TOOLS FOR THE SOFT PATH

by the staff of International Project for Soft Energy Paths

JIM HARDING, ELYSE AXELL,
CHARLES DRUCKER, FLORENTIN KRAUSE, KATY SLICHTER,
AND OTHERS

FRIENDS OF THE EARTH • SAN FRANCISCO

Acknowledgements

Tools for the Soft Path could never have been published without continuing support from friends and colleagues in 85 nations around the world who have sent us reports on the status of technologies and policies in their own nations. The *Soft Energy Notes* staff, including Charles Drucker, Elyse Axell, Cheryl Nichols and Katy Slichter have handled the herculean task of deciphering and evaluating these reports with uncompromising commitment. Nancy Austin, San Rafael's Bookman Productions, Patrick Miller, Graphic Impressions, and Laura Gest helped immeasurably to convert magazine into book. Finally, our thanks to those individuals and foundation boards who have found something unique to support in IPSEP's work.

Special thanks go to Florentin Krause who shaped the final outline of *Tools for the Soft Path* and wrote the short passages that introduce each section.—J.H.

Contents

Foreword

A few years ago, Lewis Lapham graced the pages of the *Washington Post* with an essay on energy. In it, he heaped thick ridicule upon reports "purporting to describe an unparalleled misfortune that exists, if it exists at all, at an imaginary point where six or seven lines intersect on a graph."

Not too long thereafter, the Shah of Iran (who had been described by President Carter as America's "rock of stability" in the Middle East) was unceremoniously replaced by a wondrous assortment of Ayatollahs and mullahs. A bit later, open war erupted between Iran and Iraq. The world price of oil quickly doubled—again. And for a while we were all reminded of the fragility of the energy lifeline that feeds the industrial world and the developing world alike.

As I write these words, Lapham-like sentiments are again filling the world's news publications. Talk of a petroleum "glut" has resurfaced—as it has every so often since the late 19th century. And investments in renewable energy development, and in increased energy efficiency, are tapering off. Nowhere is this phenomenon more evident than in the Reagan White House. The original 1981 U.S. budget for energy conservation was $925.6 million; the Reagan request for fiscal 1983 is $21.8 million.

To the extent that our leaders continue to address the energy issue at all, it is by supporting approaches that future historians will treat with outright derision. Nuclear power remains the central element in most "official" energy programs, and synthetic fuels loom importantly in countries possessing the requisite resource base. The awesome problems of the Greenhouse Effect and of nuclear weapons proliferation are no longer even given lip service by most high government officials. More surprisingly, economic considerations are also ignored.

The energy debate continues to turn on questions of supply. Each time a government is replaced, a new crew of dignitaries must slowly learn a painful fact: Nobody wants energy. What people want are the goods and services for which energy is needed, such as warm houses, cold refrigerators, and mobility. If these goods and services can be met using less energy, no one feels deprived. If, by insulating it, one can keep one's home just as comfortable using two barrels of oil as by using six, one generally feels no pressing desire to burn the remaining four.

I'm reminded of the man who walks into a hardware store and says, "I want a drill." He doesn't really want a drill at all. He wants a hole! If he can be convinced that a better hole can be made with a high-pressure water jet or with a laser, he will forget the drill.

The real strength of *Tools for the Soft Path* is that it helps the reader distinguish between drills and holes; between what the vested interests are selling us, and what we really want. It builds an impressive case that we are being sold a bill of goods.

The book is honest enough to admit that there are no perfect answers. Many comparatively attractive technologies can have catastrophic consequences if pursued incorrectly. This is especially true with regard to biological energy sources. Moreover, even when pursued wisely, they can have warts. The attraction of many of the current "soft" technologies is not that they will produce utopian perfection but that they are less harmful than conventional alternatives.

The strategies outlined in this volume are generally cheaper, safer, and yield more jobs than the approaches they replace. Soft technologies generally enhance the security of energy supplies while reducing negative environmental consequences. Moreover, these technologies are sustainable for as long as the earth itself remains habitable.

It is sad, but true, that few heads of state will find the time, or possess the inclination, to read this book. Instead, they will continue to try to lead the rest of us right over the edge of the cliff. That is a "given." The only uncertainty is whether or not we will follow.

H.G. Wells once wrote that "Human history more and more becomes a race between education and catastrophe." The race is continuing. *Tools for the Soft Path* provides a welcome dose of educational encouragement at a point when it's badly needed.

—Denis Hayes,
author of *Smart Energy*
and former director of
Solar Energy Research Institute

Introduction

AMERICANS ARE AFRAID of the future—and not without good reason. At no time in the recent past have the threats to our livelihood been so challenging. We face high interest rates, a stagnant economy, growing federal deficits, unprecedented numbers of Americans out of work, and the nagging fear that we are in for a future of long, hot summers. Our national defense is said to suffer from a window of vulnerability that can only be closed by adding thousands of new nuclear warheads in advanced missiles, when we can already overkill every human being on earth eleven times.

We hear that we must modernize our forces to protect strategic interests around the world, by adding new Trident submarines, the MX missile, the Cruise missile, the B-1 bomber, and the Stealth. All these add the elements of surprise and pinpoint accuracy needed to penetrate Soviet defenses in a preemptive first strike. The Soviet Union can only be expected to respond in kind, vaulting us all into a world where the first button pushed is the last button pushed.

What are these strategic international interests that are worth such a quantum leap in *insecurity*? One of the most important is energy, and the perception that continued supplies of Mideast oil are worth almost any risk. Both the Carter and Reagan administrations have announced their willingness to go to war over the Persian Gulf, which the Soviet Union is also seen to covet. A recent US Army battlefield handbook suggests the prompt use of tactical nuclear weapons within hours of a Soviet tank advance on the Arab states.

The prospect of nuclear war over oil is hastened by the complicity of many oil-dependent Western nations, including the United States, in trading potentially useful nuclear technology or delivery systems to nations capable of granting energy favors. None of the superpower confrontations that have been played out on third world stages in the post-Hiroshima age has taken place in a nation with The Bomb. But if such countries as Libya, Iraq, and Pakistan join Israel (which already has several nuclear weapons) in gaining a nuclear arsenal and delivery system, the next Middle East conflagration could begin at the nuclear level, and the superpowers may be forced to escalate from there. The shaky ethics of deterrence would disappear.

The same applies for such tension points as India, Brazil, Argentina, South Africa, Taiwan, and South Korea. All have an interest in The Bomb, though oil is less responsible for their access to nuclear technology than curried diplomatic favors, availability of strategic minerals, and the salesmanship of the nuclear industry.

This is a picture of the planet that inspires little more than taking refuge in the southern hemisphere, the Second Coming, or the downstairs broom closet filled with Atari Joysticks and a Space Invaders game. Escapes from fatalism dominate our hopeful visions, and that shortens all our lives.

Neither fatalism nor starry-eyed romantic visions are the subject of this book. *Tools for the Soft Path* presents a hopeful and practical energy strategy that can back the industrial world away from oil dependence and nuclear commerce, and the developing world away from perilous energy shortages. Based on energy efficiency and cost-effective renewable resources instead of oil and nuclear power, the approach has been called the "soft energy path." It could just as easily be called a path to a saner and more peaceful future.

During the past five years, Friends of the Earth has collected, compared, analyzed, and circulated the results of soft energy path studies developed by independent teams in nations as diverse as India, Germany, Rumania, Britain, France, Scandinavia, Japan, China, Canada, and the US. These studies are distinguished from conventional "hard" energy planning in several significant ways. Most importantly, planning begins at the point of final energy use, whether that is keeping office-workers warm in Toronto or providing dinner in a Third World village. A soft path asks what combinations of technologies and energy sources can do these energy functions at high efficiency and lowest cost, both to the bill payer and to the larger society. In Toronto, that has meant cost effective commercial buildings that use a quarter as much heating, cooling, lighting, and ventilation energy as a similar San Francisco office building, with much of that energy provided by the sun. In the developing world, a soft energy path may start with $2-3 cooking stoves that use a fifth or tenth the energy and firewood gathering time as an open fire.

Conventional energy planning approaches the problem from the opposite direction, beginning at the energy source and never bothering with how that energy is used. A "hard" energy path assumes that increases in oil, coal, natural gas, and uranium production will be gobbled up in nameless ways to yield economic growth. This approach can never show the enormous potential for improving the efficiency with which energy is used, or how local diverse flows of renewable resources can begin to supplant traditional centrally planned fuel resources. If we view the economy as an automobile, the hard energy path's logic says floor the gas pedal, increase oil company revenues, and new fuels will soon be available for the guzzling (at upwards of $2-3 per gallon). People's unwillingness to follow this plan has finally "trickled up" to policy makers in a worldwide oil glut.

The soft energy path's logic is that mobility matters, not consumption of raw energy, and for mobility we substitute a fuel efficient car, drive slower, combine trips, let our fingers do the walking, and save money and energy. A soft path extends this metaphor to the hundreds, perhaps thousands, of ways in which energy is used, asking how each of those services can be provided at lower economic cost, less environmental damage, higher energy efficiency, and greater use of renewable resources.

The unexpected miracle of many soft energy studies is the extent to which these varied objectives coincide. When the potential is added up, energy policies and problems look different, and much less bleak, though with large institutions the proof of this has been arduous.

In 1972, for example, the Chase Manhattan Bank issued a report that said no amount of conservation was possible without economic loss. Energy was already used with such perfect efficiency that economic growth in the future required equal growth in energy use. Few reports could have been more absurd. Since 1973, the US economy has grown by 21 percent in real terms and the rate of energy use has not grown—it has, in fact, shrunk by a small amount. During that period, about one hundred times more economic well being was supplied by efficiency improvements than by new drill bits. Exactly contrary to Chase's argument, we would have had no growth without the efficiency improvements of the last nine years. The same story can be told in nations twice as energy efficient as the United States; much of that story is told between these covers.

In 1977, Amory Lovins and I began the International Project for Soft Energy Paths within Friends of the Earth to rapidly circulate information on the technical and economic status of efficiency and renewable energy sources. Our audience was primarily policy makers, energy activists, and research groups. At that time we sought to encourage, through seed grants and solid technical information, soft energy studies for numerous nations.

Was it possible that nations like Japan or West Germany—both cold, crowded, and industrialized—could so improve their energy efficiency that they could come to depend largely on indigenous, renewable energy flows? What was the status of wind for electricity generation; municipal, forestry, and agricultural wastes for liquid fuel production; passive solar heat for residential and commercial buildings; low temperature solar collectors for water heat; high temperature collectors and high temperature storage for process heat; small hydropower for electricity; cogeneration for industrial processes and electricity production; plus such future technologies as advanced batteries or solar electric cells?

Could some of these alternatives be adapted for Third World energy use? Or, were China's 14 million biogas plants and Brazil's alcohol fuel program models for industrial countries?

The vehicle for asking these questions and publishing some answers has been *Soft Energy Notes,* now in its fifth year of publication. The material in *Tools for the Soft Path* has been drawn directly from those pages. We have published much material that has surprised us, disappointed us, or challenged our preconceived notions. Nevertheless, the picture that emerges is an inspirational one, showing how coherent policies can meld goals of least economic cost, environmental protection, democratic values, and higher energy efficiency into a hopeful future.

Extremely detailed studies for the US, Britain, Japan, West Germany, Sweden, the Scandinavian region, Canada, and numerous subnational regions suggests that it is not only conceivable, but definitely possible, for large industrial nations to continue to grow in material wealth and population with very substantial and cost-effective reductions in national energy use. With the best technologies available today that are cost effective against new sources of fossil fuel or nuclear energy, it is evident that the US could run on about a fifth its current rate of energy consumption, and Japan and West Germany on about 40 percent their current levels. Both are well within the potential of their indigenous resources, without any need for oil or nuclear power. Using easily conceivable technologies not currently available, these energy use rates could be cut in half again. This is the long range potential; obviously, changes like

this do not happen overnight, but over a thirty to fifty year period as a nation's capital stock, in homes and industries, is replaced with advanced equipment.

What this means is that conservation, better called improved efficiency, is not a five or ten percent solution to the energy problem, or a short term fix, but the major route for supplying future economic services as fossil and nuclear fuels are phased out. Moreover, it means that the energy solution is not "more of everything," in a rabid and expensive dash to squirrel away resources for an uncertain future. We can be more thoughtful.

Today, we know how to increase the efficiency of industrial motors by a factor of two, lighting by three, appliances by four, auto transport by five, and new buildings by ten. Much of that is being done today, in various places, at costs that are more than competitive with the fuel such efficiency replaces. The choice is not faster coal consumption or more nuclear power plants, but less of both. Over the long haul, the scope for energy efficiency and renewable resource use is sufficiently large to keep a population twice the size of that on the earth today living at better than a European standard of living and reliant solely on renewable energy flows. The technical assumptions, analytic methods, and conclusions that support such an uncommon view are included in this volume, and skeptical readers can consult the original references if they remain unpersuaded by our summaries.

All long range forecasts, no matter how detailed, are little more than sophisticated star-gazing—but they do fill one important function. By comparing alternative views of the future, analysts and citizens alike can see economic and social barriers to change. There is little question in the minds of soft and hard path advocates that we must become less dependent on oil and natural gas in the decades ahead.

One of our reports contrasts a Uranium Sweden with a Solar Sweden, and asks which would be easier to implement. Is it easy to envisage a Sweden with sixty seven giant nuclear plants providing electricity power and heat to homes, commerce, and industry? A Solar Sweden would involve the use of 6-7 percent of the nation's land for energy production from forests, about the percentage currently given to agriculture. Hydropower, wind turbines, and modest amounts of solar electric cells supply much of the remaining energy needs. Neither looks simple, but Swedish voters have chosen to phase out their current dependence on nuclear reactors by 2015 and begin the transition to solar energy. A new Swedish government study shows how the nation's economy could become twice as energy efficient as it is today by the early years of the 21st century (or about four times as efficient as the US is today). This reduces the impacts described above by about 20 percent, easing the transition process. One of the principal strengths of a soft energy path is that it relies to a much larger extent than a uranium future on popular decision making and least-cost ways to deliver energy services. A nuclear future, by contrast, demands massive institutional change; electrification of homes, industries, and transport regardless of cost; control of capital markets for centralized electricity production; and a subjugation of local values to technocratic ones.

A nuclear America, independent of oil supplies and natural gas, would probably require 800 to 1200 reactors throughout the country, as was forecast by our Atomic Energy Commission in 1973. Today, this is beyond imagining, and that says something about the possibility of implementing a Solar America. To be sure such a future requires planning, but it does not demand that people be sheep.

The major soft energy study for the United States was prepared by a number of independent research teams around the nation and assembled by the Department of Energy's Solar Energy Research Institute (SERI). The agency projected a future of material affluence, with real GNP climbing at 2.7 percent per year, 33-50 percent increases in private transport, and 55-95 percent boosts in air travel, and substantial population growth. This was a difficult test for energy efficiency, but by year 2000 energy use would still drop 17 percent from 1979 levels and non-renewable fuel use would drop nearly 50 percent. By the early to mid-90s, we would be independent of oil from the so-called "hostile" OPEC suppliers, and by the late 1990s we would be fully independent of all foreign oil. The difference between standard federal projections of our year 2000 energy use and this one is equal to all the oil, gas, and coal currently used by three quarters of the world's population, virtually everyone in the southern hemisphere. What we could win for our descendents, for our international reputation, and for the rest of the world is incalculable.

This kind of future means devising ways to encourage entrepreneurs to take larger risks than Detroit did in continuing to produce large cars. it means finding ways to increase the efficiency of rented buildings where tenants pay the energy bill. It requires the evolution of utility companies into energy service companies that more resemble Sears than a power plant engineering outfit. It means city and county planners must consider adapting the automobile to mass transit right of ways, rather than the other way around. And it means national planning to discourage bailouts for industries that can no longer compete in a world market, plus incentives for those things America can do best.

The usual free market penalty for enormous mistakes is bankruptcy, but in many industries, especially the electric and gas utilities that sell more than half the raw energy in the nation, this option is no longer available. Many other nations have state-owned utilities that understand their own mistakes even less well. A decade ago, US utilities began investing tens, if not hundreds of billions of dollars in power plants that have been or will be cancelled at public expense, largely because those plants cannot compete with newly efficient refrigerators, lights, motors, and air conditioners—the four largest categories of electricity use. This industry must adapt to a future where it understands, and positively influences, the underlying currents of change. If this means that loan officers take the place of construction engineers, so be it.

Many other industries than Detroit understand the penalty for mistakes, but make them anyway. Such Japanese manufacturers as Sanyo, Sony, and Mitsubishi have reached 10 percent of their nation's homes with mass produced solar hot water heaters that cost half what US models do, and are now eyeing the US market. Refrigerators, which are the fourth largest electricity consumers in the nation, are now made in Japan to US standards of size and convenience using one-third the power of our average current sucker. The free marketplace can tell us many things, but it cannot protect us from pervasive misallocations of national capital.

One of the primary functions of *Soft Energy Notes* has been to inspire and goad those who take risks for a living to take bigger ones. A quantum leap in Japanese auto efficiency has brought Japanese auto manufacturers quantum rewards, and so it should be for our steel, chemical, and auto industries, appliance manufacturers, and home and office builders. Instead, US Steel buys Marathon Oil and DuPont buys Conoco. If for no other reason than their own economic survival, American industry ought to be getting down with the Mitsubishis rather than keeping up with the Joneses.

Institutional barriers have a very different character in the Third World. One of the central figures in energy policy for developing countries is Amulya Reddy, an Indian physicist whose career encompasses both detailed national energy planning and the cautious integration of soft energy technologies into culturally complex, desperately poor Indian villages. A soft energy path here begins with the question of how to provide such end use energy services as cooking and lighting in the poorest parts of the country.

The traditional approach to meeting such end use needs is to build central power plants with long lines out to rural areas, but that effort has been a dreary failure. The government does not have the money to build such lines for minimal demands, and the poor could neither afford the power nor the appliances needed to connect to the central grid. Meanwhile, India's rural poor rely on heavily subsidized kerosene for lighting and woefully scarce fuelwood for cooking.

The need to subsidize kerosene prices for the rural poor has led to a dramatic recent shift in freight transportation, from rail to less energy efficient trucks. As a result of subsidized fuel prices, India's oil bill in 1981 reached 85 percent of its total foreign exchange earnings. (In the US, it was 27 percent.) The only obvious market "solutions" to decrease India's oil appetite can only hurt the poor first and further deplete firewood supplies.

Yet India's rural poor have access to large quantities of cattle dung that can be fermented to biogas in cheap units, following the Chinese example of 14 million such units. Used in this way, biogas supplies can more than meet all the cooking needs of India's villages while staving off a wood crisis. Simultaneously, small generator sets with local distribution lines can eliminate the dependence of the rural poor on kerosene for lighting. Together, these measures permit the national government to decontrol subsidized kerosene and diesel prices, so that freight begins to naturally shift back to rail. A portion of the firewood saved can be converted to charcoal or liquid fuel to further substitute for oil. The reduction in oil use nearly makes India self sufficient in fossil fuels, and the reduction in fuelwood use in open fires can prevent a rural energy catastrophe. The transition process is not simple, but not making it is very hard indeed.

Collectively, this set of approaches to the energy problem can create a very different world than we have today. It is a world that can afford to step back from nuclear commerce, and from the nuclear weapons that defend our oil aorta. Soft energy paths offer the developing world a chance to increase self sufficiency and standards of living without drastic cultural and environmental changes. For such broad prospects, this book is necessarily incomplete, but it gives us a hopeful set of choices. It demonstrates the technical and economic potential of soft energy paths for developed and developing countries. It identifies today's cost effective technologies, tomorrow's incipient ones, and rapidly developing longer range alternatives. It suggests the roles our institutions might play in the future, and what coherent policies might help move them in that direction. And it identifies the roles that activists, energy policy makers, citizens, local, state, and federal officials might play in bringing such a future about.

—Jim Harding
3/31/82
Berkeley

1. Buildings

The suburban, single-family home has come to symbolize the American way of life. They are among the most comfortable and spacious in the world, lavishly garnished with the amenities of modern civilization: two-car garages, TV antennas, and an army of appliances. Behind this affluent facade hides technological underdevelopment: eyes sensitive to infrared light would see that Americans, like people in other industrial countries, live and work in thermal shantytowns.

The average private household paid more than $700 per person for energy in 1980. A household of three people averages some $600 for space heating and cooling, $350 for lighting, appliances and hot water, and $1,200 for gasoline—$2,200 per year, all told. The same household pays about $1,000 in taxes for the US military budget, which, among other things, secures energy supplies abroad.

Even with no additional OPEC price hikes, utility bills are sure to climb: the debacle of nuclear power will have to be paid for and deregulation of US natural gas prices is proceeding.

This is the bad news. The good news is that elementary physics offers a way out.

Building energy use, for example, can be reduced to one-tenth or less of the present average, with the simplest insulation materials and construction techniques, at a fraction of the cost incurred by continuing the old dependence.

The revolution in building-shell technology is already underway. Space heating and cooling, now typically 20 to 40 percent of energy (and oil) consumption in western industrial economies, can be almost eliminated with the principles and techniques described in the articles that follow. One of the largest potential energy sources of the industrialized world lies neither inside the atom, nor outside our borders, but within our walls.

For a long time to come, our building stock will consist largely of houses already in place. The institutional and cost barriers to retrofitting existing buildings are more difficult to overcome than the obstacles to making new buildings more energy efficient—but the barriers can fall to concerted local and national policies.

Many construction practices that save energy, such as placing windows for optimum solar gain, involve no cost, or so little extra that they were already cost-effective in the era of cheap oil. Even so, they were generally ignored, which suggests that we cannot rely on the magic hand of the market alone to make buildings energy-efficient.

Buildings that are thermally tight can have new problems: more indoor air pollution, and fumes from spray-foam insulating materials that can be hazardous. Neither problem is insoluble. Air-to-air heat exchangers—themselves energy savers—can ventilate with minimum heat loss. Spray-foam insulation is only one of many insulating materials, some of which have been used for decades.

Energy-Efficient Housing

OF ALL the components in a soft path scenario, energy-efficient housing offers the quickest, least controversial, and most dramatic results for the money. Studies that focus on the near-term consequences of a soft path strategy, such as the Long Island Jobs and Energy study reviewed in this issue and the "Drilling for Oil and Gas in Our Buildings" study reviewed in our last issue, demonstrate the magnitude and rapidity of energy savings that are possible with residential conservation measures.

Soft Energy Notes has reported on energy-efficient houses and specific measures on numerous occasions in the past, as the *Contents of Volumes One and Two* (contained in this issue) attest. The present collection of articles and reviews departs from our past practice of treating these developments in isolation. Just as a soft path is more than a collection of individual events, an emerging technology is more than a grab bag of new devices and methods. Innovations are related and reinforcing; ideas collide, connect, and collaborate to accelerate the pace of change. In this issue, we have brought together a number of innovations in the energy-efficient housing field and have summarized their implications for designers, inventors, analysts, solar advocates, and policy makers.

One principal work makes conventional wisdom out-of-date. That is the normal looking, cheap, even frumpy house built by Eugene Leger in north-central Massachusetts (see photo, p. 21). Based on a computer-generated, Lo-Cal design (see p. 23), Leger's house demonstrates that a low cost residence, with no striking design constraints, can operate in Massachusetts' severe climate with a winter heating bill of less than $10 per month ($38.50 for January through April, or 11.5 GJ over 2,144 Celsius degree days). This performance level sets a high standard for other designers to match in economics, comfort, design flexibility, and environmental side effects.

What is remarkable about Leger's house is that there is nothing solar about it: it has windows on the north, east, and west sides, and no huge expanse on the south side. It does not contain tons of concrete, piles of rock, or walls of water as "thermal mass." So how does Leger perform his low-energy trick? With a superb building "envelope": extra-thick insulation combined with painstaking efforts to reduce air leaks. Leger placed a continuous overlapping sheet of polyethylene on the exterior walls and ceiling. To avoid breaking this air and vapor seal with electrical outlets and plumbing, he installed a newly developed surface-mounted electrical system called the "Gould Electrostrip" (see p. 26). This device is an important part of Leger's energy-tight design, demonstrating how innovations in one sector can spur advances in another.

Leger's house resembles the Saskatchewan Conservation House designed by Robert Besant and Robert Dumont (reviewed in August 1978 and May 1979 SENs). Leger, however, went beyond Besant and Dumont in his attention to the air-tightness of exterior walls, largely with the use of the Electrostrip. The Saskatchewan house is superior in its use of insulating shades and heat exchangers on air and water. (Leger is installing a Mitsubishi Lossnay heat exchanger this winter, and expects to halve his current heating bills.) The major difference between the two houses comes in construction costs: Leger's $55,000 house (retail) ranks with the least expensive new homes in Massachusetts; while Besant and Dumont's design cost about $4,000 more to build (conservation equipment only; the solar system added a further $15,000).

It appears that experience is lowering the cost of Saskatchewan-house construction, according to Larry Palmiter and Barbara Miller of the National Center for Appropriate Technology, who have kindly summarized for us the results of a successful low energy housing meeting held last fall in Saskatchewan. Some fifty Saskatchewan Conservation House-type houses were completed or under construction in 1979, and both builders (Concept Construction and Enercon Ltd) claim that the features were included at no incremental cost. Several Saskatoon houses were apparently "retrofitted" with Saskatchewan-type features for $3000 (labor and materials) when home additions were constructed and were cost effective against enlarging existing furnaces to meet the increased demand of a larger house.

The designers of the original prototype Saskatchewan Conservation House, Bob Besant and Robert Dumont, report that their additional experience with building conserving houses has led them to recommend a structural change: that the vapor barrier be placed two-thirds of the way into the wall, rather than directly behind the interior wallboard. This makes for easier plumbing and electrical work, and lessens the likelihood of puncturing the barrier with drill holes. Besant and Dumont have not monitored the performance of the fifty new conservation homes, but are concerned that such possible punctures in the vapor barrier may reduce their performance to a quarter, rather than a tenth, the energy use of the average new Saskatchewan dwelling. The use of an Electrostrip-like product

Comparing Efficient Houses

House Type	Space Heating Requirements: Fuel or Solar Collectors	
	kJ/m²/°C-d	Btu/ft²/°F-d
'1973' House (typical East Massachusetts housing stock)	390.4	19.1
'1976' House	312.7	15.3
ERDA Pacific Solar Handbook	163.0	8.0
AAchen House (active solar)	70.5	3.4
Trombe Wall — Kelbaugh House	55.2	2.7
Leger House	48.1	2.4
Brownell House	34.1	1.7
Lyngby House (active solar)	18.3	.9
Saskatchewan House (active solar)	4.5	.22

North America, Fuel for Single Family Residential Space Heating

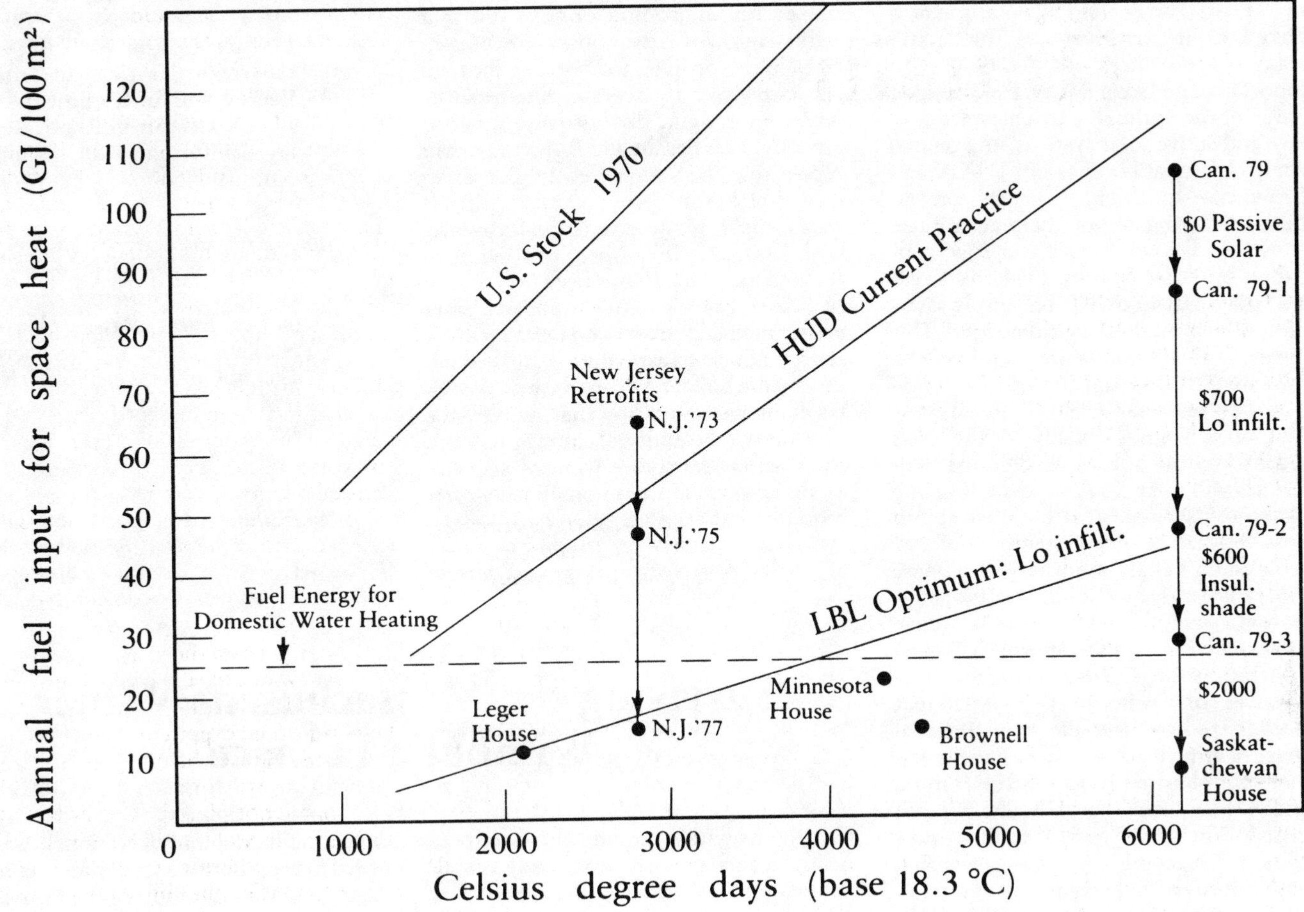

Adapted from Rosenfeld et al. 1979, fig. 4.

and better location of the vapor barrier should improve performance at the next stage.

These very tight houses pose special air quality problems. Eugene Eccli, in a personal communication, notes that problems of stuffiness and odors can be handled effectively with inexpensive air filtering devices. But very tight houses will still need fresh air, and it ought to be warm. Some reports by Lawrence Berkeley Laboratory have concluded that tight, unvented houses with gas cooking equipment can build up hazardous levels of nitrous oxides. Houses built with brick, or rock, or on some granite foundations can accumulate high levels of radioactive radon gas. The solution to these problems is an air-to-air heat exchanger, and we report here on two independent developments: one by Besant (co-developer of the Saskatchewan house) and one from Mitsubishi (see p. 24 and p. 25). According to Palmiter and Miller, Besant's heat exchanger has proven capable of exchanging 80 percent of the air in a house in an hour, a capacity much higher than we report in our less up-to-date analysis.

To compare the performance of the Brownell, Leger, and Saskatchewan houses with conventional dwellings we have adapted a figure from Rosenfeld, *et al.*, "Building Energy Compilation and Analysis." Brownell's house used 15.4 GJ/100 m² during the computed year (4,611 Celsius degree days). Gene Leger's house used 11.5 GJ/100 m² for the first four months of 1979 (2,144 Celsius degree days). These figures put both houses on the curve titled "LBL optimum: Low infiltration." This curve represents energy use for space heating that is one-ninth the level of the US housing stock as of 1970, and one-fifth the level specified by the US office of Housing and Urban Development (HUD) for new construction.

The Saskatchewan Conservation House appears in the lower right-hand corner of the figure. This remarkable performance level involved insulating shades and heat recuperators, devices Leger did not use. The series of points labeled N. J. represent houses at Twin Rivers, New Jersey, currently monitored by Princeton University and reviewed in SEN 3. They demonstrate the enormous energy-savings potential of *retrofits*. A 1977 Twin Rivers super-retrofit cost $1,250 on a townhouse and reduced energy consumption to a point below the LBL low infiltration optimum curve. This performance, combined with that of Leger's house, the Saskatchewan house, and a Minnesota design clearly set a new energy consumption standard for space heating: a level roughly *half* that of the energy used for domestic water heating! Such performance is possible for both new housing and some retrofits.

Reductions in heating loads as sharp as these imply changes in the optimum devices and designs for supplying "back-up" heat. People and their appliances have become the "furnaces" or "active solar space heaters" of the 1980s. We report here on Doug Balcomb's computer study of the optimal match between passive and active solar systems, and on two energy-efficient houses—the Lo-Cal and Brownell—which may change conventional thinking about thermal storage. Most studies of housing energy consumption treat heating loads as fixed, and the analysis dwells on supply alternatives to meet this unyielding load. This means that the results are useful only to the extent that the specified load assumptions are reasonable. The implication is that when heating loads fall to the levels reached by the houses we describe, economics will force a substantial rethinking of the requirements for both active and passive solar heating systems.

Balcomb's analysis begins with a house that falls nearly on the LBL medium infiltration curve, then adds a cost-effective passive Trombe wall to approach the low infiltration curve, and finally, a cost-effective solar water heating system that is oversized to meet the small residual space heating load. While we have found that vapor barriers, improved insulation, and a heat exchanger should precede installation of a Trombe wall in most climates, Balcomb's work suggests that tight houses with thermal uniformity can meet their water heat needs and residual space heat needs with a small solar system. The incremental costs for the small solar space system need only include a bit of extra collector, a bit of extra storage, and a fan coil in the north lower corner of the building—all rather puny fixed costs compared with the ducting, extra pumping, and control equipment needed for a traditional solar space heating unit on a leaky typical building.

Reduced energy demands are relevant to analyses at the national or regional level, such as the study of jobs and energy on eastern Long Island, the Franklin county study, the Taylor study, and the Japan analysis. Each of these studies makes assumptions about possible reductions in energy demand from retrofits of old houses and construction of new. Those assumptions, however, are behind the potential discussed here, leading to probable underestimates of the benefits derived from conservation.

To complete our overview of energy-efficient housing we report on some other technical developments: the "Transwall," a glass/water sandwich for storing heat in passive solar homes (p. 27), and the Gould Electrostrip, used by Leger to limit breaks in the building envelope.

Overall, we are impressed with the new devices and new ideas for low-cost energy efficient housing. And so, apparently, are many consumers. Eugene Eccli makes the interesting observation that tight, energy-thrifty houses are bought by wealthy people, not because they are less expensive to operate, but because, when done right, they provide superior amenity. Bob Besant and Robert Dumont report that the Saskatchewan Conservation House is still proving very difficult to monitor for thermal performance, with 1,000 *people per week* continuing to flock to its doors. (It has been on view for two full years.) Three hundred sixty people attended Besant and Dumont's low energy housing workshop. A hundred of them were builders. From the purchaser's point of view—or, for that matter, the builder's or the national energy planner's—conservation does not mean sacrifice at all. Energy efficient houses truly offer more for less.

—Mark Christensen
Charles Drucker
Jim Harding

References:

Hollowell, Craig D., *et al.*
1979 *Radon Concentration vs. Minimum Ventilation in Energy Efficient Houses.* Available from Lawrence Berkeley Laboratory, Energy and Environment Division, 1 Cyclotron Road, Berkeley, California 94720, document No. LBL 9560.

Palmiter, Larry and Barbara Miller
1979 *Report on the October 13-14, 1979 Saskatchewan Conference on Low Energy Passive Housing,* National Center for Appropriate Technology. Available from IPSEP, 8 pages, $1.50, inc. postage.

Rosenfeld, Art, *et al.*
1979 *Building Energy Compilation and Analysis.* Available from Lawrence Berkeley Laboratory, as above. Document No. LBL 8912

Surviving the Massachusetts Winter Without a Furnace

SUPERB insulation, an airtight exterior, a $30 heat exchanger, no large cement walls or water-filled barrels, no expensive south-facing windows or dark north side, no furnace, no added expense, no weird shapes, 60 percent humidity, no airconditioner, no active solar system, and it stays warm all winter long. What is it? NASA's new space shuttle? No, it's a house in East Pepperell, Massachusetts that demonstrates in northern US climates what the Saskatchewan Conservation House has done in deep snow drifts: it is possible to design and operate a normal-looking house that requires no outside source of heat. A solar house need be no more expensive than a conventional dwelling, and its solar system need be neither passive, active, nor hybrid.

As Bill Shurcliff, the granddaddy of the solar housing community, pointed out in a recent press release, the East Pepperell and Saskatchewan homes signal a new type of solar home design. This is not, by Shurcliff's definitions, an active system, a direct gain passive system, an indirect gain passive system, or even a hybrid design. "Whatever it is called," concludes Shurcliff, "it has (I predict) a big future."

Eugene Leger of Vista Homes is the architect-builder of the East Pepperell house. Its performance has been carefully monitored by Princeton University's famed Energy and Environment Group.

In a paper at the fourth National Passive Solar Energy Conference earlier this year, Leger and Gautam Dutt of the Princeton group reported that "the house is performing beyond the builder's most enthusiastic expectations. . .The combined mortgage payments and fuel bills for a house of this type will be substantially below that of a conventional house built to today's standards. . .We believe that this house will fit the needs and pockets of many families."

Clever and Thorough Insulation

Leger describes his house as an "attractive, but rather ordinary looking, 46' × 26', 1200 square feet, ranch design. . .It is my contention that any house that has to have solar added to it is an improperly designed house to begin with. I have not added 'solar' to my house, rather I utilize the sun as one of several resources that are available."

Leger's secret? Very careful attention to details and, in Shurcliff's words, "truly superb insulation. Not just thick, but clever and thorough. Excellent insulation is provided even at the most difficult places: sills, headers, foundation walls, windows, electric outlet boxes, etc." Levels of insulation are so high that homeowners and their appliances become major sources of space heat.

East Pepperell, Massachusetts is a tough climate for any structure, but the house that Leger built hoards its heat remarkably well. For the four months of January through April, 1979, when the *average temperature was* 1° *F above freezing,* Leger's natural gas bill was only $38.50 (11.5 GJ). This is about 30 percent of the energy use estimated by the US Department of Energy for a well-insulated building, and six times less heat than is used by a typical new house in the Massachusetts area. Leger and Dutt note in their report that the expense of added insulation and attention to energy details were largely offset by reduced costs for ducting and electrical wiring, combined with the large savings from avoiding the purchase of a furnace. An oversized water heater supplies hot water for domestic needs and this water, distributed through the house in copper coils along the baseboard, provides backup heat. (A solar collector for these hot water needs could have generated even greater savings.)

Double Walls and Vapor Barriers

So why is the Leger house so energy thrifty? First, it has a double wall, each wall built from 2'' × 4'' studs, which are staggered and separated by a two inch gap, so that insulation can go over and between them. The wall is ten inches thick, insulated with blown-in cellulose. Styrofoam is used in place of plywood as a sheathing material to reduce air infiltration and increase insulation. After the outside sheathing is in place, the wall is sprayed with wet cellulose, allowed to dry, and covered with a plastic vapor barrier. A second plastic sheet is stapled to the ceiling joists, with another ten inches of cellulose insulation applied to the attic floor.

All the windows in the house are double or triple glazed (two or three panes of glass within a single frame). The total window area of the house is 153 square feet, of which two-thirds faces south. Front and rear doors, both heavily insulated, open onto vestibules to limit heat loss.

Leger and Dutt attribute much of the performance of the house to the plastic vapor barrier. Other than for doors and windows, there is only one break in the plastic membrane of the walls and ceiling: a vent pipe located within a partition wall of the bathroom, and even this "penetration" is sealed with urethane insulation. Other potential sites of heat loss are cleverly handled. A trap door from the front door vestibule leads into the attic, so that the warm air in the house doesn't lead into the cold attic and cool the house down. No penetration of the outer load-bearing wall of the house occurs from electric wiring or plumbing.

Eugene Leger's house in East Pepperell, Massachusetts hides extraordinary energy-efficiency behind a plain exterior. Its price, like its appearance, raises no eyebrows in the neighborhood.

All necessary connections are made in the space between the walls. An electrical strip, mounted flat against interior walls, allows outlets to be added at any point (see *The Gould Electrostrip,* p. 24, this issue). This construction method is also substantially cheaper than drilling through outside walls to install electrical connections.

One might assume that sealing a house in plastic would give it a tropical, jungle-like atmosphere, but the humidity generally runs about 60 percent. "This makes the space comfortable without the need for a dehumidifier. Occasionally, the relative humidity exceeded 70 percent and was reduced either by opening windows or by running a dehumidifier." A further concern raised by potential owners of tight houses is inside air quality. Cooking odors, pollution from gas appliances, smoke from cigarettes, and radioactive radon gas from building components can mount to annoying, even dangerous levels in a super-tight structure. Leger says that none of these have been problems to date, but that an air-to-air heat exchanger will be installed this winter, primarily to reduce heat losses from ventilation. Air-to-air heat exchangers channel outgoing "stale" air past incoming fresh air, with only a thin plastic, paper, or plywood sheet separating the streams (see *Heat Exchangers in Energy-Efficient Homes,* p. 26, this issue). Ventilation can be adjusted to increase when cigars walk through the door, and lower again when they leave. The heat exchanger pre-warms or cools the incoming air, giving the furnace less work to do, and Leger calculates that it will save half of his current heating bill.

Comparative Costs

As of this writing, Leger's house is on the market for $59,000, very slightly more than the cost of an energy-inefficient house of equivalent size. Though calculations vary, Leger, Dutt and another energy analyst, Ron Dans, agree that the Leger house probably cost about $400 more to build than an inefficient neighbor. "Since a conventional house without the energy efficiency and solar features of the Leger house was not built at the same time, the exact cost difference is not known," Leger and Dutt explain. "Although the Leger house requires more construction material in some categories, this is largely offset by large savings in other categories. The biggest saving is in the heating system where a furnace is no longer necessary and the extent of the baseboard is drastically reduced. Additional savings result from reduced material and labor requirements for the electrical wiring. Consequently, we believe that this house is only slightly more expensive than the conventional '1976' house." (See accompanying table.) The Leger house sets standards of cost and energy efficiency that will challenge the skill and the ingenuity of other designers.

—Jim Harding

Reference:

Leger, Eugene H., and Gautam S. Dutt
1979 *An Affordable Solar House.* Presented at the Fourth National Passive Solar Conference, Kansas City, 1979. Available from IPSEP for $1.50.

SuperShell: Brownell's Low Energy House

THE CONTRIBUTION made by passive solar gain to household space heating is closely connected to the building's ability to retain heat. One of the more interesting attempts to use passive gain in a low heat loss building is the Brownell Low Energy Requirement House, constructed in 1977 by an innovative builder in upper New York State. Brownell came up with an attractive, commercially competitive house with an extremely low heating load over the long, 8300 °F-day (4611 °C-day) Adirondack winter.

Brookhaven National Laboratory became interested in Brownell's work, and collected performance data from January through March of 1978. Their measurements do not indicate how the Brownell house performs under "lived in" conditions, but this probably makes their conclusions conservative. The estimated annual heating load for the 2300 ft² (215 m²), two-story structure, assuming average solar gain of 172,000 Btu/day (181,460 kJ/day), and a contribution from lighting, appliances, occupants and other internal heat sources of 60,000 Btu/day (63,300 kJ/day), is a low 32 million Btu (33.7 GJ). This works out to 1.7 Btu/ft²/°F-day, or 34.1 kJ/m²/°C-day. The solar gain assumed here is quite modest—only three times as large as the internal heat sources—but the building envelope is so tight that the two together meet about two-thirds of the house's total space heating requirements. A wood stove or some other inexpensive unit can easily provide the remainder.

Two factors account for the building envelope's thermal integrity: high quality insulation encapsuling the entire structure, and its high resistance to infiltration. The Brookhaven calculations assume an infiltration rate of 0.5 air changes per hour, but actual performance (under test conditions, though, with no occupants to leave doors hanging ajar) suggests a rate of only 0.15 per hour. In such a tight house, low heat loss from infiltration and poor air quality tend to go hand in hand unless, as in the Saskatchewan House, air-to-air heat exchangers are installed. Brownell's house makes no such provision, but a Mitsubishi or other compact unit could easily be installed.

The low infiltration, moisture intrusion and heat loss of the building shell are products of Brownell's continuous insulation and vapor barrier envelope. Four inches (10.16 mm) of Thermax polyiso-cyanurate rigid foam insulation surround walls, ceiling and cellar. The two Thermax sheets, each two inches (5.08 mm) thick, with foil facing on both sides, provide insulation of R-32 and high moisture resistance. To reduce vapor penetration even further, Brownell added polyethylene film around the insulation on the cellar wall and floors. Careful caulking and use of foam at the joints where wall meets roof keep the building envelope tight. The post-and-beam structural arrangement used by Brownell, although not essential to the house's thermal performance, allows it to be completely sheathed in insulation attached to the exterior, thus preventing thermal "short circuits" at framing members.

The Brownell house, with R-37 walls and an R-40 ceiling, retains its heat well.

Even though solar gain from its 250 ft² (23.2 m²) of south-facing glass is not extremely large, the low heat loss can easily result in overheating on sunny days. Since Brownell's design does not include overhangs to shade the windows from late spring, summer and early fall sunlight, overheating could be troublesome during the milder months. Adirondack summers, though, are cool enough so that overheating may be prevented with blinds, shades, or the heat flushing system Brownell installed.

During winter, of course, the problem changes: how to utilize available solar gain on sunny days without undergoing unacceptably sharp fluctuations in temperature. One approach is to have less south-facing glass, but this reduces the amount of solar energy collected, and the Brownell house has a collector surface only 11% of its floor area. Brownell's solution is what he calls a "heat energy battery": a sand storage bed below the house's cellar slab (but within its insulation envelope). Warm air flows to the sand bed through a large, chimney-like thermal mass located in the center of the house. A wood stove, the house's only

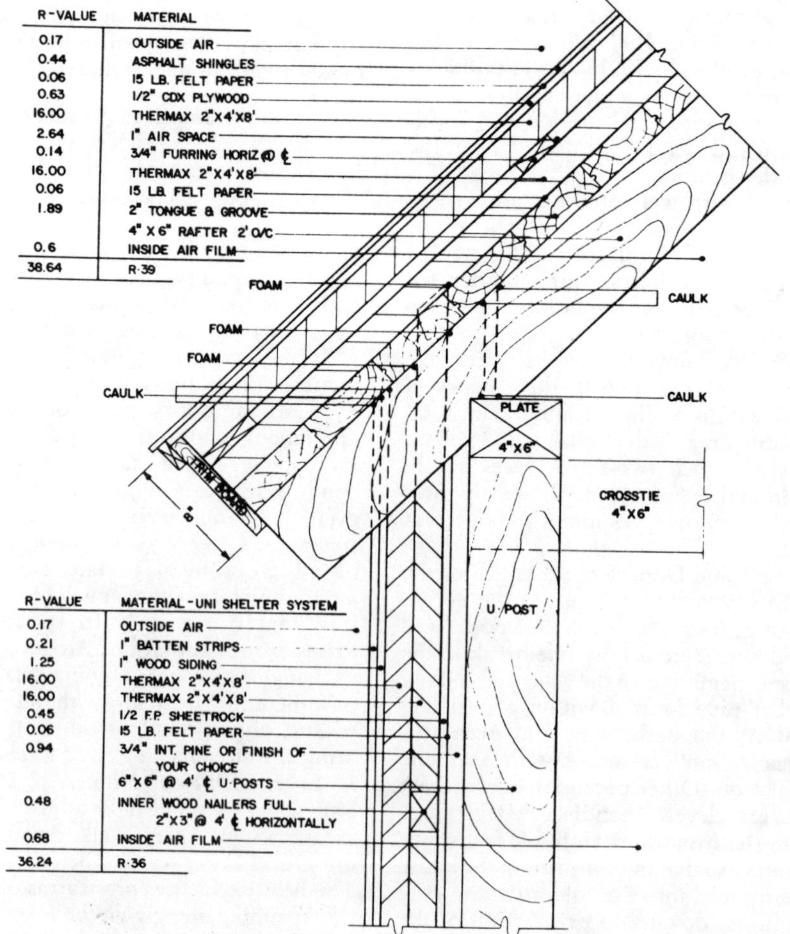

Brownell house roof system detail.

auxiliary heat source, is built into this 20-ton (18.2 tonne) mass. Heat from the sand bed is blown through vents into the living space.

The idea is a good one, but Brookhaven's test data showed the sand bed to be ineffective. Sand is a poor thermal conductor with low specific heat, and the amount of heat stored in the bed was scarcely larger than its estimated losses to the ground. The building materials themselves, however, proved to be an excellent passive storage system. The 5 in. (12.7 mm) cellar floor slab, the 8 in. (20.3 mm) concrete block cellar walls, the central chimney-like mass, the exposed framing, the post-and-beam structure and the roof deck, weighing in at 127,000 pounds (57.6 tonnes), have a combined heat storage capacity of 39,000 Btu per degree Fahrenheit temperature change (74,061 kJ/°C). Brookhaven's calculations show that 75% of this storage capacity is usable. The lesson of the Brownell house is that with a low-loss building envelope, large thermal storage systems may be unnecessary additions, and simple structural design modifications could raise the utilization rate above 75%.

There is a price on the Brownell house's performance, but it is not prohibitive. Construction costs run about $8,200 above those for a conventional house of similar size, mostly for the materials and labor needed for the Thermax insulation and the sand storage bed. Eliminating the bed, which is far from cost-effective, brings the net cost differential to the $5000 - 7,000 range. Given present heating oil prices, a healthy return on this investment is almost assured.

Adapting Brownell's construction technique to the standard mass-market stud-frame building is possible, but carries its share of problems. A tight building envelope only comes with careful attention to construction details and meticulous execution in the field. Likewise, the occupants of such a house must pay it a certain amount of respect. Doors and windows habitually left open negate the intent of a super-tight design, and bring to naught the care invested in construction. Comfort and low heating bills are possible with energy-efficient housing, but only if structure and occupants can work together. —*Steve Meyers*

Reference:

Jones, R.F., R.F. Krajewski, and G. Dennehy
    1979 *Case Study of the Brownell Low Energy Requirement House.* Upton, New York: Brookhaven National Laboratory, Department of Energy and Environment. Publication BNL 50968.

On an Energy Diet: The Lo-Cal House

THE ILLINOIS Lo-Cal house exists only on paper, but its lessons for the housing industry are very real. Developed in 1975 by architects at the Small Homes Council-Building Research Council of the University of Illinois, the Lo-Cal house presents builders with a basic design for a super-insulated, solar-oriented house. Its heating load is one-third that of a new house built to US Department of Housing and Urban Development (HUD) standards for cold climates (R-20 ceiling, R-12.5 walls, R-12 floor). The insulation technique of the Lo-Cal house, which served as the model for Leger's double-wall, super-tight construction (see p. 20), performs three functions:

● it reduces the need for solar gain from south-facing windows;

● it enlarges the fraction of total heat requirement provided by internal heat sources;

● it reduces the need for heat storage in the house's internal mass.

The basic configuration of the 1500 ft² (140 m²) Lo-Cal house involves R-33 walls, an R-40 ceiling and an R-22 floor. The distribution of windows (all triple-glazed) calls for 122 ft² (11.40 m²) facing south, 22 ft² (2.0 m²) facing north, and no windows at all on the east and west walls. The roof overhang extends 30 in. (76.2 cm) from the face of the window. Other construction details are essentially those of the Leger house.

Hypothetical Lo-Cal houses in Wisconsin and Indiana were subjected to computer analysis using Thermal Loads Analysis and System Simulation Program, developed by the US Department of the Army's Construction Engineering Research Laboratory. This simulation program reveals how variations in specific design parameters, such as window distribution or insulation levels, affect a structure's overall performance. Some of the simulation results run counter to conventional wisdom. A 25% increase, for example, in the Lo-Cal's area of south-facing glass, reduces the seasonal heating requirement by only 7% from the control configuration's. Energy needs for cooling, meanwhile, jump 14%, largely because high levels of insulation keep excess heat inside the house.

The Lo-Cal design makes no provision for excess heat storage, but simply vents it through open windows. Surprisingly, the computer analysis suggests that there may be no point in trying to capture this excess. Increasing the mass of the house by 50% from its control specification of 30,000 pounds (13.6 tonnes) to 45,000 pounds (20.4 tonnes) has a negligible effect on energy requirements. This result, however, may be specific to the Midwest winter climate, which has long periods of very low insolation. Over much of the winter, opportunities for solar gain may be insufficient to make the additional thermal mass worthwhile.

This is not to say that the passive solar design is unimportant. The Wisconsin version of the Lo-Cal house collects 14.3 million Btu (15.1 GJ) of solar energy through its windows over the course of the heating season. But the high thermal integrity of the house lowers the importance of external factors. Compared to the HUD house described above, which has 50 ft² (4.65 m²) of south-facing glass, the Lo-Cal house has only 24% greater solar gain, but heating requirements that are 66% less.

Other interesting results of the computer simulations:

● Substituting double for triple-glazed windows on the south side increases heating requirements by almost 30%. Triple-glazing also turns out to be an important factor in occupant comfort. Part of body heat loss to the environment is by radiation from the body to any colder surface, such as windows. Since the inside surface temperature of triple glass is several degrees warmer than that of double-glass in cold weather, three

panes give greater comfort at the same room temperature.

- The distribution of window area in the Lo-Cal house—85% south, 15% north, none east and west—results in summer cooling requirements 27% lower than with more standard arrangements, where one-third of the window area faces north. This is because the low window area on sides facing the summer sun prevents heat buildup.

- Infiltration of outdoor air is a very important factor. Lowering the infiltration rate from 0.5 changes per hour to 0.3 reduces heating requirements by 34% in the Wisconsin version of the Lo-Cal house, and by 50% in the Indiana configuration.

- Extreme insulation to R-60 in ceilings and R-40 in walls, with larger south glass area may be cost-effective in very cold climates with more than 8,000 °F-days (4,444 °C-days). This extent of insulation in Madison, Wisconsin, with only 7,600 °F-days (4,200 °C-days) lowers heating requirements by 22%, but makes overheating more of a problem.

- The net gain of a south-facing window is about the same for double and triple glazing, except for very cold areas with little winter sun. In extremely cold climates (more than 10,000 °F-days, or 5,556 °C-days) *quadruple* glazing may even be cost-effective on windows that do not face south.

- Double-glazed windows with night insulating panels show only slightly larger net heat gain than noninsulated triple-glazed windows.

The basic construction principles of the Lo-Cal house are applicable across the temperate zone, though exact specifications will vary with local climatic conditions. Houses with various floor plans and external appearance features can use Lo-Cal principles to obtain energy-efficient performance.

—*Steve Meyers*

Reference:
Shick, W.L., R.A. Jones, W.S. Harris, and S. Konzon
1979 *Details and Engineering Analysis of the Illinois Lo-Cal House (Technical Note No. 14).* Available for $7.50 from the Small Homes Council-Building Research Council, University of Illinois, Champaign, Illinois 61820. [Besides the computer analysis, this report includes results of special window studies, a discussion of heating and cooling systems for the Lo-Cal House, and simple plans with construction detail drawings.]

Operational Saskatchewan Solar-Conservation House Yields Further Data on Energy Efficient Building Designs

After One Year, Canadian House Demonstrates Near 100 Percent Solar System in Harsh Climate

THE AUGUST 1978 ISSUE OF *Soft Energy Notes* carried an article on the design standards used at the impressive Regina, Saskatchewan Conservation House, a model home that requires only occupants, appliances, a rather small window area, and a small solar heating system to supply all water and space heating in one of the world's least attractive solar climates. Here we will describe the house in more detail and its implications for energy analysts.

The Saskatchewan house, completed in the December 1977, has been carefully monitored since January 1978. The house is unoccupied, but not unvisited, with about 1000 visitors a week flocking to its insulated doors. The designers are two mechanical engineers, Bob Besant and R.S. Dumont. Two recently available papers (presented at the Solar Energy Society of Canada meeting late this summer) make it possible for us to pass along some further conclusions.

The house is large by US standards (let alone by European standards), with 188 m² of floor space - over 2000 square feet - and three (plus) bedrooms. Beauty is not its reason for being; but neither do its reasons for being preclude more attractive designs. The house is two stories tall and box-shaped, and faces southwest. As noted in our earlier story, Regina's climate is extraordinarily harsh, with 6003 celsius degree-days of heating in the average year. Average solar insolation is 162 watts per square meter, considerably less than that in a state like California (190-200 W/m²).

Three companies near Saskatchewan are building homes to the standards set by the "Conservation House." Enercon in Regina, Concept Construction in Saskatoon, and Permanent Construction in Prince Albert all have Saskatchewan-type super insulated homes under construction. Concept Construction is the largest of these builders, with twenty homes nearing completion. The average payback for these homes, as compared to very cheap average cost natural gas, is fifteen years.

Operational Saskatchewan Solar Conservation House

The insulation standards of the house are very strict. Walls are insulated to R-41 (7.3 m²C⁰/W), using offset double 2''x6'' walls, and a carefully caulked plastic vapor barrier. The ceiling is insulated to R-60 (60 ft²-h-F⁰/BTU or 10.6 m²C⁰/W), and also covered by the vapor barrier.

Windows are double-glazed downstairs, triple-glazed upstairs, and have insulated shutters. Also included in the design is Besant's very interesting $30 counterflow heat recuperator that captures an average of 80 percent (in August we noted 90 per cent from estimates given in Besant's design paper) of the reject heat passing through the plastic-shield baffles of a plywood box. Of course, a recuperator is of little value in a house that does not use very tight construction or a plastic shield. In that case room air would escape from so many cracks that little would pass through the heat exchanger. At a recent San Diego passive solar conference, however, Saskatchewan-style tightness was reported to have been achieved in routine series construction, and can often be obtained on retrofit too: adding exterior insulation—as is now common in Sweden—provides a good opportunity for adding a tight vapor barrier.

Besant has also designed a graywater heat recuperator into the house. This system saves about a third of the water heat energy, reducing gross water heating requirements from 22 to 15.4 GJ/y (0.697 kW to 0.488 kW).

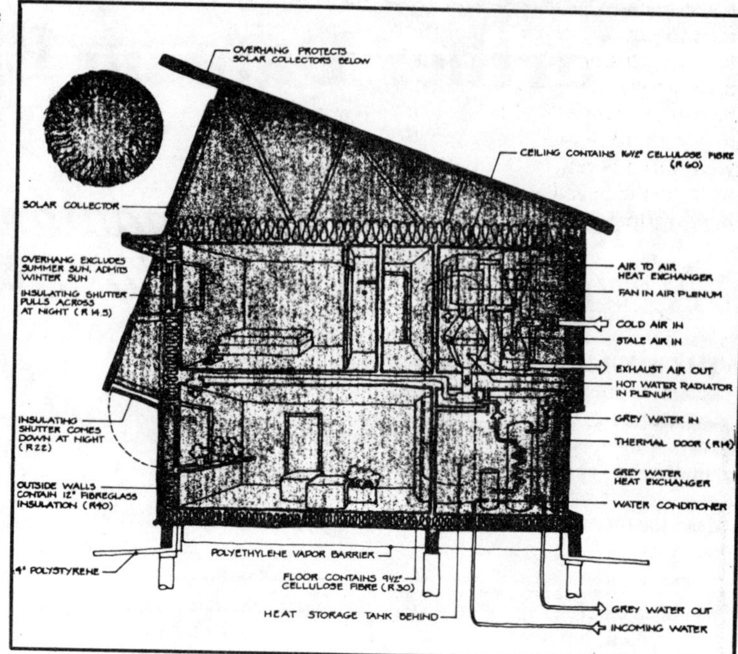

Thermal Mass

A relatively thin concrete foundation (76 mm) covered by an eight-inch thick insulation of cellufiber (ground-up newspaper), underneath a wooden floor, serves as a floor insulation corresponding to R-30 insulation. The thermal mass of the building, despite the insulation and concrete foundation, is rather small,—25 J/C⁰—of which 21 percent is from gypsum wallboard, 22 percent is from studs and wood plates, 24 percent is from the wood floor, 24 percent is from floor joists, and 9 percent is from the remainder.

Besant and Dumont claim that all these heat saving measures are cost-effective against present fuels in new construction. (Saskatchewan has electricity prices of 2.2 cents/kWh and oil and gas under world market levels.) Lovins indicates, based on recent Scandinavian experience, that most of the features are cost-effective against long-run marginal sources for retrofit into many existing houses.

Construction Questions

Several complaints have been lodged against the Saskatchewan-type house construction. One is that the vapor barrier prevents the escape of hazardous gases (NO_2 from gas cooking appliances and radioactive radon from building components). Another concern is that a barrier degrades the quality of the insulation by trapping water vapor in the walls. Some critics have also pointed to the need for careful and effective vapor barrier installation—so as not to be punctured by plumbers or electricians. Besant and Demont claim to have solved these problems with their recuperator and wall design. The recuperator can accomodate different air change requirements with a simple user adjustment. Homes with gas cooling appliances or, for example, brick (randon-emitting) basements might require one-two air changes per hour to keep hazardous gas levels low. With electric heating appliances, the air change rate, says Besant, can be as low as one-fourth change per hour, which is still enough to permit easy passage of water vapor. The Regina house currently operates on 0.6 air changes per hour.

The vapor barrier can be installed with the least amount of trouble in a two-part wall. The barrier is wrapped around the inner set of two-inch by four-inch structural studs. Additional insulation (two-thirds of the total amount in the wall) is placed outside this structural wall in a light nonload-bearing outer wall. In this way, the drywall, electrical connections, and plumbing can be installed without interference from an impermeable plastic sheet—no doubt leading to energy efficiency and improving on-site labor relations.

In frame walls, the use of a vapor barrier becomes more difficult, primarily at the important ceiling-wall juncture. Besant and Dumont have tried several local retrofits of Saskatchewan

The Saskatchewan House

The vicissitudes of management and the third worst winter on record teamed up to give the Saskatchewan house a rigorous first outing. In January and February 1979, the house required 500-700 kWh of auxiliary heating per month, much higher than design estimates. The cause of the problem was mainly mismanagement during the summer months. As the house was unoccupied (though often visited), the managers decided not to worry about turning on the active system, which is principally used for water heating. This choice burned out both the sensors and solenoid, a fact that was not discovered and corrected until well into winter, at which point the stored energy in the annual tank was quite low. Better luck next year.

Operational Saskatchewan Solar-Conservation House

house features, with some success but there has not been enough monitoring or extensive enough alteration experience to draw firm conclusions. Masonry retrofits are apparently easier, as noted above. It is a general principle that the vapor barrier should be on the warm side of the insulation to avoid water-logging. It may be hard to completely get inside the insulation in a frame wall, but it is relatively easy to wrap the outside of a masonry wall, and then add insulation to facade. This also leads to very high thermal mass.

Thermal Performance

Since monitoring began in January 1978, the house has surpassed design thermal specifications in spite of the losses introduced by the heavy visitor load. The gross heat loss from the shell is about 43.5 GJ per year, corresponding to a heat loss coefficient of 68 W/C° change in temperature with shutters closed (99 W/C° with them open, and approximately 82 W/C° with an average mix). As a result, the net space heating load has been reduced to 51 GJ per year, after deducting 14.2 GJ passive solar gain, 4.6 GJ of human gain, and 19.6 GJ of electric appliance gain.[1] This is slightly under 4 percent of the energy use in an average Regina house, and is equal to about as much heat energy as a single square meter of Regina soil receives continuously from annual averaged solar insolation.

In its first year, the Saskatchewan house did not achieve 100 percent solar heating because its late winter start-up left the heat storage largely unchanged. But based on measured performance of both the building and the solar system, it is expected to achieve 100 percent solar space and water heating with a very small system (17.8 m² of evacuated-tube collectors plus 11.7 m³

Besant and Dumont Saskatchewan House
Design Data

50½° N lat, average insolation 162 W/m², heating 6003 C°-d/y, design $\Delta T = 55C°$ two-story frame house, box shape, 187 m² floor, 13.8 m² windows (of which 11.9 m² facing 21° W of S, 1.9 m² facing 21° E of N)

***76 mm concrete slab** (no basement), R = 5.44 m²C°/W

***106.3 m² ceiling,** R = 10.6m²C°/W

***204 m² outside walls,** R = 7.3 m²C°/W (offset double 2''x6'' loadbearing walls)

3.1 m² doors (with airlock), R = 1.85 m²C°/W

456 m³ interior volume, tight 0.006'' vapor barrier with penetrations carefully calked, adjustable air-change rate (by controlled ventilation) averaging 0.6 air changes/h

double glazing downstairs, triple upstairs, both with insulating shutters (respectively 6.7 m² with R = 0.304 m²C° /W open, 2.95 closed downstairs, 5.2 m² with R = 0.489 open, 2.34 closed upstairs) (the downstairs shutters are unfortunately sheathed in wraparound sheet aluminum and thus have a higher conductive loss than they should; they are dropdown, while the upstairs shutters slide out of wall cavities)

graywater heat recuperator reducing water-heating load by about a third (nominal gross load 22 GJ/y, net after recuperation 15.4 GJ/y)

counterflow air recuperator (described fully in a Besant paper; a \$30 plywood box with plastic sheets inside, averaging over 80% efficient and peaking around 93%)

thermal mass 25 MJ/C° (21% from single gypsum wallboard, 22% from wood studs & plates, 24% from wood flooring, 24% from floor joists, 9% misc.)

It is noteworthy that all these heat-saving measures are cost-effective against present fuels for new construction (Saskatchewan has 2.2 kW-h electricity and fuels below world prices); further, all (save perhaps slab insulation, which is not so important) should be cost-effective against long-run marginal sources for retrofit into most houses.

Thermal performance has been monitored since January 1978 (operation started December 1977), with no occupants nor effort to simulate them but with the perturbations (and net heat loss) incurred by about 1000 visitors/wk. The thermal performance slightly surpasses the design basis. Measured performance implies:

gross shell loss 43.5 GJ/y (32.5 GJ from October to March), corresponding to a heat loss coefficient of 68 W/C°ΔT with the shutters closed, 99 with them open, about 82 average).

net space heating load (after credit for 14.2 GJ/y, passive gain, 4.6 GJ/y people gain, 19.6 GJ/y + 68W/C°ΔT electrical gain, 5.1 GJ/y, or an average of 162 W (under 4% that of an ordinary Regina house the same size).

gross shell loss at −34°C (interior +21°C): 3.73 kW with shutters closed, 5.45 kW with shutters open.

electrical load averages 0.71 kWe for pumps and fan (could be much reduced by better design) and 0.29 kWe for lights and appliances assumed.

The collectors have a 70° tilt, are filled with one-third ethylene glycol and two-thirds water, have a large fluid capacity—transit time is around 25 minutes—and have a thermal efficiency around 0.45-0.60 depending on inlet temperature and insolation. The pump flow—10 1/min—suffices to turn over the tank volume every 21 h of continuous operation. Data on tank stratification is not yet available, through it appears to be at least 5C°.

*This corresponds to ft²-h-F°/BTU values of 60 ceiling, 40 walls, 30 floor.

†This is the electrical load of a typical Saskatchewan house. The Regina house is 61 percent higher, but this will be much reduced by better pumps, controls, etc.

Operational Saskatchewan Solar Conservation House

of water storage that fluctuates from 40°C to 90°C).

This system—collectors equivalent to only 9.5 percent of the floor area and storage to 2.8 percent of the house volume—is an order of magnitude smaller than conventional Department of Energy-style analyses would conclude one must have for two-thirds solar coverage in a house of this size with this climate and latitude. That is because such analyses assume the solar system is added to a sieve. What the Saskatchewan house shows is that intelligent, inexpensive building design can turn an active space system into an oversized solar water heater.

A closer consideration of the physical principles of such a superinsulated and recuperated house shows why it needs such a small solar system for complete solar coverage: The insulation and recuperator have reduced the space heating load to a series of rather small blips superimposed on a water heat baseload that is three times as big as the average space heating load. The water heat load, after recuperation, is 0.49 kW; with a 21°C (70°F) interior temperature and the lowest monthly passive gain (December), the outside temperature would have to drop to -5°C (averaging over shutters open and closed) before space and water heat loads were even equal.

Further, even though the house has a relatively small thermal mass, it is so heat-tight that its time constant is 100 hours. (A normally well-insulated frame house has a time constant of about 25 hours.) This "thermal flywheel" greatly smooths out the space heat peaks, again reducing the peak heating requirements—a key dynamic effect ignored in most analyses.

Economic Implications

These physical features have important economic implications:

—**Both collector area and storage volume can be small** because the heating load is neither large nor peaky. (The storage volume, 0.66 m³/m² collector, swinging over its full 50C° range, would last 48 days at an average (0.65kW) rate of heat demand for water plus space heating.)

—**Since solar water heating is cost-effective** by itself, it can be considered to cover the fixed costs of the solar system. To cover all the space heating also would incur only its *marginal variable* costs. (Doug Balcomb has confirmed the legitimacy of this approach to the bookkeeping.) The cost of adding a little more collector area and storage volume to that which was to be installed anyway is so small that with good field-erected multiply-glazed flat-plate collectors it would probably be less than the capital cost of a gas furnace or a good wood stove.

—**Because of the long time constant of the house,** any point source of heat heats the whole house evenly, so no heat distribution system is needed.

Thus proper building design—doing the cost-effective conservation measures *first*—changes not only the size (total cost) but also the design philosophy (cost/kW) of the solar system. In this case, it has potentially[2] made a 100 percent solar heating system

cheaper than a partly solar heating system with backup (because the storage is so small that it is cheaper than backup: concrete and water or rocks are cheaper than oil, copper, stainless steel, etc.).

Indeed, the resulting 100 percent solar system is cheaper than any kind of heating system except neighborhood-scale seasonal-storage solar district heating or passive solar heating. Besant believes he can build a completely passive heated house in Regina, with more windows and larger thermal mass. This interaction of conservation with solar design is little recognized but extremely important. In one of its simpler forms, it shows up with greenhouse retrofits: if active collectors for domestic hot water are put *inside* the top of the greenhouse, they no longer need to be freeze-proofed, so a simple direct water loop (perhaps thermosiphon) can be substituted for elaborate and costly anti-freeze loops, heat exchangers, drains, etc. As with the Regina house, the design interaction is much more subtle and important

> *"What the Saskatchewan House shows is that intelligent, inexpensive building design can turn an active space system into an oversized solar water heater."*

than a mere first-order scaling effect. Many new passive houses incorporate active hot water systems inside the building, with glass substituting for roof and collector losses in heating the building.

Measured Performance

The Regina example also implies that the dynamic thermal interaction between heat gain and loss, thermal capacity of the building, storage temperature, and collector efficiency must be modelled in detail if collector performance is not be be underestimated. It may be simpler to do the experiment than to build a numerical model of the needed precision (such as the expensive-to-run Philips model used at Aachen). It is possible, indeed likely, that the design parameters of the Regina house are not optimized and that either collector area or storage volume (or perhaps both) can be reduced in the light of further operating experience. It is also clear that though the boxy Regina house is rather ugly, it need not be if the designers care about aesthetics.

In summary, the measured performance of the Saskatchewan Conservation House shows that

—**net space-heating loads** can be reduced 1½ orders of magnitude below normal North American practice by design measures that are cost-effective against present fuels for new construction.

The Saskatchewan House

Passive gain and internal heat generation supply 84 percent of the Saskatchewan house's required space heat. Seven percent comes from occupants, 33.5 percent comes from lights and appliances, and 43.5 percent comes from passive solar gain. Besant and Dumont say that 100 percent solar heating in harsh climates requires following three rules: 1) High standards of thermal recovery—a well-installed vapor barrier and an air-to-air heat exchanger, insulation levels that exceed current standards, thermal shutters on windows. 2) Use of passive gain through orientation of windows. 3) High efficiency solar collectors.

Operational Saskatchewan Solar-Conservation House

Most measures may also be cost-effective against long-run marginal sources for retrofit construction; this is a crucial area long overdue for work.

—**Once this is done, completely solar** space and water heating can be readily achieved in a very severe climate without needing backup;

—**The thermal properties** (both static and dynamic) of such a house can make the solar system not only smaller but simpler;

—**Completely solar heating** can thus become cheaper than partly solar or non-solar heating (if one shops carefully for solar equipment, preferably site-assembled, in a relatively competitive market);

—**The interactions between building** thermal efficiency and solar design are very important and deserve intensive and widespread study;

—**Once shell losses have been reduced,** air recuperators offer an attractive way of greatly reducing the residual dominant term of heat loss[3], provided the infiltration losses are low;

—**Similar techniques**[4] show even greater promise for passive heating.

—Jim Harding

Source: R.W. Besant and R.S. Dumont, "Comparison of 100 Percent Solar Heated Residences Using Active Collection Systems," submitted as a technical note to *Solar Energy;* Besant, Dumont, and Van Ee, "An Air-to-Air Heat Exchanger for Residences;" Besant, Brooks, Schoenau, and Dumont, "Design of Low Cost Ventilation Air Heat Exchangers," University of Saskatchewan, Saskatoon, Canada S7N OWO; and conversations with R.W. Besant, March 1978.

Soft Energy Notes 1:50-2 (August 1978) described briefly the Saskatchewan Conservation House at Regina Sask. Fuller data given in two papers at the Solar Energy Society of Canada meeting in London, Ontario has prompted the above semi-technical notes on the interaction between building design and solar heating design. The papers are:

R.W. Besant, R.S. Dumont, and G. Schoenau, "The Saskatchewan Conservation House; Some Preliminary Performance Results," 17 pp.

R.S. Dumont, R.W. Besant, and R. Kyle, "Passive Solar Heating—Results from Two Saskatchewan Residences," 17 pp.

---

[1] The appliance load in the house is unnecessarily high, even if its inefficiency does a nice job heating the house. The pumps and fan average 0.71 kWe and lights and appliances add an additional 0.29 kWe. The circulatory devices could probably be reduced by a large factor, as is indeed planned: the pumping energy of the much larger solar system in the Danish Zero Energy House is 0.026 kWe.

[2] The Saskatchewan Conservation House was built with a C$15,000 solar price (C$843/m[2] collector), which is not cost-effective—partly because it was experimental and partly because, pending marketing of GE and similar evacuated tubes, they are not field-erectable, but mostly because there is virtually no solar market in Canada, so one must pay over the odds.

[3] The recuperator reduces the air-exchange heat loss at the -34°C design temperature from 4.8 kW to 0.97 kW (assuming a nominal 0.8 efficiency and 0.6 air changes/h). Without the recuperator, air exchange losses at -34°C would be 52 percent of a 9.32-kW peak heating load (shutters open); with it, they are 18 percent

of a 5.45-kW load. Frost on the recuperator's plastic sheets is presently removed by running air backwards for a few minutes every few days and melting the frost into a drip-pan. It could also be prevented from sticking in the first place if Teflon rather than polyethylene sheets were used, strung slightly loose so that they flutter in the airstream.

[4] An excellent practical compendium is given in Project 2020 (Ed McGrath, Director), How To Build a Superinsulated House," 40 pp, $3 from Cold Weather Edition, Box 81961, College, AK 99708, USA. Clearly applying these techniques in climates that are only moderately severe will eliminate the net heat requirement. In many cases it will result in an effectively 100 percent-passive-solar house even though normal passive design criteria have not been followed: not only will the gross heat load be small enough to be dominated by passive gain, but fortuitous thermal mass will be effectively amplified by the long time constant resulting from the low rate of heat loss.

Superinsulation Standards Soar

Superinsulated homes first came to public attention in 1976, with the publication of the Illinois Lo-Cal House Bulletin. *Also referred to as "zero-energy," "micro-passive load," and "minimum energy" dwellings, superinsulated homes combine principles of tight construction, added insulation, and proper orientation, which are also found in passive and earth-sheltered structures. But the superinsulated design is both cheaper and simpler to construct.*

by James W. Buesing

LIKE the passive solar home, a superinsulated house utilizes the benefit of proper sun orientation. However, the design's dependency on the sun is reduced by superior insulation; the heat needed by the home, and the amount of costly storage required, are both held to a minimum.

Like the earth-sheltered home, a superinsulated structure demands careful construction practices to minimize infiltration, and thereby heat loss. But the superinsulated structure is less site-dependent, with no need for a south-facing slope or an unshaded site to save energy. And, a large portion of the building's heat comes from the equipment, appliances, lights, and people within the home.

Other advantrages of the superinsulated home:
- it may be built by local contractors with no outside consultants or engineers;
- it uses principles adaptable to virtually any home design from colonial through contemporary to earth-sheltered;
- it has a short heating season — November through April in Wisconsin; and
- it is generally the most cost-effective of all energy-efficient home design options.

> **The superinsulated structure is less site-dependent, with no need for a south-facing slope or an unshaded site to save energy. And, a large portion of the building's heat comes from the equipment, appliances, lights, and people within the home.**

Getting tighter

The principles originally put forth in the Lo-Cal Bulletin have evolved with experience and with increasing energy costs. Just as today's building codes now require insulation levels and construction practices that seemed excessive five years ago, the recommendations for a superinsulated house have become more stringent. In 1973, a house with five-and-a-half inches of wall insulation may have been called "superinsulated," but that has become the minimum allowed by 1980 code. In northern climates a superinsulated house would now incorporate a double-stud wall with twelve inches of insulation; special rafters that permit 18-24 inches of attic insulation; a fully insulated basement; and triple-glazed windows with nighttime insulation.

Recommendations for 1980 superinsulated house

Area	Amount	R	U (1/R)
• Ceiling w/ special truss	18″ glass fiber	55	.018
• Wall, 12″ double stud or	12″ glass fiber	36	.028
2x8 w/ 2″ foamboard	7″ glass fiber & 2″ foamboard	36	.028
• Basements should have 2″, R-10 foamboard which continues on the outside of the wall down to the foundation and 1″, R-5 foamboard under the floor slab.			
• Floor over crawl space	7″ glass fiber	26	.038
• Crawl space wall	2″ foamboard	10	.100
• Window, triple glass w/ R-3.5 110 ft² south, 25 ft² north	½″ night insulation	4.2	.238
• Doors, 2″ insulated w/ storm		15	.067
• Infiltration ⅓ air change per hour or less with needed ventilation and dehumidification provided with no air-to-air heat exchanger (½-¾ air changes/hr. easily obtained)			

Heat loss comparisons in 4 Madison homes

Madison, WI Design Heat Loss -Btu/hr @ 85° F ΔT †

	1974 Federal Minimum Property Standard	1980 Wisconsin Building Code	1976 Lo-Cal	1980 Super-Insulated
• Ceiling (1568 ft²)	6665	3870	2935	2395
• Wall (20% frame) (1169 ft²)	7950	5570	3680	2780
• Windows (135 ft²)	6380	6380	4095	2730
• Doors (40 ft²)	1700	485	230	230
• Floor (1568 ft²)	8125	6000	1470	1065
• Infiltration (house 12544 ft³)	19195	14400	9590	6335 (w/o heat exchanger)

Total design heat loss without considering solar and internal heat gains

• Btu/hr	50015	36705	22000	15535
• Btu/°F-day	14120	10365	6210	4380
• Btu/°F-day/ft²	9.0	6.6	4.0	2.8
• Annual loss in 10⁶ Btu	109.1	80.1	48.0	33.8

Heating requirements after considering solar & internal heat gains ‡

• Percent supplied by solar and internal gains	39%	45%	47%	71%
• Annual requirements in 10⁶ Btu	66.4	43.9	25.5	9.9
• Btu/°F-day	8595	5680	3295	1290
• Btu/°F-day/ft²	5.5	3.6	2.1	0.8
• Annual kWh heat	19470	12865	7465	2915
• Cost @ 5¢/kWh	$975	$645	$375	$145 + $12 for heat exchanger

†*calculated according to ASHRAE procedures*
‡*calculated according to solar savings fraction method*
51,000 Btu/day internal gains assumed (10 kWh/day + 2 persons)

In addition, infiltration is cut from one air change per hour to as little as one change per day. With construction this tight, air-to-air heat exchangers are needed—for dehumidification and to maintain air quality.

A home built to these specifications in Wisconsin would lose half the energy of a home constructed to meet the 1980 code. The additional construction cost would range from $500 to $5,000, for a simple payback of two to nine years.

The following chart compares the heat loss of four home designs, were they to be built in Madison, Wisconsin. All are 28 by 56 feet, with the same window and door areas.

Further information

Heat loss

- *Is Your Home An Energy Waster? How to Calculate Your Home's Heating Energy Index,* University of Wisconsin-Extension bulletin E3031.

- *How to Figure Heat Loss and Fuel Cost,* University of Wisconsin-Extension bulletin A1844.

- *Passive Variations: A Passive Solar Design Handbook,* available from Wisconsin State Energy Office, 101 S. Webster, Madison, WI 53702, (608) 266-8234.

Energy efficient homes

- "Zero-Energy House: Bold, Low-Cost Breakthrough that May Revolutionize Housing," *Civil Engineering,* May 1980.

- "Special Section on Energy-Efficient Housing: An Overview and an Outlook," *Soft Energy Notes,* 124 Spear St., San Francisco, CA 94105, February 1980.

- *Details and Engineering Analysis of the Illinois Lo-Cal House,* Technical Note #14, W. Shick, Small Homes Council, Building Research Council, University of Illinois, 1 East St. Mary's Rd., Champaign, IL 61820, May 1979, 109 pp.

- *Super-Insulated Houses and Double Envelope Houses: A Preliminary Survey of Principles and Practices,* W. Shurcliff, 19 Appleton St., Cambridge, MA 02138.

- "Superinsulation," *A.T. Times,* National Center for Appropriate Technology, PO Box 3838, Butte, Montana 59702, July/August 1981, pp. 9-12.

- "The Art of the Possible in Home Insulation," D. Robinson, in *Solar Age,* October 1979, pp. 40-46.

- "Double-Wall House: Minimizes Heat Loss and Fuel Bills," *Popular Science,* October 1980, pp. 40-46.

- *A Design and Construction Handbook for Energy Saving Houses,* A. Wade, Rodale Press.

Air quality

- "Insulation Trap: Saving Energy Can Be Hazardous to Your Health," *Mechanix Illustrated,* September 1980, p. 82.

Air-to-air heat exchangers

- *An Air to Air Heat Exchanger for Residences,* P. Besant, *et al.,* University of Saskatchewan, Saskatoon, Saskatchewan, Canada, 1980.

- Mitsubishi Enthalpy Exchanger, catalog #N780760-02, Mitsubishi Electric Corporation [Wisconsin dealer is Solar Specialists, 6701 Seybold Rd., Madison, WI 53719, (608) 271-3668].

Insulating window treatments

- *Insulating Shades and Draperies,* University of Wisconsin-Extension bulletin.

- *Movable Insulation, A Guide to Reducing Heating and Cooling Losses through Windows in Your House,* by W. Langdon, Rodale Press, 1980.

- Construction plans from Creative Energy Products, 1053 Williamson St., Madison, WI 53703, (608) 256-7696.

Construction manuals

- *How to Build a Superinsulated House,* Cold Weather Edition, PO Box 81961, College, AK 99708.

- *Low Energy Passive Solar Housing,* Energy Resource Development, University of Saskatchewan, Saskatoon, Saskatchewan, S7N 0W0, Canada, 1980.

- *Designing a Truss-Frame House for Energy Efficiency,* G. Hans, U.S.D.A., Forest Products Lab, Madison, WI 53706.

- *Reducing Home Building Costs with OVE Design and Construction: Guideline 5,* NAHB Research Foundation, Inc., PO Box 1627, Rockville, MD 20850.

- "Superinsulated Houses," *Solar Flashes,* 512 Ross Ave., Alamosa, Colorado 81101, September 1981, pp. 3-6.

James W. Buesing is an Extension Specialist in Housing and Interior Space in the University of Wisconsin-Extension and Assistant Professor, Environment, Textiles and Design, University of Wisconsin-Madison, 1300 Linden Drive, Madison, WI 53706, (608) 262-2611.

Cleaning the Air: Heat Exchangers in Energy-Efficient Houses

TIGHTLY built structures keep heat loss to a minimum by limiting air infiltration and ventilation, so they need air-to-air heat exchangers to maintain indoor air quality. The designers of the well-known Saskatchewan Conservation House (reviewed August, 1978 and May, 1979 in SEN III and VI) have developed an ingenious heat exchanger for this purpose. Their design is extremely simple and can be used by any home builder. The exchanger employs a counter flow design with .15 mm polyethylene plastic sheets to separate the entering and exiting air flows. A common kitchen ventilator fan blows air through 1.27 cm spaces between the sheets. The typical house of 140 m² needs a total heat exchanger surface area of 50 m². The University of Saskatchewan research group achieved this area using 37 polyethylene sheets, and sandwiched them into a compact plywood box measuring 200 × 60 × 45 cm.

Several factors affect the heat exchanger's efficiency. The direction of air flow through the unit is critical to its performance. The exchanger is positioned for vertical flow, with warm air travelling downward and cold air travelling upward in alternate layers. The efficiency of heat recovery also depends upon air flow rates through the exchanger, and drops with higher rates of flow.

A tightly built house of 140 m² with one exchange of interior air every two hours could require an air flow rate of 175 m³/hour. At this rate, the heat recovery ratio is about .9 (temperature change of incoming air divided by the temperature difference between inside and outside air temperatures). In the harsh Saskatchewan climate, where outdoor temperatures drop to -35° C, if indoor temperature is maintained at 20° C, the exchanger recovers sensible heat at a rate of 1900 watts. This means that under normal operating conditions, incoming air will be cooled or heated to within a few degrees centigrade of the interior air temperature.

Problems With Polyethylene Elements

Two major limitations of the Saskatchewan heat exchanger derive from its use of polyethylene sheets. Because this substance is impermeable to vapor, it can not precondition the humidity of incoming air. A separate system to humidify winter air and remove excess moisture during summer is needed to maintain comfort levels. In addition, moisture

The Lossnay Element

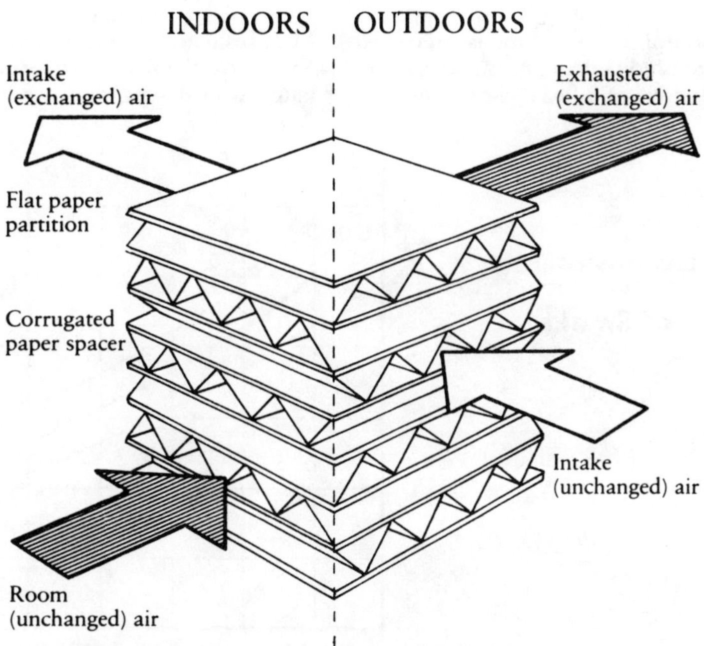

Exchange of "total heat" is the Lossnay's edge over conventional exchangers that cannot exchange latent heat. Fresh and stale air streams pass each other, completely separated by paper partitions. As the streams cross, sensible and latent heat are exchanged through the treated paper.

tends to condense within the polyethylene element and in especially cold climates ice forms, restricting air flow and reducing heat transfer across the partitions. To defrost the unit, exterior vents are blocked and warm interior air circulated through the exchanger, a procedure that may be necessary as often as once a day. An efficient method of continuous ice removal from the exchanger would improve both performance and convenience.

The Saskatchewan heat exchanger, with a materials cost of $150, is an inexpensive solution to the air quality problems of tight houses. More convenient units, though, are available fully assembled at only a slight increase in cost. The Mitsubishi Electric Corporation markets an advanced residential-sized heat exchanger for just $200, part of a line that includes much larger commercial machines. They all use an ingenious treated paper element, called a Lossnay, as a medium for transfering both sensible and latent heat. Contrary to common belief, paper makes an excellent transfer medium for air-to-air heat exchanges and suffers from none of the moisture disadvantages of polyethylene.

Stale exhaust air and fresh air pass through channels formed by the plates and fins of the treated paper. Latent heat is transferred when humidity in the air migrates across the paper partition through capillary action and is then revaporized into the low-humidity air on the other side. Because it transfers both latent and sensible heat, the Lossnay is extremely effective as a year-round air preconditioner: intake air is precooled and dehumidified in the summer, preheated and humidified in the winter. This

feature makes the Mitsubishi heat exchanger more useful than the Saskatchewan unit, as well as more convenient.

A heat transfer rate of 70% can be obtained for both sensible and latent heat, though efficiency varies inversely with the rate of air flow. Two fans drive air through the exchanger, one positioned to draw air into the system, the other to force air out. The fans draw the only power the system needs.

The Units Compared

Mitsubishi exchangers range in capacity from 80 m³/hour for residential use, to 7000 m³/hour for commercial establishments. The residential units are compact, resembling window air conditioners, and draw approximately 45 watts. On a winter day with an outside temperature of 0° C, an interior temperature of 20° C, and at an air flow rate of 100 m³/hour, the Lossnay element will recover heat at a rate of approximately 800 watts. The Mitsubishi heat exchanger, priced at only $50 above the materials cost of the Saskatchewan design, and capable of pre-

conditioning the humidity of incoming air, would seem to offer more for less.

The total costs of an air exchanger system depend as much upon structural characteristics and occupant behavior as upon unit price. In a typical house of 140 m², twelve full air exchanges per day (0.5/hr.) may be necessary if the inhabitants smoke tobacco or use incense, fry food frequently, eat beans or enjoy the company of large domestic animals. Two of the Mitsubishi or Saskatchewan exchangers would be needed to avoid atmospheric pungency. Homeowners with a less aromatic lifestyle, however, might require an exchange rate of only 0.25/hour in their homes, in which case one unit would be adequate.

The costs of air quality control through heat exchangers, as in so many of the technologies related to energy efficient houses, remain hard to pin down. The economics of energy efficiency seem to depend on both technology and human behavior, and more research is needed on how these two factors interact.

—*Craig Conley*

References:

Besant, R.W., E.E. Brooks, G.J. Schoeu, and R.S. Dumont
    1979 *Design of Low Cost Ventilation Heat Exchangers.* University of Saskatchewan, Saskatoon, Canada S7N OWO.

Besant, R.W., R.S. Dumont, and D. Van Ee
    1979 *An Air to Air Heat Exchanger for Residences.* Department of Mechanical Engineering, University of Saskatchewan, Saskatoon, Canada S7N OWO.

Mitsubishi Electric Corporation
    1979 *Mitsubishi Enthalpy Exchanger.* Catalog #N7807 60-02.

Thomas, Mike J.
    Environmental Products, Mitsubishi Electric Industrial Products, MELCO Sales, Inc., 3030 East Victoria St., Compton, California 90221 (800) 421-1132.

Sensible and Latent Heat

Sensible heat, when applied to a gas or liquid, raises its temperature. But during a phase change, such as when a liquid becomes a gas, heat may be absorbed while the temperature of the substance remains constant. The heat required for the phase change is called **latent heat.** In the case of water at normal atmospheric pressure, one calorie will increase the temperature of one gram by one degree centigrade, but 540 calories are needed to bring it from the boiling point to a vapor. This quantity, often referred to as the **latent heat of vaporization,** is released when the phase change is reversed and the vapor, or humidity, condenses back into water.

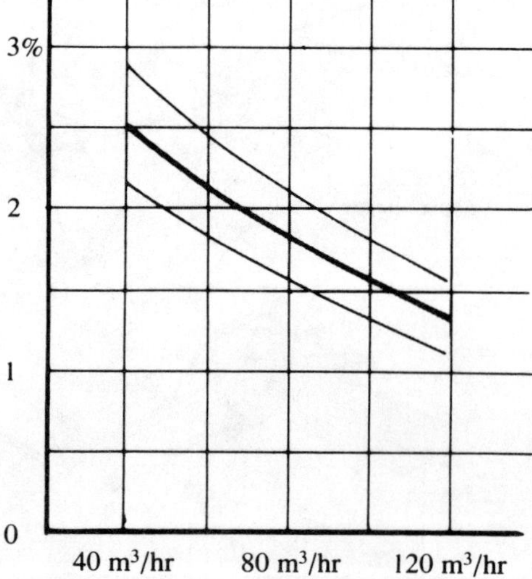

Transmission rates of gases and other impurities through the Lossnay element vary inversely with the volume of air flow. Equipped with exhaust and feed fans, the Lossnay transmits smoke at a rate of 1-3%.

Underground Movement Grows

Earth Shelters

"Earth shelter" may conjure up images of dark caves and damp leaky cellars. But, an underground house can mean great energy savings, minimal disruption of the natural landscape, and low maintenance costs.

EARTH SHELTERS, growing in popularity as the cost of energy rises, are not new. The subterranean option is a time-honored response to climate variations. For more than 5000 years, the Chinese have stored food in underground cellars, and today, in the provinces of Honan, Shansi, Shensi, and Kansu, some ten million Chinese exploit the insulative qualities of the loess—the soft, loamy soil found throughout this region.

Saving energy is the primary motive for today's move downstairs. The energy use of a building is a function of its surface area, its air leakage rate, and its insulation. Earth shelter design affects all three properties. Covering part of the structure reduces the surface area through which convection occurs and lowers the building's heat loss. Uncontrolled air infiltration through cracks and holes is also significantly reduced or eliminated through earth covered construction, and while soil itself is not considered a good insulator, it is an excellent moderator. Whereas seasonal temperature fluctuations eventually reach a depth of several meters into the soil, conduction is so slow that short-term thermal fluctuations only reach an inch below the surface.

Heat can also be transferred down through the soil by warm water percolation. But in winter, when the surface freezes, water is prevented from transporting heat and the land surface is insulated by a blanket of snow.

Thermal lag, the time it takes for climatic conditions to penetrate soil, facilitates the heating and cooling of earth sheltered housing. A minus 30° F winter chill will penetrate and cool the earth through April, May, and June. The hot summer months of July, August, and September add thermal ballast to the earth, warming it through the months of October, November, and December (Don Metz, 1979).

The thermal mass or heat capacity of the building is another important energy consideration. With a great deal of earth cover, or a large thermal mass, the temperature within the house remains almost constant. Solar energy coming through south-facing windows slowly heats the mass and is stored throughout the day. The daytime temperature inside will rise slightly; at night there will be a minimal temperature drop as heat from the thermal mass transfers to the inside air. A moderate thermal mass can not store as much energy per degree rise—incoming solar energy will become uncomfortably high dur-

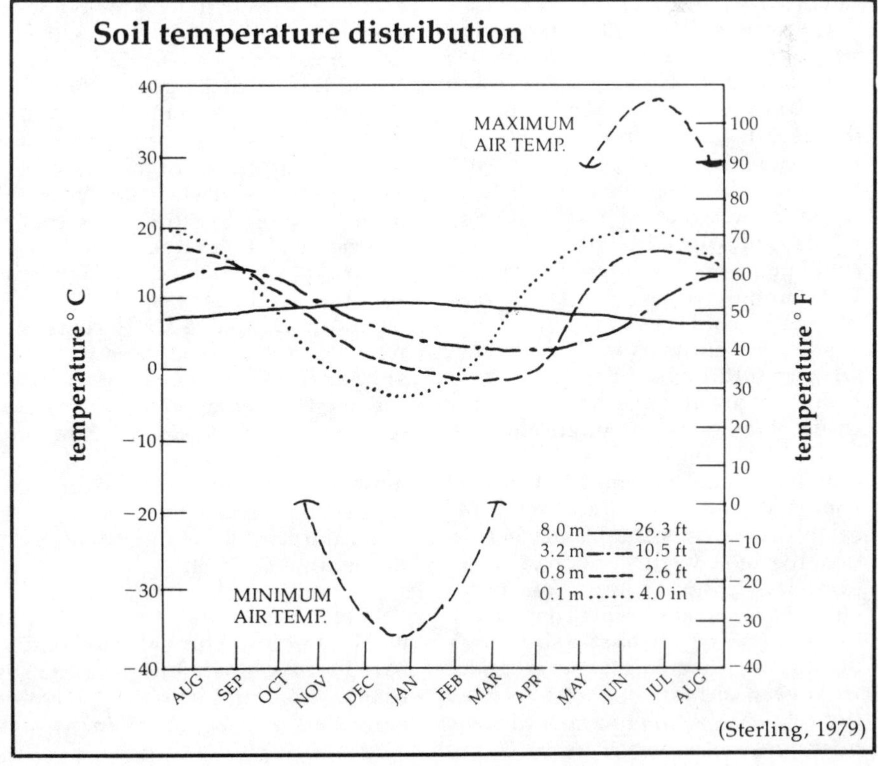

(Sterling, 1979)

ing the day, and at night, the temperature will fall rapidly requiring heating.

Three basic plans

Energy conservation is a major determining factor in architectural design for earth shelters, and in general, the larger the earth mass surrounding the structure, the more efficient it will be. "From an energy conservation view alone," points out Ray Sterling in *Earth Sheltered Housing Design*, "the ideal design would be a totally enclosed chamber well below the surface." Of course this would be uninhabitable and in violation of building code restrictions.

Practically, there are three basic plan concepts for earth shelters—elevational, atrium, and penetrational. Size and orientation of window openings account for the differences in the three concepts. The elevational model concentrates all openings on one side of the structure, usually the south, leaving the other three sides buried. The idea is to maximize earth cover around the house. Unfortunately, this limits the amount of space that can receive adequate natural lighting from one window wall. Skylights penetrating the earth over the roof may be needed to bring in more light but present potential energy drains. A south-facing skylight, on the other hand, will permit solar radiation into the house in winter, screen it out in summer, and can provide ventilation.

The atrium plan places the living spaces around a central courtyard which brings in natural light and ventilation. It is neither as compact as the elevational plan, nor does it face all its windows to the south. However, the courtyard does tend to trap sun-warmed air, reducing heat loss somewhat. And, if need be, the courtyard could be covered with plastic or glass. This model, however, may be very costly by comparison.

The penetrational plan has windows which cut through the soil in various locations around the house. Of the three plans, this one offers the best views, lighting and ventilation but usually at the expense of the best energy results since puncturing the earth's mass robs it of some thermal benefits. A few large windows, however, have less of an impact on energy efficiency than many small ones.

The roof of an earth shelter is potentially the greatest energy drain: in severe climates, heat loss may typically exceed 50 percent of the total heat losses through the building

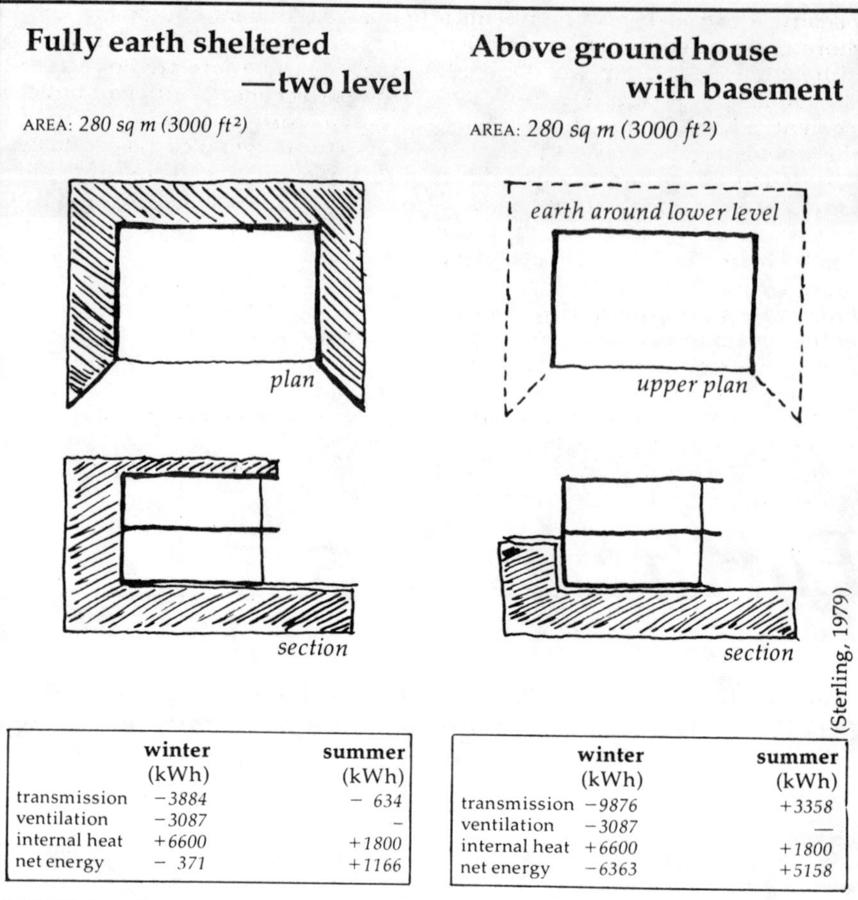

	winter (kWh)	summer (kWh)
transmission	−3884	− 634
ventilation	−3087	−
internal heat	+6600	+1800
net energy	− 371	+1166

	winter (kWh)	summer (kWh)
transmission	−9876	+3358
ventilation	−3087	—
internal heat	+6600	+1800
net energy	−6363	+5158

envelope. For this reason, a thick earth cover or a combination of soil and insulation is crucial. Structural limitations, however, determine optimal soil depth. "As the weight of the soil increases," Sterling points out, "the cost of such a structure quickly escalates beyond the benefits accrued from the additional mass. This can result in the interior spaces being subdivided into smaller areas due to the shorter spans required for high loading factors. Therefore, it is usually preferable to increase the R-value of the roof by adding insulation rather than soil once the load limit of the lighter supporting structure has been reached."

In Minneapolis, a study comparing two alternative roof designs with similar R-values, demonstrated that soil must exceed depths of 2.75 m (9 ft) to compete effectively with standard insulating materials. Increasing soil depth on the roof, on the other hand, will reduce heat losses through the walls and floor by plunging the house deeper into the earth.

In another study of insulation for earth shelters, the value of energy saved was compared to the extra cost of insulation for two roofs. Both roofs consisted of a 20 cm (8 in) precast concrete supporting structure; the first roof had 46 cm (18 in) of soil and ten cm (4 in) of polystyrene insulation, the second had an additional five cm (2 in) of polystyrene added to the layer of insulation. R-values for the two structures were 4.35 m²-°K/W (24.68 hr-ft²-°F/Btu) and 6.01 m²-°K/W (34.10 hr-ft²-°F/Btu) respectively. Based on the Minneapolis weather data, the additional five cm of insulation accrued a savings of 6.67 kWh/m² during the winter, and 0.29 kWh/m² during the summer for a total savings of 6.96 kWh/m². Using a cost of $0.03/kWh for electricity, this equals a savings of $0.21/m² per year while a projected cost of $0.10 kWh for electricity gives a total benefit of $0.69/m² per year for the added insulation. If the insulation sells for $5.38/m² per five cm of thickness ($0.25/board foot), then the payback period for the extra insulation will be 26 years at the rate of $0.03/kWh and eight years at the rate of $0.10/kWh, without any interest charged to the capital investment (Sterling, 1979).

Energy comparison

A study of earth shelter houses revealed that the plan area of a single level home could be doubled to two levels without incurring an energy penalty. A two-level plan represents a more compact configuration so that although the actual area is doubled, the surface area is increased by only 30 percent. Plus, the deeper the walls, the less susceptible they are to climate conditions. Compared with a two story, conventional above-ground home with a walkout basement, the earth shelter was far and away the more efficient. In spite of the sub-ground basement, the surface portion of the conventional house reduced its energy performance significantly (Sterling, *et al.*, 1979).

Much of the energy required in buildings is wasted due to lack of foresight during design, inadequate construction, poor operating prac-tices, inefficient equipment and un-necessary lighting, heating and cooling. Clearly, there are no recipes for the perfect energy efficient building. As with surface housing, earth shelters employ energy saving features selected on the basis of climate, orientation, planned use, and a multitude of other architectural considerations.

Earth shelters are here to stay. Although underground housing is relatively new, earth shelter advocates maintain energy and economic savings will provoke more and more people to investigate life underground.

—*Katy Slichter*

References:

Bligh, Thomas

1975 *Alternatives in Energy Conservation: The Use of Earth Covered Buildings.* Proceedings of a conference held in Fort Worth, Texas, July 9-12, 1975, pp 85-105.

Metz, Don

1979 "The Latest-Not the Last-Word in Underground Houses Design." *Solar Age,* October, 1979, pp 36-39.

Sterling, Ray *et al.*

1979 *Earth Sheltered Housing Design.* Underground Space Center, University of Minnesota, pp 31-94. Available from Van Nostrand Reinhold Company, 135 West 50th Street, New York, NY 10020, $10.95.

☐ Commercial buildings

"I have watched a new office building going up across the street. Featuring sealed single-glazed windows, it faces the summer sun on the south and west without structural shading to cut the heat. It lacks insulation; lacks a system to capture and use waste heat from Xerox machines, lights, and computers; lacks controls to manage energy; lacks a solar hot-water heater. These simple, available, and proven measures would save tens of thousands of dollars in operating costs, yet they have been left out of that new building. The builder doesn't know much about these options and doesn't care to learn about them. Why should he? He won't be paying the energy bills, the tenants will. And his commercial tenants will charge you and me more for goods and services to cover this builder's inaction."

—JONATHAN LASH, President
Energy Conservation Coalition

Reprinted from the *Energy Conservation Bulletin*, volume 1, number 1.

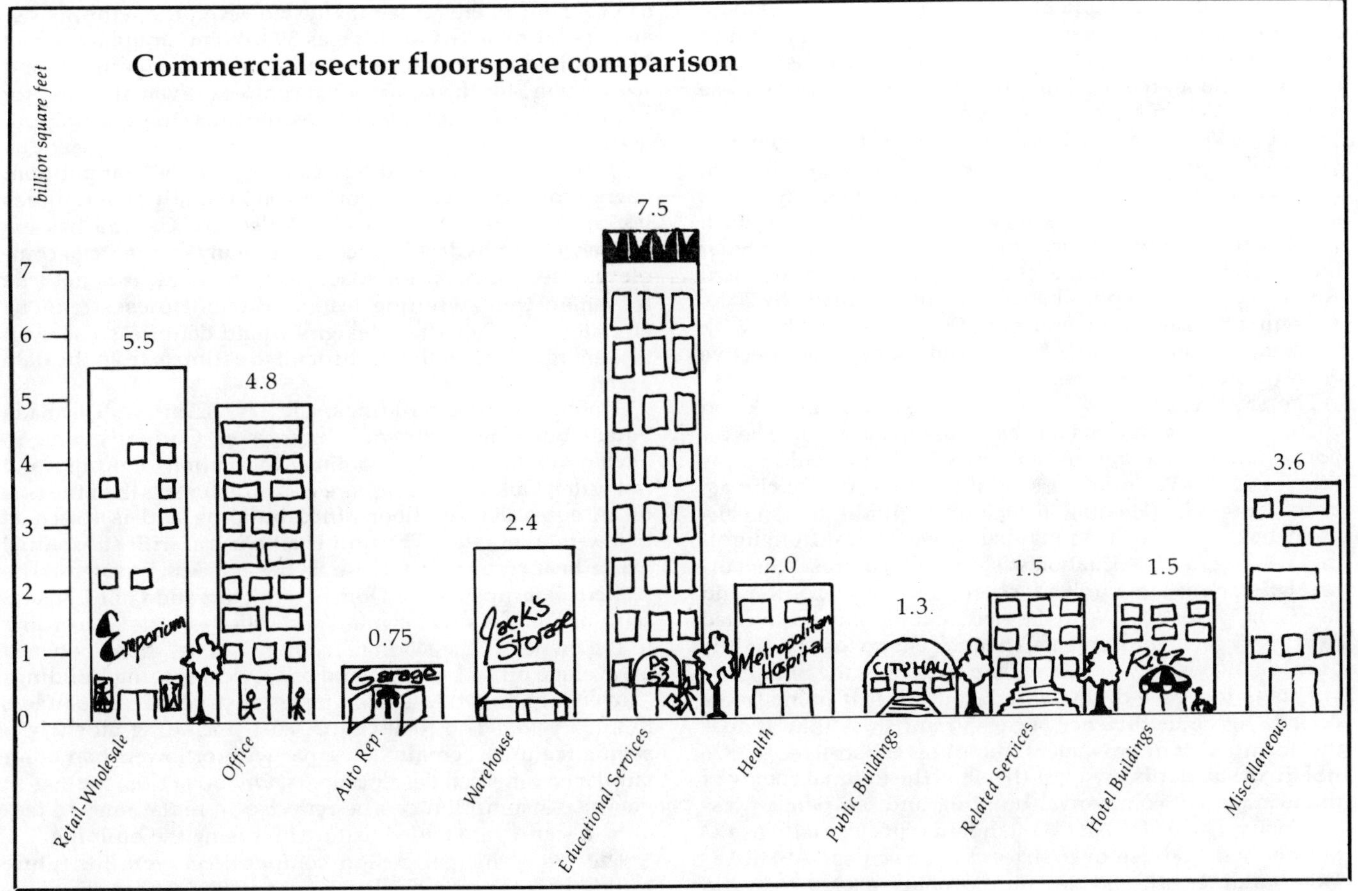

High-rise, low efficiency

OF ALL THE SECTORS of the US economy that have responded to higher energy prices by conserving, the slowest and least well understood is commercial buildings. For many years, energy forecasters made the commercial sector their "other" sector, lumping together all the energy use that didn't fit in "residential," "industrial," or "transport." Into this sector fell all government buildings, the military (this eventually fell out), schools, hospitals, large and small office complexes, and retail stores. Not only were the energy use habits of the sector unfathomed, but so too was the analysis of policy and economic measures necessary to determine whether conservation could occur.

As recently as 1978, electric utilities in California (and probably elsewhere) concealed their (then wished-for) expected growth in the commercial sector, rather than in the better understood residential or industrial categories. One such utility projected a doubling in energy use per square foot of commercial floorspace by 2000, but how that might come about was not disclosed. Even today, such basic information as floor area, building type, ownership, or end use energy utilization are barely known.

As a result, commercial buildings represent a largely untrammelled area for energy conservationists. In the US, commercial buildings use about 10.3 quads of energy, rising in government projections to about 13.3 quads in 2000, without the benefit of much conservation. The recent Solar Energy Research Institute study on an energy efficient America projected a possible decline to 7.26 quads by 2000, assuming the same economic growth in the sector, but with implementation of much better and more cost-effective energy efficiency measures.

The SERI report is not as definitive as one might like on energy efficiency options for the commercial sector. The report estimates that a year 2000 office building could use as little as 62-78 kWh/m²/yr, essentially irrespective of climate or building type. Heating in such efficient buildings is virtually negligible, as it is provided by waste heat from lighting, cooling, and ventilation systems. By contrast, the current US stock uses about 195 kWh/m²/yr in electric power and 1.75 GJ/m²/yr for heat. With the electric component converted to primary fuel requirements, the total energy use is slightly over 1,000 kWh/m²/yr, or more than five times the SERI goal.

This is indeed a dramatic reduction, but it points more directly to the inefficiency of the current stock than it does the potential for improvement. Buildings constructed to 1976 ASHRAE standards used less than half the thermal energy of the average US commercial building and ten percent less electricity. The ASHRAE 1980 standards cut electricity use 33 percent and fuel use over sixteenfold. Even so, ASHRAE-1980 buildings still require more primary energy than did the average Swedish building. Sweden is substantially colder than the US and requires all offices have windows. Nevertheless, primary energy use in the average 1975 building was less than one third of its US equivalent and twenty percent below the 1980 ASHRAE level.

Swedish buildings are also improving. The large Farsta Folksam office building uses half the space heat of the average Swedish office building. The US also has prominent examples of efficient office buildings; one excellent example is the new California State Site One office building, which uses virtually no space heat and only 40 percent of the electricity specified in the 1980 ASHRAE standards. But by far and away the most efficient office complexes are those operating or under construction in Canada.

A recently released 650 - page book, *Winning Low Energy Building Designs*, summarizes the results of an extensive commercial building design competition sponsored by the federal department of Energy, Mines, and Resources. The competition included existing buildings (constructed by 1978), plus designs for a 3,000 m² office building and 20,000 m² hotel-retail-office complex.

As Public Works Minister Paul Cosgrove notes in his introduction, "A typical (Canadian) office building that hasn't benefitted from energy conservation measures uses about 646 kWh/m² of energy annually. However, the winning designers estimate that as little as 39 kWh/m² annually would be required for the office building in the competition, and 100 kWh/m² for the commercial complex. Even allowing for some optimism, these estimates illustrate the exciting potential.

"In the existing buildings category of the competition, energy consumption is as low as 106 kWh/m². That figure is not an estimate, it is a reality." Because Canada has extremely cheap hydroelectric power, many commercial complexes use electricity for space heat, though this is not true for some of the 34 winning designs. Brought to less strenuous climates, the Canadian designs would doubtless use even less energy, though this is difficult to estimate from the data given.

Among existing buildings, the 171,602 m² Gulf Canada Square building is extremely impressive. Currently using 96 kWh/m²/yr for heating, cooling, ventilation, lighting, and hot water, Gulf Canada Square was built for less than the cost of an equivalent 20-floor office building and is leased at below-market rates. The unit is all-electric with substantial waste heat recovered for use in the hot water system. The principal occupants are Dome Petroleum and Gulf Canada, with numerous smaller retail and office tenants. According to the owners, "the savings in energy costs" alone between an average office building and energy conserving buildings "are enough to retire the complete construction costs in less than 35 years. The total capital cost, including all energy-saving features, remains on a par with or lower than other similar commercial developments. On top of this, savings in energy consumption can be reflected in rental rates to tenants, a significant added feature in leasing the building."

The two principal design competitions were for future buildings in Regina, Saskatchewan (6,000 degree days Celsius) and Sherbrooke, Quebec. Detailed computer runs on energy use and building cost were required, as were extensive architectural drawings. The total prize money for the competition, including the existing buildings competition, totalled $250,000. The winners were not necessarily the most energy efficient units. For the 3,000 m² Sherbrooke office competition, estimated energy use among the winners ranged from 32-116.5 kWh/m²/yr. Five of the fourteen winners used between 30-50 kWh/m²/yr. Eight used less than the SERI year 2000 goal, despite much more rigorous climatic limits. Most of the Regina winners fell between 100-150 kWh/m²/yr.

—Jim Harding

Reference:

Department of Energy, Mines, and Resources (Canada)

1981 *Winning Low Energy Building Designs*. Canadian Government Publishing Centre, Supply and Services Canada, Hull, Quebec, Canada K1A OS9; $19.95 in Canada.

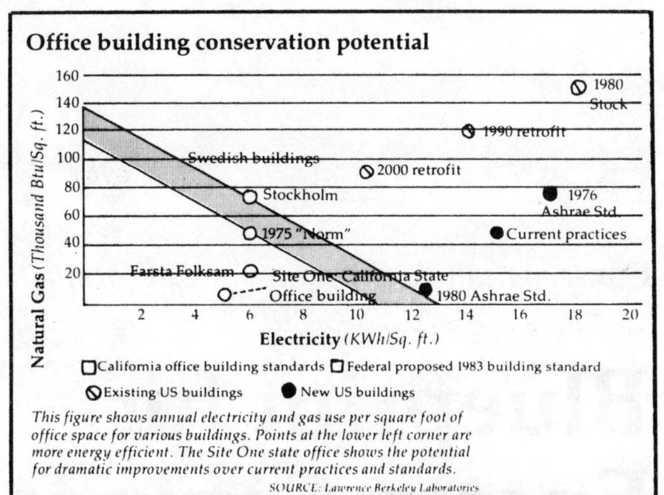

This figure shows annual electricity and gas use per square foot of office space for various buildings. Points at the lower left corner are more energy efficient. The Site One state office shows the potential for dramatic improvements over current practices and standards.

SOURCE: Lawrence Berkeley Laboratories

Building	GJ/m²/yr Electric Use (End Use)	GJ/m²/yr Heat (End Use)	GJ/m²/yr Primary Energy
1980 US Stock	0.70	1.75	3.93
ASHRAE (1976)	0.63	0.85	2.82
ASHRAE (1980)	0.47	0.11	1.58
Sweden stock (75)	0.23	0.54	1.26
Regina winner	(combined: 0.36)		1.12
Gulf Canada Sq	(combined: 0.34)		1.07
Farsta bldg	0.23	0.26	0.98
SERI goal	(combined: 0.22-0.28)		0.68-0.87
Site One (Calif)	0.19	0.08	0.68
Sherbrooke winner	(combined: 0.14)		0.44

Blueprint for Energy Savings

D ESPITE A GROWING population and increasing GNP, energy use in both commercial and residential sectors has dropped slightly since 1973. Energy conservation programs in existing buildings are largely responsible for this reduction in energy consumption; but the most effective programs begin long before the first day of construction. To minimize a building's energy consumption, occupancy scheduling, lighting needs, machinery use, the operation of windows and blinds, and other use patterns should be considered in the primary design stages. Heating, cooling and lighting needs also depend on design factors, including building orientation, layout, envelope, and glazing. California has led the way in energy-conscious design, demonstrating to business how utility bills can be significantly reduced. The first of a new generation of buildings designed to conserve energy, Site One State Office Building in Sacramento, California reflects this approach to energy conscious architecture.

Site One was designed under the direction of former State Architect Sim Van der Ryn by the project team of Calthorpe, Matthews, and Corson, and the building was completed under the tenure of current State Architect Barry Wasserman. This energy-efficient, award-winning design takes advantage of Sacramento's mild climate and fulfills its heating, cooling, and lighting needs using less than 20,000 Btu/ft²/yr. In comparison, a like-sized, nonresidential building which meets California's 1978 building standards (Title 24) would consume approximately 56,000 Btu/ft²/yr. According to the life-cost analysis by the Office of the State Architect, Site One will save an estimated $27 million in energy costs over the 50-year life of the building. At $56.60/ft², this means that Site One will save nearly twice the original $16,630,000 in construction costs through lower energy bills. And note that the construction costs for Site One nearly equalled those for a hypothetical building built to Title 24 specifications—$56.60/ft² *vs.* $56.10.

Design determines efficiency

To a great degree, Site One's energy conserving measures go hand-in-hand with its architectural strengths. A large central atrium, rising through four floors of offices to skylights in the ceiling, provides much of the building's indirect lighting through interior windows facing the central court. Exterior windows also help to reduce lighting needs to 50 percent of the lighting energy used by conventional offices. Fluorescent light fixtures with parabolic reflectors provide additional background lighting, and in specific work areas where higher lighting levels are necessary, more efficient individually controlled task lighting is used.

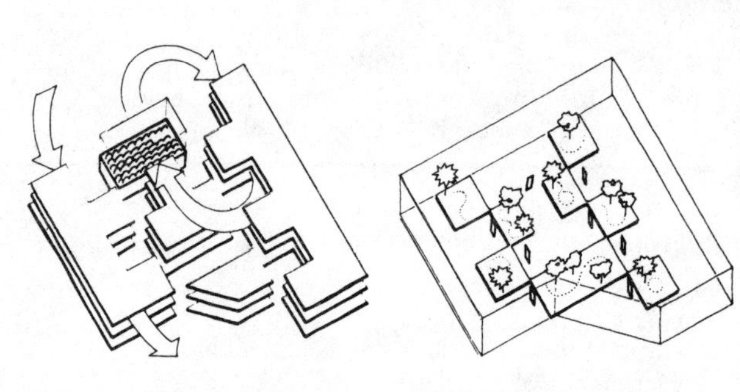

ROCKBED & THERMAL MASS

The concrete floor system acts as a thermal sponge to soak up heat generated inside the building by lights, machines and people. A thermal rockbed amplifies this role by cooling off the supply air mix before it is delivered to the office spaces. Cool nighttime air is used to flush the heat from these masses and prepare them for use the next day.

COURTYARD CHECKERBOARD

A checkerboard of courtyards organizes the 22 different state agencies around separate outdoor territories. This geometry eliminates the double loaded corridor with a landscaped open space always opposite an office block. Office entrances also face courtyards which become landmarks to identify each agency or department.

Heating and cooling of the building is moderated through several systems. One of the most significant design strategies is the extensive use of bare concrete to capture and store both heat and cold. Heat from lights, people, machinery and sun is absorbed by the concrete mass. Using this system, Site One has been able to reduce its cooling load by over 70 percent compared to Title 24.

The Rock Bed Thermal Storage System, another major temperature regulator, consists of two rock beds, each filled with 660 tons of river rock. Set beneath the courtyard floor, these rock beds store coolness or warmth until it is needed. Cool night air blown through the rocks provides 75 percent of the air conditioning needed during the day while reducing peak loads by as much as 50 percent. Fans circulate interior air through the cool rocks and blow it back into the building. This process is reversed in winter when warm afternoon air is drawn into the rock for use in the early hours of the next day to pre-warm the interior.

In addition to these systems, computer-controlled fans and louvers shade the atrium's skylight and double glazed windows. Trellises on the south side regulate heat infiltration blocking out summer sun, but allow the low winter sun to enter. Exterior shades on the east and west sides lower automatically to shade windows from direct sunlight, but rise as the sun moves off the window.

San Jose/Sacramento strides

While Site One is winning much of the publicity for energy efficient commercial buildings, two other buildings are worthy of note—the San Jose State Office Building and the McClellan Air Force Base Showcase Building, both designed by Sol Arc, the Berkeley-based architectural and energy consulting firm. The San Jose State Office building, now under construction, will have a significant role in the city's core area development. As with Site One, two primary considerations influenced its design—an efficient and revitalizing working environment and a maximum reduction in energy consumption. The building's 125,000 ft² maximize daylight by alternating courtyards and offices around an open air pedestrian circulation loop. With daylight contribution conservatively estimated, energy use for lighting is approximately 1.05 W/ft². Down from the installed wattage of 1.6 W/ft², this represents a savings of over $6,000 for the first year of operation, or $430,000 over 25 years at 7.3 percent inflation. Kept to three stories in height, this checkerboard building will employ a thermal rockbed much like Site One's for its heating and cooling needs. Its 14,000 ft³ bed will mean an annual energy savings of about $2,400 and, depending on the control strategy, can allow up to 50 percent reduction in peak demand for air conditioning. Computer simulations have shown that this design will consume 27,000 Btu/ft², or 70 percent less energy per year than a conventionally designed building of the same size and function.

The Showcase Building on McClellan Air Force Base is still on the drafting table. It promises to be an exemplary demonstration of the practicality of using solar in a commercial building. A focal point for visitor instruction and demonstration, the building will house an energy library, a small auditorium, and a display and atrium area, as well as office space and the computer terminals for the base's energy management control system. The 7,600 ft² building will use 6 Btu/ft²/°F-d, or 15,000 Btu/ft²/yr as indicated by computer simulations.

Sol Arc identified lighting as the most significant energy load for the Showcase Building. Three elements will be used to control illumination in the building—1800 ft² of skylights, a continuous expanse of south-facing glass tilted at a 45° slope; a system of operable louvers on the outside of the glass; and a reflective white surface on the ceiling opposite the skylight. Eight hundred ft² of skylights open into office space providing daylighting even on cloudy days. Automatic dimming in response to available daylight is expected to reduce the use of artificial lights by well over 50 percent. Temperature within the building is moderated through the use of thermal mass and a five-foot berm on most of the walls.

Performance considerations

California leads the nation in building for a sustainable, secure energy future. In 1976, the Office of the State Architect, under Edmund G. Brown, Jr., launched a major building program to provide the state with one-and-a-half million square feet of energy efficient office space. The result: a master plan for downtown Sacramento (the Capitol Area Plan or CAP), a major architectural competition (Site Three), one office building designed by the Office of the State Architect (Site One), and six projects designed by private firms.

Four performance measures for these new buildings were set forth in the state publication, "Building Values: Energy Guidelines for State Buildings." The first, a real estate investment measurement, revealed that the State's current leasing policy was costing 20 percent more than an owner-operated policy. The second, energy investment, demonstrated the unpredictable nature of energy futures and took account of its ambiguous but central political realities. The third, cost of operation, was found to be correlated with worker performance rather than construction or energy con-

siderations. Calthorpe notes a standard building's cost breakdown in *Progressive Architecture*: two percent for acquisition, six percent for maintenance, and 92 percent for personnel. The State emphasizes: "A better work environment that improved worker effectiveness by only six and one-half percent would be cost effective even if it quad-

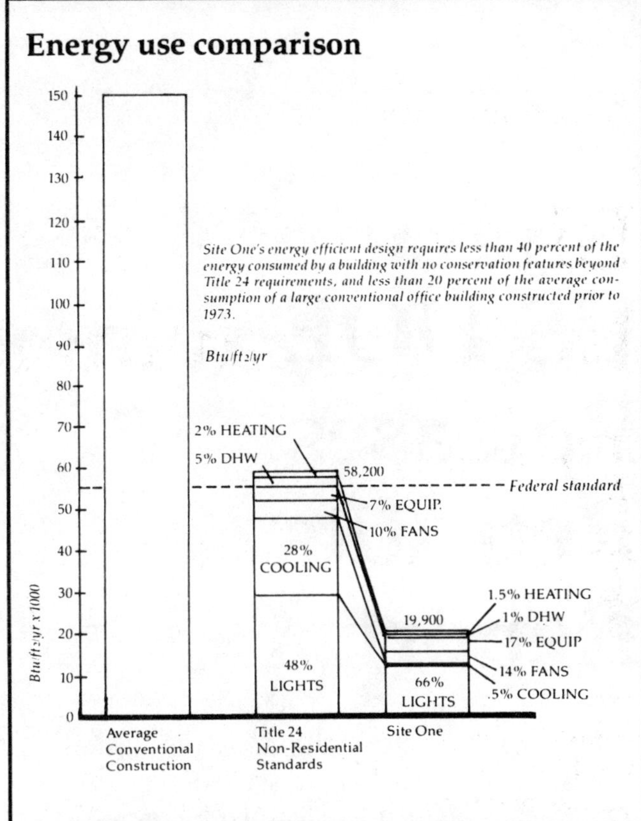

rupled building costs." The fourth measure, infrastructure costs, looks at the tangible functions of state buildings.

The mild climates of both San Jose and Sacramento are particularly amenable to passive conservation strategies. By virtue of this climate, the energy used for lighting usually exceeds heating and cooling loads combined. Since there is no practical way to store light, lighting loads cannot be displaced to off peak hours. Daylighting can reduce these peak electrical loads thereby defraying commercial penalties which are based on the peak electrical demand.

–Katy Slichter

References:
Baker, David
1980 "Daylighting Design for a New State Office Building in San Jose, California." Reprint from Proceedings of the Fourth National Passive Solar Conference, pp 411-415.

Bazjanac, Vladimir
1980 "Architectural Energy Analysis." *Progressive Architecture*, April 1980, pp 98-101.

Calthorpe, Peter
1980 "More than just energy." *Progressive Architecture*, April 1980, pp 117-120.

Holland, Elizabeth
1981 "The Showcase Building." *Solar Age*, March 1981, pp 34-37.

Ridgway, Jeff
1981 "New Site 1A: Designed For Energy Efficiency." *Energy Outlook*, pp 12-14.

Energy use per square meter varied by a factor of 10 and savings varied by a factor of 50.

Ross and Whalen contend that the source of funds for commercial sector conservation has been profits rather than loans, suggesting that these investments must be more pro-

US DOE Releases Retrofits Report

A RECENT SURVEY of progress in commercial building energy conservation suggests that much remains to be learned before potential energy efficiency gains can be reached. Authored by Howard Ross and Sue Whalen of the US Department of Energy, the draft report reviews before-and-after metered experience in 222 separate commercial building retrofits. The average amount saved in the sample was 22 percent, but a disquieting warning is raised that commercial building owners are unlikely to invest in improvements that pay back their cost in more than three years.

The Department of Energy sample contains only those buildings for which monitored information is available at least one year before and after retrofitting. This, in itself, dates the sample and may prejudice the results. The sample mainly consists of schools and office buildings and the retrofits are of a limited nature. (Very few were performed by firms whose profits depend on their ability to reduce energy use, *e.g.* Scallop Thermal Management.) Virtually no innovative measures, such as automatic daylighting, waste heat recovery, passive solar, window coverings, or night insulation, were included. Only three of the 222 buildings installed task lighting and only 25 switched from existing fluorescent tubes to more efficient ones.

The average cost of the retrofits was $0.65 per square foot (or about $7/m²), so that the "cost of saved energy (is) usually a fraction of supply costs." Payback periods were approximately two years. Oddly, the data indicate that there is no clear relationship between dollars spent and energy saved, at

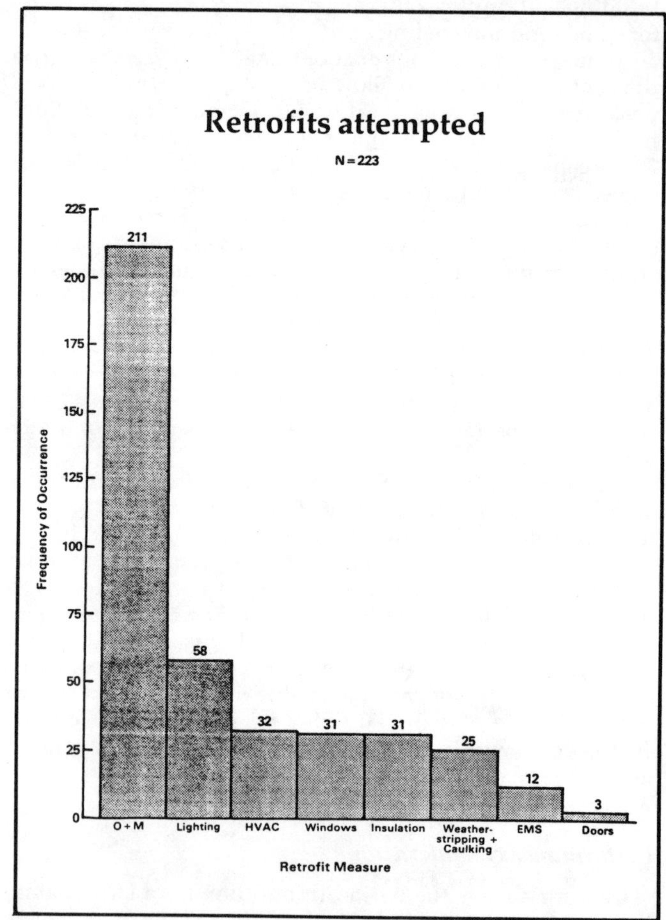

fitable than normal business investments. No consideration is given to 'capital transfers' (see *Notes* 3:5:19) that would compensate such investors that save the electric system more than they save themselves in lowered power bills. (At a current $4/million Btu for natural gas and 5.5¢/kWh for electricity, the average square meter of commercial floorspace uses $17.33 per year for energy. Assuming escalation and interest payments cancel each other out, and gas and electricity are equally saved, a $7/m² investment that saves 22 percent pays back in two years. At marginal system costs of $7/million Btu for gas and 10¢/kWh for power, the same energy would cost $31.06 annually. The utility—and its customers—could well afford to transfer at least half this difference and make the entire investment.)

Of course, not all commercial buildings are susceptible to economic pressure. Many are publicly owned or non-profit: schools, hospitals, government buildings, and the military. Capital and operating budgets for such entities are often segregated, and the concept of rate-of-return is simply alien. In such cases, heavy utility involvement may be most welcome. Still another problem in the commercial sector is the high proportion of renters, most of whom pay for their energy as a separate line item in the lease. This removes any incentive from the landlord for improving building performance, except that associated with keeping overall leasing costs down. Again, utility investments appear especially interesting, but most utility regulators are uncomfortable with the idea of "subsidizing" energy use through higher residential rates (even if these rates are lower than they would be with no commercial sector investments).

Finally, the physics of energy use in a large commercial building remains mysterious. Ross and Whalen found that within the commercial sector, energy use per square meter varied by a factor of ten and savings varied by a factor of fifty. It'll be many more designs, data collections, and programs later before most of the answers are known.

—Jim Harding

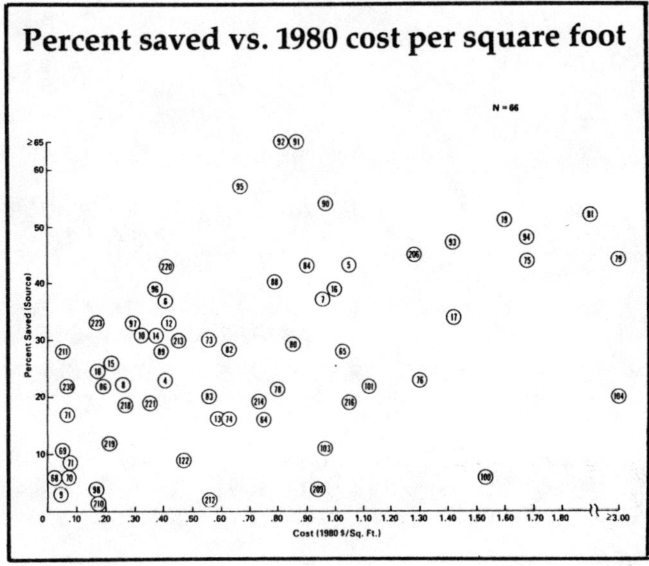

2. Appliances and Lighting

It seems that we have been too lenient towards our electric housemates. Just because they were modern and ran on a modern form of energy, we assumed that refrigerators and clothes dryers were a bargain. But household appliances are a big energy drain, as Jørgen Nørgaard shows. His exemplary engineering analysis of Danish appliances (which are representative of those commonly used in Europe) shows that there is potential for cost-effective electricity savings of 75% when averaged over all devices. His proposals for redesign apply to the larger American appliances as well.

This large potential comes from three elements.

● Within an appliance, electricity is often used to do something that is not electricity-specific but thermal in nature: to provide coolth (refrigerator, freezer) or warmth (range, washing machine water, dryer). As we know from buildings, appropriate attention to heat insulation and temperature is the key to drastically improved thermal applications.

● In many appliances, less expensive forms of heat—solar heat, district heat or natural gas—can be substituted for electricity without undue inconvenience.

● Even where electricity is best for the task (lighting, TV, pumps, fans, compressor drives, etc.), new technologies, improved electronics and careful component matching offer major efficiency gains.

If the inefficiency of electric appliances and motors surprises, so does that of gas- or oil-fired heating furnaces. With name-plate efficiencies of up to 75 percent, their actual efficiency in operation is typically less than 50 percent. Furnaces, like car engines and electric drives, suffer from the "part-load syndrome": designed for a condition of high load that rarely occurs, they are very poor performers under normal load conditions.

Can we go out and buy highly efficient appliances? Not yet, unfortunately. The refrigerator tells the story: While Larry Schlussler's prototypes have come close to Nørgaard's target, US manufacturers over the last five years have improved their products by a mere 20 percent. Meanwhile, Japanese refrigerators show an improvement of up to 50 percent over the same period. With all those taxpayer bail-outs and worker give-backs, why should General Electric learn anything from the demise of Detroit? American consumers, meanwhile, might soon be putting their eggs into a Toshiba.

Danes Design Efficient Household Appliances

Savings of 23, 53, and 70 percent Possible

Much of the famed 7 percent historical electricity growth rate came from purchases of fancy new appliances in residences: central air conditioners, frost-free refrigerators, freezers, and pong games. As householders began to stumble over these contraptions on the way to the garage, it was a logical conclusion that electric growth rates would begin to slow down. That, of course, has happened. Households are "saturated" with most major appliances.

Now as energy prices rise with no end in sight, technicians have begun to look for ways to build appliances that do their job with less electricity. Jørgen Nørgaard, a physicist at the Danish Technical University, has recently completed a review of technical prospects for improving the efficiency of important household appliances. He concludes that, for a house with all the major appliances and plenty of minor ones, reductions of

TECHNICAL POTENTIAL FOR ELECTRICITY CONSERVATION

APPLIANCES	NORMAL USE	"MODERATE MEASURES"			"STRONG MEASURES"			"RADICAL MEASURES"		
Range	950 kWh/yr	845	$0.00	0	540	$63.00	6	440	$99.00	7
Refrigerator	550	345	9.00	1	200	23.00	5	90	115.00	9
Freezer	800	480	14.00	2	270	42.00	4	145	126.00	8
Washer	575	460	0.00	0	200	25.00	6	71	25.00	6
Clothes Dryer	625	440	45.00	4	260	142.00	11	130	187.00	9
Dishwasher	650	480	0.00	0	285	27.00	5	95	27.00	8
TV (b&w)	165	120	0.00	0	120	0.00	0	120	0.00	0
TV (color)	275	130	0.00	0	130	0.00	0	130	0.00	0
Stereo	45	45	0.00	0	45	0.00	0	45	0.00	0
Elect. for Furnace	480	480	0.00	0	190	45.00	4	140	45.00	4
Sm. Appliances	115	115	0.00	0	100	0.00	0	100	0.00	0
Lighting	750	675	0.00	0	525	0.00	0	350	0.00	0
HOUSEHOLD W/ALL	5815	4495		-	2745		-	1736		-
HOUSEHOLD AVG.	3590	2888	18.00	-	1888	146.25	-	1300	319.50	-

Marginal costs are marginal from base cost, not from the neighboring column.

23 percent are possible with moderate design changes, savings of 53 percent are possible with strong stuff, and savings as great as 70 percent can be achieved with radical techniques. In an economy like California's, where fully 25 percent of electric sales end up in appliances, such improvements are not to sneeze at. The costs: $7.20 per capital for all the moderate changes, $58.50 for the strong ones, and $127.80 for the radical alterations. With Danish electricity prices at 5.4 cents per kilowatthour, payback periods range from zero for many of moderate measures to nine years for a six fold reduction in refrigerator electricity usage.

All of these reductions, of course, are from technical design changes that in no way detract from functions performed by appliances. TVs are solid-state, refrigerators have additional insulation and more efficient motors, and freezers are horizontal, so that the cold air doesn't fall out whenever someone opens the door.

The chart below shows all the major Danish household appliances, their current (1975 model) use, the cost and payback periods (in years) for technical improvements, and the resultant energy consumption (in kilowatt hours per year). Of course Danish refrigerators and freezers are somewhat smaller than American models, but the percentage improvements can still be translated to larger models. In fact, larger refrigerators would benefit more from efficient design than smaller units, so a straight percentage reduction is conservative.

It also seems that Danish models are more efficient to begin with, allowing only televisions as an exception. An average Danish refrigerator (7.8 cubic feet of refrigerated space) with a freezer cabinet uses about 550 kWh per year, frost-free. European producers are now down to 400-450 kWh in new models. By comparison, the average California frost-free refrigerator uses over 1700 kWh per year (16-17 cubic feet). A refrigerator of the same size as the Danish model uses about 1100 kWh per year. Recent standards applied by the California Energy Commission reduce these amounts to 1260 kWh for large units and 900 kWh for one of the Danish size.

Danes also use central furnaces that provide hot water and space heat, using oil as a fuel and electricity to move the heat around. Consumption by this appliance is much higher in Denmark than in the US.

To take a California example, because numbers for Califor-nia are at hand, we can assume that regulatory commissions will require stiffer appliance standards in the years ahead, standards that approximate the Danish "strong" measures. If all new households purchased efficient appliances, beginning in 1980, electricity consumption by 1990 would be reduced by the equivalent of two 1000 megawatt powerplants. And this assumes that no existing appliances, whose lifetimes are 15-20 years, would be replaced by new, efficient machines. Use of the moderate measures would reduce demand by the equivalent of 1200 megawatts and use of radical stuff would reduce it by the equivalent of 2300 megawatts. Tacked onto the Energy Commission's forecast of peak requirements, which includes less than moderate standards for freezers and refrigerators plus standards for air conditioners and heat pumps which the Danes didn't consider, these improvements reduce growth in peak demand from 2.3 to about 2.0 percent per year through 1990. This only goes to show that a few tenths of a percent in the growth rate makes a big difference in the need for megawatts thirteen years out.

Economic savings from conservation are also great. Each new household built without strong conservation equipment costs consumers over $2700 in unneeded electric plants. This is a cost that $146 worth of improved appliances would make unnecessary. With about a million and a half new households added between 1980 and 1990, according to current projections, consumers would pay $215 million on appliances instead of watching electric utilities collect about $3800 million for redundant powerplants.

Sources: Jørgen Nørgaard, "Husholdninger og Energi: Tekniske Elbesparelser," chapter 13 of Demo Project Report number 4, 1978; California Energy Commission, "Electricity Forecasting and Planning," volume 2 of the Commission's Biennial Report, 1977; Berman, et al., "Electrical Energy Consumption in California: Data Collection and Analysis," Lawrence Berkeley Lab report, July 1976. Savings in cost are at the California Energy Commission's 1977 dollar estimate of $2100 per kilowatt for a new coal or nuclear plant, including dedicated transmission. Conversion from dollars to Danish krøne has been done at $0.18/kr. Nørgaard's address, for those who wish to contact him, is Physics Laboratory III, Technical University of Denmark, DK-2800, Lyngby, Denmark.
                                                                —*Jim Harding*

Improving Furnace Efficiency

ONE ELEMENT, commonly taken for granted in the performance of conventional gas and oil heating systems is the efficiency of the furnace. Although most standard residential forced air furnaces have nameplate rated efficiencies in the neighborhood of 80%, field tests performed by Dr. Jay McGrew and his colleagues at Applied Science and Engineering (a private research and development firm) indicate that actual performance is dismally lower. The researchers report that the cumulative efficiency—a measure of furnace efficiency over a complete cycle—averaged only 46% in the eight furnaces tested. They further note that differences between their test procedure and actual operating conditions could reduce the real efficiency to around 35%. Several factors contribute to this low rating: short burn time, oversizing of furnaces, improper air flow rates, narrow thermostat ranges, and misadjusted control systems, to cite the most significant.

Their report suggests several cost-effective measures to minimize these problems. One step is a reduction in the existing burn rate to increase the total burn time. This can be accomplished by turning down the main gas valve until the furnace burns almost continuously in the coldest weather, or by reducing the size of existing gas jets while blocking part of the burner assembly to insure proper flame height.

The air flow rate can be optimized by increasing the fan speed. This often involves the mere flick of a switch, but could require rewiring or changing pulley sizes. The authors suggest that shutting off hot air registers will not always conserve heat; it may actually reduce furnace efficiency.

Additional losses can occur at the thermostat. By adjusting the thermostat for a greater span between on and off temperatures and by shielding the thermostat from drafts and flows of hot air, the number of furnace cycles can be lowered, significantly improving efficiency.

It is highly unlikely, the authors assert, that furnaces will ever reach the nameplate performance rating of 80% without major modifications in design. But simple improvements such as those outlined could significantly reduce winter fuel bills.

Finally, a note of warning: consult a professional before attempting any type of furnace adjustment or alteration. Inexperienced tampering can be extremely costly.

—*Craig Conley*

Reference:

McGrew, Jay L., *et al.*
1979 *A Field Study of Furnace Efficiency in Homes.* Available for $6 from Applied Science and Engineering, 3244 South Platte River Drive, Englewood, Colorado 80110.

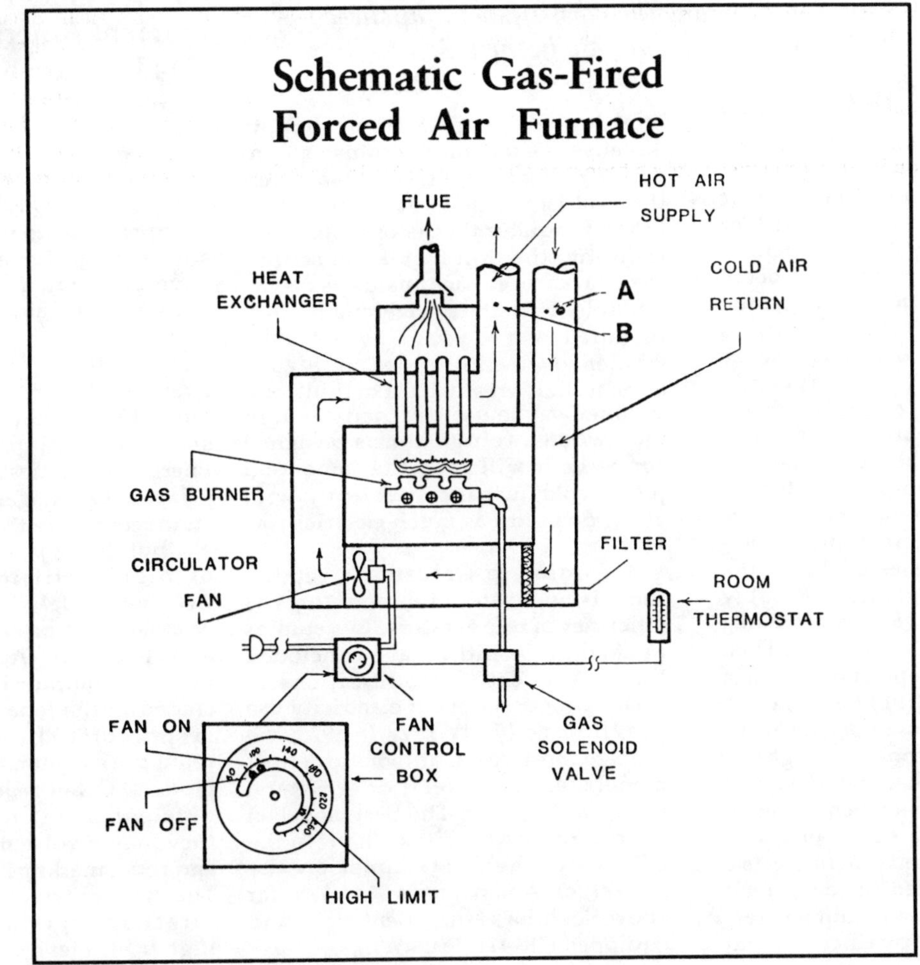

Efficient Refrigerators: The Japanese Challenge

A recent report by David Goldstein of the Natural Resources Defense Council concludes that Japanese appliance manufacturers have developed energy-efficient refrigerators that use 70 percent less electricity than the average equivalently-sized American model, scrimp on no features, and cost about as much as less efficient Japanese models. In a recent trip to Japan, Goldstein found more than a 50 percent drop in energy consumption since 1975 among the most efficient Japanese refrigerators, compared with a 20 percent drop in the US.

The Cold Falls Out

Refrigerators are among the largest single users of power in the US, ranking behind industrial electric motors, lighting, and air conditioning, and consuming about six percent of the nation's power. As a number of our prior reports have suggested (*Notes* 1:1; 4:3), this appliance is among the most notorious energy guzzlers. Amory Lovins recently described the state of the art in refrigerators: "Pre-war refrigerators," noted Lovins, "had about a 90 percent efficient electric motor mounted on top. Nowadays, the motors are more like 60 percent efficient and underneath, so the heat goes up where the food is. Your refrigerator probably spends close to half of its effort taking away the heat of its own motor. Then the manufacturers kept trying to make the inside bigger without making the outside bigger, and skimped on insulation so the heat comes straight in through the walls. Then they designed the whole thing so that when you open the door, the cold air falls out so it frosts up inside. Intermittent electric heaters were installed inside to melt the frost. Then some manufacturers installed electric strip heaters around the door to keep it from sticking because it's too simple to use a non-stick coating as in a frying pan. Then the heat is pumped out the back to a kind of radiator that is often pressed right into the thin insulation so that the heat can get back inside as fast as possible. Then the refrigerator is often installed next to your stove or dishwasher so when that goes on, it goes on as well. It's really hard to think of a dumber way to use electricity and, in fact, when a refrigerator is designed properly, it will keep the same food just as cold, just as conveniently, with about one sixth as much electricity as now."

According to Goldstein's report, there is no mystery to improving the efficiency of refrigerators. Two studies by Arthur D. Little have concluded that cost-effective modifications of existing models can cut electricity use to as little as 450 kWh/yr. In 1975, the average top-freezer, automatic defrost model (14.5 cubic feet, or 425 liters) used 1,800 kWh/yr. The best model on the market used 1,130 kWh/yr. In the US today, the best model on the market, an Amana, uses 914 kWh, far above Toshiba's equivalent sized and equipped GR-411 (550 kWh).

Even a $300 price differential would be cost effective to consumers who pay 7¢/kWh for electricity. In fact, the present value of savings from substituting the Toshiba GR-411 model for the most efficient American model (914 kWh/yr) would be $504 *larger* than the extra cost.

Improvements in the best available and average Japanese refrigerators have been far more impressive. In both traditional small (170-235 liter) refrigerators and larger modern ones (400 liters), energy consumption per liter has fallen by more than half since 1975, according to both Goldstein and the Japanese Institute for Energy Economics.

Japanese refrigerators, finds Goldstein, cost more than American models, but this is probably not the result of higher efficiencies. "The 550 kWh/yr model," writes Goldstein, "carries a list price of 235,000 yen, or about $1,025. Adjusting for a 15 percent commodity tax that is included in the Japanese price yields a net price of $890, compared to a typical retail price of about $600 for American models. Other reasons for this price differential are unknown at present; they may involve different wholesale and retail markups, or other practices. The price differential is not subject to direct interpretation as a cost of higher efficiency," adds Goldstein's

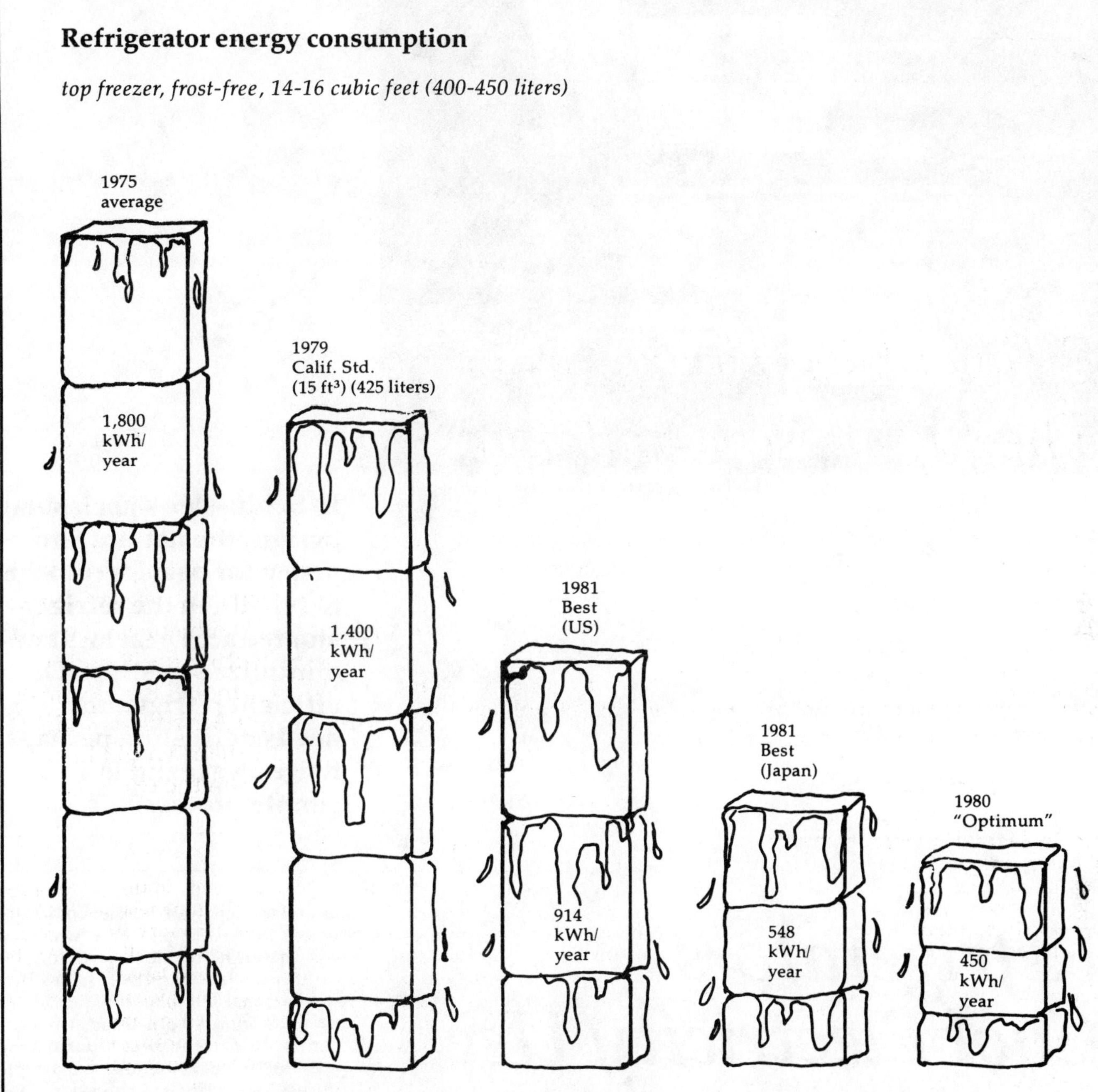

report, "because low efficiency Japanese refrigerators are also that expensive."

Savings Exceed Cost

Even a $300 price differential, calculates Goldstein, would be cost effective to consumers who pay seven cents/kWh for electricity. In fact, the present value of savings from substituting the Toshiba GR-411 model for the most efficient American model (914 kWh/yr) would be "$504 *larger* than the extra cost." (emphasis added)

One possible vehicle for accelerating the development of a better market in energy efficient refrigerators is a voucher, or rebate, on proof of purchase. (This concept has often been mentioned in *Soft Energy Notes*.) The idea is that the purchase of a highly energy efficient refrigerator not only saves customers on their electric bill, but the utility from the need to build more expensive new capacity. A portion of this savings could be applied to the purchase of energy efficient refrigerators, increasingly the visibility of electricity conservation options, improving the efficiency of the marketplace, boosting appliance sales, and saving everyone money.

—*Jim Harding*

References:

Goldstein, David
    n.d. "Efficient Refrigerators: Market Availability and Potential," preprint available from Natural Resources Defense Council, 25 Kearny Street, San Francisco, CA 94108.

The Institute for Energy Economics
    November 1980 "Facts and Figures of Energy Consumption in the Residential and Commercial Sector," Japan.

The efficient refrigerator looks like a kitchen cabinet.

> In Schlussler's horizontal design, the natural tendency for heavier cold air to remain in the refrigerator reduces heat loss and minimizes frosting. The efficient refrigerator needs defrosting perhaps twice a year and is a simple process.

Refrigerators Cool Electricity Use

Across the U.S., household refrigerators use 5.7 percent of the country's electrical energy. In a home where electricity is used efficiently, the refrigerator may easily account for 50 percent of the total electrical energy consumed—half the monthly bill. An efficient refrigerator, developed at the University of California's Quantum Institute, uses one-fifth of the energy of a conventional refrigerator—one-third that of the most energy-efficient model now available.

Funded by the University Appropriate Technology Program, Larry Schlussler designed and constructed a top-opening, 13 ft³ (.37 m³) refrigerator that looks like a kitchen cabinet. Standard counter height and 4.5 ft (1.37 m) long, it has two refrigerated compartments on either side of a center freezer section. Two drawers, including a vegetable crisper, are placed at the bottom; all stored food is easily accessible.

The horizontal design greatly reduces the infiltration of warm air. Opening the door on a conventional, vertical refrigerator releases the dense, cold air and allows warm, moist air to replace it, requiring additional energy to keep the inside cold. In Schlussler's horizontal design, the natural tendency for heavier cold air to remain in the refrigerator reduces heat loss through the door gaskets and minimizes frosting. Defrosters in a conventional model use 30 percent of the total energy consumed; the efficient refrigerator needs defrosting perhaps twice a year and is a simple process, according to Schlussler.

Placed underneath or behind a conventional refrigerator, the condenser and compressor—the coil and pump-motor assembly—make a lot of noise and transfer heat back into the unit. The metal case increases heat gain; poor air circulation around the condenser makes a fan necessary, as well. Poor insulation, metallic thermal bridges, and ineffective door gaskets also increase heat gain.

In Schlussler's design, the freezer's central location minimizes heat transfer from the room air. Three inches of urethane foam insulate the unit, with an additional inch around the freezer section. Hiding the compressor under an adjacent cabinet reduces heat transfer to the main body and eliminates compressor noise, a serious problem for some refrigerator owners.

Schlussler designed the evaporating system for passive heat transfer—not possible on a conventional refrigerator. On self-defrosting models, the evaporator is too small and improperly placed. For sufficient cooling, a fan must blow air over the evaporator. The temperature is unnecessarily low: so low that the air and food dehydrate, the evaporator needs frequent defrosting, and odors disperse readily. In some models, the fan blows cold air out the door when it is opened. Schlussler's central freezer allows for efficient passive heat transfer between the evaporator and all refrigerated sections. Moving the condenser above the unit along the wall also increases efficiency. It is an unobtrusive metal plate, one ft high and five ft long (.3 m by 1.5 m).

Schlussler recognizes that traditions like vertical refrigerators die hard. Nevertheless, his model, scaled up to 16 ft^3 (.45 m^3), requires at most 24 kWh/month—a savings of 70-88 percent from the 80-200 kWh/month of a conventional refrigerator. Considering savings on electric bills over 15 years, as well as initial cost, Schlussler concludes that the horizontal refrigerator could be competitive with conventional models.

—Elyse Axell

3. On-Site Direct Solar Energy

Self-sufficiency is part and parcel of the American dream, though the private auto is about all that is left for many US citizens. Solar energy offers an opportunity to recapture this vision, but it is a vision clouded by questions. How can I catch the sun? Will it be enough to live by? Will I be able to afford the technology?

On-site direct solar technologies fall into three categories: passive thermal, active thermal, and photovoltaic (PV) systems. The "active" flat-plate solar collector has received the widest attention.

In the doctrine of official expertdom, solar space heating has been quickly relegated to a supplemental role. In most climates, a 100 percent solar-heated house, so the argument goes, would require a seasonal hot water storage tank as large as the house itself, moving the residents out to live in the garage.

The problem of seasonal storage can be overcome by superinsulation techniques, and the costs of collector systems, as shown by the "solar sandwich" developed at Brookhaven Laboratory and other innovations, are in for a substantial decline. Today, a cost-effective, 100 percent solar-heated home, based on hybrids of passive designs or super-insulation and flat plate collectors, is not science fiction, but a real item in the process of commercialization. Even in existing buildings, on-site solar systems can supply a very substantial fraction of residential and commercial thermal needs—even without all the conservation measures that are economical.

There is another reason why solar skeptics have been far off the mark: they have uniformly applied criteria of economic cost-effectiveness adequate only for those who have money to approach solar technology as consumers. This bias fails to account for one of the features that makes collectors a soft technology in the first place: unlike a large power plant or pipeline, solar systems can be built by the do-it-yourselfer. He may seek the satisfaction of craftsmanship or accomplishment, or, like a growing sector of the US population, may be poor, be lacking marketable skills and be under- or unemployed. The costs to such "investors" are a factor of ten or so below the costs of buying a commercial system. As demonstrated in the San Luis Valley of Colorado, "middle class" solar devices can spread rapidly in poor communities.

Where solar thermal collector economics have been contested, there used to be general agreement that solar photovoltaic cells for generating electricity were still too expensive except for such applications as satellites, ocean buoys, remote military bases or northern California dope farms. The federal photovoltaics program has been uniquely successful in meeting its schedule, set in 1973, for cost-reductions. As early as 1986, roof-top panels could easily supply enough electricity to run appliances and lights at their present levels of inefficiency. Thus electrical self-sufficiency, the second part of the solar dream, may well be within reach—but we may end up buying our solar products from Japan, Italy and West Germany, if the Reagan administration successfully derails the American research effort. So large is the on-site potential for photovoltaics, that the powerline to homes may some day export electricity back to the grid.

Sunshine as Power

The motor turned toward the sun—the boiler white with reflected heat.

MANY ATTEMPTS have been made at various times to use solar heat as a source of power. A century or more ago great burning glasses were constructed both in France and England which developed a heat intense enough to melt iron, gold and silver. At the Paris exposition of 1870 an exhibit that attracted much attention was a sun engine which furnished the power for a printing press. This device was improved by the American inventor Ericsson. But though these contrivances concentrated the sun's rays, generated steam and furnished the power to engines, they failed for various reasons to be practically available for ordinary use. They were too fragile, too delicate, too costly or what not.

At last a practical solution of the problem seems to have been reached. There has been constructed and set up on Edwin Cawston's ostrich farm at Pasadena, California, a contrivance which performs its work regularly and with certainty. From one hour and a half after sunrise to half an hour before sunset it drives a ten horse-power engine, raising fourteen hundred gallons of water twelve feet per minute. This is enough to irrigate about five hundred acres of deciduous trees, or three hundred acres planted with orange trees.

In outward appearance the solar motor—for that is its name—partakes of some of the characteristics of a windmill, a steam engine, a mirror maze and a merry-go-round. It is in shape like a section of a huge umbrella of very substantial construction, having a diameter of thirty-three feet at its widest part and of fifteen feet at its narrowest. The whole inside surface is covered with mirrors, each two feet long by three and a half inches wide. Nearly two thousand of these long, narrow mirrors catch the sun's rays and reflect their heat upon a slim boiler, which is just where the handle of the umbrella would be. The great reflector is set like an astronomical telescope, the axis being north and south, and the movement from east to west. It is so nicely adjusted that one person can easily move it in either direction. The boiler is thirteen and a half feet long and holds one hundred gallons of water, with eight cubic feet of steam space to spare. It is made of fire-box steel covered with lamp-black. When the reflector is not working the boiler is quite inconspicuous, but when the concentrated heat from the mirrors is focused on it, it gleams like polished silver and is the most striking feature of the contrivance. In a little while it becomes so hot that a stick held against it smokes and bursts into flame. In almost an hour steam is generated and is conveyed from the head of the boiler through a flexible metal pipe to the cylinder of the steam engine, being thereafter used in the ordinary manner.

Though the solar motor at Pasadena develops only ten horse-power, it is already clear that plants able to develop one hundred horsepower are practicable. Many improvements are likely to be made in the present machine, especially in the direction of lessening its weight and cost, thus bringing it within the reach of people of moderate means.

The immense advantage possessed by the solar motor over most other sources of power is that it requires no fuel, which in arid regions is a scarce and expensive commodity.

[This technical review was first published by *Sunset* magazine seventy-seven years ago, and is reprinted with their kind permission. Who says solar technology is new and untried?]

The Solar Sandwich

CHEAP FILM LAMINATES
REPLACE COPPER AND GLASS

THE BEST MATERIALS for a solar collector may not be expensive copper and glass; they may be just around the corner at the fast-food chain. Well, not exactly: polymeric thin films, similar to those used in the packaging industry offer strength and low cost for solar flat-plate collector glazing and absorbers. So report the researchers at Brookhaven National Laboratory (BNL) who are experimenting with plastic and aluminum laminates to design solar flat-plate collectors with a manufacturing cost as low as $1 per square foot for an energy value of 100,000 Btu/ft²/year for solar space heating applications. The low cost insures attractive profits for the manufacturer and a short payback period for the consumer—8 years or less is the design goal.

BNL's design encourages the salesman as well as the manufacturer. The marketing plan targets small local factories as the point of final assembly.

photo courtesy of BNL

The glazing and absorber components, as well as the metal frame stock can be shipped in bulk. Thus, says William Wilhelm, principal investigator of the development effort, the materials for 10,000,000 ft² of solar collector could be shipped in a single railroad box car.

Installation, at $5/ft² total cost, can be done by one person—folding 100 ft² panels, strong enough to bear a man's weight, are only 50 pounds each. Work is still progressing to de-velop simple and efficient collectors, allowing fast and economical installation with a minimum of roof penetrations.

In the BNL design, plastic films are attached to a light, galvanized steel frame with pressure-sensitive adhesives acting as fastener and weather seal. Water entering the collector through manifolding integrated into the frame flows down between the two absorber films. Aluminum foil, laminated to the plastic layers in the absorber, provides improved lateral heat transfer and greater strength.

Wilhelm describes one absorber film which he expects will cost about $0.25/ft² and last for 15 years: it consists of two sheets of 0.002-inch aluminum foil laminated to polyacrylonitrile film. Two sheets of the laminate, attached on their polyacrylonitrile sides with adhesive to form long thin channels to encourage full wetting of the surfaces, will permit water to flow down between the sheets. Polyacrylonitrile appears to be a promising low-cost candidate for absorber films; it can withstand temperatures greater than 400°F.

High-speed fabricating machines, like those that make plastic bags for potato chips, could be easily adapted to produce the absorber, including the entire process of laminating the films, printing the optically selective absorber surface and sealing the two films with the proper water flow pattern. The fabrication cost, says Wilhelm, "can be measured in pennies per square foot."

Work at Brookhaven is still underway.

—*Elyse Axell*

References:
Andrews, John W. and William G. Wilhelm

1980 Draft: *Thin-Film Flat-Plate Solar Collectors for Low-Cost Manufacture and Installation.* Department of Energy and Environment. Brookhaven National Laboratory, Upton, New York 11973, March 1980.

Conversation with William G. Wilhelm, Brookhaven National Laboratory, 1980.

Solar Scavengers Scrounge for Cheap Heat

Today, hundreds of people in the San Luis Valley are building homemade solar systems for as little as $35. And although these systems may not look as sleek as the relatively expensive, commercial systems or keep the home as warm, they do work. With owner-supplied labor the biggest investment, these systems heat Valley homes comfortably despite sub-zero temperatures, saving each household more than $400 per year in fuel.

SOMETIMES REFERRED to as the Appalachia of the West, the San Luis Valley in south-central Colorado gets as cold as 30°F to 50°F below zero in winter (-34 to -45°C). The region has no wood fuel, higher than average fuel costs (due to transportation costs to the valley), and almost 50 percent of its 40,000 citizens fall below poverty level.

With no end to rising fuel costs in sight, many low-income Valley residents have found that solar energy — with its long-term savings — poses an appealing, non-fuel alternatives. However, the initial investments in a commercial system, combined with insulation to minimize heat loss, cost more than they can afford all at once. Rather than give up, Valley residents have taken an unconventional approach to solar. They scavenge what materials they can to build their own solar devices and will worry about maximizing system efficiency *later*.

The word spreads on cheap heat

It didn't take long for the word on homemade collectors to get around; in just a few years, the San Luis Valley has become the most "solarized" region in the nation. "We had eight solar systems in 1976 and now we have nearly 1,000," says Bob Dunsmore, founder of the San Luis Valley Solar Energy Association (SEA), a non-profit group that provides information on how to build solar systems effectively and cheaply. An average of one system every day is completed, he added. Dunsmore projects that the Valley could expect to see 1,000 solar systems built per year by 1985 and a full-blown solar industry that can help solve the region's unemployment problem.

Tom Richardson, director of the Solar Energy Association, attributes the solar boom to Valley residents' practice of sidestepping the high-priced ($3,000), commercial solar systems by building their own, simpler, and less-expensive models. Nearly 400 homes in the Valley use the homemade "North collector," built according to plans supplied by the SEA.

> The solar movement has generated a commercial solar industry that promises to revitalize the Valley's economy.

Starting out simple but warm

The North collector, a shallow wooden box, has a plywood back, a fiberglass reinforced plastic front, and black, corrugated aluminum behind the plastic to absorb the sun's heat. A fan draws cool air from the house into the collector, and blows heated air back out—either directly into the house or into a crawl space below the house to rise through the floor boards. According to the Association, a typical 64-square-foot (6 m²) North collector, complete with electric fan and store-bought materials, can be assembled for less than $250; with scavenged materials, for as little as $164.

Experts on solar energy would probably shake their heads in disapproval at such a collector because it does not include a thermal mass (such as barrels of rocks or water) to store the daytime heat and release it at night. Without storage, and in the absence of effective insulation, the North collector is not able to heat the home as efficiently as it could. But the Valley people put such niceties of system design aside.

Even without the frills, a slightly larger 94 ft² North collector saves a 1,500 ft² home between 15 million and 18 million Btu/year. Tom Richardson told *New Shelter* magazine, "They're proud of the systems they've built, so they tighten their homes to help the systems work as efficiently as possible a few months later when they've scraped the money together. After a thermal mass, insulation and weather stripping have been installed, the same 1,500 ft² home could expect to save up to 40 million Btu/year or $385 at the current fuel price of $1.30/gallon.

The association projects that by 1985, with energy at $17 million Btu, the total number of solar systems could save the region more than 266 billion Btu. Thus, with rising fuel costs, that would mean $4.5 million savings for the Valley or $935/year per family.

Savings generate industry

The solar movement in the San Luis Valley has been so successful that it has generated a commercial solar industry which promises to revitalize the Valley's economy. By 1985, solar architect Akira Kawanabe speculates, the demand for low-cost commercial solar systems will generate 165 to 250 jobs and create a $3.3 to $6.4 million industry. Kawanabe has already started a program to train unemployed people for solar jobs to ensure that the Valley will have enough trained solar builders as the demand grows over the next few years.

What is most important for commercial systems, however, is that they must stay affordable for Valley residents. Kawanabe designs new homes complete with North collectors, thermal mass, and insulation for just $33,000. And like many of the homemade systems, these commercial collectors may not look so pretty from the outside but they *feel* pretty warm from the inside.

–Angela Gennino

References:

Kawanabe, Akira

1981 "Impact of Solar Development on a Rural Community." *Solar Flashes*, March 1981, p.14.

Rawlings, Roger

1981 "The Most Solar Place in America," *New Shelter*, May/June 1981, pp. 71-83.

With a Grain of Salt

Miamisburg, Ohio. At the end of February, when the Midwestern winter is at its worst, and a layer of ice floats on the surface of Miamisburg's solar pond, the water at the bottom is still 83° F (28° C). By the first day of summer it will be warm enough to heat the City's outdoor swimming pool. When that closes in early October, the pond will still be used for another three months to heat a recreational building nearby.

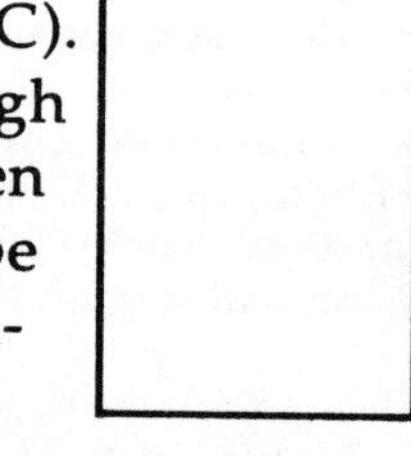

The cost for this energy? Only $9.15 per million Btu ($9.65/GJ), less than the cost of burning dollar-per-gallon heating oil at 75 percent furnace efficiency. (*See* Notes, *vol. 3, no. 4.*)

Solar ponds similar to the one built by the City of Miamisburg are now operating in Israel and a handful of US sites, while numerous others are being planned. Their popularity is easily explained: quick and inexpensive to build compared to other energy sources, modular in design, reliable, and environmentally benign, solar ponds supply heat, electricity, or both almost instantly with low-maintenance, proven technology.

The physical principles of the solar pond are just as readily understood. Anyone who steps into an undisturbed tide pool at the beach learns rapidly—and sometimes painfully—how much heat sea water can store. Large, naturally-occurring brine ponds with salt gradients hold even more energy and reach much higher temperatures, providing a model for the artificial solar pond. Typically, solar ponds are one to three meters deep, and contain water in three layers of differing densities.

The top layer is relatively fresh, and is the least dense; the thin middle layer increases in salinity (therefore density) with depth; the bottom layer is the heaviest and saltiest of the three.

Solar radiation penetrates the water, where the light is converted to heat, and the heat is trapped in the bottom, saline layer. Two factors prevent the heat from rising and escaping. First, it is held by the salt in the densest layer, which limits its movement. Second, the middle boundary layer—the density and salinity gradient—acts as an insulating blanket. Consequently the pond is both solar collector *and* storage medium.

Saline Gradient Crucial

The saline gradient is the crucial layer of the pond, because it is the only one in which convection does not occur. The top layer, exposed to air and wind, is affected by changes in atmospheric ambient temperature and surface movements. Some of its heat is lost through convection. The bottom layer loses a small amount of energy to the ground by conduction, creating a temperature difference, which generates a small convection pattern. But because the middle layer—the saline gradient—is non-convecting, it acts as an insulator, and no heat is lost from the pond through bottom layer convection.

The saline gradient forms naturally if the pond is constructed by placing a layer of fresh water on top of a briny one. Precise layer boundaries can be maintained by periodically injecting concentrated brine at the pond bottom, and fresh water at the top. The gradient is stable, and will slowly reform if it is disturbed, though with some energy lost in the process.

One simple way to tap the heat stored in the saline pond is by running a heat exchanger through the bottom of the pond; the extracted heat can then be used to condition space or water. This application requires minimal supervision and works well in small ponds, such as the 180 by 120 foot (2,000 m²) pond in Miamisburg. First tested in 1978, and put into operation over the winter of 1979, it is the largest such facility in the US. Total construction costs ran about $70,000 ($3.20/ft²), with the plastic bottom liner and some 1,100 tons of salt taking the largest share of the capital expenditure. If the pond performs as expected for ten to twelve years, its average energy cost will run below that of conventional fuels. From October to December, the pond is expected to provide 780 million Btu (823 GJ) for heating the nearby recreational building. This would replace from $7,600 to 8,200 worth of dollar-per-gallon fuel oil in a furnace at 70-75 percent efficiency. The same energy from the pond, at $9.15/10⁶ Btu, would cost as little as $7,100.

Other thermal applications for small to medium solar ponds include heating multi-family dwellings, collections of single-family houses, large commercial buildings, greenhouses, and industries using low- to medium-temperature heat. Ponds could also run absorption chillers during summer months. The restriction on thermal energy from ponds is that they should be near the point of use, since laying insulated piping is quite expensive.

Electricity From Ponds

A second way to tap a solar pond's heat is by piping the saline layer through an external heat exchanger, rather than one within the bottom layer. This method requires occasional attention on the part of trained personnel who can avoid disrupting the gradients, but it opens the possibility of using the heated water in organic Rankine cycle engines to generate electricity. As a rule of thumb, ponds can produce about one MW (peak) of electricity per 50 acres of pond area (12.4 kW/m²). The best dollar value for electricity-producing ponds comes with large areas, although small ponds generating five MW can be constructed without a major cost penalty.

The Israelis, who were the first to experiment with solar ponds, began operating a 150kW pilot power station in 1979 at Ein Bokek on the Dead Sea. The plant collects heat in a rubber-lined, 70,000 ft² (6,500 m²) pond, and sends the heated saline water through a heat exchanger, where it is transferred to a high-pressure organic fluid. The heated fluid expands through a Rankine cycle generator to produce electricity. Another heat exchanger in the pond's upper layer cools the organic fluid for recirculation.

Experimental saline ponds are now producing electricity in New Mexico, Nevada, and Virginia. Several sites are being considered for a project that could serve as a prelude for commercial-scale US production. The Great Salt Lake in Utah is a possibility, but the most likely location is the Salton Sea in southern California's Imperial Valley. This saline body of water, which has no natural outlet, was created by an accidental overflow and diversion of the Colorado River.

Now in the feasibility and design stages, the first pond in this commercially-oriented project would cover about 250 acres and generate five MW. Permission to construct this plant has yet to be obtained from the US Navy, which controls the land presently under study. Southern California Edison, the utility behind this venture, is investigating how much of the power could be provided to the nearby naval base, and how much might be put into the grid for wider distribution.

While the five MW plant was under construction, the utility would conduct feasibility studies for building a 600 MW Salton Sea plant in 20 to 50 MW modules. This ambitious project would require 15 percent of the sea's surface, or about 50 square miles (12,950 hectares), and would provide power for 500,000 to one million people.

The Salton Sea project would use a system of dikes to form three kinds of pond. Shoreline ponds would store water waiting to pass onto evaporation ponds to concentrate the naturally occurring salt. Finally, the briny water would be pumped into the solar ponds themselves.

Though solar ponds show promise, they have their technical limitations. The energy they provide fluctuates with the seasons, and pond output is lowest when heating requirements are

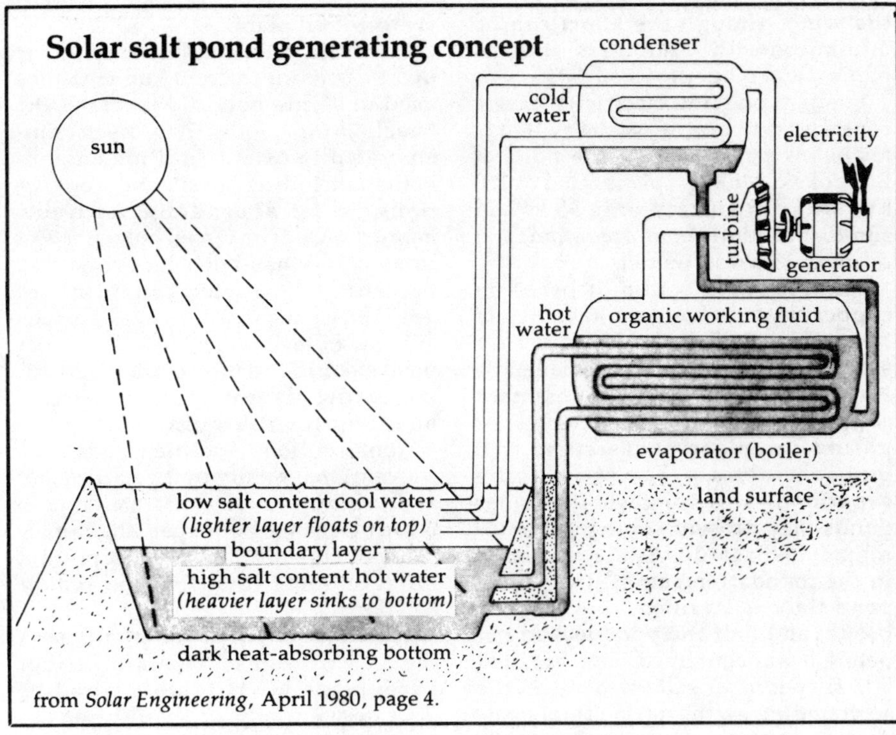

from *Solar Engineering*, April 1980, page 4.

Technical problems and approaches

problem: **Prevent convection**

approaches: 1. NaCl salt gradient
2. Other salt where cheap (e.g., bittern)
3. Saturated solutions (new research)
4. Gels (no promising candidates at present)

problem: **H₂O clarity**

approaches: 1. Copper sulphate (for algae)
2. Chlorine (for bacteria)
3. Selective precipitation for minerals
4. Fences for debris

problem: **Heat extraction**

approach: 1. Optimize hot brine withdrawal for large ponds

problem: **Slow migration of layer boundaries**

approaches: 1. Model pond and full-scale experiments
2. Theoretical hydrodynamic studies

problem: **Wind driven instabilities**

approaches: 1. Wave breaks may prove adequate
2. Problem needs theoretical hydrodynamic study

problem: **Scale up to many acre pond for IPH or electricity**

approach: 1. Field experiments including design studies

problem: **Salt pollution**

approaches: 1. Liners for small ponds
2. Natural saline environment or impervious soil for large ponds

problem: **Pond lifetime**

approach: 1. Test and develop improved materials

at their peak. Ponds are also subject to the wind, though the effect can be minimized with windbreaks, screens, baffles, and other devices.

Ponds producing electricity have a continuous operating capacity that is far below peak output. The pond at Ein Bokek, for example is rated at 150 kW, but can sustain only 35 kW in summer at continuous operation, and drops to 15 kW in winter.

The corrosive action of brine on copper heat exchangers has already been observed at the Miamisburg solar pond. However this problem is eventually solved—more resistant materials, intensive maintenance, or periodic replacement—costs will probably increase. The Miamisburg experience also suggests that solar ponds may present some local environmental hazards. As the sand used in the foundation settles, the plastic pond liner is strained. It eventually broke, and half the pond leaked out before it was contained. Five hundred and fifty tons of salt were left in the surrounding earth. Agricultural lands or local water supplies near other ponds could be contaminated if they, too, prove leakage-prone.

The problems and potential hazards of solar ponds are not serious obstacles, and are clearly outweighed by the obvious advantages of this nearly-commercial technology. A principal virtue is that the stored heat is available either for its thermal energy or for electrical production at any time, and on a few minutes notice. Because the peak power of the pond is up to ten times its continuous operating level, utilities might well consider using them for peak load, replacing expensive oil-fired plants.

Reliability is another important utility consideration. The turbines used in saline ponds, especially the Israeli Ormat model, have been demonstrated in over 30 million engine-hours (including fossil-fuel applications) to be dependable with low maintenance. The forced outage rate at solar ponds has been less than two percent. The combination of availability during peak load periods and low outage rate should give solar pond electricity a high capacity credit, increasing its economic value in a mixed-source utility grid.

Construction of saline ponds uses conventional earthmoving equipment and techniques, so it can be done in most places without specialized personnel. This simplicity is important for developing countries and remote locations.

The modularity of the ponds, and their flexibility of size are further advantages. Construction could be done sequentially, with additions to capacity made as required in small increments.

Finally, the construction components are readily available, relatively easy to use, and even inexpensive in some locations. The main expenses are for the pond liner and the salt, which costs very little to secure from briny bodies of water such as the Dead Sea. Electrical ponds add the expense of a Rankine cycle generator, but this component is expected to drop in cost. With conventional energy sources growing more expensive, salt water looks like a sweet investment.

—*Mary Lou Van Deventer*

References:

Nielsen, Carl E.

1980 "Nonconvective Salt Gradient Solar Ponds." Presented at the Non-Convecting Solar Pond Workshop, July 30-31, 1980, held by the Energy Systems Center at the Desert Research Institute, University of Nevada System, for the US Department of Energy.

Ormat Turbines, Ltd.

1980 "The solar pond development program in Israel." Presented at the Non-Convecting Solar Pond Workshop, July 30-31, 1980.

Sargent, S. L.

1979 "An Overview of Solar Pond Technology." *Proceedings of the Solar Industrial Process Heat Conference,* October 31-November 2, 1979, Solar Energy Research Institute Site Office, Golden, Colorado 80401.

Shurley, Lucian A.

1980 Letter of March 11, 1980 from Lucian A. Shurley, Contract Manager for the Salton Sea Solar Pond Project to Janis Jacobson, California Office of Appropriate Technology.

Wittenberg, Layton J. and Marc J. Harris

1980 "City of Miamisburg heats pool with salt gradient solar pond." *Solar Engineering,* April 1980, pp. 26-28.

Mary Lou Van Deventer is a writer and environmentalist who has worked with the California Office of Appropriate Technology, and is now an editor for Sierra magazine.

Double Envelope Conquers Arctic Cold

Would you expect a solar house in Alaska to produce 65 to 70 percent of its energy needs and be cost-effective, too?

photos by Bruce P. Maeda

I N HOMER, ALASKA, the passively heated home of Steve and Noko Yoshida accomplishes that and more. The Yoshida home should save at least $1,200 annually, according to its designers, based on 1977 energy costs. This house is the first of its kind in the state, and early indications are that it is living up to its promise of being a tremendous energy saver.

The house uses a "double envelope" solar design to maximize heat gains and minimize heat storage losses. The outer envelope and attached greenhouse enclose the inner shell living space area. Inside, at the crest of the house, a one-foot air space or "plenum" extends along the roof and down the north wall to the crawl space below the lower floor. Heat from the sun is collected in the attached greenhouse, rises to the crest of the roof and is circulated by natural air convection currents around the outside of the living space. This cocoon of solar heated air drastically reduces heat loss and maintains mild temperatures within the outer envelope. Situated in the Alaskan climate, the house demonstrates the feasibility of passive solar construction in even harsh north-

ern climates. Its high performance design has been creating comfortable passively heated environments in other cold US climates as well — Lake Tahoe, Claifornia; Newport, Rhode Island; and, Cincinnati, Ohio. In Alaska, the home's performance is truly delightful, according to the Yoshidas. Early last spring, "there were 30 to 40 mile per hour winds outside, but inside it was cozy and warm," said Noko. "It was like California."

Double Envelope Comfort

The performance of the Yoshida solar home is the result of a number of basic design features, and other less obvious elements of its construction. The double envelope feature stabilizes minimum living space temperatures and also provides a "comfort effect" from outer shell air temperatures, even when the relative inside temperature is between 58 and 65 °F (14.5 to 18 °C). Heat from the outer shell surrounding the living space gives a greater level of comfort than in conventionally designed houses that operate at somewhat higher temperatures (62 to 70 °F, 16.5 to 21 °C). Basically, this comfort effect is treated by the outer envelope

giving up heat to the inner shell living area, and the higher inside surface temperatures. In addition to the double envelope design, two-thirds of the energy savings are realized by using substantial insulation and house alignment for maximum heat gain from south-facing windows. The insulated crawl space area under the house also acts to stabilize temperatures within the outer shell, and it gives up heat to the inner shell. Overall, the Yoshida house provides a high level of performance for relatively low cost.

This solar-heated house of 3,000 square feet (280 m²) was constructed with conventional materials using conventional building techniques for about $38 per ft² ($400/m²). Like other costs, housing construction costs are higher in Alaska than in most other areas of the United States. Yet, the Yoshida house may save $25,000 to $35,000 on heat alone over its useful lifetime, thus making its construction more than cost-effective. The extra initial costs for this solar design over and above a conventional house of comparable size are estimated to be about five to eight percent of the total construction costs. Based on this Alaskan experience, it's difficult to see why more passive solar homes of similar design are not being built in northern US climates.

Alaskan Passive Development

The Alaska passive solar home was originally conceived by a group of solar thinkers who felt that high performance solar principles can work to provide comfortable living, cost-effective operation, and simplicity in style, even in a climate like Alaska's. Its designers, Tom Smith, Davis Alternative Technology Associates (D.A.T.A.), and Alaska Renewable Energy Associates had been interested in accelerating the rate of passive solar development and dispelling the myths that solar technology is too expensive or impractical except in the Sunbelt. As effective federal action to promote solar development was lacking, they saw the

> ## Passive heating has been termed "impractical" by many skeptics.

need to demonstrate dramatically the viability and cost-effectiveness of passive solar construction.

The project began in August of 1978 after the Yoshidas contacted Smith, owner of the Lake Tahoe double envelope house (see *Soft Energy Notes,* July 1979, p. 57). D.A.T.A. performed a computer evaluation of the proposed solar home — patterned after Smith's house and Lee Porter Butler concepts — using Alaska's weather conditions. The evaluation clearly showed the project was feasible.

The Alaskan affiliate of D.A.T.A., Alaska Renewable Energy Associates (AK.R.E.A.), is currently working on passive solar home designs for other parts of Alaska, and other climates where the use of natural heating has been termed "impractical" by many skeptics. This and other passive solar projects in the United States paint a bright future for similar solar developments. D.A.T.A. and AK.R.E.A. are especially optimistic. They contend that houses built anywhere in the United States can be cost-effective and achieve at least 70 percent of heating and cooling needs through use of solar heating and natural cooling design principles. This is surely one way of easing our domestic energy problems.

— *Eric Woychik*
& the staff at D.A.T.A.

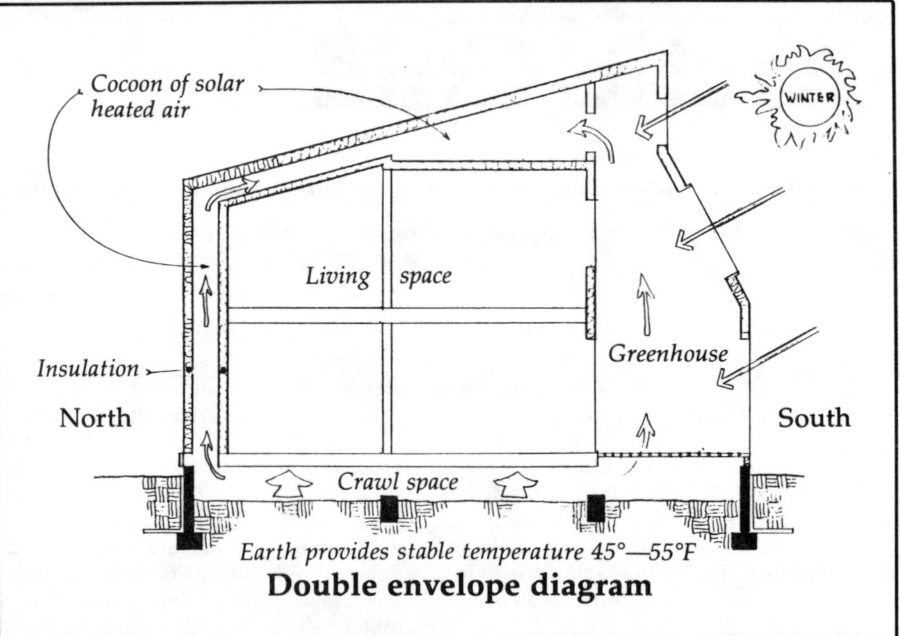

Double envelope diagram

Economical and Elegant Solar Electric System

HAILED AS A POTENTIALLY CHEAP and practical renewable energy technology, a solar electric power system being developed by Texas Instruments could be suitable for both home and commercial use. Texas Instruments (TI), a pioneer and giant in hand-held calculators and fancy circuitry, looks to have its system available for purchase by the mid-1980s. "The TI system is important, innovative, and more than any other solar energy conversion system we have heard of to date, commercially practicable," commented a recent report by the Morgan Stanley investment firm.

Featuring integrated energy storage for nights or cloudy days and a rugged collector design, the TI system is efficient, reliable, and could operate independent of utilities.

On the leading edge of the breakthrough is the company's star inventor, Jack Kilby (1958, the monolithic integrated circuit; mid 1960s, the basic patent on the hand-held calculator; 1970s, the thermal printer). "The father of the hand-held calculator may be ready to sire the nation's first affordable home power system to produce electricity from the sun," announced a first page article in the expensive trade paper, *Energy Daily*.

Over the next four years, the US Department of Energy (DoE) is putting $14 million into TI's demonstration program —an amount which TI will return to the US Treasury if the effort is profitable.[1] Texas Instruments, in turn, will wager $4

million on the program's ability to generate profits by 1983. Long term goals include electricity production for five to ten cents per kilowatt-hour on individual rooftops, with a bonus of substantial free heat for water or space in buildings.

The goal of R & D efforts is to reduce photovoltaic costs from the astronomical to the affordable. Space solar cells which deliver electricity to a spaceship cost well over $1.00 per kilowatt-hour since they use very high-grade, "semiconductor" silicon. Cells must also be thick enough to withstand meteor showers in space. On earth, however, the maximum credible accident is probably a direct Frisbee hit, so silicon production expenses need not be as high.

Much of the current interest in photovoltaic cells was stimulated by a 1977 report done for the US Federal Energy Administration by the BDM Corporation of McLean, Virginia (*Soft Energy Notes*, volume 1, number 1). The BDM report concluded that the federal Department of Defense could purchase

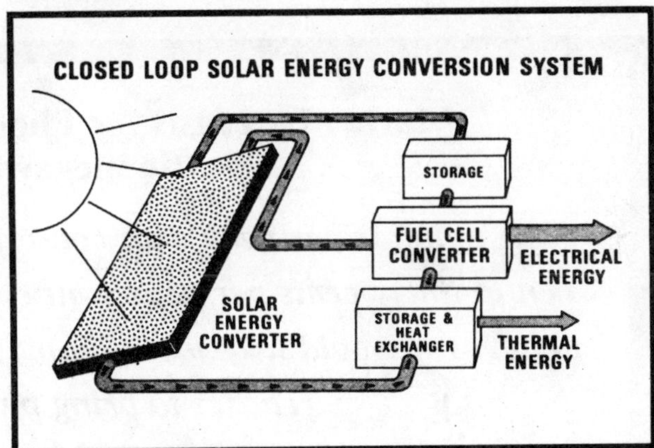

photovoltaic cells to generate cheaper electricity in remote locations than existing diesel engines and flown-in fossil fuels. These purchases, BDM asserted, would lead to mass production methods (better slicing, less waste, lower silicon grades) which would drop costs to 5 or 10 cents per kilowatt-hour by 1982— easily competitive with nuclear power. DoD need only purchase 152 megawatts with an investment of $430 million to save $500 million and reach this national cost goal. Photovoltaics could then expand into a vastly larger market.

The BDM report struck a responsive chord in many: it argued that no new technical breakthroughs were necessary for inexpensive solar electricity, only government will; and suggested a time-honored "market-pull" strategy—already successfully demonstrated on integrated circuits, transistors and diodes—to achieve commercial use. Barry Commoner, in his recent *New Yorker* series, indirectly cited BDM results, claiming that "the technological improvements in the production of photovoltaic cells which are required in order to achieve these sharp reductions in cost are almost ludicrously simple."

TI's Answer

Ludicrously simple or not, political will or no, the Texas Instruments breakthrough will no doubt speed things along. So what's so new about the system? It has four principal advantages over conventional solar electric cell approaches:

- it produces not only electric power but useful waste heat
- it has integrated energy storage for cloudy days or nights
- it is scaled and designed for small end-uses as well as large ones
- it is extremely forgiving of manufacturing defects or reliability problems.

Photovoltaic Array Prices for Several Market Projections

NOTE: With the exception of the "DOD sales" curves, it has been assumed that the Federal Government spends $129 million for photovoltaic systems between 1978 and 1981, of which 40 percent is actually used to purchase cell arrays (the remainder being spent for design, installation, storage, control systems, and other supporting devices). The forecasts assume that cumulative production of cells through 1975 was 500 kW and that average array prices at the end of 1975 were $15/Watt.

The Texas Instruments system begins with a novel "collector" composed of tiny beads of silicon bonded to a backing. The collector is immersed in a solution that decomposes when exposed to electricity ("electrolyte") generated by the silicon beads. One of the products of this decomposition is hydrogen, which is stored and burned on demand in a "fuel cell". The waste heat from the fuel cell and from other gases released in the collector passes through a heat exchanger to heat home water and air. TI currently operates a small-scale working system.

The principal cost saving measure in the TI system is the collector. Kilby has invented a way to produce tiny beads of doped (chemically treated) silicon 100-200 micrometers in diameter, with a "shotting tower just like manufacturers used in World War II to produce lead shot." The technology of this process if familiar to manufacturers of buckshot, if not to users of adding machines. The doped, melted silicon is fed through a nozzle and falls eight feet, solidifying into tiny crystal spheres (this can be considered the enrichment step in the solar fuel cycle). The spheres are then sorted for size, insulated with a thin coating, and applied to a substrate with metal contacts. Basically, TI is making fancy electric sandpaper.

Tiny spheres of silicon are inherently less wasteful than large slices. Even with silicon, the most common solid material on earth, waste is a key cost factor. In a flat, 3 inch diameter circular or hexagonal photovoltaic cell, only 12-30 percent of the silicon actually generates current. The reason for this is that most of the photoelectric conversion takes place in the top two mills (equal to 50 micrometers—a mill is a thousandth of an inch) of a cell that is twelve mills thick for durability and handling. Much of the Department of Energy's current solar cell work is focused on reducing silicon requirements in cell manufacture, but tiny spheres appear to solve this problem elegantly: TI claims 95 percent of the silicon in its device generates current.

Market Forecasts for Photovoltaic Devices at Different Prices
(in megawatts of annual sales)

Half a dollar per peak watt corresponds to electric power at 10¢/kilowatt-hour. Ten to thirty cents per peak watt corresponds to 3-5¢/KWh. Only the RCA study believes it would take more than 1000 megawatts (one nuclear power plant) of purchases to bring photovoltaic costs down to a level competitive with nuclear or coal.

	Array prices in dollars per peak watt				
Marketing study	10	3	1	0.5	0.1-0.3
1. BDM/FEA					
DOD market	10	75	100	—	—
Worldwide commercial market	1.5	20	70	100	—
2. Intertechnology Corporation	0.5	13	126	270	—
3. Motorola	1.5	20-30	—	—	—
4. Texas Instruments	0.4	2.6	30	100	20,000
5. RCA	0.8	13	200	2,000	100,000
6. Westinghouse	—		—	—	96,000
DOE planning, objectives	1.0	8	75	500	5,000

SOURCES: ERDA briefing on the "Photovoltaics Program," Sept. 9, 1977. *The Motorola Corporation.*
BDM Corporation. DOD Photovoltaic Energy Conversion Systems Market Inventory and Analysis, Summary Volume, p. 17.
BDM Corporation. "Characteristics of the Present Worldwide Photovoltaic Power Systems Market," May 1977, p. 1-1.
Intertechnology Corporation, "Photovoltaic Power Systems Market Identification and Analysis," Aug. 23, 1977.
P.D. Maycock and G.F. Wakefield (Texas Instruments), "Business Analysis of Solar Photovoltaic Energy Conversion"
The Conference Record of the Eleventh IEEE Photovoltaic Specialists Conference-1975. Scottsdale, Ariz., May 1975.

The "electric sandpaper" sheets are placed in long elliptical tubes open to the sun and bathed in an electrolyte, either hydrogen iodide or hydrobromic acid. Both solutions are cheap and continuously recyclable. In an aqueous solution of hydrogen iodide for example, iodine, water vapor, and hydrogen are produced when sun strikes the collector. The iodine and water vapor are cycled through a heat exchanger to extract hot air and water for home use, while the hydrogen is stored temporarily as a hydride.[2] Hydrogen gas is released from the hydride on demand to be burned in a small fuel cell (several hundred watts to a few kilowatts) to generate electric power. The cool iodine and water re-enter the fuel cell and combine with the hydrogen to produce electricity and an aqueous solution of hydrogen iodide. Thermal losses can be kept miniscule in this reversible process. TI's current photovoltaic panel is 7 percent efficient in converting sunlight to electricity. However, says DoE, achieving "10 percent electric efficiency is easy." The design goal is 13 percent electric efficiency which will depend on TI's ability to produce mainly single, rather than multiple, crystal spheres from the shotting tower. The entire system could be as much as 30 percent efficient (15 percent, according to TI, is more likely) because the electrolyte is heated *and* chemically separated in the collector. Thus, more than half the energy collected in the system could be used for space and water heat, with the remainder available as electric power.

Performance

An advantage of the TI collector approach is protection against individual cell defects. In a large flat cell array, the efficiency of the entire plate is limited by the lowest efficiency of an individual 1-3 inch cell. Individual cells are electrically attached in series, rather than in parallel. These cells can degrade in performance over time and can vary considerably in original performance. A major federal goal is to avoid these problems; the Texas Instruments system circumvents them. Individual silicon beads are not joined by electrical connections, but immersed in the same electrolyte. If cells short out or decay in performance, they do not make hydrogen gas, but neither do they interfere with the performance of their neighbors. The system is "fault tolerant," meaning lower grades of silicon can be used with less precise manufacturing tolerances. This is an important potential cost advantage.

Research goals of the TI-DoE contract include testing the performance and corrosiveness of various metal hydrides, improving the electrical efficiency of silicon produced by the TI shotting tower, and developing tiny fuel cells. As a top official at the Department of Energy told IPSEP, "We feel pretty comfortable with this project. You've got a first class engineering house, a good concept, and an inventor with so many credits to his name that you can't not fund it."

Sources: *TI First Quarter Report*, 1979; *Morgan Stanley Electronics Letter*, May 15, 1979; US Patent 4,021,323; conversations with SERI, DoE, and Texas Instruments officials, June 27-29; *Energy Daily*, June 15, 1979.

—Jim Harding

[1]DoE's proposed 1980 budget outlay for photovoltaics is $117 million, for fusion is $367 million, for all solar technology is $381 million, and for all nuclear fission is $987.6 million.

[2]Metal hydrides have mainly been studied to store energy rather than hydrogen. Hot hydrogen gas combines with the reactive metal. The gas is released when the hydride is heated. Details of re-heating are not available from Texas Instruments, but a small battery charged by the fuel cell would probably do.

The Distribution of Costs in the Manufacture of Silicon Photovoltaic Devices Using Contemporary Technology (all costs in $/peak W) ✶

	Representative data for 1976	Ingot technology			Noningot Sheet Technology	
		1978	1980	1982	1984	1986
Polysilicon	2.01	1.16	.76	.28	.14	.06
Crystal growth and cutting	4.18	2.50	1.43	.67	.23	.10
Cell fabrication	5.97	1.87	1.01	.58	.34	.19
Encapsulation materials	.78	.22	.12	.09	.07	.03
Module assembly and encapsulating	3.99	1.25	.68	.38	.22	.12
Price	16.96					
Goal	20.00	7.00	4.00	2.00	1.00	.50

★Does not include inflation — prices in 1975 constant dollars.

NOTE: The detailed price goal allocations are for silicon ingot technology up through 1982 and for silicon sheet technology in 1984 and 1986. These allocations for use within the LSSA Project are subject to revision as better knowledge is gained and should not be construed as predicted prices.

It is expected that after 1982, ingot and sheet technology modules will be cost-competitive, with the possibility that the $.50/W goal can be achieved by either technology. Technology developments and future production cost parameters will be major factors in determining the most cost-effective designs.

SOURCE: Jet Propulsion Laboratory, Low-Cost Silicon Solar Array Project. Received by OTA, June 1977.

Solar Power and Suburbia: Does It Add Up?

Using Space Satellite Photographs, California Study Demonstrates Potential for Rooftop Solar Cells

USING HIGH RESOLUTION multispectral space satellite photographs, the California Jet Propulsion Laboratory (JPL) has concluded that over 120 percent of existing electric power demands in the dense west San Fernando valley could be met with rooftop solar electric cells. Completed in cooperation with the Los Angeles Department of Water and Power, the study also shows that less than half of the region's usable roof area would be needed to generate power by this method.

In some areas, as much as 60 percent of annual electric needs could be met with rooftop solar electric cells. Older residential neighborhoods rate lowest, with only 2.7 percent of their land area available for solar cells, while commercial/industrial zones were able to use 33.7 percent. A forty square mile (104.2 km²) urbanized area (8 persons/acre average—3.2 persons/ha) near Los Angeles was selected for analysis because of its proximity to the Jet Propulsion Lab and the availability of many different types of aerial photography. Using the Lab's Image Based Information System and Video Image Communication and Retrieval Systems—both offshoots of the space program—JPL scientists were able to identify old and new residential, multi-family residential, mobile home, agricultural, and commercial-industrial land areas and dwellings, suitable for rooftop solar electric cells. The systems screened out parks and wooded areas, water, streets, and other unusable areas, and found south facing or flat rooftops. By superimposing the series of images and making some simple assumptions, JPL identified the amount of power available from urban rooftops, which was compared with the Los Angeles Department of Water and Power's data on existing power uses.

JPL assumed half the flat and south-facing rooftop area could accommodate solar cell arrays and that 20 kilowatt-hours per square foot per year could be generated by the solar cells. This is equivalent to a photovoltaic conversion efficiency of about 10 percent. The group also found that the forty square mile region could generate about 385 megawatts of peak power from rooftops, not counting any space and water heat from cooling the cells. This is about a fifth of the power output per unit of land area of a nuclear power reactor, if the plant's low population zone is included.

Jim Harding

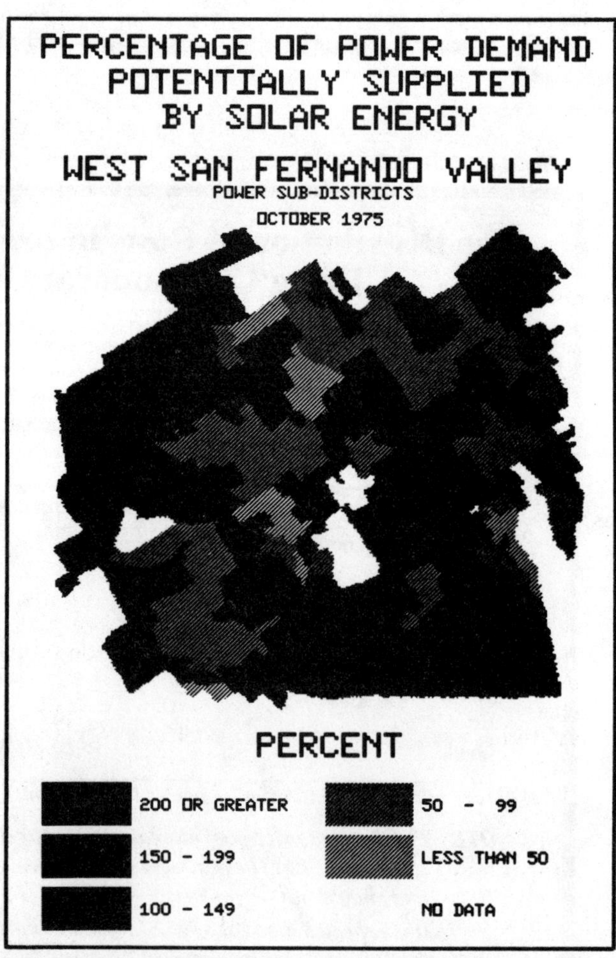

Sources: Solar Age, June 1978; JPL Feasibility Analysis, "Solar Photovoltaic Potential in the Western San Fernando Valley, Los Angeles," undated, available from Nevin Bryant, JPL, 213/324-7236.

Debunking the Myth of Solar Sprawl

T HE RESULTS of recent research, conducted by R.H. Twiss and his colleagues at the University of California at Berkeley, question the notion that the solar city needs to be spread out. "Contrary to some speculation," the authors assert, "reliance on decentralized solar technologies does not mean sprawl."

Comparing energy use among land use patterns in American cities, the UC researchers found that a shift from today's predominately suburban pattern to more compact residential patterns would achieve substantial energy savings for transportation and urban construction without unduly sacrificing opportunities for solar energy use.

The authors maintain that the critical measure is neither how much solar energy can be generated on-site, nor how much energy can be saved by conservation, "but how much energy demand remains after these techniques have been implemented." Put another way, the urban planning strategy should use as little energy as is absolutely necessary, and keep the collection effort and cost to a minimum.

High and Low Density Living

Working from aerial photographs and estimates of local energy use, the UC researchers explored the potential of solar systems for certain urban neighborhoods. In a low density (8 units/acre; 30/ha) single-family development located in Denver, the researchers found that active solar systems with short term storage could meet 35% of aggregate thermal requirements. Note that this assumes the use of all unshaded roofs as well as other sites suitable for collectors. The conservation efforts considered in this calculation are rather modest: a 19% reduction from present heating and hot water consumption to 10.5 Btu/ft^2/°F-day (214.6 kJ/m^2/°C-day).

In a Baltimore high density (30 units/acre; 75/ha) multi-family development, one might expect to find inadequate space for solar collectors. The ratio of total roof and site area to living space (0.6 to 1) is less than one-twelfth the ratio found in the low density area (7.6 to 1). Nevertheless, compact row houses of the multi-family

development can meet 40% of heating requirements with solar systems on roof tops only. Occupying other feasible sites raises the solar fraction to 50%. Once again, the conservation efforts associated with this level of solar contribution are modest: demand need drop by only 22% to 8.7 Btu/ft^2/°F-day (177.8 kJ/m^2/°C-day). The authors add that the north-south orientation of the streets turns out to be less of a detriment than the conventions of solar design would indicate. The long axis of the houses is perpendicular to the street; south-facing roofs provide plenty of suitable collector sites.

Since the climates of Denver and Baltimore are not identical, the solar percentage figures are not strictly comparable. Still, it is evident that solar energy can meet a larger fraction of heating needs in the compact residential community, even with its smaller proportion of collection area. One reason for this is that dense areas with shared walls lower heating requirements by reducing heat loss.

To drop the heating load further, both single- and multi-family dwellings can boost the solar percentage by investing in tighter structures. The major problem in the single-family development, however, is lack of solar access. Trees shade ninety percent of the south-facing roof area in the Denver neighborhood. Providing access for every dwelling's solar collectors would require removal of as much as 35% of the tree canopy.

Contrasts in the Commercial Sector

All urban sectors, not just the residential communities, need to balance increased density and on-site solar opportunities. The UC researchers examined the solar potential of the commercial sector, and found it to be a study in contrasts. The central business district in Denver has everything working against it. Solar access is greatly restricted by tall buildings; structures are oriented 45° from due south. Even after a conservation effort, the potential solar fraction remains low.

At the opposite extreme, the warehouse district can produce more energy than it needs, as can the commercial strips along the city's major streets and roads. The warehouse district study area (which includes miscellaneous other commercial uses) has a low demand for energy. In addition, little variation in building height minimizes roof and site shading. Using only roof tops, this area can produce almost double the amount of energy it presently needs for its own heating and cooling.

The commercial strip study area— composed of office buildings, multiple-family residential units, department stores, restaurants, and motels — can collect the energy to provide 70% of its space heat, 80% of its hot water, and 70% of its cooling needs with solar systems on roofs and on a small fraction of the adjacent site area. Using all the area suitable for collectors would produce a considerable surplus, especially in winter though the visual impact would be somewhat greater.

Given the potential for a surplus of solar energy in some places and a marked deficit in others, transporting energy between buildings or even between sectors becomes an attractive possibility. The warehouse district, since it receives little public use, would be a low impact source of supply; large-scale deployment of collectors would be less obtrusive here than elsewhere. The central business district, with the lowest on-site capability, is a likely recipient. The researchers found that the surplus from 3.2 acres of Denver warehouse district could make up the deficit of one acre of the central business district.

Sharing the Solar Wealth

Shared systems with annual storage, allowing effective transport of surplus solar energy within and between districts, open up new strategies for community self-reliance. One problem of solar systems with short term storage is that the inevitable extended overcast periods lower the overall capability to only 70 or 80% of typical heating requirements. Equity is a second problem, since buildings with restricted solar access have no choice but to rely on non-solar systems. Sharing collector and storage space improves both system capability and equity. If numerous individual collectors feed into a single storage facility, then long-term heat storage is possible at reduced cost. Surplus solar energy can be saved during those months when demand is low and insolation high.

A shared system with annual storage also diminishes restrictions on siting. In Denver's low density, single-family development, block-wide shared systems would make the community more self-reliant with less of an impact. The area's entire space heat and hot water needs could be met without off-roof collector siting and without any removal of the tree canopy. In the high density, multiple-family development, the improvement over short-term storage systems is less spectacular — in part because winter collector shading is less of a problem. Still, a shared, long-term storage facility boosts the solar fraction of heat demand from 40 to 70% assuming use of roof and other site areas.

The UC researchers note that medium density — townhouses and low-rise apartments (16-20 units/acre; 40-50/ha) — reduces overall energy consumption for heating and transport, while also having enough room for fairly substantial solar collection. Medium to high density is also best suited to the reliable shared solar system, since distribution costs are not prohibitive.

The UC Berkeley study also examined the use of cogenerating photovoltaics to meet community demand for both thermal and electric energy. This hybrid system is

The commercial strip area can provide 70% of its space heat, 80% of its hot water, and 70% of its cooling needs with solar systems on roofs and on a small fraction of the adjacent site area.

assumed to convert sunlight directly to electricity at 9% efficiency, while collecting heat from the array at an additional 32% efficiency. Combined with long-term heat storage, and if enough collector area is available, this system can meet as much as 100% of a district's total energy demand and then some. Even with very modest efficiency improvements in electric appliances and lighting, all the land use types studied (except the central business district) meet or come close to this goal. Use of available site area in the commercial strip and warehouse districts could produce a substantial surplus of electricity year-round. Since a grid already exists, transport of electricity to other areas is possible.

In addition to the calculation of solar supply potential for specific case study areas, the UC report includes a general discussion of environmental constraints on community solar energy development: vegetation shading, street orientation, lot configuration, residential density, and variations in building height. These constraints become

Percent of energy demand supplied by cogenerating photovoltaics
(with long term storage)

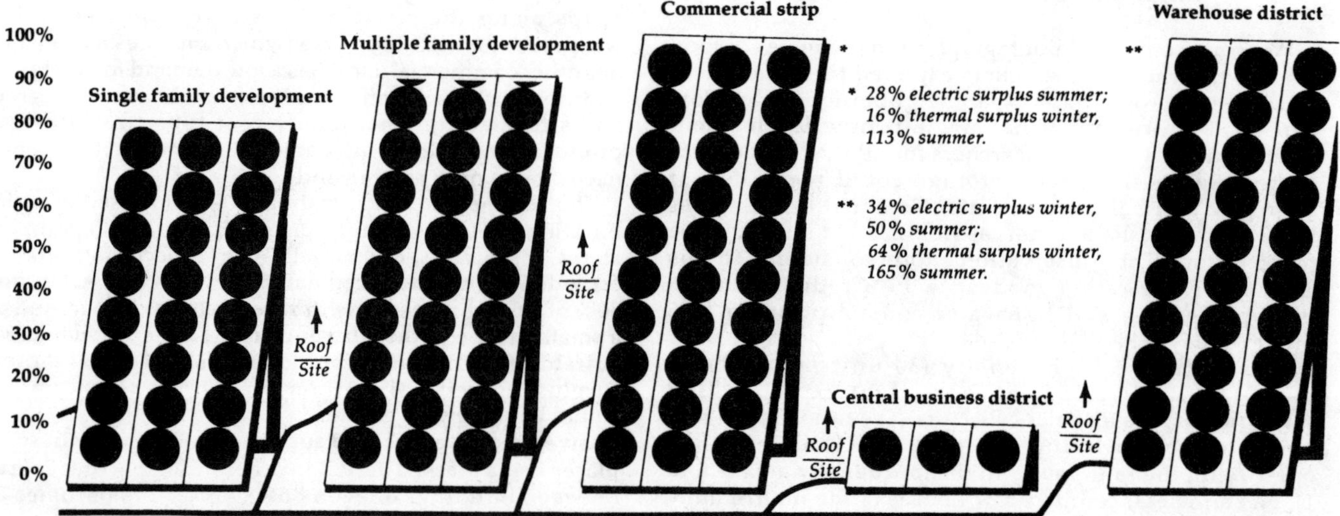

less critical as the amount of energy needed from solar technologies is reduced. Conservation to reduce demand below the levels assumed in the study places a much smaller burden on the community's endowment of land and capital. And, it is likely to occur before large numbers of collectors go on roofs and yards. With reduced energy needs, fewer collectors are able to provide a greater fraction of solar reliance.

Companion Pieces

The UC work is one of three studies sponsored by the Department of Energy's Office of Technology Impacts to evaluate the community-level effects of widespread solar energy use. *Three Solar Urban Futures* (Milne, *et al.* 1979) takes a somewhat longer cast into the future. It describes the structure of a hypothetical, average American city of 100,000 people in the year 2025 under varying solar growth assumptions. Two of the urban futures are extrapolated from the energy supply scenarios for the year 2000 developed by the President's Domestic Policy Review of Solar Energy in 1978.

With the supply mix and quantity as givens, Milne, *et al.* build each sector of the hypothetical city so that it consumes its specified amount of oil, gas, electricity, and renewables. This approach causes some incongruous

3 residential sector supplies 80% of the heating load, 50% of cooling, and 80% of domestic hot water. This leaves enough roof area for photovoltaic collectors to supply auxiliary backup as well as lighting and appliances. The commercial sector has enough collector area to achieve 60% energy self-sufficiency, while industry can meet only 18% of its needs. Demand assumptions for industry, however, are based on the Domestic Policy Review scenario, which are probably too high for 2025. The study does not look at the land use implications of the heavily passive Future 3 city in any depth, nor does it consider possible transition paths; both are areas worthy of examination.

Community Impediments to Implementation of Solar Energy (Armstrong 1979) considers some of the institutional problems associated with the transition to widespread solar use. It discusses the results of two workshops attended by representatives from those sectors most closely involved in solar implementation: utilities, finance, community planning, construction, environmental protection, special consumer groups, and legal and insurance interests.

—*Steve Meyers*

References:

Duffey-Armstrong, M., and J.E. Armstrong
1979 *Community Impediments to Implementation of Solar Energy.* Lawrence Berkeley Laboratory, Energy and Environment Division. Available for $6.50 or $3 microfiche from the National Technical Information Service (NTIS), 5285 Port Royal Road, Springfield, Virginia 22151.

Milne, M., M. Adelson, and R. Corwin
1979 *Three Solar Urban Futures.* Lawrence Berkeley Laboratory, Energy and Environment Division. Available for $7.25 or $3 microfiche from NTIS.

Twiss, R.H., P.L. Smith, A.E. Gatzke, and S.T. McCreary
1980 *Land Use and Environmental Impacts of Decentralized Solar Energy Use.* Lawrence Berkeley Laboratory, Energy and Environment Division. Available for $11 or $3 microfiche from NTIS.

Solar potential of residential building types

	Heat/HW Demand $(10^7$ Btu)	Available Collector Area/Unit (ft²)	% Shaded in Winter	Usable Energy Produced $(10^7$ Btu) Short Term Storage	Long Term Storage
Single Family Detached 1695 ft²/unit	7.00	1181	22	4.95	7.00
Townhouse 1300 ft²/unit	6.04	713	35	2.82	6.04
Row House 1300 ft²/unit	4.68	388	32	2.47	4.68
Low-rise apartments 1140 ft²/unit	3.71	316	35	1.90	3.71
High-rise Apartments 850 ft²/unit	2.65	78	32	0.88	1.60

NOTE: All dwellings are located in Baltimore (4111 °F-days).
Long term storage solar systems are "improved performance"; collector efficiency is 60%, reflectors increase output by 33%. For both systems, distribution and storage losses are 20%, with an additional 20% loss incurred for heat collected during the non-heating season by the long term storage system.

1 ft² = 0.0929 m²; 1 Btu = 1.055 kJ.

---

results. In the Future 2 city (based on a year 2000, 95 quad national scenario), 38% of the homes are uninsulated; in the high demand Future 1 city, only 29% are uninsulated. This discrepancy stems from the larger quantity of passive solar energy in the Future 2 supply scenario; the buildings are so energy frugal that *more* uninsulated structures are needed for demand to meet the given supply.

The Future 3 city is free of such scenario constraints. It represents a city built with maximum use of on-site solar energy. Passive design and tight construction reduce building energy consumption considerably. Per capita electricity use is virtually identical to 1974 levels, which seems unlikely for 2025. Still, passive design in the Future

4. Transport

A conspiracy lies behind today's lopsided transportation system. In 1949, several major oil, auto and tire companies were convicted of criminal antitrust violations for their part in the demise of American mass transit.

Although the American public was not indicted, it, too, played a role. The lure of personal mobility and privately owned transport proved stronger than caution and common sense. Cities were built around the automobile, with subsidies paving the way, and train systems were pushed into economic derailment. The only transportation system that thrived was the resource extravaganza of air travel.

The result is that two-thirds of the US labor force drives alone to work each weekday. Commanding 52 million steering wheels, Americans are ruled by a structure that has largely removed their choice to do otherwise. The dream of mobility has turned into a giant traffic jam.

Few perceived the transport trap. Drivers lived in a state of "speedometer warp," believing that they were going as fast as the speedometer indicated, ignoring the time they work just to afford their vehicles. Ivan Illich has shown that the actual speed of auto travel is equal to the miles driven, divided by the time spent behind the wheel plus the time spent working to earn the car's fuel, repair and purchase costs. Calculated this way, a bicycle is about as quick as a car.

This section shows the wealth of technical options that could make transportation more fuel-efficient, economical and environmentally benign. (Detroit has failed to make timely use of these options, though its resistence is not the only barrier to progress. Fuel-efficient cars won't have the impact they should as long as many people cannot afford to buy them,

locked into a vicious cycle of poverty and dependence on V-8-powered rejects.)

What, then, are the technical alternatives? Passenger cars consume more than half of all transport energy, and the equivalent of about half of all US oil imports. Well-demonstrated technical refinements can make them run on one-sixth the gasoline they use at present. Following the lead of foreign carmakers, US manufacturers are finally designing high-efficiency prototypes, though even these do not approach the full technical potential. Airlines, with fuel accounting for 40 percent of operating budgets, are grounding older jets in favor of more efficient models and are increasing payload capacity.

More efficient motors can bring emissions problems with them, as illustrated by the move from Otto (conventional, four-cycle) to Diesel engines. From the point of view of urban air pollution, electric vehicles and clean-burning fuels such as hydrogen have appealed—but whether they truly involve net energy savings, or carry environmental advantages, depends on the primary energy source and on the losses incurred in conversion.

Beyond technical improvements for each type of transportation, energy can be saved by redistributing the share of total transport volume carried by each mode. Specific energy needs differ by as much as a factor of ten between transport modes. Mass transit offers opportunities amid well publicized mistakes. In Europe, the tendency has been to mix trolleys and buses on city streets, with autos adapting. In the US, much mass transit has been burdened with all the development costs of a new roadbed; whether in San Francisco, Washington, Atlanta or Boston, mass transit has been forced to adapt to cars, not the reverse. So why not put the trolley tracks back in?

□ Transportation

New Fuels and Efficient Cars

*F*ROM THE CLOGGED AIRWAYS of Los Angeles to the choked, noisy streets of Rio, today's transportation system faces a dim future. With people travelling greater distances than ever before, the global appetite for dwindling energy resources has become prodigious. Transportation requires fuels with high energy value for their weight and volume. Today that means oil, a fuel with three and a half times the energy density of coal, but a fuel with a bleak outlook.

This special section in *Soft Energy Notes* examines four promising alternatives to a fossil fuel-based transport system in addition to improvements in the efficiency of transport. Among the supply alternatives are electric power, fuel alcohols (ethanol and methanol), hydrogen, and hydrocarbons from green plants. On the demand side, we review advances in electric car technology, and improvements in the fuel-based automobile. Trains, trucks, buses, and planes await analysis in a future issue.

The world's auto fleet has tripled in the last two decades and currently exceeds 300 million in number. In the United States, 73 percent of our refined gasoline feeds cars. Reducing that use is America's greatest single chance to plug the balance of payments drain. The rest of the world is not so lucky: for all the 1 billion people in the People's Republic of China, there are fewer than 40,000 cars. Three quarters of the world's cars are in the US and Western Europe.

The frantic drive for something to put in the gas tank has, at various times, brought us nuclear power, the Alaskan oil pipeline, and synthetic fuels from coal— all to be subsidized with "public risk capital." By ignoring the demand side for so long, we merely increase the agony of the eventual transition to a sustainable future.

Soft energy advocates generally side with the use of liquid fuels and efficient automobiles, rather than more electric power plants and electric autos. It is almost without question that an 80-90 mile per gallon auto (see *Reluctant Revolution* article on innovations in combustion auto efficiency) burning alcohol is more acceptable on social, environmental and economic grounds, than hooking up an electric car to five or ten cent per kilowatt-hour electricity. We think the jury is still out; though the verdict in favor of direct combustion is nearly unanimous.

Today's US transportation sector uses 19.2 quads (20.3 EJ) of oil, uncomfortably larger than the total amount we import. According to our review of improvements in combustion transport, the sector could provide equivalent service to a larger population, using only one quarter this amount, or 6 or 7 quads, 5 quads of which would be for cars. Several recent studies show that biomass can certainly provide this much energy by 2000. A new report by the US Office of Technology Assessment (June 1980) estimates that biomass from wood and plant materials could supply 12 to 17 quads of energy per year in the US by 2000. The Department of Energy's alcohol fuels policy review (June 1979) estimates a maximum potential alcohol fuels production level of 15.3 quads by 2000, 4.8 Q of which would come from ethanol and 10.5 Q of which would be methanol. A recently hotly-debated evaluation by the DOE's Energy Research Advisory Board (April 1980) pegs the potential of ethanol alone at 2.5 Q, though many of the assumptions regarding feedstock potential and conversion efficiency appear flawed.

Each of these studies expresses concern about the impact of a large fuel alcohol program on soil fertility and crop prices. The criticism is more properly directed at ethanol (grain alcohol) than wood alcohol (methanol). Even more to the point, the criticism should be directed at our unsustainable agricultural system, rather than fuel alcohol production. A new Friends of the Earth book, *New Roots for Agriculture,* by Wes Jackson documents the issue. A few snippets from Jackson's book: the US is today losing topsoil faster than it was during the "Dustbowl" years. One dumptruck's worth of topsoil, built over eons, passes New Orleans *each second.* Wind and water erosion remove an average of 9 tons of soil per acre per year. In some midwest areas, particularly those planted with corn, the loss is as high as 25 tons per acre per year. An alcohol fuels program must be based on the same policies that will save our agricultural sector, because without these policies we will be without food or energy. The current emphasis in the US on carbohydrate-based alcohols must shift to the more plentiful, less environmentally damaging cellulose feedstocks, though even there the potential damage is great.

None of this argues that we should drop our fuel alcohols program, but that we should use it as a good excuse to reform agriculture. Nor does it argue for electric automobiles, though a good hedge they may be. Our review of electric automobiles cites a suggestive, though far from definitive, analysis by the California Energy Commission, that a 50 percent shift in vehicle miles in the state to electric vehicles would increase peak demand requirements only 1000 megawatts. These vehicle miles today burn about 0.77 quads of oil, no small sum. And analysts completing a major study of the potential for efficiency improvements and solar technologies in the US economy are facing the problem of chronic oversupplies of electricity in their scenarios of the future. These oversupplies could do worse than protect topsoil.

In the end, we'll probably travel less in the years ahead. We've only one report on such a prospect; it discusses innovations in communications technology that could substitute for so much wasted motion.

—Jim Harding
Steve Meyers

The Reluctant Revolution

IN DETROIT, SMALL IS BIG AND EFFICIENCY IS IN

*C*hrysler Corporation, once known as the big car company, now stands on the verge of bankruptcy, and federal bailout funds, not fancy tailfins, get the headlines. "Small cars, small profits" has always guided the Detroit automaker, and old habits die hard. In May, Chrysler announced the closing of its Lynch Road plant in Detroit, bringing North American production of full-size cars to an end; but by June, the demise of the big car plant had been postponed.

The US auto industry, though reluctant to revamp and retool, realizes that survival means major changes. The race for fuel economy is on, but it has become largely a catchup game with Japan and Western Europe. To compete, American cars will have to consume less energy in the future—how *much* less it the question.

By law, 1985 US autos must have an EPA fleet average fuel economy of 27.5 miles per gallon (kilometers per liter = mpg × 0.42), double the 1975 figure. Although this standard will bring American cars in line with imported models, it is only a first shot in the efficiency revolution. While automakers debate with Congress over a mandatory 1995 fuel economy standard of 40 mpg, the Transportation Task Force of the Solar Energy Research Institute's Solar/Conservation Study proposes a fresh look at the actual energy requirements of passenger vehicles. The SERI panel, directed by Charles L. Gray of the Environmental Protection Agency, strips the car to its essentials. Using only demonstrated technological improvements they put together a fleet for the future that could attain as much as *100 mpg.*

Just Enough Power

The SERI 100 mpg fleet assumes no technological miracles; it reaches new heights of fuel economy by matching efficient vehicles to well-defined end use transport needs. The Transportation Task Force began remodeling the automobile by establishing the minimum performance features that would satisfy US highway drivers. They assume: 55 miles per hour (km/hr = mph × 1.6) up a 5 percent grade with a full load of passengers; acceleration from 0 to 50 mph in 20 seconds or less; and enough power to operate all the conventional accessories (heater, radio and the like) US passenger cars now provide. Motorists often haul trailers or payloads, so one-third of the full size (5-6 passenger) automobiles and light trucks in the SERI fleet need an extra 30 horsepower. One-fourth of all compact (4 passenger) autos will similarly need an additional 20 hp.

Engines even smaller than the powerplants in today's subcompacts can meet these performance requirements. The SERI design starts with a powertrain that delivers fuel energy to the wheels more efficiently: small displacement stratified charge and diesel engines, coupled to 90 percent efficient continuously variable transmissions. Similar transmissions, developed jointly by Fiat and Borg-Warner, increase fuel economy by up to 20 percent, and will soon be adopted by the Italian automaker.

A second aspect of the SERI fleet design is lower movement resistance. Advanced radial tires limit rolling resistance—the main problem at low speeds; improved aerodynamics cut the heavy wind drag a car faces traveling over 45 mph. These resistance improvements alone, according to project leader Gray, could reduce cruising power requirements by half. Finally, substituting light materials like plastics for heavier ones reduces vehicle weight. Using composite materials for certain body, suspension, drivetrain, and (low temperature) engine components can save up to 25 percent in fuel. Putting all these improvements together, the transportation panel designed a four-passenger car that meets performance requirements with a 3-cylinder, 27 horsepower engine and delivers fuel economy of 93 mpg.

Manufacturers Make a Start

Today's car manufacturers work toward more modest efficiency goals. General Motors, the world's largest automaker, recently announced that its 1985 auto fleet would average 31 mpg, well above the required minimum standard. Ford and Chrysler are moving in the same direction, with small, front-wheel drive cars. Chrysler's domestically produced *Omni* and *Horizon* models overcame initial steering stability problems to become highly popular, though fuel economy was not on a par with Ford's smaller import, the *Fiesta*. In 1980, Ford plans to unveil its second front-wheel drive model, this one manufactured at home, and with a combined EPA mileage estimate of around 40.

While American carmakers pick up momentum,

Future fleet possibilities

	Vehicle Class	Vehicle Test Weight (lb)	Cruise hp 55 mph	Extra hp, 55 mph, 5% grade	Acceleration power (hp)	Average Engine hp	cylinder	Projected mpg
2 Passenger	Available Now	2250	14.8	—	—	—	—	—
	Available 1982	1500	9.0	11.0	16.9	23	2	110
	Test demonstration	1050	7.2	7.7	12.2	18	2	140
4 Passenger	Available now	2000	13.0	—	—	—	—	—
	Available 1982	2000	9.8	14.7	21.4	32.5	3	78
	Test Demonstration	1400	7.8	10.3	15.4	27.1	3	93
5-6 Passenger & light truck	Available now	2500	13.5	—	—	—	—	—
	Available 1982	2500	10.1	18.3	25.9	43.4	4	58
	Test Demonstration	1750	8.1	12.8	18.6	35.9	3	70

Source: Congressional testimony of Charles Gray, SERI Transportation Task Force

manufacturers in other countries are not sitting idly by. Volkswagen, whose diesel *Golf* (the US *Rabbit*) already leads the industry in fuel economy, is developing a 3-cylinder supercharged diesel car capable of 65 mpg or better. The gasoline and methanol versions will be less spectacular, but cleaner. Conventional changes enhance engine efficiency: lower internal friction, direct fuel injection, and reduced pumping losses. An electronically-controlled start-stop system is more innovative. When the accelerator is released, the clutch disengages the engine and the fuel supply is cut off. Depressing the accelerator restarts the engine within 0.2 seconds. Fuel savings in stop-and-go traffic are expected to run 20-25 percent.

The new VW will also have greatly improved aerodynamics. The present VW *Golf* has a drag coefficient of 0.42, whereas the theoretical minimum is about 0.16. Wind tunnel tests of VW's new design show it to have a drag coefficient of about 0.30; this might drop as far down as 0.24. In addition, Volkswagen has put the *Rabbit* on a diet: designers look to reduce car weight by substituting lightweight materials and reducing body surface (without giving up useful interior space). Plastics and aluminum will take the steel content down from 69 percent to 50 for a weight loss of 15 percent. Even the rear axle may be non-metallic. VW's present test car weighs in at a slender 650 kg (1,430 lb).

To date, the weight loss champ is not VW but British Leyland. They have pared the already tiny *Mini Metro* down to a 510 kg (1,120 lb) Energy Conservation Vehicle. In tests, this vehicle delivers 60 mpg at 62 mph, and 100 mpg driven at a steady 35.

The Cost of Saving Gas

Auto manufacturers concede the possibility of such high fuel economy, but they argue that cars capable of 75 mpg and up would require large capital investments, and cost the consumer more than conventional vehicles. A recent study by Gorman and Heitner of TRW, Inc. concludes, however, that most efficiency improvements save more in fuel than they add in price. To confirm the cost-effectiveness of various modifications, Gorman and Heitner calculate fuel savings over a vehicle life of 123,000 km (76,000 miles) and compare this to the cost of the improvement. The resulting cost per unit of energy saved can then be directly compared to fuel prices at the pump. A diesel engine, for example, adds about $4 per horsepower over the price of a gasoline engine; the energy saved would cost about $7.50/GJ. This is easily competitive with the current cost of gasoline— $10/GJ from $25/bbl crude oil. Turbochargers are even more attractive at $3-5/GJ, and 10 to 20 percent of 1985 vehicles are likely to include them. Internal engine coatings like Xylan, which reduce friction losses and improve mileage by up to 10 percent, cost $4/GJ or less.

Other small improvements, such as constant speed accessory drives (for powering the alternator, pumps, and fans) and valve resizing, can be incorporated into engine designs at similarly low costs. Some fuel-saving moves, like aerodynamic improvements and the use of lightweight composite materials, add little or nothing to the cost of the new car. Efficiency improvements that are not cost-effective with crude oil at $25/bbl may compare favorably with synfuels, which are certain to cost more per barrel than today's crude oil.

Charles Gray suggests that radical vehicle changes may, in the long run, cost less than incremental efficiency improvements. "The redesign of an existing large car for a five-year production run, to hang onto a rapidly dwindling market, is extremely costly compared to the one-step introduction of a new car." Bit-by-bit downsizing puts high efficiency goals further into the future. Gray's rough estimate of the new capital necessary is about $60 billion, but even assuming this goes as high as $100 billion, the average capital writeoff per vehicle is only $500. The cost of delaying high fuel economy levels is high, too: in 1980, the US will spend $90-95 billion for oil imports, over one-fourth of this for automobiles.

Frank von Hippel of Princeton notes that if the US fleet of 100 million cars could get an average of 40 mpg instead of the present 15 mpg, oil imports could be cut by almost three million bbl/day, or some 40 percent. Rather than continuing to ship money abroad, or sinking it into the production of expensive synfuels, it may be cheaper to subsidize the rebuilding of the auto industry. Gray calls for an average fuel economy target of 80 mpg by 1995, to be backed up with a fuel inefficiency tax for vehicles that fall short.

Matching Cars to Needs

The technology to build an 80 mpg auto fleet is nearly ready for commercialization, but the cars would be quite different from today's. Almost half the fleet would be two-passenger urban models. Only one American production car — the sporty, high-powered Corvette — is a two-seater, while 63 percent are 5-6 passenger cars or light trucks. The present fleet is badly matched to transportation needs. Vehicle use statistics show that a two-passenger car would suffice for three-fourths of all trips; the 5-6 passenger model is actually required for less than 10 percent of all trips. And this percentage is likely to decrease as families become smaller. The SERI panel estimated that a fleet more closely matched to needs would contain nearly 50 percent 2-passenger cars and 85 percent four-passenger compacts. Incorporating the best technology already demonstrated in test equipment, such a fleet would have an average fuel economy of 103 mpg.

If new car sales are any indication, most Americans are willing to trade their over-sized gas-guzzlers for efficient small cars. Four-passenger cars claimed a 44 percent market share during the first quarter of 1980, and production of some models is lagging behind demand.

Two-passenger urban cars aren't in the showrooms yet, but they are already being planned. In early 1983, General Motors' Pontiac division hopes to introduce a mid-engine, two-seat commute car with an initial production of 200,000 per year, and Ford may follow suit. But Pontiac's two-seater is a long way from the 100+ mpg vehicle that the SERI panel says is feasible. Overpowered with a 2.5 liter engine, the Pontiac commuter will weigh 1,000 kg, more than some of today's four-passenger cars. Gray concludes that for high fuel economy, "the American public must be willing to accept energy conservation over performance potential."

—Steve Meyers
Katy Slichter

References:
Gorman, Richard and Kenneth L. Heitner
    1980 "A Comparison of Costs for Automobile Energy Conservation *vs.* Synthetic Fuel Production." TRW Energy Systems Group, 8301 Greensboro Drive, McLean, Virginia 22102.

Gray, Charles L., Jr.
    1980 "The Potential for Improved Fuel Economy Between 1985 and 1995." Statement before the US Senate Committee on Energy and Natural Resources, 16 pp. Available for $1.50 from IPSEP.

Where There's Smoke...

As YOU WAIT on the streetcorner downtown for the light to change, the bus en route across town leaves you in a cloud of hot, grimy dark soot. If current trends persist, incidents of this kind will increase, and buses will not be the primary sources of diesel exhaust.

It was in 1922 that Peugeot came out with the first experimental diesel-engine passenger car. Mercedes-Benz was quick to follow suit and began production in 1934. Today at least six car manufacturing companies produce diesel automobiles and an equivalent number have plans to do so. EPA has estimated that by 1990, 10-25 percent of all cars on the road will be diesels.

Less costly fuel, infrequent maintenance and better mileage all contribute to the diesel's appeal. The diesel Volkswagen Rabbit, for example, averages 45 miles per gallon, and its fuel costs 5-10 percent less than gasoline. Aside from financial benefits, it is interesting to note that this alternative engine emits less carbon monoxide and unburned hydrocarbons than spark-ignition engines.

Particulate emissions, however, may be a big problem. The average diesel automobile puts out approximately 1.3 pounds of sooty particulates per 1,000 miles. By 1990, if diesel emissions remain unregulated, 160,000-400,000 tons of particulate matter will be added to the air each year.

With more diesels on the road, environmental and health consequences may become more prominent. Once inhaled, the minute particles are difficult for the hairlike cilia in the lungs to expel. And, these particles absorb unburned hydrocarbons, increasing their carcinogenic potential. In *Mother Jones*, February/March 1980, Robert Rapoport claims that "these particulate emissions contribute to emphysema, asthma, and chronic bronchitis, and now it appears likely that some of the compounds attaching themselves to this matter cause genetic cell mutations."

Studies conducted by the Environmental Protection Agency have shown that diesel exhaust:

- increases susceptibility to respiratory infections;
- increases levels of arylhydrocarbon hydroxylase in the lungs, testes and prostate, which may be correlated with the growth of tumors;
- increases levels of total protein in the lungs, suggesting that longer exposure might result in fibrosis and lung destruction.

Even more alarming are the Ames test results for organic extracts from diesel exhaust particulates. These organic materials were found to be consistently mutagenic in tests conducted by EPA. Laboratory tests on rats, cats, mice and guinea pigs continue. Should they establish a link between diesel exhaust and cancer/lung disease, EPA may begin a two-year rulemaking process to further tighten recently proposed particulate emission standards of 0.6 gm/mi from 1982-1983 and 0.2 gm/mi after 1985.

Meanwhile, diesel emissions, like diesel production figures, continue to soar. General Motors now builds more than 800 diesel-engine automobiles daily; within a year, this will climb to 1,250. By 1985 both Mercedes-Benz and Peugeot expect 75 percent of total production to be diesels. Volkswagen's figures are similar. At the same time, the auto industry questions the need for diesel particulate regulation. GM, in particular, is making an all-out effort to demonstrate that diesel exhaust is harmless.

—Katy Slichter

References:
Budiansky, Stephen
    1980 "Diesel Exhaust: Regulations and Health Effects." *Environmental Science and Technology*, February, pp. 135-137.

Rapoport, Roger
    1980 "The Dark Side of Diesel Chic." *Mother Jones*, February/March, pp. 62-65.

Combustion Engines: Driving Toward Fuel Economy

NO TECHNOLOGICAL DEVELOPMENT has had a greater influence on social change than the internal combustion engine. It permeates almost every aspect of life in industrialized nations and its use is growing rapidly in less developed countries. Given the importance of the internal combustion engine as a primary source of motive power, one of the most effective near-term methods for reducing oil dependence in transportation is by basic improvements in engine efficiency. Many unique and innovative engine designs, such as the Stirling Cycle, have been proposed to replace the automotive internal combustion engine. While all promise high fuel economy and flexibility, none proves to be as practical as the internal combustion Otto cycle and its sister the diesel. The widespread application of exotic engines to transportation lies far in the future. The challenge is to improve the fuel efficiency of existing engine technology without reducing, perhaps even improving, performance. Turbocharging and the stratified charge answer this challenge.

One of the major factors determining engine performance is the ability to pump enough air into the combustion chamber to achieve optimal fuel/air mixtures over a range of engine speeds. In the turbocharged engine, hot gases from the exhaust manifold drive a turbine compressor, which increases the pressure of atmospheric air and feeds it into the engine intake manifold. The compression ratio of the turbocharged engine is the same as its naturally aspirated counterpart. Maximum cylinder pressures, on the other hand, are as much as 50% higher in turbocharged engines; combustion occurs at a higher temperature which significantly improves efficiency and performance.

Turbocharging technology is ideally suited to diesel engines and increases their already high fuel economy. Many of the drawbacks found in automotive diesel engines, such as low power density, limited speed range, and poor acceleration, are virtually eliminated by turbocharging. The first production turbocharged diesel, the Mercedes Benz 300SD, gives an impressive preview of what is to come.

Turbocharging the three liter (110-hp) engine increases the power output by 43% while adding only 7% to engine weight. A 1765 kg sedan achieves zero to 90 km/hr (0-60 mph) acceleration in 14 seconds. This compares to 21.2 seconds in a lighter (1400 kg) sedan using a conventional engine. Fuel economy is not sacrificed for increased performance; the heavy sedan achieves a respectable high-way-city combined fuel economy of 9 lit/100 km (26 mpg). An uprated version of the same engine (200-hp), installed in a C-111 coupe, was driven at continuous speeds in excess of 240 km/hr, challenging the speed limitations associated with naturally aspirated diesels.

The potential for fuel savings from increased power plant efficiency is enormous. With greatly improved power outputs, smaller turbodiesels can be used for the same tasks as their naturally aspirated counterparts. Combining a high efficiency turbodiesel with a 5-speed transmission and a lightweight Rabbit chassis, Volkswagen has created a car which exceeds 2.93 lit/100 km (80 mpg) at a constant speed of 50 km/hr. While impressive, the cost of turbocharging the already expensive Rabbit diesel makes it uncompetitive in the automotive market. No plans exist for marketing the turbo-diesel Rabbit until production costs can be considerably reduced.

Turbocharging has two main problems: high costs and increased emissions. Lower costs will result from more production experience and improved materials, such as inexpensive, low inertia turbine blades. A standard turbocharger design with turbine trim variations suiting a range of engine sizes could also reduce costs.

Emissions of unburned hydrocarbons and CO have already been considerably reduced. NO_x emissions, a problem with naturally aspirated engines as well, are produced by the high temperatures of the main cylinder. NO_x emissions are currently the focus of a vigorous research effort by Oldsmobile, Mercedes Benz and Volkswagen.

Spark-ignited gasoline engines are not as dramatically improved by turbocharging as their diesel counterparts. Porsche and Saab, among others, have installed turbochargers in several production vehicles. A major difficulty is turbocharger lag which results from the relatively slow response time of the turbocharger machinery compared to the quick response time of the engine. This turbocharger inertia is small compared to the inertia found in diesels. Higher exhaust temperatures found in gasoline engines also create additional stresses on the turbocharger materials. Turbocharging may not be the ultimate solution for improving the performance and efficiency of spark-ignited gasoline engines, but the stratified charge is somewhat closer.

The basic concept of stratified charge is the stable combustion of an overall lean fuel/air mixture. The most developed of several current strategies is the Honda CVCC (compound vortex, controlled combustion) configuration. The combustion area of the cylinder head in the

CVCC is divided into an auxiliary and a main combustion chamber connected by an opening called the torch opening. During the suction stroke, a relatively rich fuel/air mixture is fed into the small auxiliary chamber while a much leaner mixture is fed into the main chamber. At the torch opening, a controlled amount of rich auxiliary mixture combines with the lean main chamber mixture to produce an intermediate mixture cloud.

During compression the mixture cloud is forced into the auxiliary chamber making it leaner to insure ignition. Thus, the fuel/air ratio becomes stratified into three distinct zones in the combustion chamber. The mixture in the auxiliary chamber, ignited by the spark plug, propagates a stable explosion throughout the main chamber. The overall effect of this rather complex mixing strategy is that less fuel can generate the same power output as premixed carburetion.

By adjusting the relative area of the torch opening with the volume of the auxiliary chamber, an optimal combustion environment can be created to minimize exhaust emissions of CO, hydrocarbons and NO_x. Under the proper conditions, a high average temperature of 1000-2000° C is generated in the main chamber, promoting HC and CO oxidation and minimizing NO_x production. The quantity of emissions produced is load dependent. In a comparison of five different stratified charge mixing strategies, NO_x production was found to increase with load but decrease as full load was reached. Hydrocarbon and CO emissions were also found to be proportional to load, probably a result of imperfect mixing at high speeds.

The goal of stratified charge research is to produce a gasoline engine with the fuel economy of a diesel but retaining the performance characteristics of the gasoline engine. Although research has fallen short of this goal, the CVCC stratified charge has been quite successful in achieving both high fuel economy and low emissions. It is quite possible that other stratified charge configurations under development by Ford, Mitsubishi, Peugeot, and Texaco may improve fuel economy and lower emissions even further.

The internal combustion engine is a vital element of modern society. As liquid fuels dwindle, it will be necessary to push engine efficiencies to their limits. Turbocharging and stratified charge provide two currently available technologies for getting more with less.

—*Craig Conley*

References:

Society of Automotive Engineers
1974 "Why the Honda CVCC engine meets 1975 emission standards." *Automotive Engineering* 82:9.

1977a "Volkswagen develops a diesel." *Automotive Engineering* 85:6.

1977b "Can diesel specific power be increased?" *Automotive Engineering* 85:8.

1978a "First turbocharged diesel." *Automotive Engineering* 86:1.

1978b "The Mercedes turbocharged five cylinder diesel." *Automotive Engineering* 86:6.

1978c "Stratified charge mixing strategies compared." *Automotive Engineering* 86:8.

1979 "Turbocharging: What does the future hold?" *Automotive Engineering* 87:6.

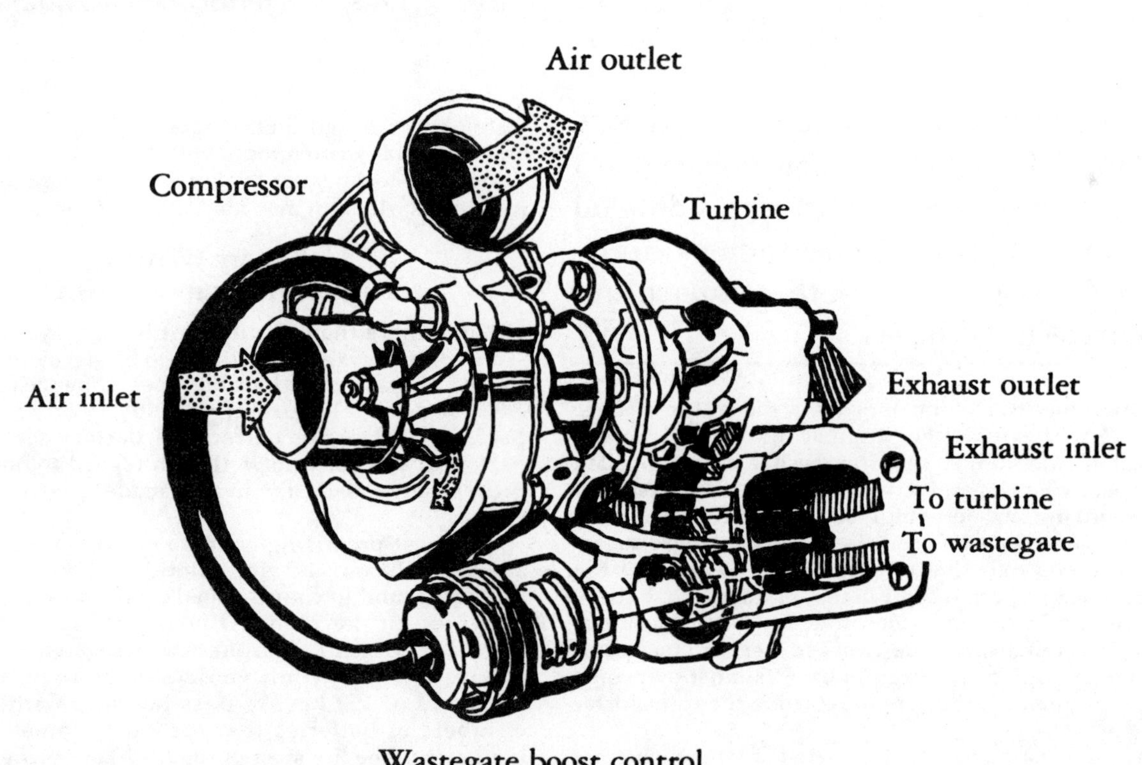

The Garrett Turbocharger adds just 15 kg to an engine's weight, but increases output by 43%.

Reproduced courtesy of the Society of Automotive Engineers, Inc.

Electric Vehicles Taking Charge

"The electric vehicle is convenient and may be well worth $500 for taking a man to his office..."
—M. C. Krarup, 1901
Prospects for Economic Automobiling

HIGH PERFORMANCE AND CONVENIENCE make the internal combustion engine a hard act to beat. Still, the drive to reduce oil consumption may replace today's high-powered gas guzzlers with compact and efficient electric vehicles.

No newcomer to the transportation scene, the electric vehicle (EV) has a longer history than the internal combustion automobile. The first mechanically propelled taxi cab was electric, and an electric car set the first land speed record in 1895. Between 1900 and 1915 the US market supported more than 100 manufacturers of electric autos; they were soon overshadowed by Henry Ford and his mass-produced internal combustion engine. The battery cars at the dawn of the automobile age were edged out because of their high price, limited performance, and expensive operation. Contemporary EVs must overcome the same problems if they are to return to the transportation picture.

Even with existing technology there is a large niche that EVs could fill. Surveys conducted in Washington, D.C. and Los Angeles during the late 1960s revealed that one-third of all urban cars were rarely used for trips longer than 55-75 km (1 km = .62 mile). To meet 95 percent of all urban driving needs, a car needs a range of 110 km in compact cities like Washington, and up to 220 km in sprawling urban areas like Los Angeles. Today's electric vehicle technology falls within this spectrum, though at the lower end, and many consumers will demand more flexibility. General Motors has set 160 km as the minimum range goal for the electric car it may market sometime after 1985.

Battery Wanted:
High Performance, Low Cost

While more than 30 different battery types are under study, only a few will contribute to EV development during this century. The need for high performance (particularly extended range), durability, and low cost in a single package makes the perfect EV battery elusive. Some battery types perform well with regard to one or two of these criteria; no design has yet made a clean sweep of the field.

The most promising candidates are lead-acid, nickel-zinc, nickel-iron, and zinc-chloride batteries. Lead-acid—the type found in conventional cars—powers most of the EVs now on the road. Although cheap and relatively durable, present designs are sluggish and rapidly exhausted. They limit vehicle range to 80 km or less; distances of 150 km are possible only with such large numbers of batteries that the car becomes heavy and inefficient. The life span of the lead-acid power pack, like most aqueous batteries, depends on the depth of discharge—how far down it is drained in use. The typical lead-acid battery can sustain only 100 charging cycles at 100 percent depth of discharge (total drain). If the depth of discharge is held to 70 percent, life span jumps to 700 cycles, and the total energy throughput increases by a factor of five.

Theoretical maximum energy density of the lead-acid battery is 171 watt-hours of capacity per kilogram of weight. Real life batteries, at about 40 Wh/kg, fall far short of this ideal. Half the weight of an actual battery consists of inert components that play no active role in storing or discharging electricity: the supports, connectors, terminals, and cell containers. With high-strength, low-weight components of aluminum and plastic, battery weight drops by as much as 15 percent. The specific energy of lead-acid batteries can probably be doubled through improved packaging and more efficient use of active materials.

Nickel-zinc batteries have a theoretical maximum energy density of 321 Wh/kg, yielding greater range for the battery's weight than lead-acid types. Unfortunately, nickel-zinc power units are more expensive and shorter-lived. Prices, now at about $180 per kilowatt-hour of battery capacity, will have to drop to $70/kWh before nickel-zinc is competitive. A 66 Wh/kg battery from Gould, Inc. will come close, if it can be mass produced for $75/kWh as the company claims. The primary research task is increasing cycle life which now rarely exceeds 200 rounds.

Nickel-iron batteries are rugged but low in power. Even advanced designs are not expected to perform much better than present-day lead-acid types. Only for the long haul does nickel-iron have any advantage. Capable of 100 percent depth of discharge for well over 1000 charging cycles, nickel-iron's long term operating costs are low. The most likely vehicles are trucks or buses where durability is more important than light weight.

The zinc-chloride battery also lasts for 1000 charging cycles or more, yet its theoretical energy density is a high 465 Wh/kg. Test vehicles, devised by Gulf & Western's Energy Development Associates (EDA), attain a range of 150 km at a speed of 90 km/hour while burdened with four passengers. The design features that provide this performance create additional technical problems. First, the battery stores chlorine in the form of chlorine hydrate at temperatures below 10° C (50° F), and the refrigeration equipment to maintain this temperature brings power losses, extra weight, and added volume. Second, the EDA battery increases cycle life by employing an actively flowing acid electrolyte, to inhibit the deterioration of the electrodes. This entails a complex plumbing system, which makes the battery more difficult to manufacture and service. Third, the acid electrolyte gives the chlorine constant access to the electrode, so the self-discharge rate is high. Because its power drains out every few days, the battery needs to be recharged even when the car is not in use.

These technical problems are not insurmountable. Gulf & Western plans to begin pilot plant production of its battery system in January 1981. Output will be one battery per day until full scale production begins in 1984. The cost stands at $100/kWh, though it should drop to $70 with mass-production.

With aqueous solution, ambient temperature batteries, electric vehicles are adequate for most transport purposes. However, for the range and performance of internal combustion automobiles, EVs may have to use high-temperature molten salt batteries. Operating in the 300-500° C range (600-900° F), batteries using sodium sulfur or lithium-aluminum-metal sulfide can reach energy densities above 170 Wh/kg. Though still experimental, they promise to store large quantities of energy while remaining small and light.

While concentrating on battery improvement, electric vehicle research proceeds in several other areas:

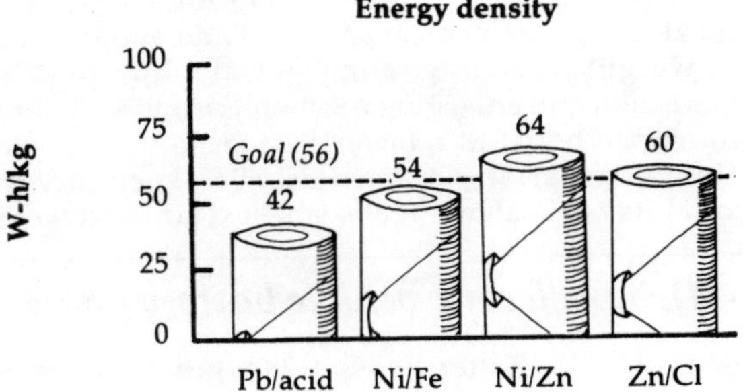

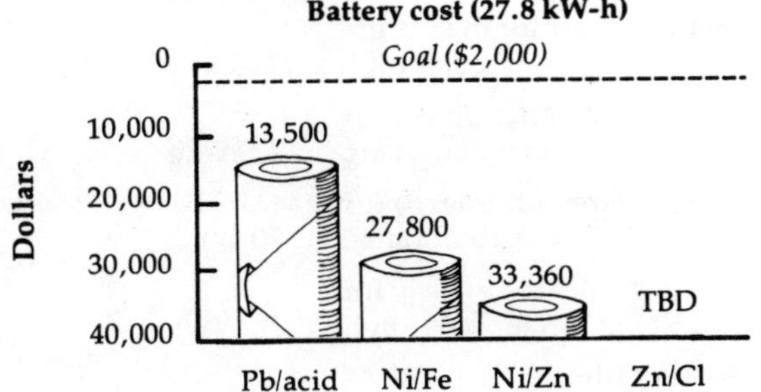

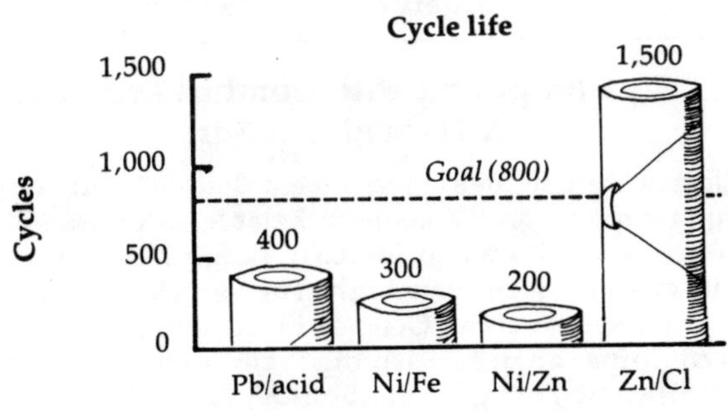

SOURCE: R. S. Kirk, P. W. Davis, "A View of the Future Potential of Electric and Hybrid Vehicles," US DOE, March 1980.

Battery technology development project 1980 performance

- *Energy efficient motors.* Direct current series motors are most common in EVs, but AC induction motors with solid state inverters could prove lighter, more efficient, and easier to maintain. Vehicle range could increase by as much as 12 percent.
- *Regenerative braking.* The EV motor can serve a double purpose by becoming a generator during braking, and converting the car's momentum back into electricity. By recharging the battery, regenerative braking can increase range by about 15 percent in stop-and-go urban driving. A drawback at present: current surges can disrupt the battery's chemistry and shorten its life.
- *Cold climate performance.* Charge and discharge capacities for aqueous batteries drop with low temperature, posing a challenge for EV design. Passengers also require space heating during cold weather, and no thermal discharge is available as in an internal combustion engine. As electric resistance heating would drain significant amounts of battery power, small gas heaters are under consideration.

- *Transmission options.* An EV transmission would improve performance under acceleration and up hills. It would also increase vehicle price and add about 45 kg (100 lb) of weight, reducing range slightly. Eliminating the transmission and enlarging the motor might give the same performance boost at a lower cost. Another option is a continuously variable transmission to keep motor efficiency at its peak, allowing the smallest size and weight.

Urban electric vehicle battery goals

Battery mass	25% gross vehicle mass
Specific power for maximum cruise	60 W/kg
Specific power for maximum acceleration	110 W/kg
Average specific power for urban driving	32 W/kg
Specific energy for 100-120 km urban range	70 Wh/kg
Specific energy for 160 km range @ 60 km/h	70 Wh/kg
Battery life, 100% depth of discharge, urban driving	300 cycles (35,000 km)
Battery cost	70 $/kWh

Competing with Combustion: A Hybrid Answer

Electric vehicle designers face a double bind: without adequate range, an EV is unmarketable; at the same time, there is no easy way to increase range without making initial costs prohibitively high. The vehicles produced are compromises, such as General Electric's lead-acid ETV-1 that combines an urban driving cycle range of 120 km with an optimistic cost goal of about $8,000 (1980 $). At this price, the electric vehicle seems to be less car for more money.

One way to make the EV more competitive is by including a small combustion engine. A range-extension hybrid proposed by W. F. Hamilton adds a small 16 horsepower gasoline engine and a clutch to the basic electric vehicle drive train. Under urban stop-and-go driving conditions the battery provides all the power. The gasoline engine takes over on the highway, which constitutes an estimated 20 percent of all travel. The hybrid gasoline-engine/battery vehicle manages to unite the advantages of both: the mobility and highway cruising of internal combustion with the low-noise, emissionless urban performance of battery cars. Projected lead-acid and hybrid EV life-cycle costs are about equivalent to a moderately efficient conventional car. The additional range of the hybrid version could eventually tip the market scales in its favor.

Changing Electricity Demand

If EVs capture the affections of future drivers, consumer demand for oil will turn into increased demand for electricity. Even with substantial market penetration, however, this doesn't necessarily mean more large power plants. EV batteries will likely recharge at night, during the off-peak hours when much generating capacity is otherwise idle. (Indeed, without favorable off-peak electricity rates, electric automobiles may never be competitive.) According to recent forecasts by the California Energy Commission, the power drawn by EVs should cause no more than a minor ripple in electricity consumption patterns. Assuming that EVs requiring 0.27 kWh/km (0.45 kWh/mile) provide 5 percent of the vehicle miles travelled in the state, total electricity demand would increase by 2 percent. Night-time recharging would prevent any change in peak load, so no additional generating capacity would be needed. If EVs became popular enough to displace 50 percent of internal combustion driving in California, an additional 1000 MW, or 2 percent, would have to be added to the projected peak load.

Overseas, the demand picture is similar. In a 2025 low energy scenario for the United Kingdom, Gerald Leach conjectures that 30-35 percent of automotive traffic will be electric powered. Assuming advanced sodium-sulfur batteries and a low EV energy use of 0.11 kWh/km (0.18 kWh/mile), 8 or 9 percent of the total electric demand would be for personal transportation.

EV Outlook

The United States is not the only country with an active electric and hybrid vehicle program; German, French, British, Italian, and Japanese governments also support EV research. The most likely EV markets, at least in the near future, are light trucks and vans. But by the mid- to late-1980s, the US expects to have a four-passenger, electric subcompact in production that can provide a range of 160 km in urban driving, consume 0.35 kWh/mile, and cost about the same per mile as a conventional internal combustion car. Acceleration and performance equal to conventional automobiles will take a bit more time. Department of Energy battery research program leader Robert S. Kirk considers it "highly plausible that a full-performance electric or hybrid vehicle could be commercialized in the late 1990s."

Efficiency gains in combustion engines, however, are so rapid, EV development may have trouble keeping pace. "We would have a chance," one EV researcher wistfully commented, "if GM and Ford and the others stood still for about ten years. But when you're looking at 50-70 mpg figures for prototype urban vehicles perhaps as early as 1985-90, things do not look so good."

With energy efficiency the fashion, EVs will have to keep current if they are to take charge.

—Craig Conley
Steve Meyers

References:

Hartman, John, Elton Cairns, and Earl Hietbrink
1978 "Electric vehicles challenge battery technology." General Motors Research Laboratories document #GMR-2654, Warren, Michigan 48090.
1978 "Battery and fuel cell technology surveyed." *Automotive Engineering* vol. 86, no. 7.

Hamilton, William F.
1978 "Prospects for Electric Cars." General Research Corporation, Santa Barbara, California, for the US Department of Energy Division of Transportation Energy Conservation. Available for $15, $3 microfiche, from the National Technical Information Service, US Department of Commerce, 5285 Port Royal Road, Springfield, Virginia 22161.
1980 "The Potential of Range Extension Hybrid Electric Cars." General Research Corporation, P.O. Box 6770, Santa Barbara, California 93111.

Rand, D.A.J.
1979 "Battery systems for electric vehicles — A state of the art review." *Power Sources* vol. 14, no. 2.

Hydrogen: Soft Gas or Hard?

*H*YDROGEN, a clean-burning, efficient, potentially abundant fuel, could replace petroleum in our transportation systems. The future of hydrogen-based transport, though, may be tied to that of nuclear power.

Hydrogen's early role in transportation came to a disastrous end one morning in 1937, when the German zeppelin, the *von Hindenberg*, exploded on a Lakehurst, New Jersey airfield. Later dirigibles made passengers and cargo lighter than air with helium gas—almost as buoyant as hydrogen, but inert. Now, decades later, hydrogen may be in line for a transportation comeback as the primary fuel of the post-petroleum era.

Unlike fossil fuel resources, which inevitably grow scarcer and more expensive, hydrogen is inexhaustible. It is the universe's most abundant material, the building block from which all other elements have been formed. Hydrogen's principal virtue as a fuel is that it is essentially non-polluting: its main combustion product is water, the familiar H_2O. Hydrogen is also versatile enough for use in both today's internal combustion engines and more advanced and efficient energy converters like fuel cells. Horsepower buffs will applaud hydrogen's high energy density, which, at 51,600 Btu/lb (24.8 GJ/kg), exceeds that of any other chemical fuel, and out-muscles gasoline by a factor of 2.7 to one. Gaseous hydrogen, after all, is the simplest, lightest molecule possible: two protons sharing two electrons. Nothing could be more compact, or release more energy for its weight, than this unstable union.

High reactivity makes hydrogen an attractive transportation fuel, but one that falls short of perfect. Hydrogen is so easily oxidized that it is rarely found in free form in either the earth's atmosphere or crust: it appears chemically bound in water and other stable compounds. Freeing hydrogen from its chemical entanglements takes energy in the form of heat or electricity—slightly more energy than the hydrogen gas provides as a fuel. Because it is unavailable in any active form, hydrogen can never be a *source* of energy like sunlight or fossil fuels, but it could be a means of *storing* energy in a form that is easily converted and conveyed. Hydrogen's transportation future, then, depends on two factors: economical, reliable vehicles, and enough energy *in some other form* to produce the hydrogen needed to keep people moving.

Cylinders and Cells

TWO DISTINCT HYDROGEN PROPULSION SYSTEMS are serious contenders for the role of 21st century power pack. One of these, the hydrogen thermal engine, is a present-day internal combustion engine look-alike. The other draws power from a fuel cell that generates electricity from hydrogen. It shares a drive train and other features with today's experimental electric vehicles, but is missing the battery.

A hydrogen thermal engine is a conventional gasoline engine, slightly modified to burn a more volatile fuel that is higher in energy density by weight, but lower by volume. An engine runs leaner on hydrogen than it would on gasoline, but is more likely to pre-ignite, or knock. Performance is best when hydrogen is injected into the cylinders under pressure (compensating for low energy/volume ratio). At the same time, flow should vary with the load, while air intake remains constant (much as in a diesel). But even these minor modifications are not essential. Conventional carburetion produces only 25 percent less specific power than complete alteration, and it also allows the vehicle to run on both hydrogen *and* gasoline.

Except for the occasional "ping" of detonation, motorists in hydrogen thermal engine cars would notice little difference in performance, but they would pay for the privilege of driving an unconventional vehicle every time they had to refuel. At 1978 prices of $0.21/liter ($0.80/gallon), premium gasoline carried an energy cost of about $6.00/gigajoule, while the dealer cost of hydrogen (assuming national markets) was estimated at about $0.20/m³, or $18-20/GJ. This direct comparison, though, does hydrogen an injustice, since engine efficiency is 30 percent higher than with gasoline. The energy cost *per equivalent GJ* is between 14 and 15 dollars, which is still about twice that of premium. Gasoline would have to climb above 46¢/liter, or $1.75/gallon (1978 $) for the two fuels to be competitive in a thermal engine.

A hydrogen vehicle powered by a fuel cell, in contrast,

would be just as economical as a typical gasoline-burning automobile with premium at only 24¢/liter ($0.91/gallon), a difference due primarily to its greater efficiency in energy conversion. The fuel cell process is just the reverse of electrolysis: fuel is supplied to an electrode where it reacts with atmospheric oxygen to produce a low voltage electric current. Whereas a thermal engine is limited by Carnot efficiency, the fuel cell can reach efficiency levels of 80 percent, with 30–50 percent possible in automobiles. This is two and a half times the efficiency of gasoline engines (Breele *et al.* 1979). The fuel cell vehicle bears a superficial resemblance to today's battery-powered electric cars. Both use high-efficiency motors to provide traction, and both can be engineered to dispense with transmissions, but the two differ in performance. The power/weight ratio of a fuel cell is seven times that of present-day batteries. Though no match for a high-compression, large-displacement engine, racing fans should keep in mind that the fuel cell is quieter, more efficient, and less polluting than any form of combustion.

Hydrogen Storage:
Hiding in the Hydrides

TODAY'S CONSUMERS look for more than performance in a car. They also consider comfort, depreciation, maintenance costs, and, especially when fuel shortages and gas lines threaten, a vehicle's driving range. To make 95 percent of the round trips for which US cars are presently used, an automobile must have a range of at least 150 miles. This distance presents no challenge to gasoline or diesel cars; liquid fuels generally lend themselves to high on-board energy storage capacity since energy density by volume is great, and they can be safely stored in lightweight tanks. Range *is* a problem with battery-powered and hydrogen vehicles, whose capacity for on-board energy storage is limited by weight. Storage systems are so heavy that electric and hydrogen cars with adequate range now outweigh conventional vehicles by a considerable margin. And the price one pays for excess weight is energy efficiency. The heavier the storage system, the more difficult this trade-off of range and weight becomes. Most hydrogen vehicle research therefore concentrates on finding a lightweight energy storage system that sacrifices neither range nor efficiency.

Industrial storage of hydrogen is a poor match to transport requirements. As a compressed gas, hydrogen needs large and heavy containers: a typical steel cylinder, filled to a pressure of 136 atmospheres, takes up 24 times the space of an equivalent tank of gasoline, and is 30 times heavier. The hydrogen fuel contributes a scant 1 percent of the weight. Cryogenic (low temperature) storage of liquid hydrogen at minus 217° C (−423° F) allows a lighter and less complex system, but insulation must be superb, and such low temperatures present serious, probably unacceptable hazards. In addition, commercial liquefaction entails an energy loss of up to 65 percent.

Any gas can be compressed or liquefied. Unique to hydrogen is its ready ability to move in and out of combination with a variety of metals, and this chemical property may hold the solution to the storage problem. Safe, compact storage at ambient temperatures is possible in hydrides of metals such as magnesium, or alloys like iron-titanium. These substances respond to hydrogen gas under pressure by adsorbing it into their surfaces. The gas molecules dissociate into hydrogen atoms, enter the crystal lattice of the metal, and at the right pressures and concentrations, form compounds, or hydrides. The storage capacity of some metal hydrides is impressive; even though total weight increases by only a few percent, there may be two or three times as many hydrogen atoms as metal. More hydrogen can be packed into the crystal structure than into the same volume of liquid. The energy density by weight of existing storage systems is still low, but not so low as to rule out vehicles with restricted range requirements, like urban buses.

Thermal engines make a better match to hydride storage systems than do fuel cells because heat is needed to decompose the gas and the metal. (An equivalent amount of heat is given off when the metal is charged with hydrogen and the hydride is formed.) Exhaust heat from the thermal engine can be used to discharge hydrogen from its hydride reservoir, while the fuel cell would have to expend some of its current for resistance heating. Several vehicles have already been built that fuel a hydrogen thermal engine in this fashion. The Daimler-Benz 12-passenger bus ingeniously incorporates hydride reservoirs operating in different temperature ranges, allowing the engine's waste heat to "cascade" from one to the next. Three storage beds are used: a high-temperature magnesium-nickel hydride bed receives exhaust gas directly from the engine and feeds hydrogen back; just behind it is a low-temperature, iron-titanium bed that

TOOLS for the SOFT PATH

takes the exhaust after the first bed has cooled it somewhat; and a third bed, also of iron-titanium incorporates a heat exchanger to provide air conditioning for the bus's interior, as well as hydrogen for the engine. One West German demonstration project proposes a fleet of 20 hydride storage vehicles for use in congested, urban areas where their low pollution levels would be most advantageous.

Hydride storage systems are expensive at present, and still extremely heavy. A projected 1985 iron-titanium hydride storage vehicle would have to weigh 5,200 pounds to achieve a range of 200 miles, more than *four times* the weight of a gasoline automobile with the same capability (Reilly & Sandrock 1980). But if non-automotive applications for hydride beds, such as off-peak electricity storage by utilities, can generate market breadth, both technological advance and price reduction seem likely.

Soft Gas

COMMERCIAL QUANTITIES of hydrogen gas are not about to appear out of thin air, and present-day production methods seem to push this almost perfect fuel into an alliance with nuclear power. Seizing on hydrogen's storage potential, some energy futurists claim that nuclear reactors and hydrogen make each other more attractive. Hydrogen storage would make large electric generating facilities more efficient, while expanding the role of nuclear power would pave the way for hydrogen as a transportation fuel.

Electrolysis, with an 80–85 percent conversion rate, is the most efficient technique known for evolving hydrogen. Nuclear proponents suggest that the excess and off-peak output of nuclear reactors could be stored in this form making the plants more efficient overall. The electricity for hydrogen production could actually come from almost any source: hydropower, wind, or solar. (In the case of coal, direct gasification loses less energy.) Yet hydrogen is generally linked to nuclear plants because electrical and thermal outputs could both be used to produce the gas. The waste heat can "reform" natural gas (CH_4) into hydrogen at about 75 percent efficiency, or derive it from coal at 57 percent (Donnelly *et al.* 1979). Hydrogen production makes the nuclear plant more appealing by raising its efficiency and lowering thermal pollution. Moreover, if breeder or fusion reactors were to supply all our electric demand, they could also supply enough hydrogen to the transportation sector to displace much, if not all, of US petroleum imports at present-day fleet efficiencies. The nuclear/hydrogen collaboration could eventually phase out the use of all fossil transport fuels—a strong argument, it would seem, in favor of the nuclear option.

Critics point out that these energy systems still rely on fossil fuel feedstocks (coal and natural gas) that will someday be exhausted; they consume large volumes of water; and they carry a heavy pollutant load. The alliance between nuclear power and hydrogen has a third and not-so-silent partner: chemical processes that diminish the very environmental benefits that make the user of hydrogen so attractive in the first place.

A hydrogen economy could, for the same investment, be powered by electricity from photovoltaic sources. AEG-Telefunken has proposed a 60-year plan for photovoltaic/hydrogen "plantations" to replace fossil fuels entirely. With solar cell efficiency of 10 percent and electrolysis into hydrogen at 70 percent efficiency, Telefunken claims that hydrogen will someday cost no more than gasoline does now. Hydrogen from nuclear reactors, they note, would be more expensive. The recently announced plan "pyramids" the photovoltaic/hydrogen plantations by using the energy generated by one installation to produce the solar cells needed to build others. A total investment of $50 trillion would be needed to replace world oil consumption of 15 billion tons with hydrogen. Because solar energy is so diffuse, 100,000 plantations are envisioned, occupying nearly 2 million km², a land mass as large as the states of Washington, Oregon, California, Nevada, Montana, Wyoming, Idaho, and Delaware combined.

On the horizon are several other processes that could produce hydrogen using only renewable energy resources. Noting that plants, in the first phase of photosynthesis, use light to split water into its component elements, laboratory researchers have attempted to mimic this natural, biological process to evolve hydrogen. They employ membranes that contain chlorophyll—the same oxidizing agent that is found in plants—with enzymes acting as a catalyst. The reaction is hard to sustain, reflecting the instability of biological components taken out of their natural environments. At the time of their discovery six years ago, photolytic systems lasted no more than a few minutes; now they have a lifespan of ten hours or more, and the rate of gas production is up ten-fold. Researchers are presently trying to replace the short-lived biological compounds with more stable, synthetic chemicals that do the same job (Hall *et al.* 1980). An alternative approach uses microorganisms to convert carbohydrates, such as crop residues and other forms of biomass, into hydrogen gas (Zajic *et al.* 1979).

Still on the drawing boards at the Systems Development Branch of the Solar Energy Research Institute is a hybrid hydrogen/methanol thermal engine that circumvents some of the production and storage problems associated with hydrogen alone. On-board energy storage is in the form of methanol, which can be derived inexpensively from coal or biomass, and is easily stored. Methanol's two main problems are low energy per unit volume and high aldehyde emissions. The SERI design proposes to decompose methanol catalytically into hydrogen-rich gases in a reactor that uses engine exhaust heat. These gases have a heating value 22 percent higher than methanol, and they burn cleaner. Because the catalytic reaction ($CH_3OH + heat \rightarrow 2H_2 + CO$) emits carbon monoxide, the hydrogen/methanol hybrid would have to incorporate a further oxidative reaction to be as pollution-free as hydrogen.

Is hydrogen a soft gas or a hard one? Will it have an important 21st century role in transportation? Much depends on whether hydrogen research can keep up with the progress made by the battery-powered electric vehicles and the alternative liquid fuels which are hydrogen's main competitors.

—*Charles Drucker*

References:
Breele, Y., P. Gelin, C. Meyer and G. Petit
1979 "Techno-economic study of distributing hydrogen for automotive vehicles." *International Journal of Hydrogen Energy (IJHE)* Vol. 4:297–314.
Donnelly, J.J. Jr., W.C. Greayer, R.J. Nichols, W.J.D. Escher and E.E. Ecklund
1979 "Hydrogen-powered vs battery-powered automobiles." *IJHE* Vol. 4:411–443.
Hall, David, Michael Adams, Paul Grisby and Krishna Rao
1980 "Plant power fuels hydrogen production." *New Scientist*, 10 April, 72–75.
Reilly, J.J. and Gary D. Sandrock
1980 "Hydrogen storage in metal hydrides." *Scientific American*, Vol. 242:2:118–129.
Zajic, J.E., A. Margaritis and J.D. Brosseau
1979 "Microbial hydrogen production from replenishable resources." *IJHE* Vol. 4:385–402.

Airlines Beat the High Cost of Flying

*I*N 1903, WILBUR AND ORVILLE WRIGHT took to the air at Kitty Hawk in the world's first sustained and controlled powered flight. The Wright Flyer weighed 750 pounds (341 kg), had a 40-foot (12 meter) wingspan, and was propelled at 35 mph (56 km/h) by a 12 horsepower engine. Today's jets, in contrast, have wings five times as long, may carry 500 passengers and fly at 600 mph (960 km/h) using 100,000 horsepower engines. The cost of fuel is much greater, too.

Jet fuel, just 13 cents a gallon (3.4 cents/*l*) before the 1973 Arab oil embargo is up to 99 cents a gallon (26 cents/*l*) today and rising. In response to these skyrocketing fuel prices, Congress initiated the Aircraft Energy Efficiency Program which calls for a 15 percent reduction in fuel consumption, a five percent reduction in direct operating cost, and reduced noise and emission levels. But sales competition in today's market has forced manufacturers to emphasize fuel economy even beyond the program's requirement. There have been significant advances in fuel-efficient engine design, aerodynamics, the use of weight-saving materials, and electronic flight management. The president of Grumman International, Michael Pelehach, illustrates the effects of these technologies for *National Geographic*: "Take an 80,000 pound (36,400 kg) tactical aircraft of the 1960s and redesign it with the technology of the 1980s. With advanced composite materials you could get the weight down to 53,000 pounds (24,100 kg); with better aerodynamics, to 44,000 (20,020 kg); with new engine technology, to 37,000 pounds (16,800 kg)."

New lightweight aircraft materials under development could help curb fuel consumption. The most promising material for planes flying at less than 700 mph (1120 km/h) is a composite of graphite and epoxy. Graphite fibers in an epoxy matrix is much lighter than aluminum and four times as strong. By using the composite material for its 767 set for delivery in 1982, Boeing plans to save 1,100 pounds (500 kg), equal to five more passengers. Dr. Robert Leonard of NASA told *Science Digest* he expects composite aircraft to fully replace current comparable aluminum aircraft by 2000. The result would be a 20 percent reduction in weight and a ten to twelve percent increase in fuel savings. Minor cosmetic changes will also help reduce the weight factor. For example, less paint will be used on the aircraft, new materials will be used for carpeting, and improved seats will replace old heavier ones.

Propellers turn fuel efficiency around

Advances in engine dynamics will increase fuel efficiency significantly. The return of the airplane propeller, although modified, is one such advancement. The new "propfan" has six to ten boomerang-shaped blades and is expected to increase fuel efficiency by 20-30 percent without sacrificing speed or creating undue cabin noise. *Aviation Daily* quotes Walter Olstad of NASA: "Combining the effects of advance propellers and improved technology in the turboshaft drive engine would mean that airplanes powered by advanced turboprops would have a 30-40 percent fuel savings over current aircraft. The importance of fuel savings of this magnitude in a time of skyrocketing fuel costs and increasing scarcity is generating a rapidly growing interest in early availability of advanced turboprop technology." The advanced turboprop will have three times the loading capacity of conventional 1950s propjets and will be used primarily on commuter routes.

In new aircraft models, such as Boeing's 757 and 767, aerodynamic design features are of primary importance. The major aerodynamic improvement common to the new transport planes is longer wings. By making the wing thicker and longer, aerodynamic efficiency can be increased by 10-15 percent.

Another significant improvement, laminar flow control systems, will streamline airflow over the wing reducing turbulence and fuel consumption by as much as 30 percent. Laminarization of the turbulent flow is achieved in either of two ways. "One is to cut very fine slots in the wing surface .005 centimeters to .03 centimeters wide," reports Robert F. Sturgeon, Jr., of the Lockheed-Georgia Company in *Science Digest*. "The other is to use either a perforated or a porous surface on the wing." Insect build-up in the small slots hasn't been worked out yet.

Electronic flight management complements aerodynamic advancements. Computer controls will automatically adjust the plane's controls to reduce stresses at various points on the airplane's frame. Shifting of the ailerons, for example, is done by sensing devices which position them for least resistance; a smoother ride and greater fuel savings are the results.

Load management and operation improvements such as reduced cruising speeds, improved maintenance, use of cruise climb, reduced holding, and grounding of inefficient aircraft, are crucial to fuel conservation. Passenger carriers can no longer afford to fly half-empty planes or turn away people seeking seats on overcrowded flights. In fact, many airlines now overbook their capacity to be sure of filling the plane.

Tightening the pitch

To combat the rising cost of fuel, airlines need to "tighten the pitch" or squeeze as many seats as possible into as small a space as possible. What is needed in many cases are planes specifically tailored to certain routes. Boeing's 767 is sized for medium-ranged trips. It will carry 50 more passengers than the 727 without burning more fuel. Boeing claims that replacing older medium-range planes with ten 767s could save an airline up to $12 million a year.

By considering weight and content, computers determine the most fuel efficient altitude and speed for the aircraft to fly. According to *Avaiation Daily*, airline jets flying in the North Eastern corridor will climb higher sooner and remain there longer as part of an FAA plan to conserve fuel. FAA Administrator Langhorne Bond estimates the new procedures will save four million gallons (15.1 million *l*) per year. Other measures will include eliminating circuitous routing where possible, route changes to reduce bottlenecks, and modification of some speed restrictions. In 1977, delays in air traffic systems wasted 700 million gallons (2,649 million *l*) of kerosene and cost an additional $800 million in compensation costs, lodging fees, and additional salaries.

Deregulation

The Kennedy-Cannon deregulation bill, enacted March 21, 1977, allows the airlines wide latitude in setting fares and in starting up and discontinuing routes. It is responsible for more competition, lower fares, higher load factors, and re-evaluation of numerous routes previously dictated by the Civil Aeronautics Board (CAB). Deregulation has caused the carriers to take serious measures to ensure economic stability rather than to rely on protective federal regulations to guarantee profits.

The cost of fuel is one of the principal factors determining routes. It governs the type of aircraft selected to be flown, the load factor, the number of scheduled flights and their destinations. One of the biggest worries regarding deregulation was the abandonment of uneconomical routes, especially to sparsely populated communities.

However, even under regulation, service to many small airports currently unprofitable to the airlines would likely be stopped by the CAB. As Michael Pustay noted in *Land Economics*, "A carrier's decision to serve a small community rests on the profitability to the carrier of the service, not on some vaguely defined benefit to society of preserving the service. As communities become unprofitable to serve, the airlines petition the board to abandon them."

The Air Transport Association determined, in a computer simulation, that 1,198 routes would be vulnerable to abandonment, assuming federal subsidies were cut. Many routes were indeed dropped, but commuter airlines replaced major carriers in many of those cities. United Airlines for example, halted service to Bakersfield, California because jets are inefficient to operate on short hops. United's successor there is Air Pacific, a commuter line begun in 1978 which has picked up four additional San Joaquin Valley towns, all of which were dropped by United. United's representative Ferris speaks of deregulation effects in *Business Week:* "There is a necessity to move toward more efficient aircraft. This could affect the mix of our 727-200s and 767s in our fleet." In September 1979, Western grounded six unprofitable 720Bs. Likewise, TWA grounded many of its inefficient 707s to be sold or used for spare parts. It also sold its entire fleet of 707 freighters.

Judging by the breeze

Mapping the flow and intensity of the jet stream permits another form of route improvement. Says *Aviation Daily:* "A weather research satellite launched in 1978 could save airlines five percent of onboard fuel because of its newly determined ability to read atmospheric ozone levels. The National Center for Atmospheric Research (NCAR) said the Nimbus satellite measures atmospheric ozone so efficiently that it can map the distribution of ozone around the globe." Ozone patterns indicate the location and force of the jet stream, which can provide a tailwind boost of up to 200 mph (320 km/h) or an equal headwind restraint, depending on flying direction. Minor alterations in an aircraft's flight path to make use of the jet stream can result in an across-the-board fuel savings of five percent on trans-oceanic routes.

It is crucial that the airlines cut their fuel consumption as much as possible. Even a three percent fuel savings on a wide-body jet can mean a savings of more than 200,000 gallons (757,000 *l*) of fuel per year. American Airlines will spend over $31 million in aircraft fuel modification improvements over the next two-and-a-half years with annual fuel savings expected to reach $22 million. Boeing is seeking to decrease fuel consumption per passenger by cramming more seats into the plane without increasing fuel costs. One solution employed in 1977 was to seat ten abreast in the main cabin. Eighty-five percent of today's 747s have been so modified. Another variation will transform the first class upper deck lounge into additional seating for 38 coach passengers. In the not-too-distant future, the upper deck will be stretched out to increase passenger capacity to 500.

— *Katy Slichter*

References:

Aviation Daily
  1981, vol. 253, no. 30, p. 239.

Business Week
  1979 "One Year After Deregulation: The Airlines."
November 5, no. 2610, pp. 104-112.

Chase, Victor D.
  1981 "Far-Out Flying Machines." *Science Digest*, January/February, pp. 58-63.

Labich, Kenneth
  1981 "The Big Company That Could." *Geo*, vol. 3, March, pp. 96-120.

Long, Michael E.
  1981 "They're Redesigning the Airplane." *National Geographic*, vol. 159, no. 1, pp. 76-103.

☐ <u>Transportation and energy</u>

Perpetual Motion

THE US TRANSPORTATION NETWORK is
unmatched in its capacity to move people
and goods, but it is starting to collapse under
its own weight. Streets, roads, and highways
are deteriorating more rapidly than they can
be repaired. Thousands of railroad track
miles are no longer used, and more are
scheduled to be abandoned as government
subsidy cutbacks force already troubled
companies into further retrenchment. Air
traffic around the nation's busy airports has
become so dense that not even the
sophisticated computers and nerveless flight
controllers can claim a perfect record for
averting disaster.

The system has grown too big, too complicated, and too
expensive to continue travelling in the same direction for
much longer. In 1980, the national transportation bill totalled 500 *billion* dollars in government and private expenditures, about 20 percent of the total gross national product.
Between 1975 and 2000, the transportation pricetag is expected to surpass $14 *trillion* (1975 $), or about $60,000 *per
capita*.

Most of the transportation money buys energy in the
form of ever more expensive petroleum. Fifty-four percent
of the US petroleum consumption in 1976 was used as
transportation fuels — it was more than we imported. Total
energy consumption that year was 76 quads, with 26.1
percent going directly for transport fuels. (Indirect energy
for transportation, including vehicle manufacture, road
construction, and maintenance, adds about half of the direct amount to the total transportation bill.) Back in 1950,
when the national energy budget was 33.6 quads, the
transport fuels share was virtually identical, at 26.3 percent. And even in energy-conscious 1980, transport fuels
gobbled a consistent 24.4 percent of the 76.2 quad total.
(Only one-fiftieth of the drop in total transport energy use
was the result of increased efficiency; the change was almost all in passenger miles travelled.)

Take to the highway

The energy share of transportation has remained virtually unchanged for three decades because we are stuck with
a 1950's plan for the American future that was based on
cheap gasoline. The commuter beltways and ring roads
built after 1952 as part of the Interstate Highway System
spawned sprawling suburbs that now circle our city centers. As the satellite bedroom communities grew, in their
G.I. bill-financed abundance, the middle class moved out
of the cities to single-family frame homes (poorly insulated), with attached two-car garage, on a tree-lined street
only forty minutes' drive from the downtown office.

The drive takes longer now that the roads are jammed
and poorly patched, and costs six or seven times as much as
it did when gasoline was 26 cents a gallon.

Unfortunately, all the other transportation options have
been closed. Rail commuter lines offer reduced service,
where they have not been abandoned. The trolley lines that
could have been the core of a light rail revival were bought
up and torn out years ago by bus companies conspiring to
eliminate the competition. And today's municipal bus systems are so poorly supported that they are unable to expand beyond the urban centers they now just barely manage to serve. We have literally torn up the tracks behind us.

Americans depend almost totally upon their highways
for passenger movement: 93 percent of the 2.7 *trillion* miles
travelled in 1977. Freight traffic, which consumes about 30
percent of transportation fuel energy, is not so one-sided.
Trucks hauled only one-fifth of the roughly 2.5 *trillion*
ton-miles moved that year — but they burned four-fifths of
the freight fuel budget to perform that much smaller fraction of the work.

A new, more energy-efficient transportation infrastructure will take decades to build, about as long as it takes for
the capital stock of the cities to turn over. Until then, we
must find more efficient and less environmentally damaging ways to use the massive transportation system we
already have: the 191,200 miles of railroad tracks, the
293,800 miles of principal domestic air routes, and the
3,166,200 miles of paved streets, roads, and highways.

Technical improvements in automotive fuel consumption, and non-petroleum based liquid fuels are perhaps the
two most discussed strategies for reducing the enormous
energy and environmental impact of transportation. (These
subjects were examined in the previous *Soft Energy Notes*
transportation issue, volume 3, number 4.) A sustainable system of transportation, though, requires other measures as well. Railroads, too, need to be improved, so they
can entice both passengers and freight away from the
highways. Light rail systems can be pieced together from
existing trackage and abandoned commuter stations. Air
travel can consume considerably less fuel than it does at
present.

But the most dramatic and immediate reductions in
transportation energy consumption come from changes
not in technology but in consumer attitudes and values.
American drivers must learn to share their cars with other
travelers — particularly when they commute — and they
must be willing to share the streets with pedestrians, bikes,
vanpools, trolleys, buses, and other highly energy-
efficient ways of moving people about.

— *Charles Drucker*

Back on Track

REVITALIZING THE RAILROADS

*Railroads are the oldest operating industrial infrastructure in the US, a
vital network that sustains a high proportion of industrial activity.*

IT IS PERHAPS the most underutilized transportation system in the world. Since World War II — and particularly in the past twenty years — while the world's railroads have generally expanded routes and services, the United States constructed highways and a dense air travel network instead. The railroads deteriorated.

Until recently it seemed we could afford to ignore railroads, placing them beside sailing ships as archaic tools of little use in the modern world. Energy and resources were cheap, the railroad industry was hide-bound by traditional regulatory and union agreements, and it had long reached its peak in market saturation and profits. All these factors contributed to a widespread belief that railroads were failing, neither a good cultural, nor economic investment.

Today, rising energy and resource costs are rapidly changing transportation's economics. Automobiles capable of over 40 miles per gallon, airplanes soon to be flying using 40 percent less fuel, and new diesel locomotives capable of 30 percent reductions in their already miserly use of fuel have appeared in the past decade. And, they portend still further refinements in the name of energy conservation. Simultaneously, the way we use transportation is changing. Recreational drives to distant parks are becoming too expensive, and smaller cars take the comfort and some of the pleasure out of a long trip. Short jet commute flights are less attractive as airline fares rise. We are far more careful in how we use transportation.

Subsidy: the real fare

Practically every railroad in the world, whether it hauls freight or passengers or both, from urban centers to rural fringes, is subsidized to some degree. Many are wholly owned by national governments, others are partially owned and some are semi-private companies receiving sporadic infusions of government funds. Unlike most of the world's railroads, US rails have always been privately owned.

Over the past thirty years there have been countless proposals calling for drastic revitalization of US railroads. Many of these plans assume railroads are minimally profitable, passenger trains cannot be profitable, and the whole network is in need of very expensive reconstruction. And, there is little evidence suggesting US railroads would improve if government owned. We might just be trading a private for a public bureaucracy, risking economic and energetic efficiency in the process.

As operators abandoned passenger trains in the late '60s, Congress established Amtrak, a government corporation chartered to keep the trains rolling. In the same period it became apparent that the Northeast Corridor, between Boston and Washington DC, was a grossly underutilized rail network beneath a sky crowded with commute flights. So Congress established experimental high-speed rail service and in 1976 appropriated the first incremental funds for the Northeast Corridor Improvement Project — a massive project involving substantial reconstruction of hundreds of miles of track, dozens of stations and new trains. Conrail, also a government corporation, consolidated the decayed network and focused on freight service. This flurry of activity has resulted in nationalizing a substantial chunk of the railroad industry.

The problem with a subsidy is not the design but in the delivery. Curiously, in the discussion of railroad nationalization, the nation's road network — an obvious parallel — is rarely mentioned. In the US, automotive transportation is subsidized by an enormous government complex of highway construction, maintenance and policing agencies, substantially financed by user fees — such as gas taxes. And, automobile transport is financed with our time — we all take the wheel instead of a pilot or engineer. Air transportation is directly subsidized by government maintained air traffic control networks costing about $2 billion a year. Airports are primarily constructed and maintained with public funds, and indirectly subsidized by countless billions of dollars spent on military research — the technological foundation of commercial aviation. Although railroads were often subsidized with land grants in the 19th century, particularly in the western US, until the last decade they received a negligible portion of subsidy funds.

The road network demonstrates an effective balance between the public need for diverse road systems and private needs for profitable vehicle operation. Subsidization is local, regional and national: the source of funds and administration tends to be closely tied to the needs of users, and where the use occurs. By contrast, the operation of Conrail and, to a lesser degree, Amtrak demonstrate a total subsidy where the source of funds and administrative functions are highly centralized and often distant, both physically and socially, from where the service occurs.

Although Amtrak receives about 40 percent of its operating revenue from fares, most of its enormous capital expenditures for equipment are Congressional appropriations.

While this may be an adequate arrangement to maintain and gradually expand the skeletal passenger train system, it would not permit a drastic expansion of the system: the costs would be enormous. Furthermore, there would remain a distinct tendency for decisions to be made not out in the marketplace but in bureaucratic offices.

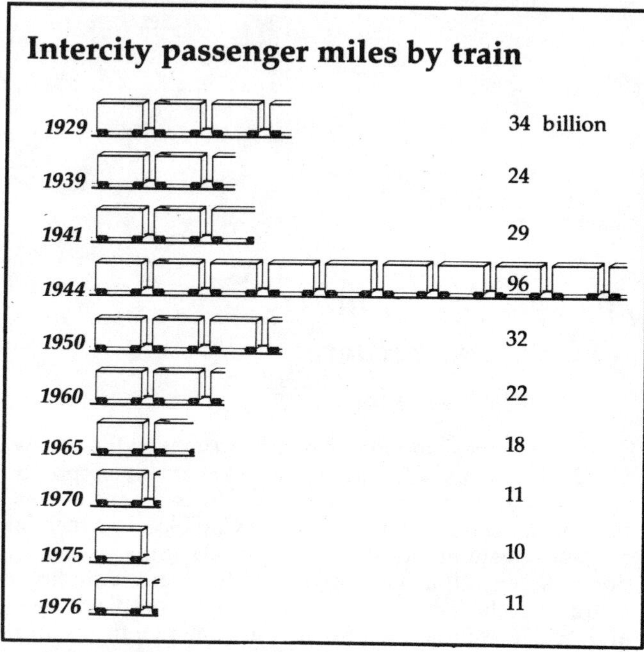

Intercity passenger miles by train

1929	34 billion
1939	24
1941	29
1944	96
1950	32
1960	22
1965	18
1970	11
1975	10
1976	11

Jets on rails

Along with nationalization, high-speed trains have repeatedly been touted as a transport panacea. It is assumed the efficiency of a train is possible at the speed of a small plane: on paper a fairly sensible proposition. But, it often proves to be a wasteful means of intercity transportation.

Japanese ''Bullet'' trains, the British Advanced Passenger Train (APT) and various French developments in high-speed rail service all demonstrate how rail service might evolve on some US intercity rail corridors. While Northeast Corridor Metroliners are capable of 120 mile per hour operation, approaching the 130-160 mile per hour range of Japanese, French and British equipment, they can only attain such speeds on highly sophisticated track. Track capable of supporting high-speed operations must withstand higher loads, it must be free of excessive switchwork, road overpasses are often required instead of track level crossings, curves must be precisely engineered, signalling must be more sophisticated and the whole system requires a high level of maintenance. The British APT and a similar Canadian train developed recently both feature special suspension systems designed to absorb shocks caused by rough track during fast operation and cant the cars as the train passes through curves. These trains were designed as a means of developing high-speed capability without extensive track modifications, but they still cost from two to four times more than conventional equipment.

Developing high-speed railroads, as the State of Ohio is now planning, requires a very high investment in either track or trains or both. Given the materials used in construction. the energy consumed in both construction and operation — a train at 120 miles an hour uses more than twice the energy than one at 60 mph — coupled with the often lengthy construction times, it seems questionable whether high-speed rail is worth the investment. When we already have an extensive but decaying railroad network, an increasingly limited budget, and a need for widespread efficient service, it seems imprudent to concentrate so much attention, money and energy on a few high speed routes.

Many existing Amtrak trains hover on the edge of meeting their operating costs from fare revenues. While the margin is slight, the potential for profitable operation is often close. It may be possible to define a profitable operation if equipment capital costs were lower, if track use was subsidized, if stations did not have to be supported just for train use, *and if the* often archaic union and management regulation were reformed.

The simplest way to decrease the capital cost of railcars is to make them smaller, lighter and with cheaper materials. US railcar construction companies now use heavy steel equipment, a Cadillac on rails. In essence, we need a lightweight railcar that can be built with the skill and materials common in the automotive and RV industries — both of which have been shrinking of late and could use the work.

The first step: own the tracks

While highways may not always be energetically efficient, the network taken as a whole can engender a high degree of economic efficiency. In effect, it is an open network free to any individual or organization able to meet basic safety rules. By contrast, railroads are an intensely private network useful only to the corporations that own them. Typically cargo bound cross country is handled by many separate companies and travels through innumerable union and railroad jurisdictional boundaries — a complex arrangement

TOOLS for the SOFT PATH

tending to discourage a free flowing, highly adaptable and economically vital relaionship between railroads and the marketplace.

Opening the rail network toward the objective of a publicly owned system could begin on the network's fringes. In many cases, particularly in the northeast, rarely-used trackage serves decaying or bankrupt industrial areas where carloadings are not sufficient to justify rail service as it might be defined by a large railroad. Many of these lines could be purchased — in some cases for the equivalent of salvage prices — by counties or towns, or local industries. Such branchline service could become the focus of industrial redevelopment — the restoration of old industrial buildings to house cottage industries, for example. While such prosaic service would probably not be exceedingly profitable, it is possible the lines could pay for themselves. Existing short line railroads, many of which are owned by municipalities for switching work in industrial areas, are often operated on shoe-string budgets with every employee performing multiple tasks. Subsequent steps would involve municipal or regional governments purchasing tracks within their jurisdiction, states purchasing major intrastate routes, and the federal government buying major interstate lines. To some extent, maintenance and management procedures for the operation of highways and railroads are similar. It is not difficult to imagine the operation of both under the common umbrella of public works agencies now in existence and the US Department of Transportation.

A "free" railroad network would allow practically any private operator — existing railroads included — to move a train anywhere through the system. There need be no more restraints than now accompany trucking operations: special safety restrictions in certain areas or different tonnage limits on certain tracks. The rails would then be managed parallel to highways, resulting in greater economic and energetic efficiency as the costs would be more visible to a wider range of the population.

Open tracks could encourage a far wider range of operating techniques. We might see shorter trains assembled to fit specific market needs such as the shipment of toxic chemicals. Train crews could become specialists in the shipment of certain cargoes for specific industries. Or a whole new industry composed of independent train operators could emerge.

Expanding passenger service

Smooth connections to other trains, buses, taxis, and air service, are a primary key to the operation of efficient passenger trains. Amtrak is engaged in a program of station reconstruction or new station development with the primary focus being intermodal use. A number of intermodal stations has already been built and some, notably the Union Station in Los Angeles, promise to be economically self-sustaining through the payment of user fees.

Over 22 million passengers rode Amtrak's 325 daily passenger trains in 1980. Despite aging equipment — new railcars are now just coming on line — and often decrepit stations, Amtrak's passenger volume has generally increased since 1971. It still doesn't even come close to the level of service available in the 1920s when over 15,000 trains were in operation. It is practically impossible to know what the market might bear, though we might take a hint from other systems: Japan operates over 26,000 passenger cars and Great Britain about 18,000 to Amtrak's 1,900.

— *Christopher Swan*

Piggybacking Improved

Truck trailers riding on railroad flatcars, called piggyback operations, are a fast-growing part of the railroad freight business. A new freight trailer design, called the RoadRailer, would get rid of the flatcar. Marketed by Bi-Modal Corporation, of Greenwich, Connecticut, it can be pulled by truck tractors over highways, or hauled by locomotives in special railroad convoys. The RoadRailer has a double running gear: standard highway tires, and a set of steel railroad wheels mounted just behind them. The trailer can be driven on streets like any other truck trailer. But instead of being loaded piggyback on a flatcar, it can be positioned over the rails, its steel wheels lowered, and its front end hitched to another RoadRailer to create a train of trailers which a standard locomotive can pull.

Bi-Modal asserts that this design cuts fuel consumption by up to 50 percent compared to trailer on flatcar operations, because the train's weight is so much lower for a given payload. And RoadRailers can save 64 percent of the fuel consumed in door-to-door intercity freight movement via highway alone. The flexibility and low investment relative to railroad cars may prove a tremendous incentive to expanded railroad freight operations.

illustration by Christopher Swan

Transport Energetics Elusive

Passenger transportation energy intensity vehicle performance in Btu per passenger-mile travelled

(1 Btu/p-m = 0.1575 kcal/p-km)

11,000	Single occupant automobile (luxury)
3,500	Compact car (25 mpg), 1.4 passengers
1,010	Carpool (40-mpg VW Rabbit, 3 passengers)
1,220	Vanpool (10-mpg, 10 passengers)
4,290	All personal passenger
2,960	Transit bus
1,010	Intercity bus
3,030	Transit Rail
3,500	Commuter rail
3,230	Intercity rail
8,320	Local air
6,620	Domestic air
4,340	All transportation modes, 1977

ARE TRAINS really more energy efficient than automobiles, trucks, buses, and airplanes, or is that belief part of the romance of railroading? The answer is not simply a matter of statistics, but takes us deep into the world of assumptions and system boundaries.

Most studies of transportation energetics and economics examine specific systems as if they were in competition with one another, rather than part of a larger, integrated system. How do we factor in a trip where a person walks to a taxi, drives to the airport, flies to another city, walks to a train, and finally rides the distance home?

To assess how railroads fit into this system, compare four typical one-way trips between two cities, 300 miles apart and connected by a four-lane freeway, a two-track railroad, and a pair of airports with frequent commute flights. On the first trip we drive from the suburbs to the airport 30 miles distant, using one gallon of gas. Then we board a fully loaded jet and fly to the other city consuming five gallons of fuel per person. Finally, we ride downtown in a diesel bus which carries 40 passengers and burns four gallons of fuel.

On the second trip we rent a car and drive the entire distance at 25 miles per gallon, or 12 gallons consumed. On the third trip we taxi to the downtown train station 15 miles away, using three-quarters of a gallon of gas, then ride with 500 other passengers using two-thirds of a gallon per person for the journey. A cab trip to the final destination five miles from the station takes another quarter of a gallon. On the fourth trip we ride a bike fifteen miles to the train station, put the bike aboard, ride to the next city, and bike to the final destination.

The first trip used a little over six gallons of fuel, cost 65 dollars, and took under three hours. The second consumed twelve gallons, cost $20 for gas, another ten for food, and involved eight hours of travelling. The third took just under three gallons, $45 on cab and train fares, and lasted seven hours. The fourth involved less than one gallon of fuel, took nine hours and cost $35 for train fares and food. Which trip was the most efficient?

Jet planes permit short travel times but high fuel consumption and operation costs mean high fares. Au-

tomobiles are the most convenient, with no transfers, but fuel use is great. Single-passenger trips gulp energy. And, we subsidize automobile use by not valuing our own time as driver. The train, in comparison, uses little fuel, but requires a crew to operate. Relative to a jet plane it carries fewer passengers per unit time over an identical route. Thus the train's fare reflects a high labor expense, and fuel costs averaging only eleven percent of the operating budget. Plane fare includes a lower proportion of labor costs, but fuel averages 40 percent of total operations.

Considering the subjective value of time we are faced with three choices: fly and take the least time while using the most fuel; drive and take more time and fuel but enjoy the convenience of door-to-door travel; or take the train, use the least fuel, and commit about the same time as driving, but spend that time reading, sleeping, working, or in any of the other ways that train travel makes possible.

From the traveler's point of view, the efficiency of a train is a matter of the relative values we assign to time, money, and convenience. The strictly energetic efficiency of trains is no easier to judge, if we include the energy involved in manufacture and maintenance. We can calculate the passenger-miles per gallon consumed by various vehicles, but computing the embodied energy of their manufacture, spread over their respective lifespans, is much more complicated.

Finally, we must include the energy costs of building and maintaining the infrastructures upon which travel depends: the airports, highways, stations, tracks, service centers, and related paraphernalia of complex systems.

Most analyses retreat from this complexity to straightforward statistics on individual vehicle performance. Yet by scrutinizing the parts, they fail to see the system as a whole.

— Christopher Swan

Christopher Swan has been a transportation writer and designer since 1970. His publications include Cable Car *(1973) and* YV 88 *(1977). He is currently developing a design for a solar train, and writing a book on transportation.*

TROLLEYS

Crawling through seventeen West German cities, they may look like caterpillars, but the trolleys in Hannover run at *four minute* intervals and that's not something for a San Francisco BART commuter to snicker at.

Hannover's largely underground light rail trains replace a tram network strangled by traffic congestion. And light rail is more attractive: one line carried 70,000 passengers per day as a tram line; now light rail, it carries 115,000/day. Klaus Scheelhase, chief civil engineer, estimates that of the more than 60 percent increase, 35 percent are passengers who switched from buses and 25 percent are from cars or are people who would not have travelled at all. In Europe and North America, light rail is making a comeback.

Light rail is a new name for an almost forgotten system: the trolley, introduced to America in the last half of the nineteenth century. In 1930, thousands of US cities had trolleys, but by the late 1950s, 40,000 miles of track had been taken up and 60,000 trolley cars junked. Today only San Francisco, Philadelphia, and a few other communities still have trolleys on tracks. Laments Gerald Haugh, general manager of the Long Beach, California bus system: "It would be great if we had it all back again. It could have been modernized. You'd have tried to extend the rails out into those areas where people were buying. It would have been a hell of a lot cheaper than to do it today. It was a damn shame they took up the tracks."

Buying trolleys to sell cars

Long Beach, California was once part of a showcase urban light rail system: extending through Los Angeles, Pasadena and out to the San Fernando Valley, this system was called the best in the country. Then during the 1940s, "the conspirators bought and dismantled it," writes Jonathan Kwitny. "Taxpayers now are faced with building a similar system at a cost of billions."

Kwitny spells out the transportation oligopoly — General Motors, Firestone Tire & Rubber, Phillips Petroleum, Mack (Truck) Manufacturing, Standard Oil of California, and others — that conspired for the internal combustion engine. Electrified-rail mass transit systems, which carried millions of riders, were bought; tracks literally were torn out of the ground, sometimes overnight. Overhead power lines were dismantled and valuable off-street rights-of-way were sold. In 1949, a Chicago jury convicted the corporations of criminal antitrust violations for their part in the demise of mass transit.

The conspiring companies replaced the trolleys with gasoline and, later, diesel-fueled buses, which, unpopular with the public, merely added to the allure of the private car. The investing companies "expected their profits to come not from bus operations at all," Kwitny explains, "but from the sale of their products after electrified transit was destroyed. An internal memo at Mack, for example, spoke of a 'probable loss' on the bus-line stock, but said it would be 'more than justified' by 'the business and gross profit flowing out of this move in years to come'.

"Nor does it appear that GM expected to make its principal profit from the sale of buses, the new form of mass transit. If it did, there is no satisfactory explanation in the trial record for why GM gave half the prospective bus business away to Mack, its supposed competitor. Another explanation, of course, is that the real profits were going to be made from the sale of cars (in Mack's case, trucks) after the destruction of mass transit opened the way for a huge public network of streets and highways. That this is what happened offers some justification for the explanation that it was intended."

That was 1940 — thirty-five years later, cities everywhere were backtracking. Edmonton, Alberta converted the most-travelled bus route to trolley and as a result it is attracting more riders. Denver had a network of 19 trolley lines in 1930 — two reached suburban Golden, twenty miles away; it has done some rethinking, too. With probably the highest number of cars per capita, Denver has one of the worst air quality problems in the US. Mayor W.H. McNichols says: "A rubber tired fleet cannot move the number of people that need to be moved on the city streets . . . light rail seems to be the only way to go."

Pittsburgh to Portland

Light rail transit can be constructed at lower capital cost than heavy rail, and it has lower operating costs than bus systems. Its track can run down the center of a roadway and it need not have rights-of-way. Electrically-operated railcars may run singly or in trains with maximum speeds ranging from 37 to 62 mph (60-100 km/hr) in North America. As such, light rail is adaptable and appealing to even lower-density, medium-sized communities. "Cities now looking at LRT (light rail transit) may be large cities like Denver or medium-sized cities like Memphis, Sacramento, and Anchorage, Alaska," writes Tom Kizzia in the slick *Railway Age*. "They may be booming cities trying to keep on top of growth, like Edmonton, Alberta, or San Jose, California, or older cities trying to strengthen corridors and arrest declines, like Cleveland and Detroit. The light rails may be descended from older streetcar and interurban lines, as in San Francisco and Pittsburgh, or they may be built from scratch, as in San Diego and Buffalo."

When Buffalo, New York decided to put in mass transit, it went out and spent $450 million — most cities these days are spending about $100 million. "We light rail advocates kind of turn our back on it," says Stewart Taylor, head of the Transportation Research Board's light rail study group. Buffalo's light rail started out as heavy rail, but city planners switched for cost savings. The savings are expected when Buffalo extends the line beyond the current 17 miles using street reservations and railroad rights-of-way. "The longer the line becomes," says Kenneth Knight, former general manager, "the greater the savings will be."

While Buffalo has been called a threat to light rail's reputation for economy, San Diego, California has, perhaps, an ideal light rail project. It has been built in just four years and it is economical — $86 million for 16 miles — all without federal money. At $5 million per mile, this new rail system has the lowest price tag in the US. Most US transit systems are 80 percent federal money, but San Diego financed 90 percent of the project with the California gasoline tax — the rest came from the state sales tax. To keep costs low, plans incorporated rehabilitated rights-of-way from the old San Diego and Arizona Eastern Railway, fare collection on the honor system, off-the-shelf Siemens-Düwag cars from West Germany, and spartan stations. "We made a great effort *not* to get anything that was new and exotic," boasts Bechtel project manager John Coil. In fact, passengers in Frankfurt, Cologne, and Edmonton and Calgary in Canada ride in the same trains.

Most transit systems are made inefficient with one-way commuters: into the city every morning and outbound each evening. Making geography economical, San Diego planners ran the tracks to another downtown— San Ysidro on the Mexican border. "With a downtown at each end of the line," explains San Diego MTDB (Metropolitan Transit Development Board) general manager Tom Larwin, "we'll have a much more dispersed patronage pattern than a typical commuter operation has." Able to preempt traffic signals, the San Diego Trolley will run at 15-minute intervals during rush hours and at mid-day. The trains are expected to carry 30,000 riders, including 5,000 crossing the border, each day. Conservative estimates place capacity at 36,000 passengers/day by 1995, but, Larwin notes, "in 1977-78 when we were doing our planning, we didn't even foresee $1-a-gallon gasoline." Or, April 1981 prices at $1.35.

Imagination & compromise

While San Diego may buy German trains and Boston may sell to San Francisco, no two light rail systems are alike. Every city's particular geography, budget compromises, and, indeed, imagination make for custom trolleys. The Portland, Oregon $146.9 million project (which includes $78.6 million for highway improvements) will be running by 1985. It will move along highway medians and reserved lanes in city streets marked off with rumble strips. Project manager Donald MacDonald says while subways are right for Edmonton, surface operation is the answer for Portland: "There were very few streets available in downtown Edmonton, and the situation required subway. Here there are multiple routes." Pittsburgh, too, is tunnelling underground with a $60 million subway to relieve downtown congestion. Also routing track over bridges, planners have left only a few grade crossings and a portion of less than a mile operating on the streets. Calgary took the opposite approach: *traffic* will be restricted downtown, leaving only trolleys on a 12-block street section.

Finally, there is the awkward compromise. Working with a tight budget, San Francisco was forced to join to Boston in finding a light rail vehicle suitable for both cities. The cars were built with a rounded shape to negotiate Boston's narrow tunnels and tight corners. Once in the San Francisco tunnels, which were designed for the rectangular heavy-rail BART trains, the Boston cars' front doors are useless. From a platform built for square-cornered cars, riders cannot get to the doors in a round-cornered car.

— *Elyse Axell*

References:

1980 "San Diego light rail: On track for '81." *Railway Age*, vol. 181, no. 11, June 9, pp. 34-8. Single copy $1, Subscription Dept., Railway Age, PO Box 530, Bristol, Connecticut 06010.

Kizzia, Tom
1981 "Light rail: Where the action is." *Railway Age*, vol. 181, no. 18, September 29, pp. 40-6.

Kwitny, Jonathan
1981 "The Great Transportation Conspiracy." *San Francisco Chronicle*, March 1, pp. 16-26, *This World*.

Scheelhase, Klaus
1980 "Hannover pinpoints light rail objectives." *Railway Gazette International*, vol. 136, no. 4, April, p. 283. IPC Transport Press Limited, Dorset House, Stamford Street, London SE1 9LU, England.

5. Fuels from Biomass

The prospect of a transport system running entirely on fuels from biomass is the least proven aspect of a renewable energy future.

Much of the confusion has its source in a dominant orientation toward supply, similar to that which created the early disputes over solar home heating. The early discussions on biomass gave little attention either to the potential for vehicle efficiency or to the potential for greatly improved conversion processes. Agricultural and forest "farms" for fuel production were the dominant concept. This touched off another debate: the competition between food and energy production for arable land and the effect of that competition on food prices. The proposal of soft technologists has been to use agricultural and forestry wastes for liquid fuels. This would bring more income to farmers while leaving food production untouched.

Yet even this proposal has its snags. For one thing, determining what is agricultural waste depends on whether the farming system is sustainable. Most analysts agree that our present agriculture is not. Before we think about exporting energy from the farm, we should put agricultural practice on a solid ecological footing.

Once it is clearer what form of recycling will maintain soils in the long term, energy production can be fitted into agriculture with a variety of schemes, but the quantities of slash, stubble, etc., available for conversion to liquid fuel, are bound to be less than early proposals indicated.

There is yet another variable. Our present, meat-oriented diets are biomass- and therefore land-intensive, just as our car-oriented transport system is fuel-intensive. Meat production alone uses 88 percent of all crops while supplying only a fraction of our nutritional needs. This suggests that the competition between food and fuel could be ameliorated by dietary shifts. Such shifts are considered desirable by many nutritionists on independent grounds of health. Even a minor adjustment could set free a large amount of agricultural land for energy farming. This must be balanced against pressures to use land to alleviate world food shortages.

Studies by the Solar Energy Research Institute and the Office of Technology Assessment of the US Congress indicate that a highly efficient vehicle fleet could conceivably be supplied with biomass fuels primarily from municipal wastes and forestry sources that do not compete with food production. Ultimately, cheap solar cells or wind turbines could provide electric power for electric or hybrid vehicles using electricity and biomass fuels.

All this speculation is no excuse for not acting on what we do know. As this section shows, internal combustion technology can thrive on such biomass-derived fuels as methanol, ethanol, and oils from plants, or even wood gas. Those were the major fuels in earlier days of scarce petroleum. While these fuels often contain less energy per gallon than gasoline, some can increase the efficiency of the motor and save energy at the refinery.

There are significant biomass wastes available now for fuel production. High efficiency wood gasifiers, ethanol and methanol plants and pyrolysis units of small and medium scale need to be developed and commercialized. Local communities stand to gain greatly from such efforts, in added employment, revenues and independence.

Biomass Energy:
Bonanza or Boondoggle?

*B*iomass energy is renewable only within limits; it could be exhausted through mismanagement. Widespread enery conversion from farm and forest land could threaten the soil, water, and natural biota that contribute to plant growth. Soil erosion and water runoff problems increase when crop and forest residues are removed. And when manures are converted into biogas, rather than plowed under, soil quality deteriorates unless some other organic matter is added to maintain soil humus and structure. Take too much from the land, and its productivity declines.

Moreover, if crops like grains or rapidly growing tree species are cultured specifically for biomass conversion, they will compete with food crops for land and water. Attempts to intensify production on marginal land will only quicken the pace of soil degradation and erosion.

Our soil and water resources determine the amount of biomass energy that might safely and perpetually be harvested. Exceeding these limits depletes the resource base, and diminishes the available biomass energy supply.

Competing for land

Fifty-five percent of all US land—an estimated 1.2 billion acres—is devoted to livestock and crop production. Forest land, mostly commercial timberland, takes up an additional 539 million acres, for a total of 80 percent of the nation's land area covered by some form of biomass. Twenty years ago, the total percentage was about the same, but the various components have since changed. Urban areas and transportation rights-of-way have increased their share of the land; active cropland is up, and less is kept idle; grasslands and forests available for grazing are lower by a tenth, despite a growing need. These trends are likely to continue, with their cumulative effect an increasing pressure on US lands.

Three hundred and eighty five million acres of cropland is not enough. The population is still growing, per capita food crop consumption is up, and food exports are on the rise. An estimated 50 million additional acres will be needed in this decade to meet export and increased domestic demand. But at the same time as more agricultural land is needed, some of our best land is being removed from production. Erosion takes an ever heavier toll. Urbanization and highway construction consumed about 45 million acres between 1945 and 1975, or about 12 percent of our current cropland. Strip mining directly disturbs at least 153,000 acres per year. Rapid growth of coal use and the development of other energy resources will increase the amount of land taken out of agriculture.

Despite these increasing land pressures, it has been proposed that 40-50 million acres of land might be put into production for biomass energy. In addition, some 10 million acres of the present 385 now in cropland would be devoted to the production of 22 million tons of grain and other products for conversion into ethanol.

To meet the higher demand for cropland, some grazing and forest land could be brought into production; marshes could be drained (with added environmental impact); and perhaps more land could be irrigated (assuming that water is available). The United States Department of Agriculture estimates that 33.5 million acres of land might be brought into production by these means. But such marginal lands will not be as productive as current croplands, may be more susceptible to erosion and degradation, and will require at least the same heavy investment of energy inputs.

Soil Erosion and Environmental Degradation

Cultivation of the soil for grain and other energy crops will increase soil erosion, the rate depending on land, crop, and management techniques. The removal of any vegetation from the land creates erosion problems; annually about 3 billion tons of topsoil are carried to streams and other environments by runoff water from the nation's agricultural land—and an additional 1 billion tons is eroded by the wind. Roughly one-half of the original topsoil on one-third of the croplands has been lost since agriculture began in the US. During the past 200 years, land degradation by soil erosion forced farmers to abandon at least 100 million acres of cropland. In 1973-74, when the price of grain doubled in some areas, farmers intensified their agricultural efforts and soil erosion increased—a 22 percent jump in Iowa alone.

Sediments running off the land have many adverse environmental effects: they deplete reservoir storage, silt harbors and navigation channels, fill drainage and irrigation ditches, impair recreational use of streams and lakes, increase hydroelectric power generation costs, increase flood damage, and raise water treatment costs. The removal of one year's sediments from the reservoirs would cost about $1 billion. Dredging of the harbors and waterways costs the federal government $240 million annually. Sedimentation also reduces the ecological life in streams and lakes.

Water Resource Shortage

Water is the limiting factor in US agricultural production. Requiring 83 percent of the US water supply, agriculture is the single largest consumer, with industry, power generation, and domestic use sharing the rest. The total amount of water being withdrawn from lakes, reservoirs, and streams has increased steadily since 1900. Since 1950 alone, water use has mo e than doubled, from some 200 billion gallons per day to 425.

An acre of corn under arid conditions requires 1.5 million gallons of water during the growing season, one-third of which is lost in transpiration. With current high energy prices, many farmers can no longer afford to irrigate low value crops like alfalfa. In some water-scarce regions, land is being abandoned as farmers cannot afford the energy to pump from 700-1,000 foot depths.

In the arid regions of the Western states, agriculture, urban populations, industry, and fossil energy mining clash over scarce water supply. Coal and oil shale producers may well out-bid farmers, and if agriculture receives less water for irrigation, more rainfed land will be needed to meet food demands.

Livestock Manure as Biomass Resource

If livestock manure is used for biomass energy (as in methane gas production), the value of the manure after fertilizer should be considered along with the value of the biogas. Unprocessed manure's value as a fertilizer is worth only about $3.20 per ton, 65 to 82 percent water, it is difficult to handle and has a low nutrient yield. Processed through a biogas digester, the manure retains its nutrients, but its physical structure is altered, and its volume reduced. Digester effluent emerges as a thick soup; particle size in the slurry is too small to maintain soil structure and prevent water erosion.

Biomass Energy From Crop and Forestry Residues

Producing methanol, ethanol, or direct heat energy from crop and forest residues looks expensive, partly because of collection and transport costs. Corn residues, for example, reportedly cost between $2.40 and $3.50 per million Btu delivered, compared to coal at $1.00/million Btu.

Nutrient loss is an additional cost, as crop and forest residues are essential to maitaining the productivity of agricultural and timber lands. Taking 4,900 pounds per acre of corn stover from the land removes 49 pounds of nitrogen, 4.9 pounds of phosphorus, 45 pounds of potassium, and 29 pounds of calcium nutrients. To replace these nutrients with commercial fertilizer requires an input of 821,000 Btu/acre, or eight gallons of oil equivalent. Forestry residues, averaging 11 tons/acre, also contain valuable nutrients in nearly the same proportions: one percent nitrogen, 0.01 percent phosphorus, and 0.5 percent potassium.

Crop and forestry residues left on the ground help control erosion by reducing the impact of water and wind on soil particles, lessening the chance of their being carried away. If biomass residues—the soil's protective cover—are taken from agricultural and forest lands, soil erosion would increase significantly. This would, in turn, lead to lower soil fertility, as nutrients are lost in soil sediments. Every ton of soil lost carries with it 4.5 pounds of nitrogen and 0.9 pounds of phosphorus.

For centuries, agriculturalists have known that the soil's potential for crop production is a function of its organic matter content. Organic materials improve soil structure, increase water holding capacity, heighten cation exchange, and stabilize the mineralization rates of nitrogen. The minimum crop residue cover needed to convey these benefits and prevent erosion varies with climate, soil type, rainfall, wind, slope, and previous erosion history. But on the average, only about 17 percent of crop residues can be removed with current agricultural technology and still maintain soil quality. Even on the relatively flat lands (zero to two percent slope) from which residues are likely to be removed for biomass energy conversion, great care will have to be exercised to maintain cropland productivity.

Energy Crops

Fuelwood plantations have the potential to supply significant quantities of heat energy, and it is estimated

---

Major uses of US land
(millions of acres)

Agricultural Land	1959	1969	1974	1977
Crops	391	384	382	385
Cropland used for crops	(358)	(333)	(361)	(377)
Idle cropland	(33)	(51)	(21)	(8)
Livestock	944	890	860	857[a]
Grassland grazed	(699)	(692)	(681)	(678)
Forest land grazed	(245)	(198)	(179)	(179)
TOTAL	1335	1274	1242	1242
Forest Land				
Private, Federal, State	483	525	539	539[a]
Urban and Other Lands Used				
Cities, roadways, recreation and public installations	146	172	182	182[a]
Unused Land				
Deserts, rock areas marshes, tundra	306	293	301	301[a]
TOTAL LAND AREA	2271	2264	2264	2264

[a]Estimated

---

that 140 million tons could be produced annually. The land needed would be 40 to 50 million acres, 10 million of which would be taken from grazing land, and most of the remainder converted from forests. Even with the pressures on US land increasing, it was felt that energy crop production on this much land was economically justifiable.

Rapidly growing trees like poplar and sycamore have several advantages on fuelwood plantations. First, they can be mowed and harvested every second or third year, much like a hay field. New vegetation would coppice, or sprout from the stumps left in place. The land would not have to be tilled, and the roots, leaves, and twigs would remain on the soil surface. These two practices help protect the land from water erosion and reduce water runoff rates. In addition, the organic matter from roots and leaves contributes to soil quality and structure.

Ethanol Pollution

The production of ethanol from biomass can increase competition for cropland, accelerate soil erosion, and poses several other environmental problems as well. Liquid stillage (not wet solid stillage) from a farm alcohol plant must be recycled or safely disposed of if it is not to become a pollutant. On small farms, the liquid stillage can be pumped or drained into a sewage lagoon, and it can be used as a soil conditioner or fertilizer.

This technique, however, will not work for a large, 20 million gallon per year ethanol plant. The stillage from such a plant may have a BOD_5 concentration (Biochemical Oxygen Demand measured over a five-day period) of 25,000-30,000 mg/liter, which is ten times the ethanol produced. This is equivalent to the BOD_5 load of the domestic waste from a city with *one million* inhabitants.

Other environmental impacts of an accelerated ethanol fuel program include: increased air pollution from coal-fired power plants (particularly sulfur oxides, nitrogen oxides, and fly ash); carbon dioxide from the fermenters; dust from handling and processing dried stillage; and some venting of odors. All these wastes and pollutants are subject to existing federal and state health and environmental regulations.

Possible Climatic Effects

Ethanol production from grains involves the combustion of oil, gas, and coal, all of which release carbon dioxide into the atmosphere. Significant quantities of this gas are also released during grain fermentation. Processing wastes from livestock, food, urban, and industrial sources into energy does not effect the CO_2 level. But removal of crop and forest residues does increase atmospheric carbon dioxide because their use for biomass energy oxidizes them more rapidly, and because intensified soil erosion exposes humus to oxidation into CO_2. In addition, the reduced rate of plant growth on degraded soil will probably slow the incorporation of atmospheric CO_2 into plant biomass.

Crops and forests annually remove about 15 percent of the carbon dioxide produced by fossil fuel combustion. Even so, it is projected that the total quantity of CO_2 in the air will double its preindustrial level by the middle of the next century. There is concern that such an atmospheric change could result in a "greenhouse effect" capable of altering the global climate.

Atmospheric Pollution

When wood is burned at relatively low temperatures to increase the efficiency of energy conversion, the pollutant level is higher. Where there are numerous installations, especially in valleys, air quality suffers. Pollution from wood burning is already quite serious in cities such as Aspen, Colorado.

Reduced Ecological Diversity

The conversion of some natural forest and grazing lands into intensively cultivated energy-crop plantations will reduce the natural biotic diversity of the landscape. The wide array of species associated with natural habitats will be reduced, and the normal use of these areas for wildlife and recreation will be severely restricted.

There are many ways in which the use of biomass as a source of energy could have major environmental impacts. Water, soil, and air resources could be affected, and the competition for land could increase food prices. If biomass energy is to be a fully renewable resource, it is essential that the environmental limits on biomass use not be exceeded.

—*David Pimentel*
College of Agricultural &
Life Sciences
Cornell University
Ithaca, N.Y. 14853

David Pimentel is Professor of Insect Ecology and Agricultural Sciences at Cornell. He has investigated biomass energy potential and the environmental consequences of biomass sources from agriculture and forestry.

References:

ERAB. 1980. Gasohol.
Report of the Gasohol Study Group, Energy Research Advisory Board, Department of Energy, Washington, D.C. April.

GAO. 1977.
To protect tomorrow's food supply; soil conservation needs priority attention. Report to the Congress No. CED-77-30. General Accounting Office, Washington, D.C. 59 pp.

OTA. 1980.
Energy from Biological Processes. OTA-E-124. Office of Technology Assessment, Washington, D.C. 195 pp.

Pimentel, D., E. C. Terhune, R. Dyson-Hudson, S. Rochereau, R. Samis, E. A. Smith, D. Denman, D. Reifschneider, and M. Shepard. 1976.
Land degradation: effects on food and energy resources. Science 194: 149-155.

Pimentel, D., D. Nafus, W. Vergara, D. Papaj, L. Jaconetta, M. Wulfe, L. Olsvig, K. Frech, and E. Mendoza. 1978.
Biological solar energy conversion and U.S. energy policy. BioScience 28: 376-382.

Pimentel, D., S. Chick and W. Vergara. 1979.
Energy from forests: environmental and wildlife implications. pp. 66-79 *in* Trans. 44th N. Am. Wildl. Nat. Res. Conf.

Pimentel, D., P. A. Oltenacu, M. C. Nesheim, J. Krummel, M. S. Allen, and S. Chick. 1980b.
Grass-fed livestock potential: energy and land constraints. Science 207: 843-848.

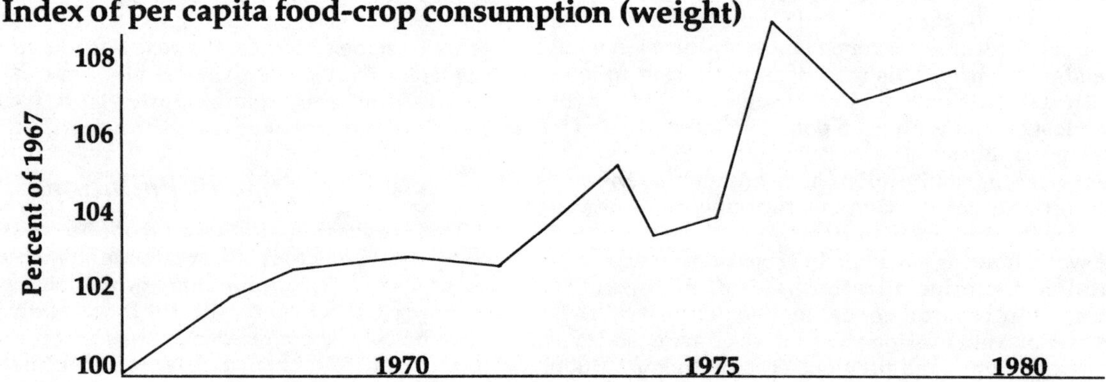

Index of per capita food-crop consumption (weight)

Fuel Farms

WHY DRILL FOR OIL WHEN WE CAN GROW IT JUST AS CHEAPLY

Melvin Calvin, Nobel Prize-winning biochemist, suggests that agronomy, rather than geology, may point the way to our future petroleum supplies.

Nonrenewable fossil fuels are not the only sources of petroleum we have available. Thousands of plant species, growing throughout the world, secrete hydrocarbons that could substitute for crude oil. At least one variety, *Euphorbia lathyris,* may contain enough hydrocarbons to produce a high quality, low molecular weight crude oil right now, at forty dollars a barrel.

In comparison to the fossil fuels they could replace, oil-bearing plants are environmentally benign. Coal, natural gas, and underground oil all increase atmospheric carbon dioxide when they are burned, with unknown, but possibly dangerous, effects on climate. Hydrocarbons from green plants, however, restore to the atmosphere only as much carbon dioxide as they recently removed via photosynthesis, and the contemporary CO_2 balance remains unchanged. Other biomass fuels, such as alcohol from grain or corn, have much the same effect, but hydrocarbon-bearing plants are grown without artificial fertilizers. Even more important, they can be grown without irrigation in arid areas unsuitable for food crops. Food and fuel need never compete for land or other agricultural inputs.

The plants that *don't* produce hydrocarbons are "doing only half the job," according to chemistry professor Melvin Calvin of the University of California, Berkeley. The "job" of a plant is chemical reduction (removing oxygen from a compound). The materials are carbon dioxide and water—fully oxidized substances having no chemical potential energy. Using the energy from sunlight, plants liberate oxygen from its chemical bonds and synthesize carbohydrates: wood, sugar, and starch. Re-united with oxygen (burned), these half-reduced products release some 7,000 Btu/lb (16.3 MJ/kg) of stored solar energy. The more a substance has been reduced prior to combustion—the less oxygen it contains—the higher its energy potential per unit weight and the greater its value as a fuel. One way to reduce substances further is to ferment the plant sugars into ethanol. The oxygen, in leaving, combines with some of the carbon to form CO_2, but little energy is actually lost. The energy density of ethanol, at 13,000 Btu/lb (30.2 MJ/kg) is nearly double that of wood. The ideal plant would take this process one step further within its own cells, and reduce carbohydrates all the way to hydrocarbons. Doing the whole job, and leaving little or no oxygen in chemical combination, gives an energy density of 19,000 Btu/lb.

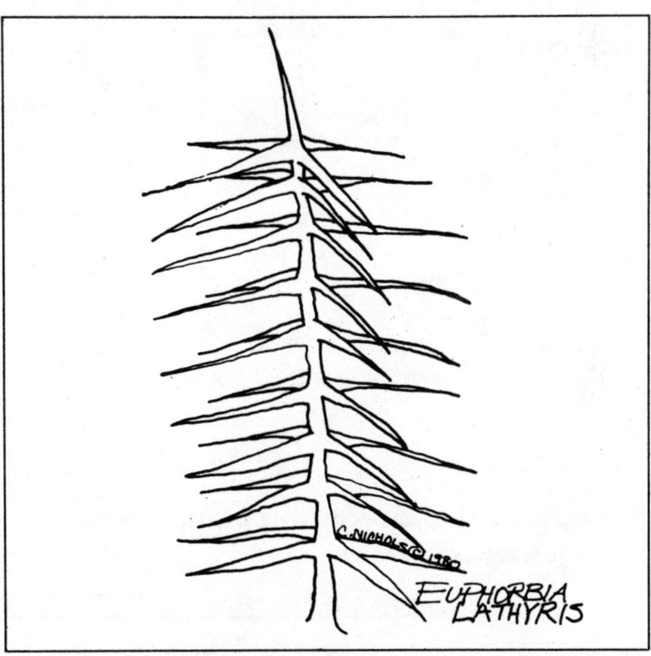

One plant capable of complete chemical reduction, *Hevea braziliensis,* has been used commercially for decades, but the hydrocarbon it produces is a rubber rather than an oil. *Hevea,* better known as the rubber tree, is a genus of the family Euphorbiaeceae, another genus of which is *Euphorbia.* All of its more than one thousand species secrete latex, which is an emulsion of roughly 30 percent hydrocarbons in water. The latex from *Hevea,* however, has such a high molecular weight (2×10^6 gm/mole) that when the water is removed, the remainder forms a solid elastomer, or rubber. *Euphorbia* plants, on the other hand, produce latexes with molecular weights ranging from 2,000 to 20,000 that remain liquid after the water from the emulsion is gone. What is left after extraction, Calvin notes, is "a black oil, which looks like crude oil and behaves like crude oil when it is cracked" (molecularly broken down in refining). He is convinced that with selection of plant species and some minor agronomic improvements, *Euphorbia* could supply refineries with a chemically

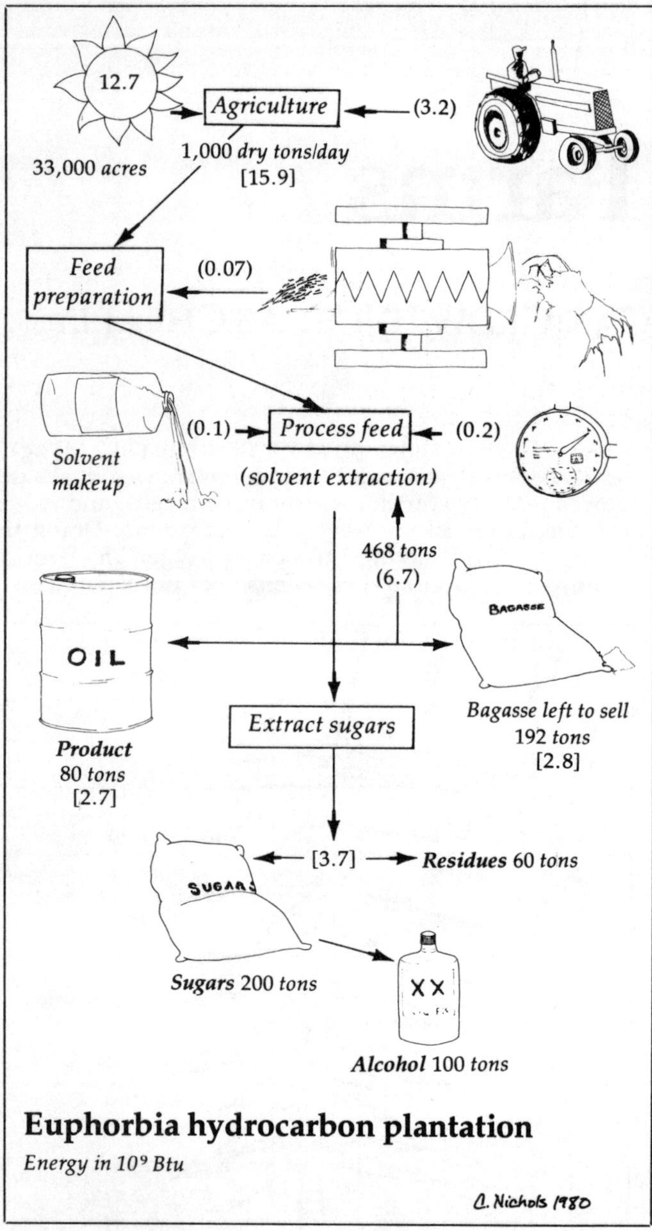

Euphorbia hydrocarbon plantation

Energy in 10⁹ Btu

a. Nichols 1980

suitable, economically competitive, environmentally benign, and perpetually renewable source of crude petroleum.

Euphoria over Euphorbia

The idea of using *Euphorbia* species as a source of oil is not original with Calvin. Italians in Ethiopia conducted a short-lived gasoline production project with the local *Euphorbia* variety, *E. abyssinica*, over forty years ago. Their experiment, according to one observer, deserved to be "watched with interest since geologists inform us that the world's petroleum supply is limited and diminishing rapidly, and who knows but that the answer rests with the Euphorbiae." During the Second World War, the French in

Morocco also tried to remedy gasoline shortages with a Euphorbia project, and in their one harvest obtained an oil yield of 3 metric tons per hectare (18 barrels/acre). But cheap oil pushed hydrocarbons from green plants completely out of the picture until just a few years ago, when petroleum prices and atmospheric carbon dioxide levels both began rising rapidly enough to warrant giving the Euphorbiae another look.

Calvin's work began in 1977 with experimental plantings of two species particularly well suited to commercial production. *E. tirucalli*, the African milkbush, is a perennial that can grow in one year from a 3-inch cutting to a 2-foot bush. In the lush rain forests of Brazil it matures into a 20-foot tree. But *tirucalli* is sensitive to frost, and its range is limited to mild climates. *E. lathyris* is somewhat heartier and matures more rapidly. In only 7 months it spurts from seed to harvestable 4-foot plant. Both Euphorbia species need only 10–20 inches of rain (or irrigation water) per year; neither requires good soil, and they grow well in rocky areas where the only other commercial potential is grazing at one head per square mile. *Tirucalli* and *lathyris* also grow upright and are adaptable to mechanical harvesting, which is essential because an irritant in latex makes hand work impractical.

The 1977 plantings produced oil at a rate of 8–12 percent of total dry weight, which Calvin extrapolates to 8–10 bbl/acre for *lathyris* and 15 for *tirucalli*. Because he was using wild seed from local California plants, Calvin insists that with selection and plant breeding, a yield of 20 bbl/acre can be attained soon, and 50 bbl/acre eventually. Seed companies agree that simple seed selection can double yields within a single year, and Euphorbia's close relative, the rubber tree, provides ample precedent for expecting much greater improvements in the long run. Prior to World War II, the annual rubber harvest from *Hevea braziliensis* in Malaysia was 200 lbs/acre; since 1946 this has risen tenfold—2,000 lbs/acre/year.

Once the *lathyris* plants are harvested, extracting the oil is a fairly straightforward procedure. The plant matter is dried, ground into a fine powder, and then subjected to two separate solvent extractions. The first, with hexane or methylene chloride, removes the hydrocarbons; a subsequent application of methanol removes a mixture of sugars and amino acids. The residual plant material, the lignocellulose, or bagasse, can be burned to supply the necessary heat for recovering the solvents. The resulting hydrocarbons can be cracked at 500° C with a zeolite catalyst into an array of compounds much like that from fossil oil. "There is no question," Calvin concludes, "that the oil from plants such as *E. lathyris* can be used as the basis for materials for the petrochemical industry."

Energy Balance and Costs

To calculate the cost and potential quantity of energy from *Euphorbia* oil, Calvin and his associates have designed a hypothetical petroleum plantation. With a processing plant capable of handling 1,000 dry tons (.45 tonnes) of *E. lathyris* per day, and an expected yield of 12 tons dry matter per acre per year, the plantation would have to keep some 33,000 acres in annual production. Capturing solar energy at about 1% efficiency (fractionally less than wheat or corn), *lathyris* would contribute 12.7 billion Btu (13.4 trillion joules) each day to the plantation. Agricultural inputs, notably fertilizer, water pumping, and fuel for farm machinery, would require 3.2 billion Btu/d (3.4 trillion joules). Energy input for the processing plant itself is a relatively low .37 billion Btu/d (390 million joules). Each day, the plant would produce 80 tons of hydrocarbons, with an energy density of nearly 17,000 Btu/lb, for a total of 2.7 billion Btu. This is less than the 3.57 billion Btu input, suggesting a negative net energy balance for *Euphorbia* oil. However, the processing leaves 660 tons per day of bagasse, and even after some of this is used for heat to recover the solvents, 192 tons remain, with an energy value of 2.8 billion Btu. On top of that, there are also 200 tons of sugars, readily fermentable to 100 tons of alcohol, for another 2.6 billion Btu/d. Discounting solar input to the plantations, which is free, the energy balance equation gives a net gain of 4.53 billion Btu commercially usable energy for this size of operation.

The exact price tag on this energy is hard to determine from experimental plantings and chemical processes tested only in the laboratory. Independent estimates by the Stanford Research Institute, however, point to agricultural and operating costs around $200/acre, or $20/bbl, together with $40/bbl capital costs for a pilot plant producing 1,600 bbl/day. This adds up to a hefty $60/bbl, but the bagasse residue can defray some operating costs by recycling part of it into the operation and selling the rest. Valued at $1 per million Btu, the bagasse drops the per barrel oil cost to about $40—a level which the world oil spot market reached not long ago, and which OPEC contract oil will probably hit within the next year or two.

Oil Company Optimism

Diamond Shamrock oil company believes these cost estimates are much too high, and it has already invested several million dollars in a research project, conducted by the University of Arizona, to obtain better economic data. Test plots of *E. lathyris* in seven Western and Southwestern states should reveal optimum planting and harvesting routines. With Calvin, Diamond Shamrock believes that agronomic research can easily increase extractable hydrocarbons two- to three-fold, up to 20 percent by weight, or 25 bbl/acre. Full commercialization would also lower per barrel capital costs through process development and large-scale production (100,000 bbl/day and up). The company estimates an eventual cost of only $20/bbl, well *below* present market prices for oil.

This cost per barrel amount includes a credit for bagasse, but it does not consider the commercial value of other *lathyris* processing byproducts. For every barrel of oil extracted, there are enough residual sugars to ferment 52.5 gallons of ethanol, using considerably *less* energy than is available from the bagasse left over *after* solvent recovery. The market value of ethanol is hard to assess since alcohol fuel subsidies vary so much by state and over time, but 50-plus gallons may well be worth more than the accompanying barrel of crude oil. Calvin speculates that there could be other, equally profitable ways to use bagasse residues. Lignocellulose, under heat, can be made into levoglucosenone, which the petrochemical industry can use as a feedstock for plastics and a host of other products. Even the amino acids and other bottom-of-the-barrel processing residues could have economic value as a chemical raw material.

The high value and low cost of hydrocarbons from *Euphorbia* lead both university researchers and oil company executives to predict a substantial portion of future petroleum needs coming from these plants. Based on their yield projection of 25 bbl of oil per acre of Euphorbia, Diamond Shamrock representatives estimate that 10% of US petroleum consumption could be met with 26 million acres of arid and otherwise unproductive land. To put this vast area into perspective, they note that 84 million acres produce corn every year, and another 80 million are planted with wheat. *Euphorbia* could take an important chunk out of crude oil imports while increasing the total land area under cultivation by less than 10 percent. And if petroleum consumption continues to drop while *Euphorbia*-produced ethanol goes into transportation needs, those 26 million acres of petroleum plantations could displace an even larger fraction of our fossil fuel consumption.

—Charles Drucker

References:
Calvin, Melvin
1979 "Petroleum plantations for fuel and materials." *BioScience* 29:9:553-538.
1980 "Hydrocarbons from plants: Analytical methods and observations." Submitted to *Die Naturwissenschaften*. Available from Lawrence Berkeley Laboratory, Energy & Environment Division, Berkeley, California 94720, document #LBL 10912.
Johnson, Jack D. and C. Wiley Hinman
1980 "Oils and rubber from arid land plants." *Science* vol 208:460–464. Available for $1.25 from the American Association for the Advancement of Science, 1515 Massachusetts Ave. NW, Washington, DC 20005, order #0036–8075/80/0502/0460.

Does Oil Grow on Trees?

MELVIN CALVIN'S SEARCH for hydrocarbon-bearing plants has taken him to little-known parts of the globe and the botanical field. In the Ducke forest of the Amazon basin in Brazil he examined the *Copaiba* tree, which produces almost pure diesel fuel. *Copaiba multijuga* is a species of the genus *Copaifera*, which is spread throughout Brazil. Secreted in the tree's leafy canopy, the oil runs down in small pores, or canals about 0.2 mm in diameter, evenly distributed through the 3–4 foot width of the trunk. A single tap hole will produce 20 or 30 liters of oil in two to three hours. The molecular weight and volatility of the oil allow it to go directly into a diesel engine with no refining or purification in between. Since 1978, the Instituto Nacional du Pesquisas da Amazonia has run its Toyota pickup trucks on oil from the diesel tree. They tap each tree twice a year with a single hole, and Calvin is confident that a second hole during each tapping would be practical. Twenty or 25 gallons of diesel fuel per tree per year may be enough to interest the oil companies in forestry.

Diesels on a Vegetarian Diet

Rudolf Diesel wrote, in 1911, "the Diesel engine can be fed with vegetable oils and would help considerably in the development of agriculture of the countries which will use it. This may appear a futuristic dream but I can predict with great conviction that this use of the Diesel engine may in the future be of great importance."

STREETS AND HIGHWAYS filled with vegetable-powered diesels remain a vision of the future. Today a "fill-up" at the local supermarket with sunflower, peanut, maize, soya, or palm oil would prove an expensive way of getting around, but it may not be so for long. Engineers, experimenters, and farmers around the world are now examining vegetable oils as a direct substitute for diesel fuel. They have suddenly reawakened to the fact that vegetable oils have a high energy content, and can be burned in diesel engines, either pure or blended with other liquid fuels, such as diesel or alcohol. This represents another very promising way, besides alcohols, for obtaining liquid fuels from biomass. The relative simplicity of extraction and use of vegetable oils, compared to ethanol and methanol, is generating a lot of activity and excitement in the field of renewable fuels.

Rudolf Diesel would find little of this surprising. He himself tried all kinds of alternative fuels for his engines, and correctly concluded that any injectible material which could be ignited at the temperatures generated by high compression might serve as a diesel fuel. The energy content (heat of combustion) of sunflower and peanut oil, for example, is 90 percent that of diesel fuel. The fuel efficiency, however, is only about four percent lower, and if the oil is esterified, can be even better than diesel.

China's early experiments

During the Second World War, China developed an industrial batch-cracking process for producing motor fuels from vegetable oils, mostly tung oil from commercial tung nut plantations. From one ton of crude tung oil the China Vegetable Oil Corporation of Shanghai could obtain 0.6 tons of "veg-diesel oil," 60 gallons of "veg-gasoline," and 40 gallons of "veg-kerosene," plus about 0.06 tons of tar. Besides tung oil, rapeseed and peanut oil were also used. The Chinese had previously found that vegetable oils could be used directly in diesel engines, but because of higher viscosity they required increased injection pressures. And there was a further problem: vegetable oils have a high carbon residue, and a tendency toward gum formation, particularly with tung oil. So the Chinese opted for their vegetable gasoline program, in which the products were close enough to diesel and gasoline to run engines with little or no modification.

Oils can be extracted from an impressive range of plants, in addition to the tung nut, rapeseed, and peanut examined by the Chinese. Sunflower, palm, olive, and several others are already in use for industrial and culinary purposes—many more are less well known: desert plants such as jojoba and quayule; copaiba and malmeiro, both indigenous to N.E. Brazil; castor bean, milkweeds, eucalyptus trees, and squashes. Whether any of these (or the many others being investigated) can be recommended for diesel use for a given situation or country depends on costs and energy ratios. These factors, in turn, follow from the productivity of the plants (oil per hectare), and the nature of the production system.

Some important trials with sunflower and soya, now underway in South Africa and in the US Midwest, show that yields of one tonne of oil per hectare are readily obtainable with unsophisticated extraction processes, while yields of two tonnes per hectare are not unusual. This oil costs about two dollars per gallon, and the reported energy output/input ratios vary between three and ten to one. Energy ratios have always been controversial in the biomass-for-ethanol business, but not with vegetable oils. It is claimed that if a mechanized maize farmer devoted only ten percent of his land to grow sunflowers he could become completely fuel-independent. It is this possibility that has captured the imaginations of farmers and governments.

Brazilian barometer

The Brazilian experience, as with their ethanol PROALCOOL program, is an important barometer. In 1980, Brazil initiated a new Probiomass program to look at a diverse range of biomass feedstocks for the production of fuels. Oils from soyabean, sunflower, peanut, rapeseed, dende palm, and castor are being tested, along with oils from indigenous species such as pinhao (*Jotropha* spp.), *Copabeira, Toloume,* etc. The tests involve the use of straight oil, and various blends with diesel, such as a 30-70 mixture. Numerous large-scale trials are underway, notably the conversion of the bus fleet in the capital of Brasilia. It is hoped that this year vegetable oils will replace six percent of the nation's diesel requirement, rising to 16 percent by 1985.

Using African palm oil as a source of vegetable oils is a well-developed industry in many tropical countries. The Brazilians, and others, are looking closely at the economics of palm oil as a diesel substitute or extender. Yields of four tonnes/hectare/year of oil have been claimed—far in excess of alcohol yields from sugarcane or cassava of two to three tonnes/hectare/year. Oil from palms as a fuel source offers further advantages over ethanol from sugar or starch crops: simple fuel extraction, year-round production, minimal polluting by-products, whole-plant use for a variety of end products, favorable net ratio, no water requirement for processing, and the production of a protein-rich meal as a processing by-product. Other indigenous palms under examination in Brazil are dende and peach; coconut oil is being eyed from Fiji, the Philippines, Samoa and Papua, New Guinea. Australia is leading this Pacific research in its collaborative R & D program, and displays a growing internal interest in peanut, soyabean, sunflower, and rapeseed oils.

In South Africa, the search for liquid fuel substitutes is an important part of the fuel diversification program. The significant research is mostly at the Division of Agricultural Engineering, Pretoria, on the use of sunflower oil in diesel tractors. The Division's R & D group has studied the properties and use of 100 percent sunflower oil, as well as various blends of ten to 90 percent sunflower oil in diesel. The calorific values, combustion efficiencies, and cetane numbers of various sunflower oil grades were compared: crude after simple extraction, degummed crude, and refined oils. Nine different tractors have been tested, their engines run continuously up to 58 days on a one-fifth sunflower oil blend, with only minor deposits accumulating in the combustion chamber, cylinders, and piston ring grooves. Starting on pure sunflower oil presented no problem, and after 100 hours at maximum power, no adverse effects had been noticed. Clogging of injection nozzles was somewhat of a problem with high percentages of sunflower oil—especially under part-load—but this could be overcome by proper tuning and a filter. Contamination of the lubricating oil with partially combusted fuel oils could become a problem if oil changes are not performed at regular 200-hour intervals.

Esterification to mimic diesel

With these possible problems in mind, the South African group has modified the sunflower oil by esterification to the ethyl or methyl esters, using ethanol or methanol, respectively. Heating for a few hours at 30 to 40°C in the presence of acid produces fatty acid ethyl esters, some unreacted oil, and ethanol. This fuel mixture has distillation and viscosity characteristics very similar to diesel fuel. However, it causes less coking and exhaust smoke than diesel, and increases engine thermal efficiency. Tractor engines which run on this sunflower oil ester mixture achieve better results than with conventional diesel. The Brazilians, though, have found that esterification with methanol, rather than ethanol, is preferable. The methyl ester so produced can be naturally separated, unlike the ethyl ester which requires a distillation step to obtain the pure product.

South African farmers are fortunate in that over half a million hectares are now planted in sunflowers (triple from a decade back), and extensive trials and selections have increased yields 2.5-fold since 1970 to an average of 1.2 tonnes/ha in 1977. Top yields of four tonnes of seed/ha with 45-50 percent oil content are obtainable with new hybrids.

Work on vegetable diesel oils is also progressing in the US, spearheaded by the US Department of Agriculture, North Dakota State Extension Services, and the American Soyabean Association. One set of trials showed that a 75 percent sunflower oil-diesel mixture gave fuel efficiency similar to straight diesel.

Already, the extraction process is simpler for farmers than the production of ethanol from sugar or starch crops. Cost comparisons, however, are complicated by fuel taxes, exemptions, and subsidies. On a heat of combustion basis, diesel fuel at $1.13/gallon is competitive with soy oil at $0.14/pound; but tax exemptions of $0.12/gallon for diesel fuel use on farms, and the absence of such tax advantages for renewable diesel fuel substitutes, make further comparisons difficult.

The US, like South Africa, also has a well-developed sunflower production regime, with yields varying from 500 to 3,000 pounds per acre (0.56 to 3.34 tonnes/hectare) within three months of planting. Sunflowers can be grown commercially in all North American agricultural areas, often on marginal lands, and with minimal water. Oil yields range from 0.3 to 1.6 tonnes/hectare, or about 100 gallons (800 pounds) per acre when seed yield is

---

The Mediterranean region produced a surplus of olive oil, which would make a good diesel fuel substitute. Another source is the Eucalyptus tree, whose leaves contain oil, as the pharmaceutical industry has known for ages. Extraction yields are quite low, but a recent report claims that selected species carry close to 25 percent oil.

2,000 pounds. This is equivalent to 0.7 to 4.4 barrels of oil per acre.

The Mediterranean region produces a surplus of olive oil, which would make a good diesel fuel substitute. Another source is the Eucalyptus tree, whose leaves contain oil, as the pharmaceutical industry has known for ages. Extraction yields are quite low—about one percent—but a recent report claims that selected species carry close to 25 percent oil. Japanese researchers claim that eucalyptus oil can be used straight, or blended at 70 percent oil in gasoline, for tractors and small cars. Cold starting is apparently a problem, but horsepower is up while carbon monoxide emissions are decreased.

The day of diesel buses and trucks on a vegetable diet may not be too far away, if crude oil prices keep rising, and vegetable oils come in with high

---

Bus Fuels

THE NATIONAL INSTITUTE of Technology in Rio de Janeiro recently completed a series of studies, running local buses on a vegetable-diesel mixture. One trial involved an 80-20, diesel-peanut fuel, and the other a blend of 73 percent diesel, 20 percent palm oil, and seven percent ethanol. In the latter trial, the bus ran 500 hours (10,000 km), having previously clocked 60,000 km on diesel alone. After the trial, the bus went back to its normal diet of diesel, with no readjustments. The consumption rate of the mixture was 3.4 percent *lower* than with the diesel alone.

yields and low costs. But there will be problems associated with vegetable fuels, as there are with all biomass-for-energy programs. They could compete with food production, they could contribute to soil depletion or erosion, and there may be other environmental effects as yet unsuspected. But it is certain that imported petroleum will not become less expensive, and will increase in scarcity, while vegetable oils can be produced locally, and they are renewable.

—*David O. Hall*

David O. Hall is Professor of Biology at the University of London, King's College, 68 Half Moon Lane, London SE24 9JF, United Kingdom.

References:

Alvim, P. de T. and R. Alvim
1979 "Energy from plant carbohydrates, oils and hydrocarbons." *Oleagineaux* 34: 465-472.

Bruwer, J. J. *et al.*
1980 "The use of sunflower seed oil in diesel engined tractors." Division of Agricultural Engineering, Post Bag X515, Silverton 0127, South Africa.

Erickson, D. R. and P. Dixon
1980 "Soy oil as a diesel fuel: Economic and technical perspectives." American Soybean Association, PO Box 27300, St. Louis, Mo. 63141.

Hall, D. O.
1980 "Renewable resources (hydrocarbons)." *Outlook on Agriculture* 10:246-254.

Hofman, V. *et al.*
1980 "Sunflower oil as a fuel alternative." North Dakota State University Cooperative Extension Service, Fargo, North Dakota 58105: Paper no. 13 AENG-5.

Johnson, J. D. and C. W. Hinman
1980 "Oils and rubber from arid land plants." *Science* 208: 460-464.

Trindade, S. C.
1980 "Energy crops—the case of Brazil." In *Energy from Biomass*, P. Chartier *et al.*, eds., London: Applied Science Publishers.

photos by Jim Taulman

Energy Role For Forest Waste In Sweden

Once again, energy inefficient countries can take a lesson from less wasteful neighbors. Sweden, already more efficient than most countries, aims to cut oil imports by 30 to 40 percent during the 1980s. Weaning itself from a petroleum-based energy diet will be no mean feat, for imported oil makes up nearly 54 percent of Sweden's total energy supply.

THE PARLIAMENT is looking away from conventional supplies for alternatives to oil. A 1980 national referendum calls for a phase-out of nuclear power—now nine percent of the total energy supply—by the year 2010. Current plans also restrict the growth of hydroelectric power (now 215 petajoules or PJ), with a production level of 240 PJ (65 TWh) expected by 1990. In addition, the government has increased taxes on oil to encourage industry and homeowners to switch from oil to other energy sources. Grant schemes have supported oil substitution in industry since 1974.

The "complete transfer to a domestic, renewable energy system in Sweden" is called for in a recent release from the Swedish Consulate. Authors Jane Summerton and Thomas Tyden review Sweden's transition to renewable energy sources from an oil-based system in the September 1981 circular. Though they broadly cover the range of soft energy options, the authors emphasize forest waste as Sweden's energy fortune.

Forestry's Oil Appetite

Forests cover more than half Sweden's land area, with 23.5 million hectares of productive forest land. The forestry industry earns the largest share of Sweden's export revenues and ranks fourth among the world's pulp producers with 21 percent of the market.

David Brown, writing for the industry magazine, *Sweden Now*, describes the energy consumption of Swedish forestry: "With the explosion in oil prices over the last decade the Swedish forestry industry needs to be especially careful about its use of energy. It is, in fact, the country's largest energy consumer, but it has been able to achieve considerable savings by burning lignin and spent liquor as well as the bark from the trees. In such a way, the industry manages to satisfy 60 percent of its own fuel needs and then obtains much of its electricity from its own hydroelectric plants." According to the Pulp and Paper Association, an oil-conserving trend has begun among Swedish foresters, and is expected to continue: from 1973 to 1979, the industry's oil requirements dropped 17 percent.

Wood Wastes—A Resource

With over 70 million m³ of timber harvested annually by the Swedish forestry industry, over 45 million m³ of wood wastes are left—an energy resource estimated at 325 PJ (8×10^6 tonnes of oil equivalent). Mechanical harvesting of forest wastes for fuel is currently only on an experimental basis. Summerton and Tyden cite the lack of clear marketing and distribution channels as a major constraint to widespread use of forest wastes. Current R&D focusses on "whole tree" harvesting with new systems available by 1987, and conversion of wastes to compact forms—woodchips and pellets—for economical transport. These authors estimate the long-term potential of forest wastes at 25 million m³/yr, a fuel value of 190 PJ (4.5 Mtoe). Present-day production costs are in the range of 9-18 Swedish kroner (SEK)/GJ (US $ 1.62-3.24/GJ). This excludes preparation and conversion costs necessary for comparison to current oil prices.

dards; these are now savings in the world market. "In Sweden we have been forced to build environmental care into the process of paper-making—and the recovery of gases and wastes have provided us with much-needed energy," reports an industry spokesman in *Sweden Now.* "In Canada the regulations are easier, but they will eventually be forced to concede to stricter environmental regulations as well."

of 2,000 kW and 3,000 kW, are presently being designed for completion in 1982. Summerton and Tyden estimate electricity costs from these at .13-.17 SEK/kWh (US $.023-.03/kWh), excluding possible reserve capacity costs.

Finally, the authors project 40 PJ (0.9 Mtoe) per year is available from peat by 1990. R&D focusses on dewatering techniques and harvesting methods for irregular marshes. Peat was used extensively during World War II for fuel—an estimated 1.3 million tonnes annually. Current peat production costs are estimated at 7-13 SEK/GJ (US $1.26-2.34/GJ) oil equivalent, excluding variable transportation costs. In northern Sweden, a pulp and paper company has rebuilt its oil burners for peat combustion. Now importing peat from Finland, the company plans full-scale harvesting of some 1,000 hectares by 1983.

—*Elyse Axell*

References:

Brown, David
1981 "Our Forests are Our Fortune," in *Sweden Now, vol. 5*, pp. 61-63. Midskogsgränd 11, Box 27315, S-102 54 Stockholm, Sweden.

Summerton, Jane and Thomas Tyden
1981. "Renewable Energy Sources in Sweden," In *Human Environment in Sweden, no. 19*, September. Available from: Swedish Consulate General, 825 Third Avenue, New York, NY 10022.

Swedish energy supply (1979)

(with hydropower and nuclear power counted in fossil fuel equivalents)

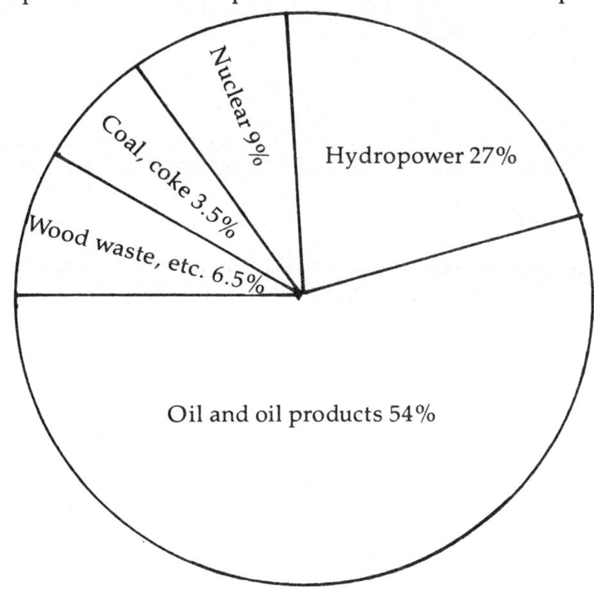

Total energy consumption (excl losses) = 1,450 PJ (35 Mtoe)
from Swedish Information Service, Swedish Consulate General.

Cultivation of energy forest plantations is currently in the early stages of development. Swedish land areas favorable for intensive cultivation include marshlands, forest areas, and abandoned agricultural soil. Results indicate that short-rotation (2-3 years) of Salix plantations is the optimal technique for most available land areas in central and southern Sweden. Two large-scale (100 hectares) pilot farms are testing selected marshland and abandoned agricultural land for harvesting methods and plant breeding. Summerton and Tyden do not report a major contribution to Sweden's energy supply from forest plantations before the year 2000. They conservatively project that three M tonnes dry substance or 50 PJ (1.2 Mtoe) may be produced by the turn of the century. Production costs vary from 6-28 SEK/GJ (US $1.08-5.04/GJ).

Ahead of its competitors, the Swedish forestry industry has already spent the money for environmental controls. At an early stage, the industry adopted high environmental stan-

Sun, Wind, and Peat

In addition to its energy fortune in forest wastes, Sweden is developing other renewable sources. Underground solar heat storage is favored in Swedish research, according to Summerton and Tyden. Pilot plants with concrete tanks and underground rock caverns ranging in size from 5,000-10,000 m³ have been built, as well as deep-level storage systems, suitable for highly populated areas. Solar collector research is well developed. Swedish companies now market complete systems for solar-heated tap water. Summerton and Tyden record 2,000 systems in present operation, primarily in single-family residences.

With favorable winds and a completely integrated national utility grid, Sweden is well-suited to wind powerplants. Wind-generated electricity production is projected at 55 PJ (15 TWh) per year during the 1990s. Two experimental horizontal-axis wind powerplants, with rated outputs

photo by David Powers

TOOLS for the SOFT PATH

When the Chips are Down...

"MARINDALE DAIRY, 3 miles ahead," reads the billboard along Highway 101 in Novato, California, "Purveyors of Milk, Cream, and Electricity." The sign isn't up yet, but it won't be long before the Marindale Dairy has its methane digester/generator system up and running, and begins to sell electricity back to the utility.

The dairy has 275 milking cows and high energy bills. Up till now, the 3,861 pounds of manure produced each day was regarded as a disposal problem, not a potential energy resource. Unprocessed, it might be sold for only about $3.20 per ton, or six bucks for the lot—not even worth the hauling fee. But run through a methane digester, and converted into 16,000 cubic feet of biogas (60 percent methane) with an energy content of 10.4 million Btu, the daily manure load is worth as much as 1.8 barrels of crude oil, or nearly $60 at current world prices.

Most of the dairy's energy use is in the form of electricity, but the daily fluctuations are extreme. To meet peak demand, Marindale would have to install gas storage facilities and a large methane generator. With recent regulations requiring utilities to purchase electricity at no less than their avoided costs of production (about $0.05/kWh in California), it becomes more efficient and cost effective for the dairy to generate electricity at a modest, constant rate, and sell it to the utility. The efficiency of the methane generator is only about 20 percent, so 609 kWh are available for sale each day. However, 90 percent of the engine's waste heat is recovered in the form of hot water, which is easily stored and put to a variety of dairy uses.

Total yearly energy savings could be as much as 220,000 kWh, and 2.7 billion Btu, or about $25,000. Operation and maintenance should run $5,000 annually, and with an initial investment of around $100,000, the system is expected to pay for itself in 7-8 years, depending on how fast energy prices rise and the cost of money.

Dairy Energy Balance

In 1978, the milk cow population of California was 842,244, and together they produced nearly 12.5 billion pounds of milk. The energy value of this milk was about one-fifth of the total fossil fuel energy expended in milk production. Dairy operations take only one-fourth of the energy, with the rest embodied in the feed. And as an energy conversion device, the cow is only 20 percent efficient in transforming feed into milk.

Much of the unused food energy is discharged in manure—40 to 50 percent of the food energy the cow consumes. Biogas digesters offer dairies an opportunity to recover more than half the energy contained in the manure their animals produce. Depending on their needs, they can then use the gas for heating, or to generate electricity which they can either use themselves or sell to the grid. The biological methane production process is also an effective waste treatment method; effluent remaining after the biogas has been extracted is lower in weight and volume than the original manure, but with most of the nutrients retained.

In the digestion tank, bacteria convert the organic matter into methane. The digestion process works best at 35° C, and heat must be supplied; unlike aerobic decomposition of organic material, anaerobic decomposition produces little heat. Biogas and spent slurry are the process' two products. Sixty percent methane and 40 percent CO_2, the gas has a heating value of 21 MJ/m³ (600 Btu/ft³). It cannot, like gases such as propane or butane, be liquified at moderate pressures and temperatures. Because storage in gaseous form is cumbersome and expensive, the biogas is best used as it is produced, either for direct heating or for electricity.

Plug Flow System

The Marindale methane digester system uses a "plug flow" system developed by Dr. Jewell of Cornell University's School of Agriculture. A collection trough along the concrete animal containment area receives manure, which is then washed into a tank and diluted with warm water. A centrifugal pump converts the mixture into a uniform slurry, which is pumped into the anaerobic digester. Each day a new "plug" of manure mix is pumped in. The cost of this system is expected to be well below that for a conventional mixed tank digester.

This project, along with a similar research and development effort on another small California dairy, is under the direction of the Landal Institute, a non-profit corporation which works to preserve agricultural land in urban fringe areas. Their objective is to develop a system that can be

Energy type		BTU (in millions)	
Electricity	455,194,200 kWh	4,551,942	
Natural Gas	9,327,750 therms	932,775	
Gasoline	310,925 gallons	38,555	
Diesel Fuel	3,233,620 gallons	452,707	
		5,975,979	Total BTU (in millions)

scaled down to suit dairies below the 300-400 animal size thought to be the minimum for cost-efficient methane production.

Methane digester systems save dairy operators the time and trouble of waste handling, and the fertilizer value of the digester effluent adds to the overall value. But the energy savings are foremost. If every dairy in California installed such systems, the energy value of the methane produced each year would be about 1.35 trillion Btu, equivalent to nearly a quarter of a million barrels of oil. And because the methane energy is right on the farm, and can be used at nearly 90 percent thermal efficiency, the actual energy displaced would be several times greater.

—Katy Slichter

Reference:
Landal Institute
1980 "A Methane Digester System for Coastal Dairies: A Two-Phase Project." Prepared for Energy Resources Conservation and Development Commission, State of California.

After this article was published, a correspondent suggested that 842,244 cows in California would yield 11.6 × 10¹² Btu of methane — a factor of 8-10 better than the estimate above.

Can Methanol Be Managed?

IN AMERICAN FARM COUNTRY, ethanol from grain is everyone's favorite fuel. Trickling from farmyard stills and shipped from giant distilleries, ethanol offers farmers independence from oil and a second market for their crops. The Congressional Office of Technology Assessment, however, recently concluded that grain ethanol's contribution to US liquid fuel supplies is likely to remain small. *Methanol* is another matter entirely. Though not yet produced in large quantities for transportation, the potential supply far exceeds ethanol from grain and sugar. Methanol can be refined from the abundant US stores of wood, crop residues, fast-growing grasses, and even coal, though at a higher environmental cost. Wood alcohol, as it is also known, avoids the food *vs.* fuel controversy since it derives from inedible biomass; it consumes less energy in production than grain alcohol, and with proper land management has less of an impact on the environment.

The limit on large-scale grain alcohol production is primarily political. OTA estimates that as production approaches 8 billion liters per year (20 times the present level), the diversion of grain to alcohol fuels would cause inflation in the farm sector and a hike in farm commodity prices. The exact production limit will vary somewhat with the acreage of new land brought under cultivation, and the amount of crop switching. (Distillers' grain, a byproduct of ethanol production, substitutes for some soybean uses, allowing grain cultivation on otherwise soybean land.)

The biomass resource base for methanol, by comparison, is much larger and more flexible. Taking competition from other uses into account, enough wood, grasses, and crop residues remain to convert some 4,000 Petajoules (1 PJ = roughly .001 quad) into 115 billion liters of methanol, with an energy value of 1,800 PJ. Present US gasoline consumption is over 400 billion liters per year, or about 12,000 PJ (the energy content of gasoline, at 31 MJ/liter, is twice that of methanol). But alcohol fuels displace more than their energy content when they serve as an octane-boosting additive to gasoline. Refineries can produce high-octane fuel by mixing alcohols with relatively low-octane gasoline, which requires less energy to produce than unleaded premium. By the 1990s, fewer cars should require high-octane gasoline, so the long-term energy savings from using alcohols as octane boosters is uncertain. Still, if one-fourth of the 115 billion liters of methanol available from biomass is used as an octane booster, and the rest as a stand-alone fuel, the resultant oil displacement would be about 1 million barrels per day.

Diverse Feedstocks

For the past half-century, methanol has been manufactured almost exclusively from natural gas, but it was originally produced by the destructive distillation of wood. Once the US's primary fuel, wood remains the country's most important biomass resource, and could play a transportation role through conversion to methanol. The forest products industry already derives 1,300 PJ annually from wood—more energy, though in a lower quality form, than the electricity generated by US nuclear power plants in 1979. Intensive management of commercial forests, including more complete biomass removal, improvement cuts, and conversion of low quality stands, could increase the energy available from wood. By the year 2000, commercial stands could supply 5–10,000 PJ per year from logging and wood processing residues, thinnings, and the removal of dead and diseased timber. The industry will use much of this additional energy, and some will go to meet industrial process and home heating needs, but substantial quantities should still be available for methanol conversion.

Crop residues represent a second major methanol feedstock. Roughly 5,000 PJ of residues remain standing in US fields after the yearly harvest, of which 20 percent can probably be removed without erosion problems. Larger amounts of energy could come from haylands and pasture in the form of grass. Farmers could take 3–5 tons per acre, for a net harvest of 1–2,000 PJ, though this would mean heavier fertilizer applications and more frequent cropping. As much as 5,000 PJ annually are possible in the year 2000, if mediocre lands which are now uncultivated turn out not to be needed for food crops and are planted with grasses.

Wood, crop residues, and grasses together can supply a great deal of biomass energy, but much of it will go to purposes other than methanol production. Direct combustion of biomass or gasification for industrial purposes are cheaper, more energy efficient, and require less capital investment. The markets involved are also more easily penetrated than consumer transportation.

Refining Methanol

The first methanol refinery in the US to use biomass as a feedstock, rather than natural gas, will be constructed in Winchester, New Hampshire. From 500 dry tons of sawmill and logging wastes, it will produce 40,000 gallons (150,000 liters) of methanol per day. Plants much smaller than the Winchester facility are unlikely since methanol refining, unlike ethanol distillation, has substantial economies of scale. The upper limit to plant size is set by the cost of transporting biomass over long distances; the Winchester plant will procure material within a 40 km radius, and a much larger refinery would face higher transport costs. Capital costs would run $95 million for a plant capable of 190 million liters per year, somewhat more, according to OTA estimates, than a comparable ethanol plant. Conversion of crop residues or grasses into methanol should not differ much economically from the use of wood, but a commercial production facility is further off.

Methanol from wood, like ethanol distillation from grain, rests on a well-known technology, but neither is likely to provide energy as inexpensively as methanol from coal. They may, however, compare in cost to some synfuels. With feedstock costs ranging from $20–60 per dry ton, OTA calculates methanol costs of $0.17–0.34/liter. Considering only energy content, methanol priced at $0.24 at the plant is equivalent to gasoline selling at the refinery gate for $0.46/liter. (This is equivalent to crude oil at $45 per barrel; marginal foreign oil costs $35/bbl.)

Running on Alcohol

Of the two alcohol fuels, ethanol is the easier to use in today's cars. Blended with gasoline in "gasohol," grain alcohol is already widely used in the US and in Brazil. Methanol makes a more problematic blend. It not only separates from gasoline when small amounts of water creep into the blend, but it attacks gas tank liner and engine gasket materials, and presents emissions problems as well.

Tests by the California Energy Commission using both alcohol blends in unmodified 1978 Honda CVCCs showed lower NO_x emissions with no real loss of fuel economy or performance. Unfortunately, evaporative emissions—which contribute to ozone and smog—increased. A 10 percent ethanol blend emitted twice as much as pure gasoline, and a 5 percent methanol mix made evaporative emissions five times worse than gasoline alone. Retrofits to control the problem are possible, but expensive; new car design modifications are comparatively easy. A stock 1979 Volvo equipped with an advanced three-way catalyst beat the tough 1984 emissions standards for *all* pollutant categories while running on a 10 percent ethanol blend. Higher alcohol percentages are possible with additional modifications. A 15 percent blend in a Honda with simple carburetor adjustments gave poor fuel economy and unacceptable exhaust emissions, while vehicles with feedback-controlled fuel injection can use up to 50 percent alcohol and still meet emissions and performance standards.

Since alcohols produce no particulates and have low NO_x emissions, they can help reduce the pollutant load from diesel fuels. Methanol is not readily soluble in diesel fuel, but a 25 percent methanol-diesel *emulsion* can be combusted with little change in engine thermal efficiency. The power loss that comes from methanol's lower energy content can be corrected with a higher fuel flow rate. Blends of 30 to 40 percent alcohol are feasible if the fuels are aspirated into the engine from separate tanks.

Pure methanol makes an excellent fuel for conventional spark ignition engines—including the stratified charge designs—provided that carburetors and fuel injection systems are altered. Methanol's high octane rating allows a higher compression ratio, which boosts vehicle efficiency. OTA reports that a methanol engine can be 15 to 20 percent more efficient than a gasoline engine of comparable size. A car getting 24 mpg on gasoline, for example, would get 14 mpg in a methanol version, rather than the 12 mpg which one would expect on the basis of energy content alone. Using pure methanol, evaporative emissions drop to near zero, and, with the exception of smog-producing aldehydes, exhaust emissions compare to gasoline. A further advantage of unblended methanol is impressive power output, since its high heat of vaporization allows a large amount of fuel and air to enter the cylinder. Pure and near-pure alcohol fueled cars are already building a market share: Ford, GM and Volkswagen now have vehicles for sale in Brazil designed for 100 percent ethanol, and two manufacturers—Volvo and VW—have tested engines capable of running on 90 percent methanol.

The main problem with methanol is cold-weather starting, which becomes impossible below 4° C without additives such as butane. An electric heating element beneath the carburetor extends the temperature range to −4° C, which is not low enough for a large number of American consumers.

How Benign?

OTA estimates for ethanol production suggest that with a four- or five-fold reduction in today's transportation energy requirements, alcohol fuels could completely replace gasoline. The problem of oil imports would then be solved, but the price of our energy independence could well be environmental degradation, unless the large-scale use of biomass resources is properly managed. "The rapidly rising use of wood for fuel in New England," OTA points out, "has raised both hopes and fears for the future of America's forests." The heightened importance of biomass resources could stimulate either more careful management, or else short-sighted and destructive exploitation. Forests under intensive management experience shorter rotation periods and greater biomass removal, and run the risk of nutrient-depleted soils. More careful management, on the other hand, could increase the yields of high quality timber from commercial forestlands, thereby relieving the pressure to harvest stands with high recreational and esthetic value. But the problems inherent with intensive management would remain: limited species and age diversity, affected wildlife, a forest of altered visual and ecological character. As wood prices climb with increased demand, so does the incentive to harvest poor quality stands on delicate marginal lands. OTA concludes, "in light of the current weak economic and regulatory incentives for practicing good environmental management and the alarming lack of information about current logging practices, the potential problems appear quite serious."

Similar management problems arise with the use of crop residues and grasses as methanol feedstocks. The stalks and byproducts that remain in the field after harvest play a part in erosion control, but their economic value, if it becomes high enough, will tempt farmers to remove them indiscriminately. Should alcohol fuels ever supply a major part of transportation needs, they will raise the familiar conflict over land use objectives: short-term economic benefit versus long-term environmental value.

—*Steve Meyers*

Reference:

Office of Technology Assessment

1980 *Energy from Biological Processes.* Available from Superintendent of Documents, US Government Printing Office, Washington, DC.

Estimated costs of alternative liquid fuels (1979 $)

Fuel source	Raw liquid $/GJ[1]	Refined motor fuel $/GJ	Refined motor fuel $/liter[2]
Imported crude ($30/bbl)	4.83	8.88	0.31
Enhanced oil recovery	2.84—6.63	5.21—12.23	0.18—0.42
Oil shale	5.59—6.92	11.85—15.35	0.41—0.54
Syncrude from coal	4.83—8.06	10.33—16.87	0.36—0.59
Methanol from coal	—	5.21—8.34	0.09—0.15
Methanol from biomass	—	9.67—19.81	0.17—0.34
Ethanol from biomass	—	10.14—16.87	0.24—0.40

1. $/MMBtu = $/GJ × 1.055
2. $/gallon = $/liter × 3.79

*SOURCES: K. A. Rogers and R. F. Hill, **Coal Conversion Comparison,** prepared for US Department of Energy under contract EF-77-C-01-2468; and **Coal Liquids and Shale Oil as Transportation Fuels,** a discussion paper of the Automotive Transportation Center, Purdue University, West Lafayette, Indiana.*

Ethanol's Balance Sheet

Ethyl alcohol appeared, at first, to be an uncontroversial energy technology; it is a clean-burning, direct replacement for refined foreign oil, produced from renewable plant and animal sources in America's breadbasket. But that optimistic view was not to last. Several recent reports—including a sharply critical one from the Department of Energy's high-ranking Energy Research Advisory Board (ERAB)—have left ethanol off the apple-pie slot on the energy menu. And President-elect Ronald Reagan's Energy Task Force comes to the same conclusion: given the high fossil fuel consumption of agriculture, "ethanol from grain... is energy inefficient using more petroleum than it displaces and presently very uneconomical." Instead, the Task Force suggests methanol, which "can be produced from coal and may be more efficient even though some engine modifications will be required."

Net Energy

An inefficient ethanol program would lose support from all quarters in no time, by diverting large areas of cropland for fuel production, boosting food prices, accelerating soil erosion, and doing little to reduce our dependence on imported oil. But this is far from the impression one gets from an analysis of ethanol technology by University of

An all-out biomass to ethanol program could produce over 90 billion gallons annually; while a reasonably efficient transport system would require only 37 billion.

Pennsylvania scientists E. Kendall Pye and John Ferchak. Their report focuses on two key ethanol controversies: net energy and resource sufficiency. Ferchak (a biochemist) and Pye (a biophysicist) conclude that "total gasohol use could be (achieved in a) short period of time... with relatively minor adjustments in the agricultural sector... (and) no evident impact on food production." Over the longer term, "the potential additional yield of ethanol from (woody) biomass appears to be well in excess of liquid fuel requirements of an (efficient) transport sector in the United States (assuming) present mileage demands."

The US currently consumes about 101 billion gallons of gasoline per year, plus another 12 billion gallons of diesel in cars and trucks, for a total of 15.6 quads. These uses of petroleum are the largest in the nation, slightly exceeding the total sum of imported oil. Against these enormous sums, the 200 million gallons of ethanol produced in 1980 are small potatoes. But the commitment to increased production is growing rapidly, and industry sources expect to turn out 500 million gallons during 1981. A *Value Line* investment survey notes that "dozens of ethanol facilities are either in planning or under construction" and that ethanol production capacity has grown 130 percent during 1980.

The oil displaced by an acre of cropland in ethanol production is a key contention in the reports by both ERAB and Ferchak & Pye. By ERAB's accounting, corn grown specifically to produce ethanol takes 45,000 Btu/gallon from seed to harvest, and an additional 59,000 Btu/gallon to refine. The sum—104,000 Btu/gallon—easily exceeds the fuel value of the product (82,000 Btu/gal.), and nearly equals its value as gasohol (assuming 1:1 volume equivalence with gasoline based on alcohol's value as an octane booster). If most of the agricultural fuel inputs are oil, the acreage required to make a dent in oil imports would be astronomical. Thus, ERAB's recommendation to concentrate on methanol from coal appears sound, and it is echoed in the Reagan report.

Ferchak and Pye reach a very different conclusion, stressing crop wastes and by-products, and "lignocellulosic substrates" or woody sources of biomass. In their view, 192 or 200 proof ethanol could be distilled "at a maximum energy cost of 18,000 Btu/gallon with present innovative technology and for as little as 8,200 Btu/gallon... as a process goal." A typical 3 million gallon/year postwar distillery used 103,000 Btu to make a gallon of "spirits-grade" ethanol. Today, advanced microbial technologies for cellulose fermentation and the use of membranes or cellulosic desiccants in the distillation process suggest that a revolution is underway in alcohol fuels production technology. Ferchak and Pye believe that within broad limits, an ethanol facility could be "totally process energy self-sufficient if necessary." On the key question of high quality liquids, Ferchak and Pye note that efficient energy plantations could produce nine times the ethanol they themselves consume. They add, "the liquid fuel is needed exclusively for the operation of agricultural equipment."

Sugar & Starch:
30 Billion Gallons

After subtracting the energy required to farm, gather, ferment, and distill sugar-starch resources, Ferchak and Pye calculate that some 30 billion gallons of ethanol could be produced with little change in current agricultural practices.

Citing Ed Lipinsky's (Battelle-Columbus Laboratory) recommendation for diverting feed corn to ethanol production, Ferchak and Pye claim that 16.2 billion gallons of ethanol could be produced with no reduction in the ruminants' standard of living. The high-protein mash that remains after fermentation could be mixed with cellulose or starch and fed to cows just as feed corn was. The authors note that the "technology is well established and could be implemented within a year," incremental energy use is small, and that the quantity of ethanol produced from diverted livestock feed would be sufficient "to make an immediate impact on oil imports and to begin the development of an infrastructure for the transition to an ethanol-based transportation sector."

Another 1 billion gallons of ethanol per year is possible from small-scale fermentation of spoiled grains. Ferchak and Pye second another approach developed by Battelle-Columbus Laboratories for expanded farming of high-yield sugar crops on "new acreage" in the Southeastern, Delta, and Southern Plains farm regions. The Battelle group estimates that 50 million "new" acres might be available in these areas for alcohol production, and that 21 million of these, if planted in sugar crops, could produce 10 billion gallons of ethanol, net of energy costs. (Sugar crops produce 1.8 times more ethanol per acre than grain.)

Idle or set-aside land offers an additional opportunity. Ferchak and Pye estimate that this marginal land may total 150 million acres in the continental US, 31 million of which the federal government pays $2.3 billion to leave idle. (For comparison, 70 million acres are now in corn production.) This subsidy, argue Ferchak and Pye, might reasonably be transferred to ethanol production. Depending on crops, technologies, and yields, the potential net ethanol production from 30 million of these acres could range from 3.8 to 8.7 billion gallons.

Finally, the US is moving toward less consumption of animal proteins in favor of grain protein. This trend increases the available cropland, because animal protein requires 8-15 times more land to produce than an equivalent amount of grain protein. Should the trend toward reduced meat consumption continue to 2020, and the lands to put to use for corn or sugar crop production, the net ethanol yield would range from 19 (corn) to 30 (sugar) billion gallons per year. (This trend is not considered in Ferchak and Pye's estimate that 30 billion gallons of ethanol per year could be produced with no substantial change in agricultural practices.)

Cellulose: 60.9 Billion Gallons

Cellulosic sources include agricultural residues, forestry residues, bagasse from increased sugar cane production, and forests cultivated for energy production. DOE's year 2000 estimate for agricultural residues is 560 million tons, half of which would be returned to the land for soil conditioning and half of which would be used for ethanol production. The net production from residues would total 16.3 billion gallons. An equal sum, using the same accounting basis, would be available from forestry wastes. Seventy percent of municipal wastes would add a further 4.1 billion net gallons. Bagasse from sugar cane would add 5.2 billion net gallons. These sources total 41.9 billion gallons.

The US has 500 million acres of commercial forests, of which 300 million might be suitable for energy crops, according to Ferchak and Pye. DOE's interim goal is that 10-15 percent of the 300 million acres be used to make fuel alcohol. Based on a projected yield of 10 oven dry tons per acre, minus energy requirements of 77 gallons of ethanol/ acre, each acre would provide 633 net gallons of ethanol per year. With 30 million acres in production, net ethanol from this source would reach 19 billion gallons. Production of ethanol from swampy areas, or from offshore kelp plantations was not considered in the analysis.

Process Improvements

New technologies for converting biomass to ethanol include a few potential "show-stoppers." Agricultural energy use in the US is quite high (45,000 to 50,000 Btu/ gallon of ethanol produced, assuming corn). Low-till farming methods, cereal perennials, and substitution of manure for conventional fertilizers could reduce agricultural energy use by over 50 percent.

Distillation currently consumes 57,000 Btus per gallon of produced ethanol, according to Ferchak and Pye. But this is for "spirits-grade" ethanol, and "fuel-grade" processes promise significant reductions. A two-grade (rather than five-grade) process pioneered by Vulcan Cincinnati cuts energy requirements to 18,000 Btu/gallon. Another process developed by Cyseweski and Wilke cuts energy needs to 16,300 Btu/gallon for 95 percent ethanol. Katzen Associates has developed an advanced distillation process that requires slightly over 10,000 Btu/gallon.

None of these systems takes advantage of some even more important technologies. Most of the energy in distillation is used above the 85 percent ethanol level. Using starch or cellulose products to dewater the 12 percent feed, anhydrous ethanol could be produced with an energy requirement as low as 8,200 Btu/gallon. Harry Gregor of Columbia University suggests that advanced membrane technology could be used to dewater the fermentation stillage using less than 5 percent of the energy required by conventional evaporation. Solvent extraction processes under development by Arthur D. Little use CO_2 to extract ethanol from mash, using as little as 8-10,000 Btu/gallon.

Co-products of ethanol processing include biogas from

An ethanol facility could be totally energy self-sufficient...
producing nine times more liquid fuel than it would use.

the fermentation mash and lignin. These sources can total three-fourths the value of ethanol from a fermentation plant. The lignin can be removed after pretreatment and is soluble in ethanol; this mixture has a fuel value exceeding that of pure ethanol, but its use in internal combustion engines has not been tested. Such a use would nearly double a facility's liquid fuel output.

On the engine side of the equation, work is being done to test whether hydrous ethanol could run cars and trucks. There is some evidence, note Ferchak and Pye, that 80-90 percent ethanol has proved satisfactory with tractors and that new engines (e.g. The Beta Engine) may effectively run on 60 percent (or 120 proof) ethanol. Pye and Ferchak also cite a commercially available dual injection system developed by M & W Gear Company of Gibson City, Illinois that uses a secondary tank to inject a 50/50 mixture of water and ethanol into a turbocharged diesel. The developer claims that 1 gallon of ethanol in this system displaces 1.25 gallons of diesel.

Policy Issues

An all-out biomass to ethanol program could produce over 90 billion gallons of ethanol annually, according to Ferchak and Pye. With new car mileage levels of 42 miles/gallon in 1995, cars and light trucks would use 49 billion gallons of fuel in 2000. At 63 mpg, year 2000 use would total 36.8 billion gallons. Both are within the reach of a fuel alcohol program, but the difference between a sector that uses over 100 billion gallons per year and one that uses a third as much is enormous in environmental impact, acreage requirements, and impact on food and lumber prices. A combined effort to increase auto efficiency levels and boost alcohol production is the best insurance (and, according to past articles, the most cost-effective route) for a biomass program that is consistent with long-term political support, an improved environment, and reduced use of foreign oil.

—Jim Harding

References:
Ferchak, John D. and E. Kendall Pye
  1980 *Utilization of Biomass in the United States for the Production of Ethanol Fuel as a Gasoline Replacement*, published in "Farm and Forest Produced Alcohol: The Key to Liquid Fuel Independence," Joint Economic Committee, US Congress.

US Office of Technology Assessment
  1980 *Energy From Biological Processes*, available from OTA, Washington, DC, 20510 or from US Government Printing Office, Washington, DC, 20402.

Bio-Energy Council
  1980 *Proceedings of the Bio-Energy World Congress and Exhibition*, available for $60.00 from the Bio-Energy Council, 1625 Eye Street, NW, Suite 825-A, Washington, DC, 20006. (Some discounts may be available.)

Calories and Controversies

Food or Fuel: New Competition for the World's Cropland, a recently released (March 1980) Worldwatch Institute paper by Lester Brown, has stirred considerable controversy over agriculturally based alcohol fuels. At issue, Brown contends, "is whether governments can encourage the production of alcohol fuel without inadvertently launching an industry that competes directly with food production."

Brown foresees prime farmland being diverted from food to energy crops, decreasing the quantity of food on the world market and, in turn, driving up food prices. Even without the added burden of cultivating energy crops, efforts to expand global food production have lost momentum over the past decade and have barely kept pace with population growth. Brown therefore questions the national priorities which place energy self-sufficiency ahead of plentiful, cheap food.

Domestic alcohol fuel production appeals to countries with ample agricultural resources as a way to decrease dependence on imported oil. This would reduce their susceptibility to price increases and supply disruptions on the world petroleum market, while improving their balance of payments.

An alcohol fuel industry would also generate new jobs. Since energy crop cultivation and alcohol distillation are far more labor intensive than petroleum extraction and refining, an alcohol fuel industry could employ more people than the gasoline industry it displaces. Most of the jobs would be located in rural areas — often areas of high unemployment — where the raw materials for the industry are grown. In developing countries, these new jobs could help slow down rural-to-urban migration.

Alcohol fuels have environmental advantages as well. Ethanol burns more cleanly than gasoline, and it can substitute for lead as an octane booster in gasohol (9 parts gasoline, 1 part ethanol). Brown recognizes all these benefits, but argues nevertheless, that the negative consequences of large-scale conversion of agricultural commodities into fuel outweigh them.

Brown contends that as an agriculturally based fuel industry expands it would compete increasingly with food production for limited agricultural resources. This competition would drive up food prices. Since the fuel produced from energy crops would substitute for gasoline, food prices would be tied to gasoline prices. Thus, as the world price of petroleum increases, so too would the prices of basic foods.

Distillers' demand for energy crops could also decrease food production. As Brown points out, "more and more farmers will have a choice of producing food for people or fuel for automobiles. They are likely to produce whichever is more profitable."

Brown warns, too, of possible environmental degradation. Higher prices and attendant higher profits would prompt farmers to cultivate more marginal land, and to use land already under cultivation more intensively. This could lead to extensive erosion and soil deterioration, accelerating the current fall in global cropland productivity.

In many countries, higher food prices resulting from alcohol fuel production could increase income inequities. Middle and high income groups tend to dominate liquid fuels use, especially in less developed countries where only the affluent can afford automobiles. These groups benefit the most from national programs promoting alcohol fuels, while the poor suffer most from the food price increases that accompany them.

International Consequences

The prospect of agriculturally rich countries turning to large-scale alcohol fuel production is most alarming in the context of the global food situation. Citing the growing world population, and diminishing returns from efforts to increase crop yields, Brown predicts more frequent and more devastating world food shortages. From 1950 to 1971, he notes, per capita grain production increased 30%, improving nutrition in both industrial countries and much of the Third World. However, per capita production since then has been decreasing. To compound the problem, the world's largest reserves — huge US grain stocks and cropland left idle under US farm programs — have been nearly exhausted in an effort to keep pace with recent world demand. Population continues to grow by 70 million annually, with a corresponding increase in world grain demand of 30 million tons. But the cropland base is expanding only one-fifth as rapidly. On this basis, Brown warns that agricultural fuel programs which raise world food prices and reduce the amount of food available for export to agriculturally poor countries would affect nutrition and health worldwide and could result in serious political conflicts.

Brown provides data demonstrating the trade-off between producing fuel for automobiles and food for people. Thirty-six people, according to his calculations, can subsist on the

---

Food and Fuel

annual per capita grain and cropland requirements

	grain (pounds)	cropland (acres)
Subsistence diet	400	.2
Affluent diet	1,600	.9
Typical European automobile[1] (7,000 miles/yr. at 25 mpg)	6,200	3.3
Typical US automobile[2] (10,000 miles/yr. at 15 mpg)	14,600	7.8

[1]Based on average world grain yields in 1978, according to the US Dept. of Agriculture.

[2]Fuel use converted at 380 liters of alcohol per metric ton of grain.

from Lester R. Brown, "Food or Fuel: New Competition for the World's Cropland," p. 27

grain equivalent of the fuel consumed by a typical US automobile. Several countries including Brazil, the US, South Africa, New Zealand, and Australia have, nevertheless, instituted programs promoting ethanol production from agricultural commodities. At a time when starvation threatens millions, the priorities underlying this trade-off deserve close examination.

National Policies

Brazil, the world's leading ethanol producer, hopes to achieve automobile fuel self-sufficiency through sugarcane distillation by the end of the decade. Massive efforts to expand sugarcane production are now underway. But in recent years Brazil has developed a chronic grain deficiency despite its vast agricultural potential. In 1979, Brazil imported 5.7 million tons of grain; in 1980, it is expected to be by far the largest grain importer in the Western Hemisphere.

Although recent droughts are in part reponsible for Brazil's large grain imports, Brown contends that government policies subsidizing energy crops at the expense of food production have contributed to the problem. Brazilian officials claim that the country's vast land resources are sufficient to support both energy and food cultivation without competition. Brown, on the other hand, contends that energy crops compete not only for land but also for management skills, capital, agricultural credits, fertilizer, water, and technical advisory services.

Low domestic food production has led to higher food prices. Brown cautions that if food prices continue to rise, energy crop cultivation could put a severe strain on the country's social and political fabric. With an average income ratio of 36 to 1 between the richest and poorest fifths of the population, Brazil already has one of the world's most dramatically skewed income distributions. The decision to fuel the rapidly growing automobile fleet with energy crops in effect increases the wealthy individuals' claim on cropland while further squeezing the nation's impoverished millions.

In stark contrast to Brazil, the US is the world's largest grain exporter. Much of the world has come to rely on US grain exports. But as demand for alcohol fuels grows, US distillers might divert domestic grain surpluses to ethanol production. According to Brown, ethanol from corn can be competitively marketed as gasohol even without government subsidies. To substantiate this claim he cites estimates by the US Office of Technology Assessment that, with US corn prices at $2.44/bushel, "ethanol produced and delivered to US gas stations at $1.20 to $1.40 per gallon (32¢ to 37¢ per liter) would be competitive as a gasoline additive when the average price of crude oil is $20 to $31 per barrel, at which point unleaded gasoline would be selling for $1.10 to $1.60 per gallon (29¢ to 42¢ per liter) on the average."

Current US policies make gasohol even more competitive. The Energy Act of 1978 removed the 1¢ per liter federal gasoline tax on gasohol obtained from non-petroleum sources. Under this tax exemption, mixing 1 liter of ethanol with 9 of gasoline exempts all 10 from the 1¢ federal tax for an actual subsidy of 10¢ per liter of alcohol used as fuel. Additionally, sixteen states have exempted gasohol from their state gas tax. In Iowa, for example, the net exemption on a liter of ethanol amounts to 16¢. Combined with the federal exemption, the total subsidy exceeds 25¢ per liter. When these subsidies took effect, Iowa's monthly gasohol sales surged from 2.4 million liters in November 1978 to 33 million in December 1979. Brown finds that distillers taking advantage of these subsidies and selling distillation by-products for animal feed might well be able to pay over $5 per bushel for corn and still sell alcohol fuels competitively when gasoline prices are at 52¢ per liter.

Alcohol fuels received another boost in January 1980 when President Carter announced a program providing guaranteed loans and other incentives to encourage growth of the alcohol fuel industry. He set a production goal of 2 billion liters of ethanol for 1981 (six times the 1979 production figure) and 8 billion liters—2% of the 1979 US automotive fuel consumption—for 1985.

Brown estimates that 800 million bushels of corn, or the equivalent, will be required to meet the 1985 goal. This amounts to 20 million tons of grain, approximately 18% of 1978 world grain exports. As more and more distilleries are built to meet the national goals, and as rising world petroleum prices improve the competitive position of distillers vis-à-vis the food industry, countries which have come to rely on the US for food will find less of it available at higher prices.

Controversy

Although Brown is alarmed by national programs which trade food for fuel production, he is actually a strong supporter of alcohol fuels. However, like many serious alcohol fuel proponents, he advocates the use of forestry products, and agricultural and municipal waste.

Although the technologies for conversion of these materials have been developed, some require refinement for wide-scale economical application. Crop distillation on the other hand, is a well-known process; the technology and even some industrial capacity are readily available. But government decisions to promote conversion of waste materials is not simply a case of pursuing safer, more conventional paths. Strong lobby efforts by powerful farm organizations, for instance, have influenced US energy policy. Farmers saw the emerging alcohol fuel industry as a potential source of steady demand at

---

World grain trade: the changing pattern

Region	1934-38	1948-52	1960	1970	1978
			(million metric tons)		
North America	+ 5	+23	+39	+56	+104
Latin America	+ 9	+ 1	0	+ 4	0
Western Europe	−24	−22	−25	−30	− 21
East. Europe & USSR	+ 5	0	0	0	− 27
Africa	+ 1	0	− 2	− 5	− 12
Asia	+ 2	− 6	−17	−37	− 53
Australia & N. Zeal.	+ 3	+ 3	+ 6	+12	+ 14

Plus sign indicates net exports; minus sign, net imports

from Lester R. Brown, "Food or Fuel: New Competition for the World's Cropland," p. 32

---

relatively stable prices for their products. Accordingly, they lobbied for government support of gasohol.

Some proponents of crop distillation counter Brown's arguments by pointing out that the fermentation by-product, distillers grain, retains the original grain's fiber and protein value. Brown does not disagree with this point. He contends, however, that there would be a loss in food value; 18 kg of starch are lost from each 25 kg bushel of corn during distillation. Since food deficient countries need this starch as well as the protein and fiber, distillers grain does not fulfill their nutritional requirements. Because livestock also need starch, Brown disputes the claim that distillers grain could substitute for livestock feed, freeing significant amounts of grain. Brown notes an additional problem with the feed market absorbing vast quantities of wet distillers grain. Due to its weight, bulk, and spoilage rate, it requires drying before being transported to feedlots or overseas. This process, according to Brown, tips the net energy balance unfavorably.

Brown's paper questions whether the benefits of alcohol fuels produced from energy crops outweigh the social costs. Is it justifiable, for instance, to trade nearly 20% of the world's exportable grains for 2% of the fuel consumed annually (1979) by US automobiles? Brown views energy crops as a short-sighted solution, especially when there are so many ways to conserve fuel. He calculates, for instance, that such a modest measure as banning automatic transmissions in the US would save more fuel by the mid-eighties than the government's alcohol program is expected to yield. Brown recommends that "sacrifices" of this nature be weighed against the worldwide social costs of diverting resources away from the production of food.

—Cissy Wallace
John Fore

References:
Brown, Lester
1978 *The Twenty-Ninth Day,* 363 pp. A Worldwatch Institute book. Norton & Co., Inc., New York.

1980 "Food or Fuel: New Competition for the World's Cropland." Worldwatch Paper #35, 43 pp. Available from the Worldwatch Institute, 1776 Massachusetts Ave. N.W., Washington, DC 20036, $2.

Chevy Purrs On 200 Proof

THERMAL EFFICIENCY *of a gasoline engine has been improved 30 to 100 percent with a modified methanol fuel system developed at the Systems Development Branch staff at SERI. Installed in the chassis of a 1980 Chevrolet Citation, the engine runs on hydrogen-rich gases produced from methanol which is dissociated on board. The year-long project shows great promise for significantly improving the thermal efficiency of conventional vehicles while greatly reducing per-mile fuel costs and pollution problems.*

Minor engine modifications make this system different in three key ways from gasoline or other methanol engines:
- the compression ratio has been increased;
- the hydrogen-rich gases have a 22 percent higher heating value than did the original methanol; and,
- the carburetor has been adjusted for ultra-lean air-fuel mixtures—achieving cleaner, more efficient combustion.

The engine has been designed to combine the best qualities of both hydrogen and methanol vehicles. On-board energy storage is in the form of methanol, which can be derived inexpensively from coal or biomass, and is easily stored. Methanol's main problems are low energy per unit volume, high aldehyde emissions, and difficult cold starts. Clean-burning hydrogen, in contrast, is difficult to store and transport. The SERI design decomposes methanol catalytically into hydrogen-rich gases in a reactor that uses engine exhaust heat.

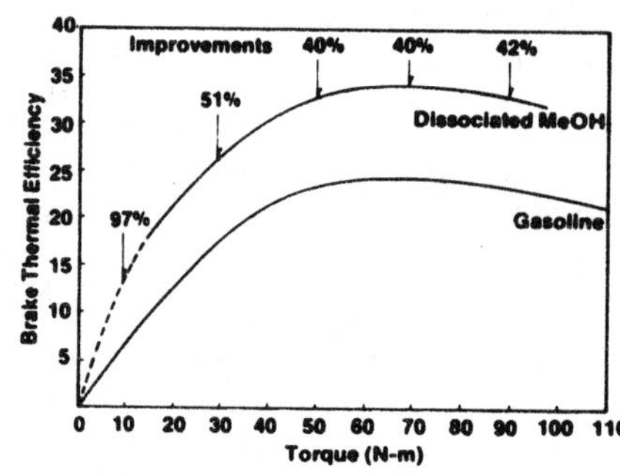

This graph shows the efficiency improvements of a dissociated methanol-powered engine over a conventional engine at 200 rpm (the equivalent of a car operating at 55 mph).

Although prior research has been done on each subsystem (see *Soft Energy Notes* 3:4:14-16), this project represents the first attempt to integrate the two. Power output, which would normally be limited by the large volume of hydrogen-rich gases in the intake stream, is being increased by inducting liquid methanol to supplement or replace these gases at wide open throttle. The results have been a peak thermal efficiency of 35 percent compared to 27 percent for gasoline engines; and a 50 percent increase in engine fuel system efficiency over that of conventional vehicles.

When other components—a catalytic reactor, superheater, gas cooler, and methanol vaporizer—are installed, the Chevy will be road-tested and run on a chassis dynamometer to determine the efficiency of the vehicle as a whole.

—*Elyse Axell*

6. Agriculture

Eating into our Pocketbooks

OF ALL THE ENERGY SOURCES used to produce heat, the last one that comes to most minds is food. What was once the nation's primary energy source has now become the most exotic. We pay an extraordinary premium for food calories delivered to us in creative ways: a home cooked meal delivers energy to its consumer at the equivalent of $2,500 per barrel oil, plus or minus quite a bit.

This figure says as much, of course, about fossil fuel's cheapness as it says about the energy cost sensitivity and inefficiency of US agriculture. In one of our principal articles, Wes Jackson of Kansas' Land Institute estimates that the food delivered to the average American's table has a daily energy cost equivalent to 24 cups of crude oil. Nationwide, the "food system"—from ground to mouth—uses 13 quads of energy, or about 16.5 percent of national use, says Jackson. Few other nations on earth could pour as much oil onto the dinner menu with such appetizing results.

We've asked two different questions in this special section on agriculture. First, is our agricultural system sustainable over the long term and, if not, what do we do about it. Are there techniques that reduce energy costs of farming and enhance long range productivity? Second, what is the impact of agricultural instability on biomass energy development. Can the two co-exist?

There isn't a clear path through this maze, but there is some good news. Respected agriculturalists have found substantial energy and environmental benefits in cereal perennials, no-till agriculture, biological pest control techniques, and so-called "organic" farming methods. The jury is still out on the economic benefits of these measures, but with higher farm energy costs, many are on their way to widespread adoption. Two percent of US cropland—3 million of the 150 million hectares under active cultivation—is planted and farmed without the plow. The percentage has doubled in the last five years, and the US Department of Agriculture estimates that 45 percent of the nation's cropland will be so farmed in 2000. According to our reports, no-till agriculture can reduce soil erosion levels by an astounding 90 percent. Cornell's David Pimentel estimates that the US loses over 4 billion tons of topsoil to wind and water erosion annually; an amount substantially greater than during the Dustbowl Years. No-till agriculture offers a key opportunity to arrest further damage.

No-till farming has side effects, including a 10-20 percent energy use reduction, increased pesticide use, and more seed use. Use of cereal perennials and biological control methods can reduce the undesirable impacts of this otherwise critical technique. Both are discussed in this section.

Studies by the Department of Agriculture, by Washing-ton University (for NSF), and others reverse the unproductive reputation of organic farming methods. These analyses find that such methods can reduce farm energy requirements by 40-50 percent without a perceptible impact on profitability and with only slightly higher labor requirements. As the USDA report comments, "many of the practices regularly followed by organic farmers... have been highly recommended to all farmers for almost 50 years by agencies of the USDA and land-grant universities for improving productivity and tilth of soils." The report acknowledges this was not always the official USDA or agricultural extension position.

Nebraska's Small Farm Energy Project is a fine example of the original goal of agricultural extension, broadened to the energy area. Rob Aiken's article notes that it was possible to reduce farm energy use by 15 percent with community cooperation and effective information dissemination. None of the big payoff improvements (no-till, organic methods, biological controls, or cereal perennials) was stressed in the first go-around.

David Pimentel's cautionary report and our analysis of an optimistic one by University of Pennsylvania scientists John Ferchak and Kendall Pye appear to square off on the potential of wringing biomass fuels from a tested agricultural system. Pimentel's report recommends that the bulk (80+ percent) of crop residues from farms and forests be returned to the soil, that intensive cultivation of fallow or marginal land for fuel crops be discouraged, and warns that mismanagement of soil resources could turn renewable biomass into an exhaustible resource. Pye and Ferchak believe that only minor disruptions of the agricultural sector would yield total gasohol use and that the year 2000 yield from all usable biomass resources would dwarf the requirements of an efficient transportation system.

These positions aren't necessarily contradictory, but they are sobering. An energy and economically efficient transport system with more cars than we have today need use no more than 30-35 billion gallons of alcohol fuel (see *Notes*, August/September 1980). Using Pye and Ferchak's estimates, this amount could probably be produced from a fifth of farm and forestry residues (6.5 billion gallons), corn starch diverted from feed (16.2 billion), spoiled crops (1.0 billion), municipal wastes (4.1 billion), and energy crops grown on six percent of the nation's commercial forest land (19.0 billion). If, as Pimentel argues, 50-60 million acres of productive cropland are intensively cultivated to produce 100 billion gallons of ethanol, the environmental damage and impact on food prices could turn an attractive idea into a tragedy. With clever minds, long range thinking, and dedication from all sides, that won't happen.

—Jim Harding

American Food: S/oil And Water

FOSSIL FUEL *calorie in for edible calorie out, the American food system is the world's most inefficient. Compared to the developing countries, which typically expend a fifth of a calorie or less for every food calorie harvested,[1] US agriculture spends two calories of fossil fuel energy for each calorie of food in the field. And the energy consumption doesn't stop there: from ground to mouth we use 9.8 times the energy value of the food we eat.[2]*

The food system in America is more energy costly than in the less developed countries because it runs mostly on oil, rather than on soil. The diet of the average American embodies, each year, the energy equivalent of 13.3 barrels of petroleum—70 percent as much as the average American car consumes.[3] But because people still outnumber cars in the US, farming uses more petroleum than any other industry.[4]

These fossil fuel inputs make land and labor highly productive, but before we congratulate ourselves for producing enough food last year to export $34.3 billion worth, we should tally up all the costs: not just the energy dollars sent abroad, but the nitrogen, the phosphorus, and the potassium shipped overseas in that food, and eroded out of the land.

Table 1. **Energy consumption within the US food system[9]**

Sector	% of US Total Energy Use	% of US Food System
On farm		
Direct Energy	1.0	6
Input Due to Chemicals and Transportation	1.9	12
Food processing	4.8	30
Distribution system	1.8	10
Commercial food service	2.8	17
Home food preparation	4.3	26
TOTAL	16.5	101

In the long run, soil and water are more important than oil. But because the American food system is increasingly dependent upon fossil energy, we ignore the declining quality and quantity of our soil and water resources. During the last 200 years, at least a third of our cropland topsoil has been lost.[5] Studies within the last decade[6,7] indicate that we are losing from 25 to 50 percent more soil per acre now than we were over 40 years ago, when Hugh Hammond Bennet—founding chief of the Soil Conservation Service—was lamenting the loss of American soils.[8]

Since that time, water has also become a serious resource problem. A full two-thirds of the ground water is pumped to irrigate our crops. About one-fourth of this withdrawal is overdraft.[9] Our agricultural system seems to have a penchant for "double mining": we now mine fossil fuel in order to mine water; long ago we became accustomed to mining soil nutrients during harvest while promoting the downward rush of remaining soil to the sea. Conserving these natural resources is at least as vital to our interests as conserving the massive fossil energy inputs that American agriculture requires.

Not counting food exports, the food system consumes 16.5 percent of the country's entire energy budget.

On-Farm: Fertilizer, Irrigation, Pesticides, Field Operations

Energy used on the US farm is only 18 percent of the total used to get food on the table. Of this initial energy expenditure, nearly half is for fertilizers, pesticides, and irrigation.

Table 2. **1976 US on-farm energy use[10]**

Component	10^6 bbl. Petroleum At the Wellhead	% of Total
Fertilizer	153	30
Field Operations	98	20
Transportation	81	16
Irrigation	67	13
Livestock	40	8
Crop Drying	29	5
Pesticides	26	5
Miscellaneous	15	3
TOTAL	509	100

For comparison, the daily US energy budget in 1979 was 39.3×10^6 barrels of petroleum-equivalent, at the wellhead.

Almost three-fourths of on-farm energy, and 88 percent of crop output is spent to provide meat. Per capita consumption is among the world's highest. To produce the 250 pounds of meat the average citizen consumes each year, we have to maintain a livestock population that weighs four times as much as the human population it feeds.[11]

Diverting much of our human-edible grain to livestock represents a major diseconomy of energy and protein in the food chain. Livestock convert food with an energy efficiency of only 5 to 15 percent. Overall, for every 5 pounds of vegetable and fish protein fed to livestock (in addition to the large forage intake), we obtain one pound of animal protein.[12]

US annual fertilizer consumption amounts to 50 million tons: 10 million tons of nitrogen and 5 million tons each of phosphorus and potassium. The production of chemical fertilizers is America's fourth largest basic manufacturing industry. (Petroleum is first, steel second, and cement third.)

The agricultural system's total energy consumption for

fertilizer "from the ground to the ground" is about 120 million barrels of petroleum equivalent at the wellhead. Of this total, 88 percent was for production, one percent was for transport of raw materials, and 11 percent was for transportation, storage, handling, and application. The major energy-consuming phases of the fertilizer system are

Table 3. Energy cost of animal protein production in the US[13]

Animal product	Feed Energy Input / Protein Energy Output	Fossil Energy Input / Protein Energy Output
Milk	30	35.9
Eggs	20	13.1
Broilers	19	22.1
Catfish	25	34.6
Pork	65	35.4
Beef (feedlot)	122	77.7
Beef (rangeland)	164	10.1
Lamb (rangeland)	188	16.2

ammonia production (71 percent of the total), product distribution (11 percent), and production of wet-process acid (8 percent for production of phosphate fertilizer). Of the energy used to produce the fertilizers, nitrogen, phosphate, and potash account for 85 percent, 11 percent, and two percent respectively.[15]

Table 4. US on-farm energy and crop input to the American diet.[14]

% of Diet	% Energy Input	% Crop Input
Meat—40%	73	88
Plant—60%	27	12

Corn, wheat (winter and spring), and hay (other than alfalfa) consume 45 percent, 17 percent, and ten percent, respectively, of the invested energy in fertilizer for agriculture.[16]

Irrigation accounted for one-third of US water use in 1974[17] and required 58.4×10^6 bbl. of petroleum-equivalent at the wellhead—or 13 percent of US agricultural energy use.[18] Ground water used on 75 percent of all pump-irrigated acreage accounted for 90 percent of irrigation energy use. Sixty-three percent of groundwater irrigation withdrawal was in regions of critical ground-water depletion.[19]

Although irrigation accounts for 13 percent of agricultural energy use nationally, that figure jumps to 25 percent[10] for a state such as Nebraska which depends heavily on irrigated crops. A center pivot sprinkler in Nebraska uses ten times the fossil fuel used for all other on-farm requirements to produce a field of corn.

The 17 western states are the most heavily dependent on irrigation. Eighty-three percent of all pump-irrigated land is in the 17 western states; they consume 93 percent of all irrigation energy. The Northern and Southern Great Plains, as defined by the USDA, consume 71.4 percent of all natural gas used for irrigation and 58 percent of all energy at the pump.[20] In the Southern Great Plains alone, the land area irrigated by natural gas totals one-fourth of all pump-irrigated land area.[20] Irrigated agricultural production in the Northern and Southern Great Plains depends mainly on the Ogallala aquifer. Groundwater mining here occurs at the rate of 14 million acre feet annually.[5] By comparison, the annual discharge of the Colorado River at Lees Ferry, Arizona is 12 million acre feet a year. If water were diverted from the Missouri at Kansas City during the pumping season to try to supply the irrigation demand above the Ogallala, it would be a dry bed at that point.

Table 5. Irrigation from pumped water in the US, 1974[20]

	Ground Water	Pumped Surface	Both	Total
Area irrigated 1000 acres	25,997	7,356	1,729	35,082
%	74	21	5	100
Quantity of Water 1000 acre ft.	51,530	14,101	3,266	68,899
%	75	21	4	100
Energy for On-Farm Pumping[21] 10⁶ bbl. petroleum equivalent at the wellhead	52.4	6.0	—	58.4
%	90	10	—	100

US farmland received 641 million pounds of active pesticide ingredients in 1974. Half were herbicides, nearly 30 percent insecticides and 15 percent fungicides.[22] Nearly as much—about 500 million pounds—was used in 1970 by government, industry, homeowners and in mosquito control programs.[23] Cotton, corn and soybeans are the big users: in 1974 they accounted for 37 percent, 30 percent and 15 percent, respectively, for the energy used to manufacture these pesticides and provide the carrier solution. Corn acreage consumed 33 percent of all herbicides and cotton land received 42 percent of all insecticides by weight in

The energy equivalent of an average American's diet embodies 13.3 barrels of petroleum, each year.

1974.[22] In 1976, it was estimated that 90 percent of the corn crop and 80 percent of the soybean acreage received pesticides. In 1971, 47 percent of the wheat acreage was treated and overall 52 percent of all cropland received pesticide chemicals.[24]

The amount of energy required for this part of our chemotherapy treatment of the countryside is really minuscule, around a twentieth of one percent of the 1980 oil consumption. Nevertheless these huge quantities of poisons could very well have a serious long term negative impact on the energy value of healthy soil, forcing technology to do what a balanced and healthy microbial flora once did.

Second to fertilizer, which requires 153 million barrels of wellhead petroleum, field operations require about 98 million barrels.[25]

Food processing & distribution

The major consumption of energy in the US food system occurs in the processing and distribution of food, which account respectively for 30 percent and 10 percent of the energy used in the total food system.[10]

S. G. Unger reported the most energy intensive food-related industries, as a whole, derived about 48 percent of their energy from natural gas, 28 percent from electricity, nine percent from coal, seven percent from residual fuel oils, and eight percent from other fuels.

Table 6. **Energy consumed in food processing and distribution in 1970.[26]**

Component	10^6 bbl. Petroleum, wellhead	% of Total
Food processing	276	36.7
Food processing machinery	5	0.7
Paper packaging	34	4.5
Glass containers	42	5.6
Steel cans & aluminum	109	14.5
Transport (fuel)	221	29.3
Trucks & trailers (manufacture)	66	8.8
TOTAL	753	100.0

Soft agriculture vs. food consumerism

The American food system needs to be re-organized, quite literally from the ground up. Changing farm technique is not the most productive way to increase energy efficiency, since this captures a relatively small share of the total energy expenditure. Priorities should be given to revamping the largest areas of end-use: the food processing industry, the ultimate consumers who prepare food in the home, and the system of marketing and distribution.

The food processers, in consuming 30 percent of the total energy invested in food, go through countless expensive

Table 7. **Energy used in 1973 by various energy-intensive food-related industries[27]**

Sector	Energy used 10^6 bbl. petroleum, wellhead
Meat packing	22.4
Prepared animal feeds	19.5
Wet corn milling	18.9
Fluid milk	17.7
Beet sugar processing	16.8
Bread & related products	15.5
Frozen fruits & vegetables	14.1
Cane sugar refining	10.0
Sausage & other processed meat	5.7
Animal & marine fats & oils	5.6
Manufactured ice	1.0

motions of taking substances out, then putting them back in again—along with other additives. It is time we recognized this wasteful effort for what it is: packaging and consumerism, which is itself the heart of the ecological problem. This is one reason the food co-op movement, so small and presently so powerless, is also so radical and so necessary. If the energy conservation values which motivate people to take on the food processing industry can carry over into the kitchen, it might be possible to chip away at the 26 percent energy consumption that occurs in the household sector. These same values would inspire food preferences and purchases that render some of the long distance transport of food unnecessary.

Soft Agriculture saves more than energy. It could also help to conserve our shrinking endowment of soil and water resources, without which agriculture itself is impossible. Our present form of plow agriculture presents our number one ecological problem (aside from the threat of nuclear disaster)—a planetary disease which has reached epidemic proportions. It will not be cured until we resolve to meet our food needs without losing and poisoning our soil.

—*Wes Jackson and Marty Bender*
The Land Institute
Route #3, Salina, Kansas

References and notes

1 Steinhart, John and Carol Steinhart. 1974. "Energy Use in the U.S. Food System." *Science* 184: 307-316.

2 Cox, George and Michael Atkins. 1979. *Agricultural Ecology*. W. H. Freeman and Co. San Francisco. p. 618.

3 Federal Energy Administration. 1975. "Energy Consumption in the Food System." Rep. No. 13392-007-001, Booze, Allen, and Hamilton, Inc.

4 Committee on Agriculture, House of Representatives. (92nd Congress, 1971). p. 20.

5 Pimentel, David, et al. 1976. "Land Degradation: Effects on Food and Energy Resources." *Science* 194: 150.

6 Hargrove, T. R. 1972. *Iowa Agricultural Home Economic Experiment Station Special Report No. 69*. This report sets soil loss as 12 tons per acre.

7 General Accounting Office. 1977. "To Protect Tomorrow's Food Supply, Soil Conservation Needs Priority Attention," CED-77-30. A calculation from the table on p. 5 suggests the average loss is closer to 17 tons than to 15 tons/acre.

8 Bennett, H. H., 1939. *Soil Conservation*, McGraw Hill, New York.

9 U.S. Water Resources Council. 1978. *The Nation's Water Resources: 1975-2000*. Vol. 1: Summary.

10 Bethal, James S. and Martin A. Massengale. 1978. *Renewable Resource Management for Forestry and Agriculture*. U. of Washington Press. Seattle.

11 Pimentel, David. et. al. 1975. "Energy and Land Constraints in Food Protein Production." *Science* 190: 756.

12 Pimentel, David. et. al. 1975. p. 756.

13 Pimentel, David. et al. 1975. p. 757.

14 Heichel, G. H. and C. R. Frink. 1975. "Anticipating the Energy Needs of American Agriculture." *Journal of Soil & Water Conservation* 30: 4853.

15 David, C. H. and G. M. Blouin. 1977. "Fertilizers and Plant Nutrients." *Agriculture and Energy: Proceedings of a Conference*. William Lockeretz ed. Washington University. St. Louis. June 17-19, 1976. Academic Press. New York. p. 315-331.

16 USDA. 1976. *Energy and U.S. Agriculture: 1974 Data Base*. Volume 1, p. 14.

17 Zaporozec, Alexander. 1979. "Changing patterns of Ground-Water Use in The United States." *Ground Water*. 17(2): 200.

18 USDA. 1976. *Energy and U.S. Agriculture: 1974 Data Base*. Volume 1. p. 7.

19 USDA. 1980. *Soil and Water Resources Conservation Act*. 1980. Appraisal. Review Draft. Part 2. p. 3-135.

20 Sloggett, Gordon. 1977. "Energy Used for Pumping Irrigation Water in the United States, 1974." *Agriculture and Energy: Proceedings of a Conference*. William Lockeretz ed. Academic Press. New York. p. 113-130.

21 Energy used by irrigation organizations to deliver water to farms is not included. The Bureau of Reclamation used an estimated 2.75×10^6 bbl. petroleum at the wellhead to pump irrigation water in 1974.

22 USDA. 1976. *Energy and U.S. Agriculture: 1974 Data Base*. Volume 1. p. 16.

23 Pimentel, David. 1973. "Realities of a Pesticide Ban." *Environment* 15(2): 19.

24 Cook, L. 1975. "Using More Pesticides Than Ever." *Chicago Sun-Times*. Sept. 16. p. 58.

25 Jackson, Wes and Marty Bender, 1980. Unpublished manuscript prepared for Office of Technology Assessment, U.S. Congress.

26 Steinhart, John and Carol Steinhart. 1974. "Energy Use in the U.S. Food System." *Science* 184: 307-316.

27 Unger, S. G. 1975. "Energy Utilization in the Leading Energy-Consuming Food Processing Industries." *Food Technology*. December. pp. 33-45.

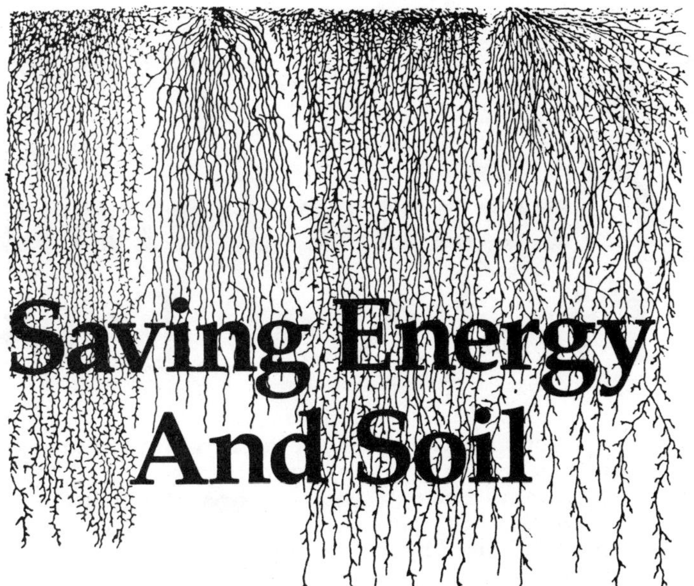

Saving Energy And Soil

T HE TOP TEN US CROPS, grown on 316 million acres, not only feed the American populace, but produce enough surplus for most of the $40 billion of agricultural exports in 1980. Enough grain remains for the most massive welfare program in history—our cattle and pig welfare program. American agriculture suffers from success, a success which blinds us to the casualties which follow from such high production.

A young continent and lots of oil, more than our ingenuity, has allowed us to evolve what must rank as about the least sophisticated form of agriculture in the world. In the name of high production we have consumed vast energy resources, and prematurely aged our young continent, with its fortuitously located soils and generous rains. These 316 million acres alone drink, each year, the energy equivalent of about 284 million wellhead barrels of petroleum. Fertilizer, pesticides, irrigation, traction, and equipment manufacture all go into this 0.9 barrel/acre. Accountants keep tally of this energy cost as a regular part of bookkeeping; unfortunately, though, the most important half of the balance sheet is not considered. Who considers the energy value of the nitrogen, phosphorus, and potassium washed seaward with the eroding soil each year?

There are four principle estimates of how much soil we lose and the energy value of replacing this loss. Last year the Soil Conservation Service set the average annual loss at five tons/acre/year. Two years ago, the Service's estimate was nine tons. In 1972, some Iowa State University researchers said it was 12, and in 1977 the General Accounting Office gave the US Congress an estimate of 15-17 tons/acre/year. The energy value, at five tons loss, is the equivalent of 99.5 million wellhead barrels, or nearly 0.6 quads (10^{15} Btu). At nine tons, the energy cost is 179 million barrels; at 12 tons, 239 million, and at 15 tons/acre/year, the replacement cost comes to 299 million barrels, or 1.7 quads. If we accept a figure between the Iowa State estimate and that of the General Accounting Office, then the replacement value of the soil nutrients lost each year per acre is the energy equivalent of 90 percent of a wellhead barrel—equal to the energy expended per acre in growing the crops.

This cost accounting doesn't consider the value of the organic matter and other elements associated with humus that are lost each year. This debt will be felt by future generations as they move from oil agriculture back to soil. It is a debt unlikely to be retired until we establish a radically different form of food production. Our bet is that in the oil-scarce future, the number of *soil* accountants at any one time will far exceed the total number of energy accountants that have ever existed.

Perennial Polyculture

The method of food production proposed by the Land Institute is a bio-technical fix for the area now under till agriculture. This involves breeding and developing herbaceous, perennial, seed-producing polycultures as substitutes for herbaceous, annual monocultures on sloping or rolling land. These plants would be developed by wide crosses and selection of certain wild grasses, legumes, members of the sunflower family, and, in all likelihood, other plant families as well. Such an agriculture, based on the principles of a stable ecosystem, would be high yielding in seed production, and would drastically reduce the energy cost for pre-planting, cultivation, and irrigation, as well as for fertilizer and pesticide treatment. There would still be costs for harvest, the occasional replanting, some fertilizer application, and a limited number of pesticides.

We have estimated the total direct energy savings from such an agriculture to be 175-195 million wellhead barrels of oil per year. There would also be savings that today's accountants pay little attention to: 99.5 to 299 million wellhead barrels to replace the nitrogen, phosphorus, and potassium that conventional plow agriculture loses each year. Combining the energy savings of food production with the energy value of the soil saved, we have a total energy value savings of 242-497 million wellhead barrels of oil equivalent per year, or 1.4 to 2.8 quads. This represents a financial savings to the country of $8.5 to 17.4 billion (1980 $) each year.

Better to Wear Oil Than to Burn it

Toxic wastes in the environment cause a great deal of legitimate concern, but industrial pollution does not compare to the ecological problem of soil loss due to till agriculture. The corn plant, as a product of technology, has in a certain sense, destroyed more options for the future than the automobile, which only consumes oil. The cotton plant is heavily dependent on pesticides and fertilizer; because of the plant's nature and the way it is tilled, soil loss is severe. Perhaps we should recognize cotton as a serious liability and boycott it in favor of polyester until we develop one or more sensible methods for growing it. We might be better off wearing oil than burning it for conventional cotton production.

A complete rethinking of till agriculture is in order. Though civilization has believed otherwise, the plowshare has destroyed more options for the future than the sword. We at the Land Institute believe that we must start immediately to develop perennial polycultures of seed producers. As we do, we can then begin the important task of beating our plowshares into tools which sustain rather than erode.

—*Wes Jackson and Marty Bender*
The Land Institute

Wes Jackson, author of New Roots for Agriculture, *and Marty Bender are with the Land Institute, a non-profit educational research organization devoted to the study of sustainable alternatives in agriculture, energy, waste-management, and shelter.*

When he was US Secretary of Agriculture, Earl Butz said that before the US could talk about organic farming, we would have to decide which 50 million Americans were going to starve. Butz was echoing the opinion often expressed by agribusiness and the chemical industry: organic farming doesn't work.

JANICE FILLIP,
Not Man Apart, September 1980

Better Living Without Chemistry

Organic Farms Find New Fans

Today's Department of Agriculture (USDA) and the National Science Foundation do not agree with Butz. In fact, it might well be said that *conventional* farming doesn't work.

USDA cites the following as adverse to a "stable, sustainable, and profitable agricultural system": a heavy reliance on increasingly costly chemical fertilizers and energy inputs; steady decline in soil productivity and tilth from excessive soil erosion and loss of soil organic matter; environmental degradation, erosion, and sedimentation from water pollution by agricultural chemicals; hazards to human and animal health and to food safety from heavy use of pesticides; and the demise of the family farm and localized marketing systems.

Farmers have economic reasons for turning to organic methods. "Their main motivation is a concern for the soil and soil life," explains Garth Youngberg, member of the USDA Study Team on Organic Farming. Chemical farming is killing the soil, compacting it so that water runs off the surface carrying topsoil with it. "Farmers say that unless the soil is healthy, the plants can't be healthy and the animals and the humans can't be healthy."

Organic farms number 30,000 in the US, ranging up to 1,500 acres in size. Organic farming trades the use of synthetically compounded fertilizers, pesticides, growth regulators, and livestock feed additives for techniques which cut conventional farms' energy use in half. Organic farming systems rely upon crop rotations, crop residues, animal manures, legumes, green manures, off-farm organic wastes, mechanical cultivation, mineral-bearing rocks, and biological pest control to maintain soil productivity and tilth, to supply plant nutrients, and to control insects, weeds, and other pests.

In the midwest corn fields, William Lockeretz has shown an organic farm's output per unit energy is twice that of a conventional farm. G. M. Berardi's study of northeast wheat fields determined conventional farms may yield 29 percent more than organic, but they use 48 percent more energy.

The first comparison of organic and conventional farms has just been completed for the National Science Foundation (NSF) by scientists from Washington University in St. Louis. The five-year study of 51 farms in the midwest corn belt found that farmers who used organic procedures consumed only about 40 percent of the energy required on conventional farms using chemical fertilizers and pesticides.

The energy savings on organic farms mainly come from the fertilizers they no longer employ, nitrogen in particular. According to Berardi's study, 3.7 percent of the organic farm's energy is for fertilizer, while conventional fertilization requires 28.9 percent.

Organic farmers use a variety of fertilization methods: manure is the most widely used. David Pimentel estimates that the manure from three cows can adequately provide 125 kg nitrogen, 35 kg phosphorous, and 67 kg potassium per year—enough to fertilize one hectare of corn. Plowing under legumes, such as sweet clover, will add about 150 pounds of nitrogen/ha as the bacteria fix nitrogen from the atmosphere into the usable form of nitrate.

Additional organic wastes which would otherwise be thrown away can be used as fertilizer, as well. A cannery's crab trash is used by the Coolidge Farm, an experimental farm in Topsfield, Massachusetts, for nitrogen. The dust of basalt rock from road stone manufacturing is used for potassium and phosphorous.

Comparing 14 pairs of farms, labor requirements for crop production have been found to be only slightly higher for organic farms (3.3 hours/acre) than for conventional farms (3.2 hours/acre), according to the USDA report. Organic farms can also be more labor-intensive depending on the requirements for manual pest control techniques.

With decreased energy input costs, an organic farmer can produce a smaller crop for the same profit. The NSF study showed that although the cash value of crops was lower on organic farms, the organic farmers' expenses were 36 percent less than those of conventional farms. That resulted in a net dollar return only two percent lower than that of conventional farms using chemical fertilizers and pesticides. Consequently, when labor input is expressed per dollar value of crop output, labor requirements appear much higher on the organic farm. USDA points out: "The use of labor was much higher for the organic group because the value of crop output per acre was lower. The organic farmers spent 19.8 hours per $1,000 of crop output compared with 17.8 hours for conventional farmers."

"Contrary to popular belief," the USDA report points out, "many of the practices regularly followed by organic farmers are the best management practices that have been highly recommended to all farmers for almost 50 years by agencies of the USDA and land-grant universities for improving productivity and tilth of soils." As fuel prices rise, organic techniques will offer farmers more attractive profits. As other institutions, especially the local extension agencies, follow the USDA's lead, American farming will become more organic and more energy efficient.

—*Matt Biers and Elyse Axell*

References:
Berardi, G. M.
1977 "Organic & Conventional Wheat Production: Examination of Energy & Economics." Dept. of Natural Resources, Cornell University, Ithaca, New York 14853.
Pimentel, David
1974 "Workshop on Research Methodologies for Studies of Energy, Food, Man and Environment." Cornell University Center for Environmental Quality Management, Ithaca, New York 14853, June 18-20.

United States Department of Agriculture
1980 *Report & Recommendations on Organic Farming.* July, USDA, Washington DC.

The Toll of No-Till

WITH THE ENCOURAGEMENT OF THE US, the exploding world population has come to depend heavily on the stability of US agricultural output. Some forecasters predict that by the year 2000, 100 percent of all the grain available for export will come from the croplands of the United States and Canada. Yet the agricultural system the world has been urged to emulate, the system that has been the marvel of our time, may not be sustainable. We are perilously close to the brink and running flat out. It would be well if we took off our blindfolds.

JOAN GUSSOW, *Progress As If Survival Mattered*

The problem is soil erosion, the plowman's folly. The US has lost more than a third of its topsoil in its first 200 years; annual loss averages 12 tons per cropland acre and is likely to increase, writes Joan Gussow. "Yet just as much of the land that blew away in the thirties ought never to have been plowed, since it was incapable of production except when climatic conditions were ideal, so much of the land recently pushed back into production in order to pay for our imported oil is blowing away again..."

More than 5,000 years ago, Middle Eastern farmers devised the plow—a labor-saving invention for controlling weeds. Today, chemists work to replace the plow with herbicides for a planting method called "no-till." No-till farmers plant without their plow or with just enough tillage to place the seeds and cover them with soil. No-till can actually out-produce conventional till for some crops on some soils; it requires smaller capital investments in equipment and tractor power; it lessens farm machinery fuel requirements; and it can reduce soil erosion dramatically.

Soil losses from water runoff can be "virtually eliminated," according to Ronald E. Phillips and others of the University of Kentucky's College of Agriculture. Harrold and Edwards found Ohio soil losses in 1964 from a conventional tillage watershed were 6,477 kg/ha, while only 134 kg/ha were lost on the watershed with five years of continuous no-tillage production. Corn yields were 7,091 and 7,909 kg/ha, respectively.

Fewer trips across the field also mean energy and labor savings. "For example, a conventional-tillage system could require 80 hours per acre and 315 gallons of fuel per 100 acres while a no-till system might require only 25 hours and 40 gallons," report Giere, Johnson, and Perkins. Darrel Hager, county agricultural extension agent in Kansas, says: "When you go past 60 cents a gallon for diesel fuel, you begin to try to cut down on the number of trips you make across a field."

These savings are partly offset by an increased need for herbicides and insecticides derived from petrochemicals at a high energy cost. "About 50 percent more pesticides are used for corn production by the no-tillage method than by conventional tillage," Phillips estimates.

Herbicides are the Hallmarks

America planted nearly three million hectares, mostly corn and soybeans, in 1977 using no-tillage cultivation. That equalled two percent of total cropped acreage and *twice* the no-till acreage of five years before. The US Department of Agriculture (USDA) has estimated that 62 million hectares or 45 percent of the nation's total cropland will be grown with no-till systems by 2000. Sixty-five percent of corn, soybeans, wheat, oats, barley, rye, and sorghum could be under no-till by 2000 and 78 percent by 2010.

Reduced plowing and increased reliance on pesticides are the hallmarks of conservation tillage—a spectrum of methods including reduced-till, minimum-till, and no-till. US farmers planted 30 million hectares using all conservation tillage methods in 1979, the No-Till Farmers Association of Milwaukee, Wisconsin estimates.

By leaving a mulch of the previous crop on the field and not churning up topsoil, no-till reduces soil erosion to "almost zero," Phillips says, without reducing crop yields. McGregor *et al.* found that in Mississippi, soil erosion was reduced 90 percent from 17.5 metric tonnes per hectare to about 1.8 tonnes/ha with no-till.

Annual energy savings, through reduced machinery manufacture and the elimination of plowing and disking, reach an equivalent of "about 36.6 liters of diesel fuel per hectare," Phillips reports. However, he cancels 2.65 liters, representing the fuel equivalent of the additional pesticides recommended to maintain crop yields. Extra seed, also recommended, offsets another liter. Phillips places the annual energy savings for a hectare of corn under no-till at about seven percent; for soybeans, which do not require nitrogen, at 18 percent.

Gambling with Soil

No-till leaves little room for error. The farmer takes a greater economic risk: "He needs to command greater technical skills to use no-till," Giere warns, "and he usually cannot go back over a field with a cultivating device to control weed problems not handled by the herbicides." Different soil types, moreover, require different techniques; no-till methods are not yet suited to all soil types. Well-drained soil does best, while poorly drained soil under the accumulation of organic matter on the soil surface won't hold necessary nitrogen.

In addition, no-till soil needs extra nitrogen. The lower evaporation rates from the mulched soil mean nitrates are not drawn up to the corp roots. Phillips also recommends lime applications to neutralize the soil surface which is likely to become acidic under no-tillage systems.

Without plowing, pests such as rodents, disease organisms, and insects are left to breed on previous crop debris, encouraging increased use of rodenticides and pesticides. "Probably the most serious disadvantage of a continual mulch cover," Giere claims, "is its enhancement of populations of some plant pests, especially insects and diseases." Plowing would have killed the pests or their host—the plant residue; without it, pests find a favorable habitat and can overwinter into the next growing season. With no-till, the simultaneous attack of a crop by several pests is more common than in conventional systems, Giere says.

When coupled with the modern system of large monocultures, no-till risks the outbreak of widespread plant diseases. "Special attention," Giere warns, "needs to be given to the possibility that no-till agriculture, if widely adopted, may create conditions which will make our agroecosystem even more vulnerable to devastating pest outbreaks than they are now."

Know Your Enemy—(Weeds)

Controlling weeds without a plow means big sales for the chemical companies. Monsanto Company literature recommends increased dosages of atrazine to control weeds with no-till practices. Chevron Chemical Company recommends paraquat, the other principal herbicide used on US no-till farms. "In no-tillage farming, herbicides are used to replace horsepower," the company literature says; "the one most extensively used for this purpose today is Ortho Paraquat CL, a product of Chevron Chemical Company."

When you switch to no-tillage farming, you switch to an often total reliance on chemicals for control of unwanted vegetation.

The extra pesticides cost the farmer, farm worker, and the public as they add to our already contaminated environment. Paraquat is extremely toxic: it is potentially fatal, requiring proper instruction, proper clothing, and well-enforced occupational health standards. Whether paraquat can cause cancer or genetic mutations is still under debate. Giere reports: "One sensitive test indicated that paraquat could induce genetic changes at concentration levels similar to those which would occur from residues of the material left on harvested crops. Other tests based on different detection methods have failed to find mutagenic activity."

Robert Rice, of Gibbs Soell, a public relations agency which handles Chevron's agricultural accounts, says: "I haven't heard anything on mutagenic or carcinogenic" aspects of paraquat. "Most complaints are about its toxicity: paraquat is a poison."

Some studies indicate atrazine, too, may be mutagenic; not all investigators agree. "Furthermore," Giere warns, "atrazine can be transformed chemically under conditions that are found in the human stomach into its N-nitroso derivative. N-nitroso compounds as a class are under strong suspicion as carcinogens. The present contamination of many Mid-western drinking water supplies with atrazine is thus a cause for concern."

Finally, paraquat may be a self-limiting chemical. Rice says paraquat kills weeds on contact and then is deactivated by binding to clay particles in the soil. It binds so tightly that only leaching the soil with strong acids will remove it. Giere adds, the time required for all bind sites to become filled "varies from 20 to 10,000 years depending on soil

TOXICOLOGICAL PROPERTIES

Fish and Wildlife:
Alachlor possesses a low degree of toxicity to fish and wildlife.
Acute Toxicity:
Rat acute oral LD50—1800 mg/kg
Rabbit acute dermal LD50—5000 mg/kg
Irritation:
Causes eye burns and skin irritation. May cause allergic skin reaction.
Toxicity Category:
Category 1 pesticide. Label carried DANGER! statement based upon eye irritation as required by the Environmental Protection Agency.
Symptoms of Poisoning
No unique symptoms.

composition: the chemical could then no longer be used; no-till technologies based on paraquat may therefore have an inherent tendency towards obsolescence." Rice says, "Even 20 years is a long time, and paraquat is not something you use every year." Chevron literature advises, however: "When you switch to *no-tillage farming,* you switch to an often total reliance on chemicals for control of unwanted vegetation." The brochure explains combining, not alternating, paraquat with other herbicides: "So a combination of both Paraquat and one of more residual herbicides is needed for complete, season-long weed control."

Insects Controlling Insects

Biogenesis, a Texas corporation, which sells biological crop pest control services, characterizes insecticides as abused chemicals, "detrimental to our ecosystems," and "economically disastrous." Company president Malcolm Maedgen lists these environmental consequences:
● some insects have developed a high level of tolerance to these toxic compounds, resulting in more frequent and

heavier applications of poisons to eliminate the pest;
- soils accumulate non-biodegradable elements harmful to people;
- plants absorb these toxic elements through the root system and foliage; and,
- diseases such as cancer have been linked to consumption of these toxic substances through foods.

Pesticide-induced pests abound all over the world, according to Carl B. Huffaker, of University of California, Berkeley's Entomological Sciences Department.

Pesticide-resistant insects are not only increasing in numbers of species, but also intensifying within the geographic ranges of the resistant pests, according to Richard Garcia and Donald Dahlsten, entomologists at the

Soil erosion

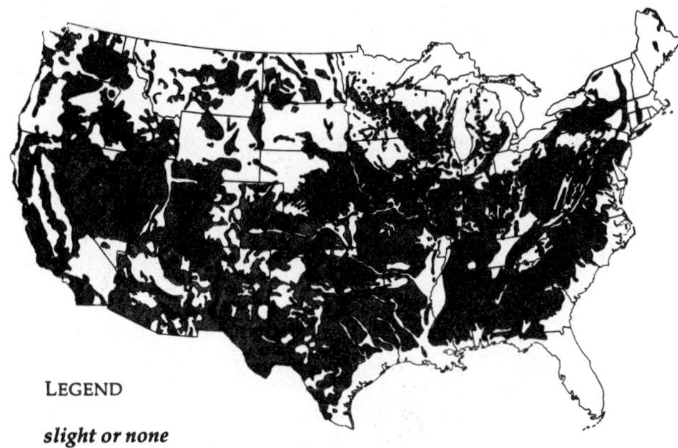

LEGEND

slight or none

■ *moderate*
(25 to 75% of topsoil lost; may have some gullies.)

map courtesy US Soil Conservation Service

■ *severe*
(more than 75% of topsoil lost, may have numerous or deep gullies, includes severe geological erosion in parts of low rainfall areas.)

Many small areas could not be shown at this scale.

Based on data from 1934 reconnaissance erosion survey of the United States and other soil conservation surveys by the Soil Conservation Service.

University of California, Berkeley. One California study, they report, found that of 25 pests, each costing farmers a million dollars or more, 24 are either resistant, insecticide-aggravated, and/or controlling them with chemicals is linked to the outbreak of *other* pests.

But in Texas, cotton yields have been increased 30 percent, costs per pounds of lint decreased 29 percent, energy use per pound cut 48 percent, and profits have jumped from $12.40 to $104.97 per acre—a 746 percent increase—all with a 27 percent *reduction* in pesticide use. Natural enemies halted the cotton boll weevil: the method is called *biological control*.

Biological control introduces parasites and predators from a pest's former native home area to a region the pest has invaded, and includes sound crop management—crop rotation, shelter belts, contour plowing, and terracing. Many of the experiments in introducing natural enemies to crop pests have taken place in developing countries.

Huffaker claims some 400 cases "of substantial or complete success" with insect control of insects. Likewise, he adds, there have been some 39 cases of substantial to complete success with insect predators for weed control around the globe in distinct major geographic areas.

Meanwhile, in California's Imperial Valley, farmers stubbornly apply insecticides to eradicate the budworm and pink bollworm. As many as 20 or even 30 insecticide treatments per season have failed to control the pests, Huffaker says. While biological control has "literally rescued" cotton production from budworms, bollworms, and economic disaster in Peru and Texas, in California, no solution has been developed. Huffaker asks: "When will we ever learn?"

Pest Control Without Poisons

As tractor fuel becomes more expensive and less available, American farmers are finding the diesel savings of no-till attractive. No-till offers fewer trips across the fields; yields on some switched-over farms will actually improve; and soil erosion becomes minimal. But, a quick switch to more chemical pesticides in place of the plow may not be the answer.

The environment and public health aren't holding up under the annual dumping of pesticides onto croplands. Concentration of toxic substances through the food chain, massive fish kills, animal poisonings and reproductive failures are all "legacies of the pesticide syndrome," Garcia and Dahlsten claim. "The application of billions of pounds of pesticides during the last 30 years, some so persistent they last for decades, has left its mark throughout the world." These chemicals poison an estimated 45,000 farmworkers and other handlers each year. Of the 1,500 active ingredients in registered pesticides, one-fourth are probably carcinogenic, according to the US Environmental Protection Agency.

Apart from environmental and health risks, others will caution against those pesticide-resistant insects. Pesticides that are not effective after 20 or 30 applications do not make economic sense. *Rather* than their impacts on human health and the environment, pests' genetic resistance "probably stands as the major obstacle to the use of these chemicals for pest control," Garcia and Dahlsten conclude.

A more sensible strategy might be to combine the soil and energy conserving principles of no-till with integrated pest management—a method of pest control that substitutes biological agents for most uses of pesticides.

—Elyse Axell

References:

Garcia, Richard and Donald L. Dahlsten
1980 "Pest Control Strategies." *The New Environmental Handbook*, Garrett De Bell, ed., Friends of the Earth Books, 124 Spear Street, San Francisco, California 94105, $5.95, 350 pp.

Giere, John P., Keith M. Johnson and John H. Perkins
1980 "A Closer Look at No-Till Farming." *Environment*, vol. 22, no. 6, July/August, p. 15-41.

Huffaker, C. B.
1980 "Use of Predators and Parasitoids in Biological Control." From: *Linking Research to Crop Production*, edited by Richard C. Staples and Ronald J. Kuhr, Plenum Publishing Corporation.

McGregor, K. C., I. D. Greer, G. E. Gurley
1975 *Am. Soc. Agric. Eng. Trans.*, vol. 18, p. 918.

Phillips, Ronald E., Robert L. Blevins, Grant W. Thomas, Wilbur W. Frye, Shirley H. Phillips
1980 "No-Tillage Agriculture." *Science*, vol. 208, 6 June, p. 1108-1113. Single issues of *Science* are available for $2 from American Association for the Advancement of Science, 1515 Massachusetts Avenue, NW, Washington, DC 20005.

7. Industry

During the post-war period, the energy-efficiency of buildings, appliances and cars has been stagnant or declining, while energy productivity in industry (the amount of energy it takes to produce one unit of value-added) has steadily increased, even when oil was declining in price. This is not surprising. Unlike private consumers, industries have direct control over the development and design of substantial parts of their energy-consuming systems; many have R&D departments that can generate innovation and information, and they are organized to minimize the costs of their operations.

These characteristics have led some analysts to believe that the potential for further technical improvement is limited in the industrial sector. In fact, the potential is dramatic, and expensive oil will lead to great changes in this sector.

To grasp the full potential of industrial energy efficiency, we must take a broader perspective. Much of the energy used in industry is embedded in materials that can be recycled, remanufactured, reconditioned and reused. Such recycled products compete with the same products produced from increasingly scarce virgin materials. Secondly, high energy prices promote growth of hi-tech industries that use as little as a tenth the energy used by basic material industries to generate a unit of GNP. Energy-intensive industries such as steel, basic chemicals and cement will contribute less to total industrial production over time. A reduction in steel production has a major effect on domestic energy use, but a minor effect on industrial output. (This also reduces the energy requirements for carrying industrial freight to and fro.)

A third factor lies in structural changes within the ''basket'' of goods produced by each industry group. This effect has been particularly pronounced in the chemical industry where high-value, low-energy products such as pharmaceuticals and pesticides, have gained in importance over basic chemicals.

Technical fixes, in turn, are of two categories: the consistent application of insulation, waste heat recovery, cascading and cogeneration, to standard equipment such as boilers, furnaces and other heat processes and the careful matching of electric drives to varying loads in the operation of pumps, compressors, fans and other equipment. Then there are new ways to produce particular materials, such as the Alcoa-process for aluminum or the replacement of

acetylene-based synthesis by ethylene-based technologies. These new processes are not being developed solely to reduce energy costs: frequently, energy gains are secondary to other cost advantages.

Despite the fact that industry is by far the most well-organized energy consumer, it has its handicaps. Even industrial managers display economically irrational behavior. They may expect conservation investments to pay for themselves in one or two years while allowing five years or more for other investments. Until recently, fuel costs typically accounted for five percent of the cost of industrial production, and were tax deductible. Even when this figure doubled, energy was not given prompt priority by many industrial managers.

Another handicap is utilities' historical hostility toward onsite congeneration as a threat to their own business.

Much major plant—steelmaking furnaces are an example—will have to be scrapped in order to install more efficient technologies; oftentimes, retrofitting of energy-saving devices is physically impossible. In other instances, retrofitting can be both easy and very cost-effective, but if the power requirements of individual process components have never been measured, waste continues unnoticed.

Recycling has been treated by industry much as independent power generation has been viewed by utilities. With capacity standing idle in production plants, little cooperation can be expected from industry for a thorough materials-efficiency program.

Taken together, the potential for increased efficiency in direct industrial energy use and in the materials cycle is comparable to that in the other sectors of energy end-use. Energy prices alone are unlikely to suffice as an incentive for this potential to be tapped. Implementation could be boosted by subsidies for scrapping and rebuilding obsolete equipment, by promoting an electricity market for cogeneration (as done by the 1978 US PURPA Act), and by making the principle of returnable-bottle laws apply to industrial products as a whole.

Not surprisingly, industry has been skeptical of renewables' potential. By contrast, almost 30 percent of US process heat requirements fall into the temperature range which can be reached with simple concentrating solar collectors (below 285° C). But solar can preheat liquids that can be boosted to higher temperatures with conventional fuels; this raises its potential contribution to about half of all process heat. Cost barriers are being surmounted for even higher-temperature collectors as a larger market develops and lighter, mass produced materials are used. Biomass can also be a significant industrial fuel, either in direct combustion as a solid fuel or after gasification. As a transitional technology, fluidized-bed combustion of coal is particularly relevant, as it eliminates 90 percent of sulfur dioxide emissions and permits the use of a variety of fuels in a single boiler.

☐ Energy and Industry

Beyond the Oil Barrel

YOUNGSTOWN, OHIO WAS ALWAYS A STEEL TOWN, until the last of the big plants there folded, leaving long lines of unemployed. And when Kaiser Steel threatened to shut down its troubled Fontana steel plant, the Los Angeles city government scrambled to keep the doors open, and save 5,500 jobs. Throughout the United States, steel is in trouble; but it is not alone. All of the energy-intensive industries with large capital requirements find it hard to keep the doors open, and as their sales fall and export markets close, they pull employment and the rest of the economy down with them.

The problem is not just that energy is now so expensive, but that it once was so cheap. Awash in underpriced oil and mesmerized by promises of a cheap energy Eden, US industry sought for several decades to reduce labor costs through energy-intensive capital investments. The real price of industrial electricity fell for thirty five consecutive years, through World War II, Korea, and Vietnam.

The energy cost escalation of the last five years is, of course, only one of the culprits in the decline of US industry. Yet the new efforts to "reindustrialize" or "revitalize" American industry will be little more than empty campaign slogans if they are not married to an intelligent energy policy.

The least expensive new "source" of energy is good housekeeping, of which industry on the whole has been a stellar performer over the past five years. Dr. Vince Taylor of the Union of Concerned Scientists states that, relative to 1973 energy output levels, industry is saving much more energy than the entire output of the Alaska pipeline. Nearly two-thirds of the "unexpected conservation" between 1973-78 (7.6 quads, had ratios of energy/$ held constant in each sector) came from improved industrial efficiency.

Many companies and some industries have active housekeeping programs to improve the thermal integrity of their physical plants and production processes. Dow Chemical Company has reduced its energy use per pound of product by 40 percent over the past ten years. In 1973, IBM set out to cut its energy consumption by 10 percent in 34 major facilities across the nation. It actually reduced its consumption by over 50 percent, despite a major increase in sales.

Many other companies or industries are less foresightful, well-informed, or able to invest. George Hatsopoulos of Thermo Electron Company, an industrial conservation and cogeneration firm, notes that many firms are unwilling to accept energy saving investments that take longer than 1-2 years to pay off.

The difference between this payoff period and those undertaken by regulated utilities or the federal government is enormous. Investments by utilities in coal or nuclear power plants operate under payback constraints of 25 to 40 years, using discount rates as low as 10 percent to justify the use of ratepayer funds.

Even with widespread industrial housekeeping steps, overall consumption of energy by industry is expected to increase. The Department of Energy, in its 1979 Annual Report to Congress, forecasts that 80 percent of the growth in national end-use energy consumption between 1978 and 2010 will occur in the industrial sector. Most of these increases would be, according to DOE, in increased coal use and larger purchases of utility electricity. Over 60 percent of the projected increases in utility generation are attributable to industrial use. The potential exists, however, for industry to use far less energy than the government forecasts, and to generate, rather than purchase, large amounts of electricity for the utility grid.

Improved use of waste heat in industry closes one important loop in the system. Many industrial activities involve the use of process steam, substantial fractions of which are discharged to the environment as waste. The DOE forecasts an increase in industrial electricity generation, but this increase, from the current 15 gigawatts (plus 9 gigawatts under construction) to 44 gigawatts in 2010 is extraordinarily pessimistic.

Increased recycling of waste materials closes a second major loop. Reuse of aluminum, copper, and some plastics, can save 95 percent of the energy required to produce virgin materials. Historically, materials costs prompted industrial reuse of steel and other primary products; now the embodied energy in scrap may represent a major resource. A smarter materials policy, in which materials are matched to structural or functional needs as we have advocated in energy, holds still greater potential rewards. All we can point to on this score is our article in volume 1, number 2, reviewing the work of Ayres and Narkus-Kramer on materials efficiency. These authors point to the 6 to 7 percent efficiency figures cited for the "second-law" efficiency of the US economy, and ask how much further one could go by trying to minimize the resources consumed in a given product. Their estimate of US energy efficiency in delivering a function is 1 to 2 percent, with the potential of reuse or less use of materials taken into account.

Ultimately, industry's principal energy requirement—process heat—will be met by renewables. (A tautology, but not an unreasonable one.) Solar industrial process heat technology has made remarkable strides in the last few years, but is finding a difficult environment in the heady world of industrial capital investment. The fact that energy bills are tax write-offs for large firms forces solar to compete against economic signals of the past: $12 per barrel oil, $4 per "barrel" coal, and $6 per "barrel" natural gas. Solar ponds for pre-heat and inexpensive concentrating collectors that have every chance of breaking into the domestic water heat market are, or will soon be, available for $20-$40 per barrel. They deserve at least as much support as synfuels that will clearly cost our bank accounts and coal deposits dearly.

—Jim Harding & Charles Drucker

Energy Efficiency Could be Doubled in Steel Manufacture

Steelmaking, an energy intensive process that consumes about 8 percent of all the energy used in Sweden, can be performed more than twice as efficiently in the future as it is in the US today, according to a recent report prepared by the Swedish Royal Institute of Technology.

On average, producing a metric ton of alloyed steel takes 9,000 kilowatt-hours of primary energy, states a recent comparison of Swedish and US energy use by Schipper and Lichtenberg. The Swedish industry, newer and more efficient, uses about 7,800 kilowatt-hours per ton.

"The technique used today in the iron and steel industry," writes Professor Sven Eketorp in a recent summary report of the group's investigations, "is based on methods which are old and have rested unchanged during centuries. Principally nothing has changed in the blast furnace process during the last 500 years. Bessemer and Kelly invented...the pneumatic process 120 years ago and our way of casting is also in principle of the same age. Unbelievable progress has, indeed, been made but only by successive improvements of old existing technique."

By successive applications of known technology to a model Swedish steel plant, Eketorp and his group manage to reduce this end use energy consumption from the current level of 6,390 kWh per ton to 5,055 kWh/ton, to 5,210 kWh/ton, and, finally, to 3,850 kWh/ton. The second stage of improvement, which looks like anything but, comes from a substitution of coal for oil, allowing a reduction of 84 percent in oil requirements.

The dense 43-page report by the five authors and six-member advisory committee considers three levels of improvement:

Level 1 reduces end use energy consumption by 21 percent, through improved blast furnace practices, re-use of process heat, re-capture of off-gas from furnaces, continuous or ingot mold casting, and design changes in sintering plants.

Level 2 replaces the blast furnace, sintering plant, and coke ovens with an integrated smelting reduction and scrap melting process, operating under simultaneous coal gasification. Heavy use of poor quality, high sulfur coal in this synergistic effort drops oil use more than 90 percent from that commonly achieved in Level 0, the technology used today. Total energy consumption drops 18.5 percent from Level 0. "The flow of material is here completely identical with that of the previous levels, and energy data calculations are based on process technology which is relatively well known, but not applied nor tested in every detail," note the authors. "The energy consumption figures should only be regarded as conservative estimates."

In Level 2, a smelting reduction process produces molten metal, with ore concentrate, coal powder, and oxygen injected into an induction-heated hot metal bath. Iron oxide reduction is simultaneous with partial combustion of coal to carbon monoxide. Carbon acts as the reducing agent. "The required process heat," note the authors, "may be supplied by either externally produced electricity or by coal powder only, with recycling

of the waste gas energy as electricity by way of a turbine."

The oxygen potential of such a hot bath is kept very low, forcing the carbon only to carbon monoxide and avoiding oxidation of the hydrogen in the coal. "The off-gas," add Eketorp and others, "is very rich, as it contains no nitrogen...If the energy of gas is converted into electrical energy, the total process energy can be supplied by coal, even the energy for oxygen production. (Such a solution would, however, seem quite impractical)."

Mechanical devices separate slag from hot metal prior to desulfurization. Such a steel plant would probably be small, with the reduction process a limiting factor. Even with recycling of the off-gas from the hot metal bath, induction heaters would have to be quite large, with 5 units of 30 megawatts each suggested by the Swedish group. Each would produce 20 tons of hot metal per 100 minutes. Limestone injected into the slag free bath desulfurizes the molten metal, with calcium sulfide dust collected above the metal surface, "a method...that can be handled by use of technology known today." Oxygen, metal oxide, and metals injected into desulfurized hot metal produce steel. Low sulfur coal supplies the necessary heat for oxide reduction, the only time this high quality fuel is needed for steel production.

All casting in Level 2 is continuous, except for very large bars, which will continue to require ingot casting. This minimizes reheating, as hot material feeds into reheaters prior to rolling, inspection, and surface conditioning. Heat is recovered from finished products.

The energy requirements for this method of steel production can vary substantially, depending on one's assumptions. The gross end-use consumption calculated by Eketorp, et al. is 5,210 kWh per ton of alloyed, cast steel. "The gross consumption," they caution, "is naturally highly dependent on the way the reduction process is operated. If only the stochiometric amount of coal is injected and external electricity applied, instead of the electricity generated within the process, (then) 4,315 kWh per ton may be anticipated.

"For further reduction of the gross energy consumption, one might consider recycling the off-gas discharged from the reduction units for some other purpose than generating electricity. Assuming that the gas heat could be used with a thermal efficiency of, say, 80 percent instead of the 36 percent when generating electricity, the gross consumption would be decreased to approximately 3,500 kWh."

Steel in Level 3 "is produced in one step passing from ore concentrate into liquid steel by means of plasma reduction, with hydrogen and possibly methane or coal powder as reductants. Ore concentrate and plasma heated reductants are mixed, and the product is injected into a steel bath which is continually deoxidized. Grading of the steel bath is done simultaneously, also by the application of plasma technique. Casting is done by atomization of the melt, with subsequent rapid solidification, and compacting of the microingots, while the metal is still hot enough. Considerable research efforts are needed," conclude the group, "both with respect to the plasma technique and the casting technique."

Provided cheap electricity is available for hydrogen production, end use energy consumption can be reduced in Level 3 to 3,850 kWh/ton.

In addition to energy benefits, continuous steelmaking offers environmental benefits, noted Eketorp in a presentation before the Third International Iron and Steel Congress, held April, 1978 in Chicago, Illinois.

"The physical environmental conditions have in many steel plants in the world been enormously improved. So much has in fact been done that costs for environmental protection inside and outside the steel plant is rapidly reaching impossibly high levels. It has also been pointed out that energy requirements for dust and heat removal require so much energy that dust and SO_2 in this primary energy production result in an overall increase in environmental hazards.

"The solution to this dilemma seems to be new technology and new thinking, where the environmental problem is attacked at the source. Today's efforts are to a large extent guided towards decreasing the effect of dust, heat, noise, draft, etc. But the environmental problem is not solved by building high stacks, installing more filters, employing stronger fans or the use of effective ear protectors (perhaps with built-in music). The real solution is to be found in complete removal of specially troublesome process steps (sintering, coking), in the exclusion of air and thus large gas volumes in all process systems, in avoiding free exposure of molten iron and steel and off-gases from furnaces.

"The scientist is not in the real forefront pointing out the road to travel," Eketorp notes earlier. "Instead he is called on to explain the phenomena and problems as they appear and is asked to suggest remedies. Naturally such work must go on. But would it not be possible for the metallurgical scientist to start the innovation of new processes by strict application of fundamental scientific and technical principles and possibilities? ...Fundamentals, however, are not only metallurgical but also economical, technological, and social."

—Jim Harding

Sources: Sven Eketorp, et al., *The Future Steelplant—A Study on Energy Consumption*, published by Styrelsen för Teknisk Utveckling (the National Swedish Board for Technical Development), Stockholm, Sweden; Sven Eketorp, "Decisive Factors for Planning of Future Steel Plants," paper presented at the Third International Iron and Steel Congress, Chicago, Illinois, 16-20 April 1978, available from Professor Eketorp, Department of Ferrous Metallurgy, Royal Institute of Technology, S-100 44 Stockholm 70, Sweden.

Efficient Electric Motors: Winding Up and Slowing Down

ELECTRICITY IS THE ROLLS ROYCE of fuels: high quality comes at a high price, which the environment must pay along with the purchaser. At 4¢/kWh (the US average), using electricity is equivalent to buying oil at $64 per barrel. At these prices, it makes little sense to electrify when cheaper, more thermodynamically appropriate fuels will do.

In industrialized countries, electricity is essential for anywhere from 7 to 11% of all end-use needs. Among those special uses that require electricity, motor drives claim the largest share. Some 50 million electric motors of all kinds are sold every year in the United States alone. Most of these are small, fractional horsepower units (i.e., less than one-hp), and their annual consumption is low, but there are still millions of motors in the 1 to 125-hp range being sold, and many are even larger. All together, motors accounted for an estimated 64% of total electricity consumption in 1972 (Arthur D. Little, 1976).

From the standpoint of energy conservation, the most important electric motors are integral hp (i.e., more than 1-hp) AC induction motors. The fractional hp motors, mostly in home appliances, are greater in number, but account for only about 15% of total motor electricity consumption. Integral hp AC induction motors, in contrast, receive an estimated 44% of *all* the electricity delivered in the US. The other integral hp motor types, such as synchronous and direct current designs, take much smaller fractions of electricity use, and they are already so efficient that there are limited opportunities for conservation.

Increasing Motor Efficiency

It is generally thought that electric motors are already quite efficient, but most commercial and industrial motors, in fact, operate at only 50-70% efficiency. Purchasers traditionally assigned efficiency a low priority, and manufacturers obliged by stressing reliability and quality control, while trying to maximize sales by keeping prices down. This trade-off typically meant minimizing the materials used, with the result that efficiency suffered for the sake of lower cost and smaller size. With the industrial cost of electricity a low 1.25¢/kWh in 1973, efficiency was clearly not a priority concern.

The recent sharp rise in electricity costs and the prospect of further jumps, have turned the motor market around. Most manufacturers now have an energy-efficient line, priced somewhat higher than their standard models, but with an average payback of the additional cost within one or two years. Their customers at first lacked enthusiasm for the new higher-priced products, but energy efficiency has become a strong selling point.

The operating efficiency of an electric motor is a function of its internal power losses—that is, the amount of electrical energy which is not converted into mechanical motion. Losses fall into three broad categories: *mechanical*, including friction; *electrical*, notably through resistance; and *magnetic*. Standard design, high-efficiency motors concentrate on electrical (I^2R) and mechanical losses, though some use higher quality magnetic materials such as silicon steel, to reduce magnetic loss. Present efficiency methods resemble the design techniques of the 1950s. The stator and rotor are lengthened, giving a larger cross-sectional area of magnetic flux, which lowers resistance and core losses. Upgrading materials decreases eddy current and hysteresis core losses. For a typical 5-hp induction motor, such modifications (particularly the use of silicon steel) can lower losses by about 25%. These are solid, but not spectacular gains. They have been possible for a long time. The only thing that is new about them is that they are finally economically attractive.

The Wanlass Wonder Machine

An innovative new approach to motor design, developed by inventor Cravens Wanlass, ushers in a new era in efficiency, and may well send motor manufacturers back to the drawing board, or into licensing arrangements with Wanlass. In test after test, the Wanlass Motor works at efficiency levels that leave run-of-the-assembly line motors sputtering. And the performance from retrofits is as impressive as that of new motors. A US Electric 30-hp motor, converted to the Wanlass design, shot from 80 to 95% full-load efficiency, for a 75% reduction in motor losses. Efficiency gains at two-thirds load were comparable, jumping from 77 to 92%. A "Wanlassized" Toshiba 25-hp motor similarly showed a gain in efficiency, from 85 to 95%. In these and other tests, the Wanlass modification improved the power factor too, in some cases to near unity.

The Wanlass Motor design is an example of precision end-use matching. Its performance comes not from changes in the core materials, but by means of a clever rewinding that allows the motor to respond to the exact load placed upon it. In a conventional induction motor, the flux density is constant, set for the motor's rated load. Yet many motors operate at partial load most of the time, at greatly reduced efficiency. The Wanlass winding technique optimizes the transfer of energy through the stator, and thus increases operating efficiency at all load levels. When the load changes, the motor automatically alters the magnetic properties of the circuit, reducing the flux density when it is not needed. More efficient energy transfer and flux density control combine to reduce eddy current and hysteresis losses in the core, and I^2R losses in the windings. The Wanlass motor is substantially more expensive than standard motors, in part because their production is so low. Yet Wanlass claims a payback period for the additional cost of 4 months to 3 years, depending on motor size, usage, and the price of electricity.

The physics principles behind the Wanlass motor's new winding design are complicated, but the results can easily be understood by anyone capable of reading a watt meter—or a utility bill. Wanlass plans to produce a line of new, three-phase motors in sizes up to 200 hp, and to license his design to other motor manufacturers. Based on tests of rewound motors, full-load efficiencies should range

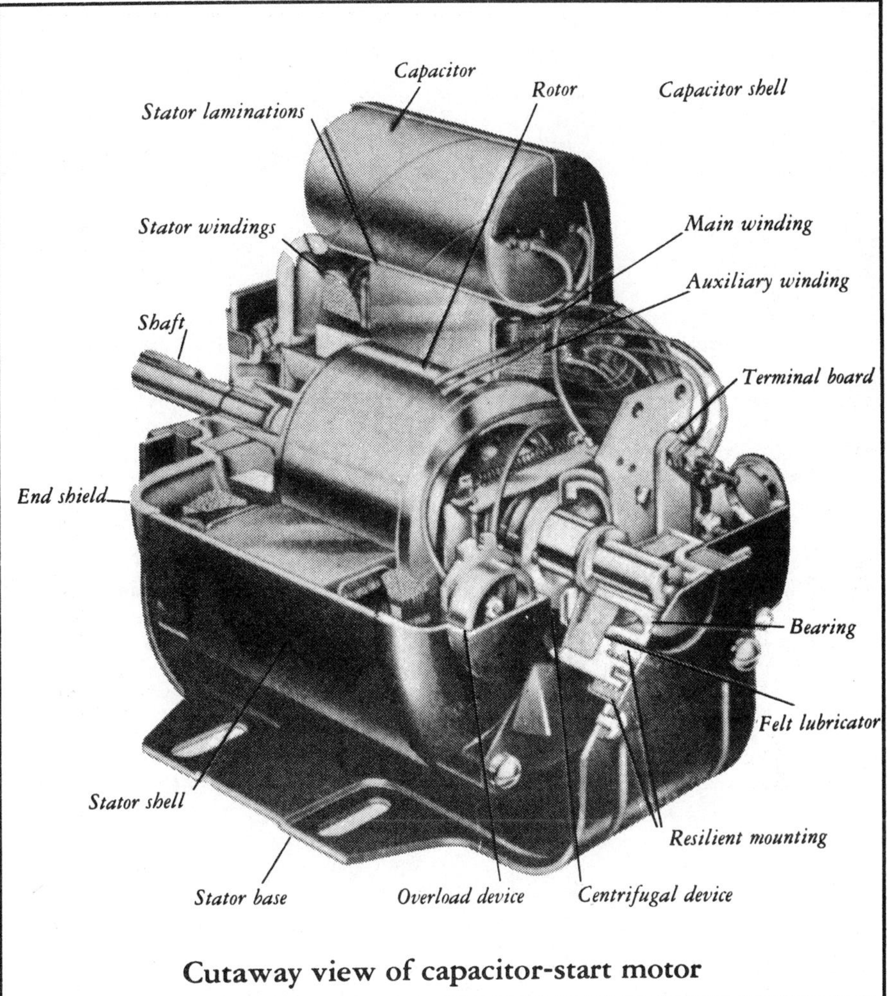

Cutaway view of capacitor-start motor

from 88% in 1-hp motors up to 96% for the largest units.

Retrofitting Efficiency

The lifetime of an electric motor depends on application and size. Large units over 50-hp are usually rewound rather than scrapped, and the stock turns over slowly. Small and medium-sized motors have a useful life of 7 to 10 years, though in some process industries this is lower. Still, the short payback period for a Wanlass rewinding makes it an attractive investment, especially for heavily used motors 5-hp and larger. Sixty-five Authorized Rewind Centers around the country serve a growing market.

Wanlass is not the only motor efficiency enterprise. Twenty-two firms have received licenses from NASA to market the Nola power factor controller, a space program spinoff developed at the Huntsville Space Flight Center. Like the Wanlass design, the Nola power saver matches the output of the motor to the load placed on it, but it does so without any alteration to the motor itself. Instead, a solid-state circuit "chops" the voltage down to a load-matched level just sufficient for the task at hand. Calibration is crucial, because if the device is set too low, there may not be enough torque to start the motor against a heavy load. The Nola design is most attractive for motor drives with irregular and light loads, and on those that idle much of the time. And since most motors are least efficient at light load, the Nola power saver can reduce electricity use by up to 35%. One small Nola licensee sells power saver units for 3-hp motors at a price of $70, and reports having received 30,000 requests for information in the last year.

Exxon's Entry: Variable Speed Control

The most intense use of electric motors is in five industries: chemicals, primary metals, pulp and paper, food and beverage, and petroleum refining, together accounting for half of the electricity consumed by industrial and commercial motor drives. Many applications in these process industries involve variable through-

Tests of Motors Rewound to Wanlass Design		
Motor	Efficiency (full load)	
	before	*after*
Dayton 1/4 HP	50	67
Lasson 1/2 HP	61	81
Gould E+ 1 HP	80	88
US Electric 30 HP	81	95
Toshiba 25 HP	85	95
Westinghouse 100 HP (70% load)	90	95

put, and thus variable speed, from pumps, compressors, fans, and blowers. Direct current motors permit simple and inexpensive speed control, but they are bulky, costly, and unsuited to many industrial environments. Process control with AC motors—the mainstay of industry—usually means mechanical throttling while the motor continues to run at nominal speed. Valves, baffles, and other throttling devices are highly reliable, but quite energy inefficient. At 40% of maximum operation, for example, the motor consumes 70% of full-load energy, and the throttling device dissipates the 30% differential as heat.

An alternative to throttling is an AC Variable Speed Drive (VSD). Varying both line voltage and frequency simultaneously, a VSD can load-match by reducing the energy input to the motor, rather than dissipating the unwanted portion of the motor's output. The heart of the VSD is a power inverter that synthesizes the required current, tuning frequency and voltage to varying speeds. Variable Speed Drives themselves are not new. Traditionally, they have used silicon-controlled rectifiers, which require complex timing and control circuitry, making the VSDs more expensive than the motors they are designed to improve. A further deficiency is that the quality of the wave forms they produce is poor, which can increase motor heat loss.

Semiconductor electronics technology has spawned an entirely new generation of VSD power inverters with neither of these handicaps. The fully transistorized Alternating Current Synthesizer (ACS) offers improved wave form quality, higher energy efficiency, smaller size, and greater reliability, all at a price low enough to stimulate widespread use. The ACS inverter promoted by Exxon eliminates bulky and costly inductors, capacitors, and output transformers by substituting state-of-the-art power transistors which directly produce the desired output wave at the correct voltage.

Size, weight, and unit cost plummet while efficiency climbs from the 60-90% bracket to between 94 and 97 percent and possibly even higher.

Tests of prototypes ACS units on pumps at Exxon refineries show substantial savings over conventional throttling control: 27 and 38% over a two-month test period. At low throughput levels, 50% savings and more are achievable. BenDaniel and David of Exxon estimate that the new VSD devices could affect half of all industrial AC motor applications by 1990. Assuming an average energy savings of 30%, this diffusion of VSDs would reduce electricity demand in the US by 110 billion kWh.

Exxon's marketing strategy for their ACS recently led them to a 2 billion takeover of the Reliance Electric Company, a move opposed by the US Federal Trade Commission on anti-trust grounds. The government would prefer Exxon to license the invention for others to produce and distribute, but the energy conglomerate argues that licensing is not the most aggressive method to commercialize their invention.

It is certain that the Exxon ACS and the other motor efficiency innovations will eventually reach their proper markets. The Wanlass Motor demonstrates that Second Law efficiencies close to 100% are not only possible, but also cost-effective. Variable speed control brings high efficiency to presently wasteful industrial processes. What effect will electric motor efficiency have on the national energy picture? At 1978 US consumption levels, an average efficiency increase of only 10 percent would save an amount of energy equal to the annual production (at 65% capacity factor) of more than twenty 1000-MW power plants.

—*Steve Meyers*

References:

BenDaniel and David
1979 "Semiconductor Alternating-Current Motor Drives and Energy Conservation." *Science* 206:773-776.

Little, Arthur D., Inc.
1976 *Energy Efficiency and Electric Motors*. Prepared for the Federal Energy Administration, August 1976. Available from National Technical Information Service, Report # FEA/D-76/381.

Popular Science
1980 "Nola's Power Saver," *Popular Science*, February 1980. Wanlass Technologies Inc.
1980 Product description packet and personal communications. Wanlass Technologies Inc., 1442 Irvine Blvd., Tustin California 92680.

Glossary

Core Losses — occur in the magnetic steel of the stator and rotor due to hysteresis effects and eddy currents; typically 20-25% of all internal losses; eddy current losses, caused by current circulating within the core steel, can be reduced by using thinner gauge steel; hysteresis losses are a function of flux density.

Efficiency — the ratio of electrical energy (input) to mechanical work (output); to estimate the power consumption of a motor at any load, divide its output by its efficiency at that load.

Horsepower — 746 watts; 550 foot-pounds of work per second.

Induction Motor — an alternating current motor in which a primary winding on one member (usually the stator) is connected to the power source, and a secondary winding on the other member (the rotor) carries only current induced by a magnetic field.

I^2R Losses — resistive power losses resulting from current passing through the conductors on stator and rotor; vary with the square of the current; constitute 55-60% of total motor losses.

Magnetic Flux Density — a vector quantity used as a quantitative measure of a magnetic field; the product of the field intensity and the permeability of the medium.

Power Factor — ratio of the total power (in watts) dissipated in an electrical circuit to the total equivalent volt-amperes applied to the circuit.

Stray Load Losses — caused by flux leakage induced by load currents; vary with the square of the load current.

Some Like It Hotter

HIGH INDUSTRIAL TEMPERATURES CHALLENGE SOLAR TECHNOLOGIES

Direct use of the sun's energy for industrial processes would reduce oil imports, remove a major supply burden from utilities, and eliminate the largest single contributor to increased energy consumption. Still, technical and economic barriers stand in the way.

No other end use of energy in the United States exceeds industry's demand for process heat. Of the 77 quads of primary energy consumed in 1977, (1 quadrillion Btu = 1.05×10^9 Gigajoule), 20 to 28 percent became thermal energy for the nation's factories and industrial plants. Not even the 120 million automobiles now on US roads captured this large a share. And the Department of Energy forecasts that process heat demands will be responsible for nearly half of the increase in energy consumption over the next four decades.

Solar energy would seem the natural candidate to supply industrial process heat in an era of ever-increasing prices for fossil fuels. All but a few percent of the energy we now use, or expect to use in the near future, starts out as solar energy, and a scant .03 percent of the solar radiation that enters the earth's atmosphere is utilized as light in photosynthesis, rather than as heat. If industry could tap this renewable reservoir of ambient and incident energy, then the fossil fuels and the hydroelectric power now going into process heat could remain in the ground, or be put to more appropriate and more efficient ends. Surrounded by thermal activity, industry should be able to meet its heating needs by direct capture.

Yet harnessing the solar flux turns out to be more complicated in the industrial sector than elsewhere. Process heat demands are quite specific as to quantity, medium, and temperature level, and the acceptable range of variation is small. Solar energy flows, in contrast, are highly variable. They change with the hour of the day and with the cycle of the seasons; they follow the caprices of weather, and disappear entirely at night. Medium- and long-term storage, or even conventional backup systems, can smooth out the solar peaks and troughs, but the major obstacle is the extremely high temperatures required by most industrial processes.

Thermal Mismatch

The great majority of processes require temperatures above 600° C, with nearly half of the demand above 1,100° C, according to a 1977 study by InterTechnology Corporation. Less than 10 percent is for heat under 100° C, the highest temperature at which water-based, flat-plate collectors typically operate. These nontracking, nonconcentrating collectors are the most common and familiar solar devices, with more than 80,000 residential and commercial

systems presently in use across the United States. Industrial processes such as cooking, washing, bleaching, drying, curing, and anodizing need heated water or air between 50 and 100° C, and readily available flat-plate collectors are well suited to these purposes.

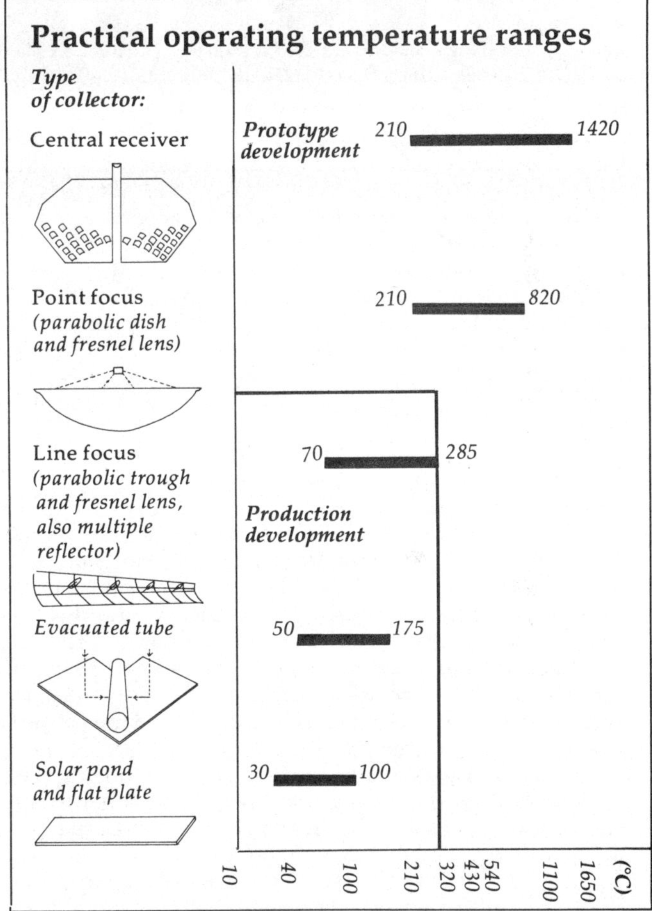

Many more industrial processes, though, require direct heat or low pressure (100 psi) saturated steam between the boiling point of water and 175° C—still in the low-temperature range by industrial standards, though hotter than any residential need. Flat-plate technology can be stretched to this level by using a heat transfer medium—air, pressurized water, or oil—that can be raised to temperatures above 100° C, then moved to a heat exchanger or low-pressure chamber so that the heat is supplied to the process in the proper medium. But for temperatures at the higher end of this range, the most practical devices are simple solar concentrators with a ratio of about 10-20. Typical of this technology is the nontracking evacuated tube collector, within a cusp or v-trough reflector.

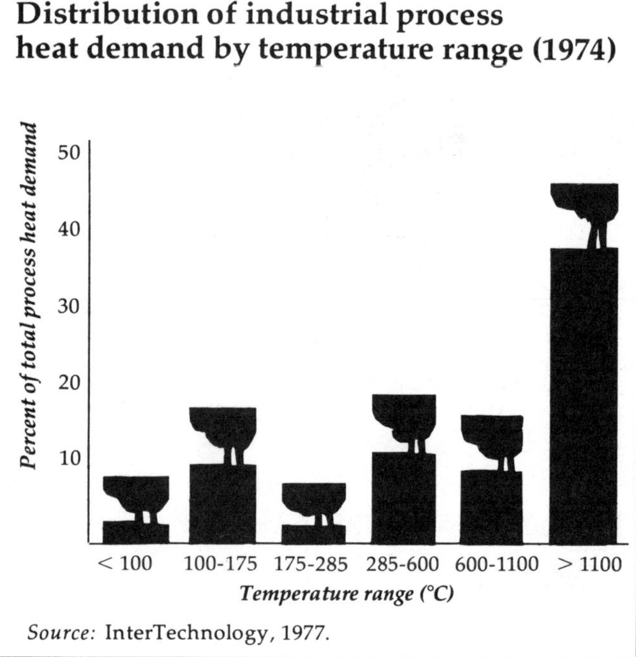

Source: InterTechnology, 1977.

Higher solar concentration ratios, up to 40 or 50, are needed to meet processing heat needs in the 175°-285° range. The most common instrument in this category is the tracking parabolic trough collector, with over 3,600 m² installed for industrial purposes since 1977. Line-focus fresnel lens systems can also generate heat at this level, and innovative materials and extrusion processes promise to bring significantly lower initial costs within the next few years (see box).

More than 70 percent of industrial process needs, however, fall in the high-temperature range above 285° C. Petroleum refining, cement, glass, and primary metals processing all call for 285° plus, but solar technologies capable of these temperatures are not yet commercially available. Development work is under way: Ken Brown, in a recent Solar Energy Research Institute report, describes point-focusing, tracking collector systems as "on the verge of industrial deployment." Parabolic dishes, with a concentration ratio of up to 1,000, achieve these high temperatures by tracking the sun on two axes. At the focal point of the dish, a receiver converts the concentrated solar energy into heat via a working fluid such as gas or steam. A heat exchanger then transfers the energy to the specified process medium. Point-focus Fresnel lens systems are also capable of these temperatures.

Extremely high-temperature processes of up to 1,350° C require solar concentration ratios of 2,000, achievable only with centralized receiver systems. Here, a large number of heliostats, or tracking mirrors, reflect light to a tower-mounted central receiver. The working fluid—liquid sodium, in some designs—transports the collected energy to the point of use or a heat exchanger.

Manufacturers interested in calculating the potential savings from solar industrial process heat systems should note that the energy delivered by a solar system actually displaces *more* than its fuel equivalent. As Ken Brown points out, "the efficiency of conversion of the solar system is implicitly contained in the energy delivery figures given; the efficiency of fuel conversion (which may vary from 65 to 85 percent in conventional boilers and furnaces) is often not included in calculating fossil energy displacement." The true fuel savings to the end user equal the annual energy capacity of the solar heat system *divided by* the conventional fossil fuel conversion efficiency.

From the larger perspective of the global energy system, the savings are even greater, as there are conversion and distribution losses not just in the end user's facility, but all the way back to the fossil fuel's point of origin.

Process Heat Prospects

Solar energy can, at least in theory, be supplied to industry in any form of thermal range, but extremely high-temperature systems are still in the prototype and design stages. Much more work will be required before their costs are established. Low- and medium-temperature collector and concentrator systems operating under 285° C are more likely to become cost-effective in the near term. At least 27 percent of industrial process heat demand falls below this level. Available solar technology can also supply a portion of higher temperature industrial needs through preheating to 285° C, for a total contribution of 51 percent of all process heating.

How much of this 51 percent looks economically attractive to industry depends as much upon the efficiency and reliability of the equipment as upon the financial terms that industry requires of its investments. The Inter-Technology study of industrial process heat estimated that tracking parabolic concentrators (with terminal temperatures of about 285°) could command 36 percent of the process heat market by 2000. Their assumptions about collector performance were fairly conservative, but they also relaxed the pretax rate of return demanded by industry to a leisurely 15 percent. (See *Soft Energy Notes* II, March 1978, for a review of this study).

The present situation may be both better than this, and worse; collectors are now, or soon will be, more efficient and less expensive than assumed, but it is highly unlikely that any financial manager would give serious consideration to an investment with only a 15 percent rate of return. As we noted in 1978, tests of Winston compound parabolic collectors at the Los Alamos Scientific Laboratory suggest that temperatures as high as 315° C are possible at an efficiency of 0.44, with a price tag of $118/m². This is at the extreme low end of the cost range that InterTechnology assumed for this collector variety.

On the other side of the ledger, expecting industry to accept a 15 percent rate of return ranks as wild optimism. Payback periods of 3 to 5 years, or 20 to 33 percent rates of return, are more realistic, given the alternative investment opportunities available for capital. This economic fact of life makes it difficult for solar process heat to compete with conventional fossil fuels. If the capacity cost for a solar system runs about $170/GJ/year (as anticipated for a number of Department of Energy-sponsored projects), then conventional fuels, under a three-year payback, would have to cost $57/GJ, or $325/bbl of oil equivalent, for solar process heat to compete. Under a ten-year payback assumption, the break-even fuel cost drops to about $12.50/GJ, equivalent to $72 for a barrel of oil.

Impact of Federal Policy

No easy way exists to make a 10 year payback period palatable to industry, given today's cost of money. There are, though, federal tax incentives that could give solar industrial process heat a substantial boost. Tax credits have historically encouraged capital investment, and since 1978, solar process heat equipment has enjoyed a 10 percent credit *in addition to* the standard 10 percent capital equipment tax credit. Legislation introduced in 1979 sought to increase the total credit to 50 percent—a level which seems justified since solar heat must compete with conventional fuels that industry often pays for with "fifty cent dollars." Fuel costs are deductible operating expenses for businesses, and corporations in the 50 percent tax bracket receive, in effect, a 50 percent "credit" for their conventional energy purchases. The final version of the industrial solar tax incentive proposal, however, raises the total credit to only 25 percent, an increase so small as to be ineffective. With a 50 percent credit, solar process heat systems installed for $170/GJ/yr with a ten year payback, can compete with conventional fuels costing $5.30/GJ, or oil at a mere $30 per barrel.

The federal government could also spur adoption of solar process heat systems by allowing a form of accelerated depreciation. Tax law presently requires a depreciation period of at least seven years to receive all the allowed investment tax credit. Industrial energy equipment, such as boilers and furnaces, is generally allowed a depreciation life of 15-23 years. The Internal Revenue Service has made no specific rulings on solar energy capital investments, but shortening the allowable tax life makes any such expenditure a great deal more attractive.

With a combination of 50 percent tax credits and rapid depreciation, low and medium temperature solar process heat could easily compete with conventional fuels in the majority of industrial settings. The potential market is vast, and so is the amount of capital required to have any effect on the energy supply mix. To meet President Carter's modest goal of 20 percent of domestic demand from renewable resources by 2000, solar process heat would have to provide 2.6 quads (assuming 93 quads total national consumption). At the output efficiencies and costs assumed by SERI, this goal would require between 700 and 900 million square meters of solar collectors in industry, with a total cost of $400 billion.

—Charles Drucker

Reference:
Brown, Ken

1980 *The Use of Solar Energy to Produce Process Heat for Industry.* Available for $4.00 from Solar Energy Research Institute, 1536 Cole Blvd., Golden, Colorado 80401 USA, document # SERI/TP-731-626.

Roll Out the Lenses

INNOVATIVE DESIGN AND NEW MATERIALS TECHNOLOGY are well on the way to making solar industrial process heat less expensive. Using polycarbonate plastic sheets costing only $3-4/m², scientists working with the US Department of Agriculture have fabricated a Fresnel-type line-focus solar concentrator capable of relatively high temperatures. At the theoretical maximum concentration ratio of 50, output temperatures would reach 550° C. Existing lenses, with ratios of 15 to 25, are roughly 30 percent efficient and easily reach output temperatures of 300° C—adequate for many industrial purposes.

The rectangular concentrators can be mass produced to even higher performance levels using plastics technology already in widespread use. A thin plastic sheet (0.5 mm) is passed, while still in a semi-molten state, between two rollers: one smooth, the other engraved with a precisely shaped ridge every half millimeter or so. The pressure on the rollers embosses a series of parallel grooves in the plastic, giving it the optical properties of a Fresnel lens. Each groove acts like a tiny prism; light rays, passing through the plastic, are bent according to the pitch of the individual grooves. As the grooves approach the outer edges of the lens, they increase in depth and slope, bending the sunlight to a greater degree. The rays thus converge on a focal line or strip.

The USDA concentrator, in a high temperature configuration, can focus the light on a blackened collector pipe within an insulating, evacuated glass jacket.

To maximize the total energy output of the collector, though, the device must track the sun's movements. In addition to seasonal compensation for changes in the sun's orientation, adjustments are needed throughout the day, because the focal length of the device changes with the angle of the incident solar rays. This is where materials choice pays off. The thin lens material weighs about two pounds for a three-by-eight foot collector, or one kilogram for two square meters, so that small motors using very little energy are adequate to track the sun.

Placing the concentrator under glass keeps the system a featherweight. Protected from wind and rain, little structural strength or rigidity is required of the lens and its frame.

The USDA concentrating device, mounted in modules under a glass roof, could pay for itself within 2-10 years, depending upon application and insolation level. The original design team envisioned rows of collectors, installed in the greenhouses of small farming communities, where they would support such light industrial activities as canning or drying. But as conventional prices rise, industry may decide to "roll out the lenses" to generate process heat.

For further information on the USDA research, contact Glen Bailey, Western Regional Laboratory, United States Department of Agriculture, 800 Buchanan Street, Berkeley, California 94710. For specifications on the lens material, contact Jack Rhoads, Director of Product Quality, Hallmark Cards, Inc., 25th & McGee Street, Kansas City, Missouri 64108.

—Charles Drucker

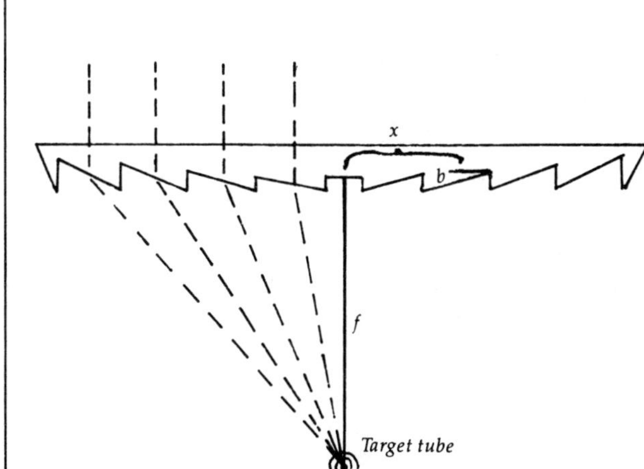

For flat lenses of refractive index n and focal length f, the equation for the prism angle b is:

$$\cotan b = \frac{n(f^2 + x^2)^{1/2} - f}{x}$$

Curved lenses have similar optical properties, but a different prism angle equation. In both cases, the focusing effect only operates on direct illumination, and not on diffuse light.

Piping Hot

Urban energy planners now look to replace on-site oil- and gas-fired heating systems with district heating and cooling—DHC. DHC carries thermal energy from a central source through a piping network to end-use equipment in buildings. It offers more efficient energy conversion—cogeneration and renewable energy sources—and low capital cost.

THE TECHNOLOGY has developed over the last century, from the small steam systems serving the dense central business and industrial districts of US cities, to the large hot-water systems that provide European cities with supplies of space heat and hot water.

District heating is slated for a major expansion in Western Europe and is being studied in cities across the United States. Recent research, development, and demonstrations focus on three system components: end uses, distribution, and heat and cold production, collection and storage.

End-use systems

Energy conservation in buildings has important repercussions for the retrofit of internal building heat distribution equipment to district heating. After conservation measures have been implemented, building internals sized to meet pre-conservation heat demand levels will be oversized. This overcapacity allows the use of lower quality (temperature) heat than the pre-conservation distribution temperature levels for which existing building internals were designed. This facilitates the use of hot water in steam building internals and the use of lower temperatures in hydronic and forced-air systems.

The lower supply temperature limits for existing hydronic and forced-air systems are approached at 49° C (120° F) and 32° C (90° F), respectively. Showers and handwashing require hot water at temperatures of about 42° C (108° F), while dishwashing typically uses hot water at about 49° C (120° F).

These minimum requirements contrast sharply with traditional district heating system designs. Typical Swedish systems supply hot water at temperatures varying between 80° C (176° F) and 120° C (250° F) and return water at a minimum of 45° C (113° F) to a maximum of 70° C (158° F). Danish systems supply somewhat lower temperature and pressure hot water to end uses at 70° C (158° F) to 90° C (203° F) and return water at about 45° C (113° F).

Danish systems typically dispense with the use of heat exchangers between hydronic building internals and the district heating system, an advantage over other systems. Heat exchangers are necessary in systems that employ higher temperatures and pressures to safely isolate the consumer's premises from the district heating system. Danish systems directly connect hydronic building internals to the district heating system. Direct connection has been shown to have a 2.5 to 1 cost advantage over heat exchangers when capital, energy and pumping costs and the capitalized value of lost electricity production are combined.

Distribution systems

Since the distribution system frequently represents as much as 75 percent of the total cost of a district heating system, much recent R, D and D has focused on reducing costs here. Reducing distribution temperatures to levels under 60° C (140° F) allows the use of plastic and concrete pipe—less expensive than steel.

Distribution system material costs are usually exceeded by installation costs. Running large-diameter district heating pipe through the utility-clogged substreets of central cities costs from $750 to $2,000 per foot. Where attached buildings are prevalent, however, insulated 6″ (15 cm) steel pipe can be run through basements at capital and installation costs of less than $50 per foot. The resulting block-scale distribution system can then either connect to a small, efficient cogeneration unit or to a larger district system employing a single connection per block. Such grouped connection practices can contribute to reduced distribution system installation costs.

If the distribution systems (and heat sources) are sized to the existing, artificially high, pre-conservation heat demand densities (measured in MW/linear km² or MBtu/h/ linear acre), district heating systems run the risk of oversizing and underutilization. After building efficiency improvements, revenues from heat sales will decline, while the system's high fixed costs force the utility to raise rates to cover costs, encouraging higher levels of consumer investment in energy conservation. Thus, the design of district heating systems to serve artificially high, pre-conservation heat demand levels runs considerable financial risks.

Energy conservation measures also reduce building heating reference temperatures. (The reference temperature is the minimum outside temperature reached before heat is required by a building.) A reduction in the reference temperature lowers the distribution system capacity factor, since efficient buildings will not require heat during moderately cold periods. (Capacity is calculated by the average annual heat demand divided by peak heat demand.) The greater the level of improvement in building energy efficiency, the higher the cost of delivered heat, if that heat is supplied from a conventionally designed district heating system. If very high levels of investment in building energy conservation are assumed, conventional district heating system designs stand a good chance of being uneconomic. It is necessary to strike a cost-effective balance between investments in improved end-use efficiency and in the district heating system; modify conventional district heating system designs; and, increase the system capacity factor by providing as many additional thermal services as possible, such as space heating and cooling, hot water, and refrigeration.

Efficient technology for DHC

By lowering both the quantity and quality of heat demand, energy efficiency reduces the scale and capital cost of the heat production, collection and storage systems. The efficiency and economy of heat supply systems is considerably improved, as well. The efficiency and economy of biomass- or fossil-fueled steam turbine cogeneration units, for example, are directly related to the design temperature of the DHC system it supplies. Electric power production in a condensing steam turbine is a function of condenser temperature: increasing condenser temperature to obtain heat at higher temperature decreases electrical production; the cost of the foregone electrical production is borne by the district heat. Thus, the higher the temperature of the heat obtained from a steam turbine cogeneration unit, the higher its cost will be. Low design temperatures lower system cost and can also allow low-capital-cost methods for retrofit of existing steam turbines to cogeneration.

Low temperature district heating systems can provide a market distribution mechanism for the many abundant, locally available energy sources that are currently wasted. Examples include: heat from urban waste, industrial surplus—"waste"—heat, low- and moderate-temperature geothermal/ground water resources, and waste (sewage) water. Heat pumps are increasingly employed to boost these low temperature heat sources to useful levels. The lower the district heating design temperature, the more sources of low temperature heat can be used and the higher the heat pump's coefficient of performance will be. Wind energy is an attractive source of electricity for heat pumps since peak wind energy availability often coincides with peak heat demand at northern latitudes.

Solar for interseasonal thermal storage

Solar space heating systems that employ interseasonal thermal storage are potentially more economical and require a smaller collector area since collector efficiency is highest during seasons of high solar insolation. The ratio of collector area without long term storage to that with has been estimated for the following US cities:

San Diego, CA, 1.96	Denver, CO, 2.65
New York, NY, 2.68	Yuma, AZ, 3.60
Chicago, IL, 3.88	Cleveland, OH, 5.22
Seattle, WA, 5.32	Spokane, WA, 8.44
Sault St. Marie, MI, 6.05	

Both the economy and efficiency of interseasonal hot water storage increase with size. With much larger scale storage capacity than is appropriate for single buildings, district heating becomes necessary. Existing communities were not planned with solar access in mind—another reason for considering the solar district heating option. Factors such as poor building orientation, shadowing by other buildings, and insufficient roof area for collectors, make centralized solar collectors necessary if many existing cities are to derive a significant portion of their low temperature heating requirements from solar energy.

Sweden's solar district heating R, D, and D program, begun in 1978, includes three solar district heating systems and two designs-in-production. These first-generation Swedish solar DHC systems include: a prototype system supplying 100 percent of the annual heating requirements of a 500 m² office building (Figure 1); a community-scale application that supplies 52 houses with 92 percent of

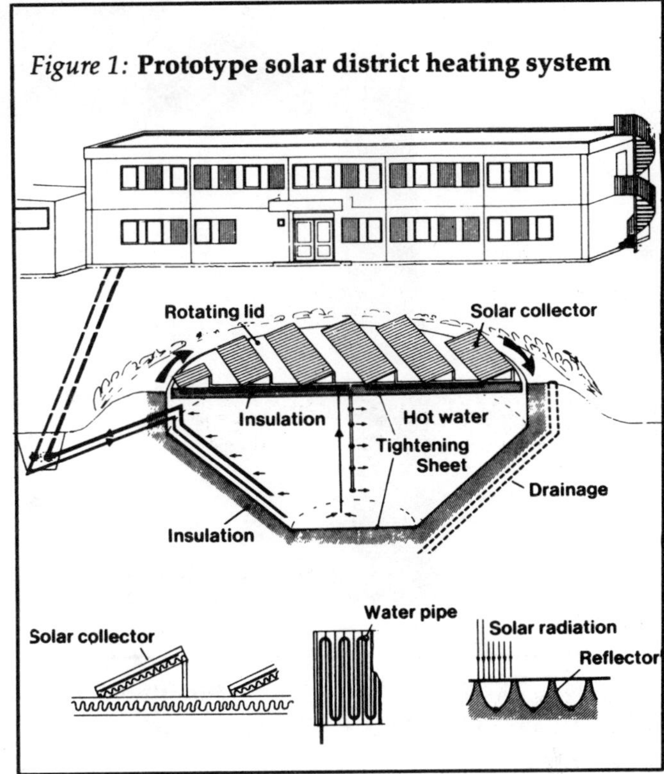

Figure 1: **Prototype solar district heating system**

annual heating requirements with the remainder provided by a heat pump (inside front cover); a system supplying 56 houses with 50 percent of annual heat demand; a solar district heating system slated to begin operation in 1984, designed to provide 100 percent of the annual heating needs of 500 homes; and, a system scheduled to begin construction in 1983 that will supply most of the annual heating needs of 500-1,000 dwellings with the remainder provided from a heat pump. Each of these systems employs a different solar collector, interseasonal heat storage techniques, and distribution and end-use systems.

Most of these systems use components that appear to be more costly than necessary. None of these systems is competitive with conventional Swedish district heating systems. US research, however, indicates that expensive solar collectors and heat storage techniques may not be necessary. The required collector area for tilted arrays, for example, is greater than for horizontal arrays when both employ interseasonal hot water storage for widely different areas of the US. If the design temperature of the district heating system is sufficiently low, very inexpensive integrated solar collectors and heat storage facilities, such as modified solar ponds, might be used. One proposed modified solar pond design uses potentially inexpensive components: a shallow water-filled, black bottom pool covered with an air-inflated, transparent plastic mat, which converts the pool to a one-window flat-plate collector. The shallow pool is supported by an insulating layer on the pond surface—an interseasonal hot water store. The pond/store is 5-10 m deep and is formed by earth berms around the sides. Its inside surface is covered with a plastic liner (see schematic).

Supplies of cold water for district cooling are most appropriately obtained from renewable energy sources using ice pond technology. The ice pond is a natural cooling process that uses abundant cold winter air to extract the heat of fusion from water sprayed by a relatively low energy-consuming recreational snow machine. The ice

Far Rockaway, Queens

Figure 2: Schematic Diagram of Fresh Water Heat Storage Pond/Collection System

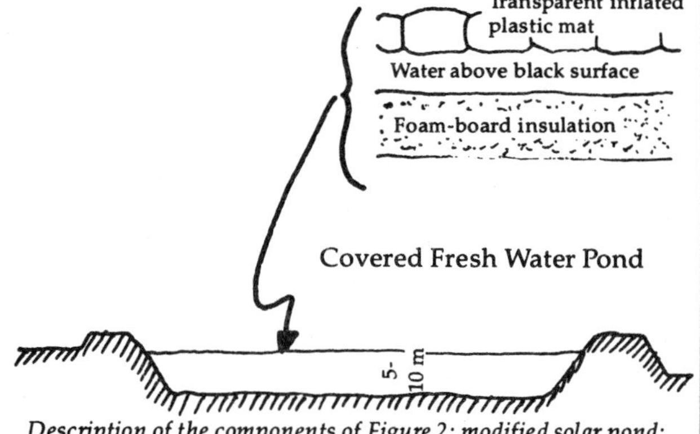

Description of the components of Figure 2: modified solar pond:

Foam Board Insulation: 10-20 cm, depending on storage temperature and climate.

Water (atop foam board insulation): 1-5 cm depending on time of day, location and desired collection temperatures.

Transparent Inflated Plastic Mat: 2 layers, 2 air gaps each 3-5 cm.

TOTALS:		
Glazing:	6-10 cm
Water:	1-5 cm
Insulation:	10-20 cm
TOTAL:	17-35 cm

produced is stored until summer when it is gradually melted to supply 0° C (32° F) water for space cooling. The ice storage facility is an excavated earth basin, with a water-tight membrane liner and an insulated/reflective cover. A prototype ice pond has been built, and the first commercial application is under construction. The ice production COP of the latter facility is estimated to be 23:1 for a typical Princeton, New Jersey winter.

Given the continued development of ice ponds, modified solar ponds and other renewable heat production technologies, it may soon be both economically and technically feasible to design efficient district systems supplying a full range of thermal services from renewable energy resources. The emerging generation of advanced DHC technology offers considerable advantages over conventional systems for efficiency, economy, and energy source flexibility. These advanced DHC technologies have a significant role in the transition to a renewable urban energy future.

—Jack Gleason

Jack Gleason is project coordinator for the Self-Reliance District Heating Group, 1717 18th St., NW, Washington, DC 20009; (202) 232-4108. He is the author of "Efficient Fossil and Solar District Heating Systems: Preliminary Report," Solar Energy Research Institute, Golden, CO, 1981.

References:

Geller, Howard S.

1980 *Thermal Distribution Systems and Residential District Heating.* Center for Energy and Environmental Studies, Princeton University, Princeton, NJ, p. V-4.

Kirkpatrick, Donald L., Theodore B. Taylor, et al.

1981 "A Unique Low Energy Air Conditioning System Using Naturally-Frozen Ice." *Proceedings of the Annual Meeting of the American Section of the International Solar Energy Society,* p. 537.

Robinson, Peter J.

1980 "Transmission and Distribution Networks and the Consumer—The Potential for Development." In W. H. R. Orchard and A. F. C. Sherratt, *Combined Heat and Power: Whole City Heating—Planning Tomorrow's Energy Economy,* John Wiley and Sons, New York, NY, p. 191.

Taylor, Theodore B.

1981 *Long Term Storage of Heat and Cold.* Center for Energy and Environmental Studies, Princeton University, Princeton, NY, p. 6.

Contacts:

Peter Margen, Swedish Solar District Heating Program, Studsvik Energiteknik, AB, S-611-82, Nykopping, Sweden and the Swedish Council for Building Research, Sankt Goransgatan 66-S-11, 230 Stockholm, Sweden.

Ostgota Byggen AB, Solarvarmcentral Lambohov, Allmogegatan 66, 583 30, Linkoping, Sweden.

Fluidized Bed Technology Nears Commercial Application

Looking for cleaner, cheaper, more efficient ways to burn coal, energy policy-makers have taken increased interest in fluidized bed technology. Although problems with calculating cost-competitiveness and overcoming institutional uncertainties have hampered the technology's development to date, a new INFORM survey of fluidized bed combustion (FBC) concludes that the time has arrived for the systems' commercial application.

Walt Patterson and Richard Griffin, authors of *Fluidized Bed Technology: Coming To A Boil*, outline a remarkable array of fluidized bed projects in the planning, engineering design or demonstration phases, indicating the systems' potential for industrial uses—combustors can be used to produce heat for manufacturing purposes and/or for electricity generation. Utilities have also expressed interest in the systems, which many say could aid in broadening national energy options by spurring coal use as an interim replacement for oil and natural gas.

Coming To A Boil profiles nineteen private and governmental institutions from the US, UK and Sweden that are involved in FBC technology development, and contains an annotated index of fifty-one other such groups. Each organizational reference includes an address and, when available, financial data. The authors obtained profile information through telephone and direct interviews, written responses to a questionnaire sent out by the authors, and public relations and technical material provided by the organizations themselves. Most of the organizations "...responded enthusiastically to the chance to describe their fluidized bed program."

Designed like a sandbox with a floor of strong, porous material, fluidized beds operate when air blown under pressure buoys or "fluidizes" sand particles until they are suspended in a turbulent mass along with the burning fuel. Balanced between the forced air and gravity and churning like a basinful of boiling water, the hot sand particles facilitate the burning of virtually any type of combustible material.

The technology's promise is based on the fluidized bed combustors' ability to burn coal—or almost any fuel—efficiently and with relatively little pollution. The fuel, burned in the fluidized bed in small quantities (as little as 1 percent of the total material) heats the sand, ash, or other particles in the turbulent bed to temperatures generally lower than conventional boilers. Because the mass is "boiling" at a constant (or controlled) rate, the temperature does not fluctuate rapidly. Tubes placed in the bed, filled with water or air, absorb heat from direct contact with the huge total surface area of the churning particles—

a heat transfer process more efficient than that of a conventional boiler.

A fluidized bed combustor's lower operating temperature reduces nitrogen oxide production during combustion. Furthermore, ash does not melt and sinter into fly-ash—both important environmental and operating advantages over conventional boilers.

FBC appears advantageous when high-sulfur coal is used as fuel. Sulfur can be absorbed during combustion by injecting crushed limestone or dolomite into the bed with the fuel. Calcium in this sorbent material (and perhaps, in the case of dolomite—magnesium) combines with sulfur; the resulting calcium sulfate drops to the bed's bottom with the coal ash. This waste, according to the many observers INFORM contacted, is both manageable and relatively benign compared with liquid sulfur waste from conventional scrubbers on coal-fired boilers. Hard evidence on this aspect of waste disposal is not cited.

Research and Development Problems

Development of an efficient, simple, environmentally tolerable and manageable coal technology appears irresistible, but *Coming To A Boil* points out that the technology does have problems. For one, research and development have so far turned up few operational statistics for pollutants. [The book gives only one reference to sulfur emissions performance standards outside test-units.] However, the project director at a fluidized bed combustor used for a district heating system in Enkoping, Sweden, says he "...expects the unit to be able to remove up to 85 percent of the sulfur from coal containing 1.5 percent sulfur."

Laboratory tests by the American Argonne National Laboratory and Britain's National Coal Board, provide the primary sources of optimism concerning the industry's ability to handle sulfur removal. They indicate that fluidized bed combustors with sorbent can remove 90-95 percent of the sulfur in high-sulfur coal during the combustion process. Most studies show that calcium to sulfur ratios of 2.5-4.0 to 1 are necessary to achieve acceptable sulfur absorption, but utility representatives surveyed by INFORM say that a 1.5 to 1 ratio would be more economically desirable. British Coal Utilization Research Association (BCURA) pressurized test beds, however, have demonstrated sulfur reductions of 85-98 percent with calcium/sulfur ratios of 1.0 to 2.0.

The scarcity of solid information on sulfur and nitrogen oxide emissions performance standards does not appear to be the fault of the book, but rather reflects the level of research to date. The authors do claim that "...a fluidized bed combustion unit burning high-sulfur coal can *demonstrably* [our emphasis] trap enough sulfur to comply with air-quality standards *without* [their emphasis] expensive and potentially unreliable stack-gas cleaning equipment."

It is difficult to conclude from data presented whether this is correct. However, if test figures can be reliably transferred, in practice, to large-scale commercial designs, even stringent Swedish and US air-quality standards should be met. In addition, IPSEP's conversation with the Enkoping project director, suggests that nitrogen oxides could be entirely controlled within the combustion chamber. Research should soon produce the definitive answers needed before industry will risk large investments in the technology.

The question of commercialization of fluidized bed combustors is closely linked to how research and development (R&D) money will be spent. The US Department of Energy looks to spend $56 million on FBC research in 1978, but heated controversy exists among policymakers concerning what research is most needed. Some FBC-related companies want money for large demonstration fluidized bed systems (and designs for several such units have been commissioned). Others want to concentrate on specific problems such as pollutant emissions, combined-cycle technology (using the fluidized bed system for both heat and electricity generation), fuel feeding mechanisms (which have given some American developers difficulty), and more prototypes of both atmospheric and pressurized test units.

Technology Establishment Within the Next Decade

Most observers surveyed by INFORM think that FBC technology will become economically competitive within the next decade, starting with small to mid-sized (under 200 MWe) atmospheric pressure units for industrial uses.

Pressurized units—slightly more efficient thermally and in trapping sulfur, probably cheaper, more adaptable to combined-cycle operation, and lower in sulfur and nitrogen oxide emissions—will probably take longer to commercialize while being more suitable for utility and electricity purposes (this conclusion is disputed within the industry). The book suggests that American decision-makers would do well to investigate carefully the European efforts that appear to be moving more quickly towards commercial development of FBC (Sam Biondo has recently completed such a trip, and is apparently much impressed).

Fluidized bed systems compete primarily with scrubber-equipped coal-fired systems. Unfortunately, according to INFORM, FBC costs are notoriously variable. Consequently, calculating comparative investment sums for the different fluidized bed systems, themselves, much less working through

ORNL-DWG 75-9227

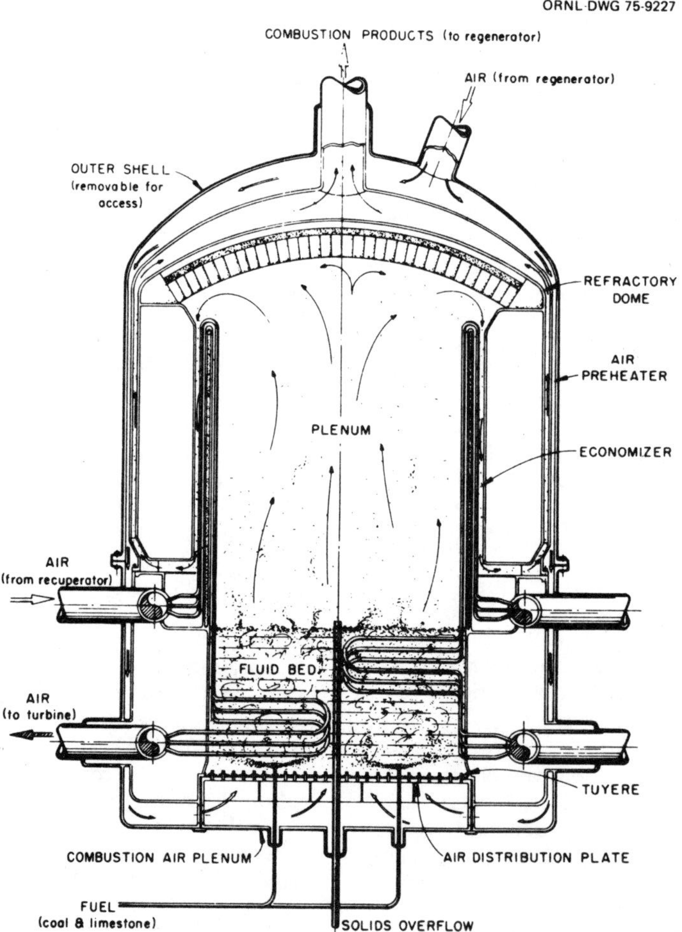

SCHEMATIC DIAGRAM SHOWING FLUIDIZED-BED COAL-COMBUSTION SYSTEM DESIGNED TO SERVE AS A HEATER FOR A CLOSED-CYCLE GAS TURBINE

This Oak Ridge National Laboratory design is unusual because it uses an atmospheric rather than a pressurized fluidized-bed combustor in combination with a gas turbine. In this system, the gas which drives the turbine is heated in tubes in the fluidized bed. Drawing: Courtesy Oak Ridge National Laboratory.

economic analyses between FBC and coal and solar technologies, have proven difficult. Indeed, INFORM quotes only R&D figures, noting that cost figures at this point, are "founded on conjecture," owing to site, company and size-specific factors.

Institutional Obstacles

Problems beyond the technology itself inhibit progress toward FBC development. In the US, bonds for pollution-control investments like scrubbers are tax exempt, but this is not the case for fluidized bed systems that control pollutants during combustion. Questions regarding future tax breaks, the general supply market, changes in environmental quality standards, applicability to electricity production without attention to combined end-uses, and uncertainty about who will make energy decisions, all retard bold investment in the technology.

Yet the surge of government and industry interest, documented by INFORM, evidences a breakthrough for FBC. The "...survey of companies active in this field suggests that fluidized bed combustion is on the threshold of becoming a major energy technology," says INFORM. For example, the largest fluidized bed combustor, developed by Pope, Evans and Robbins, Inc. (PER) is designed as a 30 MWe plant, and became operative in September, 1977. The Enkoping 25 MWt unit started using coal in March, 1978.

Designs for separate 800 MWe and 570 MWe plants are under way by Foster Wheeler Corporation/PER, and PER, Stone and Webster, respectively. Foster Wheeler is offering commercially large-scale atmospheric combustors capable of producing steam ranging up to 176 MWt, and Babcock and Wilcox Ltd. is marketing a 147 MWt unit. Fluidyne Engineering Corporation offers a FBC unit for air heating that delivers more than 3.5 MWt per hour.

A sampling of FBC activity includes work by Stal-Laval Turbin AB (Sweden) on turbine corrosion in combined-cycle units and innovative work by the relatively small Stone-Platt Fluidfire Ltd (UK) on small-scale application of FBC. One Fluidfire project focuses on a residential unit that converts about 50 percent of coal energy to radiant heat. Fluidfire also has devised a shallow fluidized bed heat recuperator unit that is fluidized by exhaust gases from an engine, instead of by burnt fuel. Such a package has been marketed and is in use in a Norwegian ship to augment steam production.

Coming To A Boil persuasively advocates more fluidized bed activity. The authors say they hope the book will help encourage the inclusion of FBC technology in energy planning discussion. The book's format and style lend themselves to that goal.
—David Chatfield

Source: *Fluidized Bed Energy Technology: Coming To A Boil*, by Walt Patterson and Richard Griffin. Published by and available from INFORM, 25 Broad Street, New York, New York 10004, USA.

Economists Stress Material Matching, Reuse

The first and second laws of thermodynamics offer only a starting point for calculating the inefficiency of national energy use, suggests a report by two economists, Robert Ayres and M. Narkus-Kramer. Most studies of national energy efficiency, notably that of the American Physical Society, roughly calculate that the American economy has a second law efficiency of 6 or 7 percent. In their novel approach to quantifying efficiency, Ayres and Narkus-Kramer look beyond the first and second laws of thermodynamics to the re-use of materials from existing capital stock and the efficiency to delivering real services (not just energy) to users. To that end, the authors pointedly argue for recycling basic materials and matching construction materials to structural requirements. Taking these factors into account—recognizing the extreme difficulties in their quantification—leads the authors to suggest the nation's energy efficiency is between 1 and 2 percent.

U.S. Energy Efficiency Only 1-2 Percent?

Ayres and Narkus-Kramer share the perspective of the economist, interested in final consumption of services, rather than physical consumption of resources. "Although the goods that provide services are somewhat difficult to distinguish from the physical materials from which they are made, the services themselves are essentially non-material in nature." Approaching the problem of increasing desires as non-Malthusians, Ayres and Narkus-Kramer argue "there is no fixed relationship between the extent to which final demand is satisfied and the quantities of physical materials or energy required along the way." Conservation potential is limitless.

Recycling Saves 'Embodied Energy'

After a brief discussion of published work on the first and second law efficiencies of energy use, the authors move into efficient use of materials. "...Consider the use of tin for plating sheet steel in cans. A minimum quantity of tin can be specified based on the thickness required to form a rustproof protective coating. Any extra use is not required by the function and represents an inefficiency in the design or manufacturing process."

The first major distinction drawn by the authors about materials use is between consumables and durables. Generally, they must be treated separately. "To estimate the efficiency of materials use in existing applications, one would have to define the 'ultimate' material for each application and determine its embodied available energy...We know of no comprehensive attempt to assess the future potential for improvement in materials performance or the implications for embodied energy. Such a study would obviously be a very difficult and challenging one.

"Altogether probably no less than 18 quads were used in 1970 for processing materials (including food) from the raw state to finished form; while no more than 5 quads were used for all other industrial activities...Of this total, most of the iron and steel, all the cement and a substantial fraction of other metals and glass go into *durables*, while most of the paper and almost all of the chemicals are used for *consumables* (including finished fuels).

"From an energy standpoint, there are three ways to reduce the energy embodied in durable materials. The first, which we frankly cannot evaluate quantitatively—but which is unquestionably of major importance—is to reduce or eliminate the amount of material that is required....The other two ways of increasing the efficiency of embodied energy use are, respectively, to increase product lifetimes and to increase the recycling rates of their contained materials....the extent to which aluminum *could* be recycled, but is not, is a measure of inefficiency of energy use.

"On an energy weighted basis, no more than 4 percent of consumable materials (mainly paper) are recycled. For durables, the figure is probably closer to 20 percent."

Services Are Rarely Matched To Needs

Ayres and Narkus-Kramer then look to the payload or "utility efficiency" of supplying services to users. Energy conversion and energy/materials transformation, they note, can be analyzed in thermodynamic terms, but these are not supplying real services to human beings, only "pseudo-services." The purpose of an automobile trip, they add, is not to move the car, but the passengers and their belongings. "The second law efficiency for supplying energy to the rear wheels of the car does not reflect inefficiencies arising from the fact that the vehicle 'shell' is larger than it really needs to be to accommodate the average payload, not the fact that it is much bigger and heavier than it has to be to serve its primary function of providing a comfortable moving environment for the passengers."

The authors suggest that space heating systems could be designed to heat only occupants of a building, not unoccupied space or the structure itself; that ultrasonic sound could be used to loosen dirt from clothes; that cooking appliances could be created that heat only the food; and that illumination could be work-specific and more efficient, with plastic light pipes and

TABLE 6. ALLOCATION OF ENERGY AMONG SERVICE-GENERATING FUNCTIONS

Service Function	Fraction of U.S. Energy (Reference 12)	"Second Law" Thermal Efficiency E (Reference 5)	Utility or Payload (b) Efficiency m	Composite Efficiency Ex.m	Total Available Energy in $\times 10^{15}$ BTU (1971)	Available Energy Loss $\times 10^{15}$ BTU (1971)
Transportation						
Truck	0.049	0.11	-0.5	-0.056	3.39	3.21
Bus	0.00196	0.2	-0.3	-0.060	0.136	0.127
Rail (diesel-electric)	0.0098	0.3	-0.5	-0.150	0.68	0.578
Air	0.0196	0.3	-0.1	-0.030	1.36	1.319
Military, Pipeline, etc.	0.0392	0.3	-0.2	-0.060	2.7	2.556
Automobile	0.127	0.11	-0.2	-0.022	8.8	8.6
Total or Average	0.247	0.163			17.14	16.438
Commercial[a]						
Space Heat	0.0605	0.06	-0.2	-0.012	4.16	4.11
Water Heat (Laundry)	0.0096	0.03	-0.5	-0.015	0.66	0.654
Cooking	0.0017	<0.2 (b)	-0.3	<0.060	0.12	0.113
Refrigeration	0.0096	0.04	-0.5	-0.02	0.663	0.6507
Air Conditioning	0.0157	0.05	-0.2	-0.025	1.086	1.059
Other (including light)	0.0149	0.05	0.1	<0.005	1.025	1.021
Total or Average	0.112	0.055			7.79	7.414
Residential						
Space Heat	0.086	0.06	-0.3	-0.018	5.952	5.84
Water heat (wash)	0.022	0.03	-0.2	-0.006	1.569	1.56
Cooking	0.0086	<0.2 (b)	-0.3	<0.06	0.595	0.559
Clothes Drying	0.0023	<0.1 (b)	-0.2	<0.02	0.162	0.159
Refrigeration	0.0086	0.04	-0.5	-0.02	0.595	0.583
Air Conditioning	0.0054	0.06	-0.3	-0.015	0.379	0.372
Other (including light)	0.016	<0.05	<0.1	<0.006	1.136	1.13
Total or Average	0.1506	0.0615			10.45	10.2

(a) Asphalt paving is omitted from this "end-use" category and included among basic materials.
(b) Author's estimate.

TABLE 9. EFFICIENCY OF THE U.S. ECONOMY AS A WHOLE

Conversion Step	Fraction of U.S. Energy in (b)	Total Available Energy in 10^{15} BTU (1971)(c)	Efficiency Definition	Estimated Efficiency of Step (c)	Fraction of U.S. Energy Lost (c)	Available Energy in x10^{15} BTU (1971) (c)
RM-FF (a)	1.00	69.38	Available Energy, Finished Fuels (Out) / Available Energy, Raw Fuels (In) (a)	0.81	0.19	13.20
FF-FM	0.30	22.25	Available Energy, Finished Materials (Out) / Available Energy, All Sources (In) (b)	0.33	0.21	14.62
FM-DG	0.055	3.81	Fraction of Finished Materials Allocated That is recycled to Durable Goods (DG)	-0.20	0.043	- 3.04
FF-DG-FS	0.51	35.36	Miminum Finished Fuels Input to Durables Required to Yield Final Services / Actual Fuels Input to Durables	-0.04	-0.49	34.24
FM-CG-FS	0.055	3.81	Fraction of Finished Materials Allocated to Consumable Goods (CG) That is recycled	-0.04	0.054	- 3.76
TOTAL	1.00	69.38		-0.016	-0.984	68.26

(a) Including Electricity
(b) Not including available energy in recycled iron and steel, chemical feedstock and wood pulp. Assuming no available energy in metal ores.
(c) Including terms omitted above, for convenience.

light-emitting diodes that convert 10-20 times more of the input electrical energy to visible rays than incandescent or fluorescent bulbs.

In conclusion, Ayres and Narkus-Kramer trace energy losses from raw fuels to final services, considering thermodynamic losses, materials recycling, and payload efficiency. The latter, they believe, "offers significant long-term opportunities for energy conservation. An exploration of the magnitude of the potential would seem to have high priority in view of widespread public misconceptions about the overall efficiency of energy use in the United States economy.

"The most surprising result (of our work), of course, is the national total. Based on our analysis, the efficiency of the United States economy is in the range between 1 and 2 percent. While estimates of utility or payload efficiency are obviously very rough and crude, we have tried to make them fairly conservative. In each case, we are satisfied that a detailed analysis would probably lead to a *lower* estimate of efficiency than that shown."

—Jim Harding

Source: "An Assessment of Methodologies for Estimating National Energy Efficiency," R.U. Ayres and M. Narkus-Kramer, reprinted from the publication of the American Society of Mechanical Engineers, United Engineering Center, 345 East 47th Street, New York, New York 10017; paper #76-WA/TS-4.

Efficiency Improvements: The Largest Energy Source

WHILE CONTEMPORARY ENERGY studies consider more efficient energy use to be important, no two studies agree on the quantitative potential of efficiency improvements. Even thoroughly-documented low-energy scenarios differ. Measures to improve energy efficiency are often selected with an eye towards short- and medium-term possibilities, and with an awareness of economic and political constraints. Analysts thus fuse their technical assumptions with economic ones about prices of energy and GNP growth rates. They also express in these assumptions their political dispositions about what they think is desirable and acceptable.

However, since energy scenarios are intended to guide policy, a uniform yardstick is needed by which political judgment and technical possibilities can best be separated. Such a yardstick would be provided by estimating the total theoretical potential of efficiency improvements. We could then compare scenarios by asking how much of this reservoir of technical fixes has been drawn upon by each scenario, by what time, and through which measures.

This approach, simply stated, treats efficiency improvements as yet another energy source. All conventional and renewable sources have a theoretical potential (e.g., solar radiation incident on the earth's surface) and a technical potential (e.g., the solar radiation that can be converted with state-of-the-art collector technology while observing land use constraints). The theoretical potential of efficiency improvements is defined in thermodynamic terms by the theoretical second-law efficiency of 1.0. The technical potential can be defined as the efficiency which would be achieved if the best available technologies were used in all energy applications in the economy.

I have estimated these two potentials for the Federal Republic of Germany based on 1973 statistical data, the last year before the so-called first oil crisis. The purpose of the calculation is to obtain an order of magnitude for the size of this energy source in an industrial country that is relatively efficient, as well as to illustrate a new methodology.

Methodology

Energy use involves a series of *conversions* of energy, followed by an ultimate *translation* into an energy service, such as a warm house. The conversions can be divided into direct uses of energy (operation of engines, motors, processes, etc.) and indirect uses (the consumption of materials in which energy has been embedded). These conversions are thermodynamically governed. Previous estimates of national energy efficiency have been limited in that they considered only thermodynamic conversions, and calculated only the theoretical potential for their improvement.

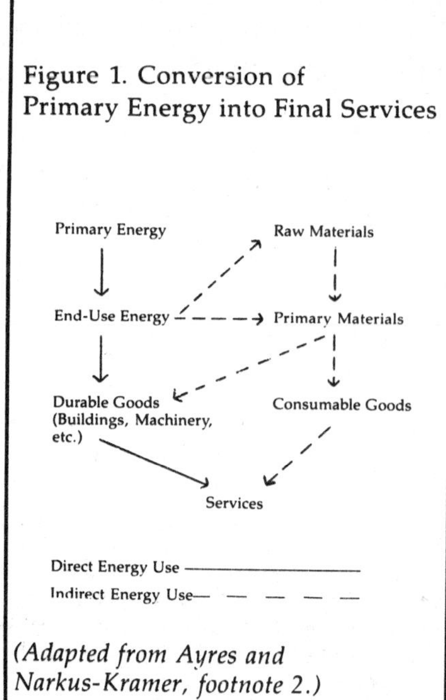

Figure 1. Conversion of Primary Energy into Final Services

(Adapted from Ayres and Narkus-Kramer, footnote 2.)

Yet the translation into an energy service is an equally decisive step in energy use. It is accomplished with specific technologies: building shells, vehicle shells, refrigerator shells, and others, all of which I will refer to as *service technologies*. Each service technology has a certain energy intensity (e.g., heat loss per degree-day and unit area, or roll and drag resistance). No theoretical minimum energy intensity can be defined for service technologies or energy services, which is why they are not included in second-law efficiency estimates. *Service efficiency* can only be compared to a relative standard, such as the best available technology.

The calculation of an overall national efficiency requires that all the steps in energy use (as depicted in Figure 1) be evaluated by such a relative standard. Analysis falls into three parts: second-law efficiencies for direct energy conversions, recycling efficiencies for indirect conversions of embedded energy, and service efficiencies.

Second-Law efficiency

Table 1 presents average second-law efficiencies in the Federal Republic of Germany for four end-use categories: space heat, process heat, liquid fuels for transport, and electricity-specific functions. The largest fraction of all end-uses is space heat (40 percent); it also has the lowest second-law efficiency, with less than two percent primary energy becoming useful energy in the room. Other applications show considerably higher val-

ues, but the overall second-law efficiency in the FRG is estimated to be barely ten percent. Estimates for the US and the British economy arrived at second-law efficiencies of six and eight percent respectively (Ayres and Kramer, 1976; Ford *et al.*, 1975).

Embedded-energy efficiency

Of the energy embedded annually in materials, only an estimated 13 percent was captured for re-use in 1973 (energy expenditures for recycling were accounted for). As a rough approximation, the embedded fraction of energy can be identified with the process heat fraction in Table 1. The combined direct and indirect efficiency of process heat then becomes two percent. This shows that the conversion efficiencies of space and process heat, when considered on a whole system basis, do not differ substantially.

Combined direct and indirect conversion efficiencies

Since heat applications account for 75 percent of all end uses in the FRG, it is not surprising that the combined conversion efficiency for all end uses is only 4.5 percent. An equivalent estimate for the US economy is about 1.6 percent.

Service efficiency

The following specifications for best available technology are employed:

● SPACE HEAT: buildings with heat loss characteristics equivalent to the Saskatchewan superinsulated house (0.1 kWh/m²/°C day). This is about one-tenth the heat loss of a typical West German building.

● TRANSPORT: cars weighing about 500 kg, with an aerodynamic drag resistance of 0.25, as exemplified by recent manufacturer prototypes. This

standard represents a 55 percent improvement over average 1973 FRG car technology. Overall, the service efficiency standard in transport is estimated at 50 percent.

● ELECTRICAL APPLICATIONS: the two major categories are appliances and industrial electrical drives. Appliance performance standards described by Nørgaard in the Danish DEMO study are assumed here, corresponding to a 65 percent reduction in the energy intensity of the 1973 West German appliance stock (see *Soft Energy Notes* I:8). Manufacturing systems and product design, however, do not lend themselves well, if at all, to the definition of a standard for best available technology. I have therefore set the service efficiencies of industrial drives at 1.0, and taken the same standard for process heat applications. In a sense, the service aspects of products are ac-

counted for in the recycling fraction included in this analysis. Weighting the industrial drive fraction of electricity-specific end uses at 1.0 and the appliance and lighting fraction at a service efficiency of 0.35, the overall value for these end uses is 0.80.

The weighted average of the service efficiencies in all applications is 53 percent, as depicted in Table 2.

Combined Conversion and Service Efficiency: The Technical Potential

A lower limit for the theoretically possible increase in energy efficiency can be obtained by multiplying the service efficiency value of 0.53 by the total thermodynamic efficiency of 4.5 percent previously derived. This yields a value of 2.3 percent, meaning that the energy thrift of the FRG could theoretically be increased by a factor of 44 or more.

Table 1 **Conversion efficiency of direct and indirect energy use in the Federal Republic of Germany** *(1973)*

Application	Fraction of end-use energy	Efficiency direct	Efficiency indirect	Combined Thermodynamic efficiency, weight
Electricity	0.06	0.17	——	0.0102
Liquid fuels for transport	0.19	0.10	——	0.0190
Space heat	0.40	0.02	——	0.0080
Process heat	0.35	0.17	0.13	0.0077
TOTAL	1.00	0.0964		0.0450

(Adapted from Ayres and Narkus-Kramer, footnote 2.)

By comparison, the combined direct and indirect thermodynamic efficiency of the US has been estimated at less than two percent. Obviously, national differences are considerable, but they are small compared to the tremendous potential for improvement that exists in all these industrial nations.

The Technical Potential

To assess present technical possibilities, this approach must be extended to the technologies whereby energy is converted and recycled. In space heating, a well-regulated system using cogeneration can attain a first-law efficiency in excess of 80 percent, compared to only 52 percent in the average 1973 heating system. The energy quality from a solar or cogeneration unit is a three-fold better match than direct fuel combustion to the very low quality requirements of space heating. Thus, the relative efficiency of the average West German heating practice, when compared to the best technology available, is estimated to be 19 percent.

In transport, the estimated first-law drive train efficiency achieved in the 1973 stock of cars was only ten percent, and not much higher for the other transport modes. With modern low-friction materials, turbo-charged diesel engines to replace conventional Otto cycle designs, continuously variable transmissions, and the elimination of idling, this average efficiency can be raised to 20 percent. On this basis, the relative conversion efficiency of transport fuels in 1973 was only 58 percent.

In electricity-specific applications, major losses are due to part-load conditions and to oversized motors in industrial drives. Estimates of the average first-law efficiency in industrial electric drives range from 40 to 60 percent for West Germany. Modern power electronics allow for average efficiencies in excess of 94 percent. In lighting, the best available lamps need less than a third of the electricity required by incandescent lamps. The overall relative efficiency compared to the state-of-the-art technologies is estimated to be 65 percent.

Finally in process heat applications the average second-law efficiency can be raised from an estimated 17 percent in 1973 to about 35 percent with modern processes and by consequent application of cogeneration, recuperation, regeneration, insulation and heat cascading. On that basis, the relative efficiency of process heat applications was 49 percent here.

| Table 2: | Comparison of status quo and state of the art efficiencies in West German energy use (1973). | | | |

Application	Fraction of end-use energy	Relative efficiency conversion	service	(1973/best) combined
Electricity—Specific	.06	.71	.80	.57
Liquid Fuels for Transport	.19	.58	.50	.29
Space Heat	.40	.19	.10	.019
Process Heat	.35	.32	1.00	.32
Weighted Average		.40	.53	.208

More difficult to define is the state of the art in recycling of materials since it is to a large degree a matter of social organization rather than one of available technologies. Clearly the potential is large. Conservatively, a 65 percent relative efficiency has been assumed for the FRG.

The overall relative efficiency of the West German economy is then 20.5 percent, as shown in Table 2. This means that the same standard of living and level of economic activity which was reached in 1973 can be provided with one-fifth the energy used by the FRG in 1973. Or conversely, a five-fold increase in the standard of living over that of 1973 would be possible without any increase in energy demand — if we were to exhaust only the efficiency improvements contained in presently available technologies.

Comparison of the technical potential (improvement by a factor of five) and the theoretical potential (improvement by at least a factor of 44) shows that even after the present state of the art in energy use becomes general practice, another tenfold improvement is possible through future innovation.

A Conservative Scenario

How much of this technical potential for more efficient use of energy has been included in the West German soft path scenario (Krause, 1980)? In this scenario, the national energy thrift increases by a factor of four, as measured by the GNP/energy ratio. This allows the FRG to increase GNP 2.3-fold by the year 2030, while lowering its primary energy use to 60 percent of 1973 levels. The effect of technical fixes alone is to increase national efficiency by a factor of two, while a further factor of two results from growth-induced structural changes in the composition of the GNP. Thus, only 40 percent of the technical potential is exhausted.

While analysts from established energy interests have described this scenario as an unrealistic program of all-out efficiency, it is really based on rather conservative assumptions. Thus, continued use of larger than desirable cars is allowed for, existing buildings are only retrofitted to Swedish thermal insulation standards, increased recycling and some improved industrial processes now under testing have not been considered, and a smaller amount of low temperature heat is still produced directly from fuels rather than in cogeneration units.

These conservatisms, which contain the "missing" 60 percent of the technical potential for efficiency improvements, underscore the pragmatism and realism of the scenario. Yet, even this modest approach would allow the FRG to shut down all its nuclear reactors immediately and for good; by 2030 the FRG could have entirely phased out oil and natural gas as fuels, and be totally energy self-sufficient; the continued use of coal (Germany's only significant fossil resource) at the present level of consumption would suffice as a transitional strategy; by 2030, 45 percent of all energy needs would be supplied from renewable sources.

Politicians and industrial leaders are finding it harder these days to explain why they have failed to significantly tap our largest energy source: efficiency improvements.

— Florentin Krause

Mr. Krause, a physical chemist, is the former director of Freunde der Erde, *and author of* Energiewende, *a soft energy*

Cogeneration: The Canners Can

A miniature mountain range of biomass wastes rises in a field beside the Modesto, California cannery. Heaped up in fourteen-foot high piles, the pits, shells, and other cannery operation byproducts used to present a disposal problem and cost the company money to get rid of. Now, the biomass materials are a source of energy and will soon be a source of revenue.

THEY ARE FED into a boiler which produces more steam than the canning operations require; the remainder can spin a turbine, generating enough electricity to handle the plant's needs, plus a surplus to sell back to Pacific Gas & Electric, the local utility. In addition to the almond and walnut shells, and the peach, prune, olive, and cherry pits left over after processing, the boiler is fed with other woody materials normally discarded: cotton gin trash, corn cobs, tree prunings, sawdust, and coffee grounds. It can extract energy from almost anything that burns.

The canning company, Tri-Valley Growers, initiated this pioneer effort when faced with the replacement of a production facility destroyed by fire. But it was not the first time they recognized the potential energy value of biomass. In 1978, one of their existing boilers had been converted to handling biomass fuels, with the encouragement of financial incentives from California's State Energy Commission. The success of this venture helped convince the company's Board of Directors to take a chance on an unproven technology, by designing and building a canning facility powered *entirely* by biomass.

Waste combustion

S & W Fine Foods (a subsidiary of Tri-Valley) Plant #9 is unlike most canning plants which direct-fire their biomass wastes. Other boilers use a "fluidized bed" of red hot sand, continually fluffed up by air blowers, to combust ground-up particulate matter. The S & W plant employs pile burning: biomass materials as large as three inches in diameter are conveyed in metered amounts to three separate chambers and incinerated in piles, a method which results in very complete combustion. The steam thereby generated cooks vegetables and bean products, cleans cans, and will spin turbine blades when the electrical generator installation is completed in June.

"Since we started up our boiler in October of 1980 we have not had to fire any oil or natural gas with it—which is uncommon," commented Dr. Peter Thor, an economist who has worked closely with the project. "In our combustion chamber we've managed to generate more steam than the manufacturer thought we would. In effect, we've been able to generate more with less. We're totally pleased with the system."

Many other companies are watching the project closely as the operating bugs are worked out. "A lot of companies have contacted us," Dr. Thor said, "a lot are hanging on our shirttails."

Among the problems to be considered was the storage of the biomass material—some 30,000 tons annually are needed to produce 26 to 28 million kilowatt-hours and supply the steam requirements of the plant. A field beside the cannery is heaped with almond and walnut shells, and various fruit pits. "We're currently storing it in piles about 14 feet high. To get it any higher than that, the material compacts and will spontaneously combust. Or it will start smoldering on you. If you have the pile lower than that, when the rain comes, it soaks the material. So there's a lot of technology that is back-to-basics which we're having to relearn and develop."

Drying the biomass materials creates its own special problems, although the S&W generator can accommodate biomass up to 50 percent moisture content. California's abundant solar energy is used for drying, but an area outside of town must be used (creating transportation expenses). The materials are spread in thin layers over an abandoned airfield, and will generally dry sufficiently for use in a short period of time even in the winter. Solar grain drying techniques being developed in the Midwest by progressive farmers may prove helpful.

Although the initial capital investment for this system is high, fuel savings are expected to pay for the cogenerator within three or four years. Tri-Valley Growers' network of farm producers supplies much of the biomass fuel within the 40-mile radius beyond which transportation costs become prohibitive. "We have about 40 percent of our fuel as internally generated waste. If we then count our nut plant, we get around 60 percent internally generated waste materials. Some of the materials we gener-

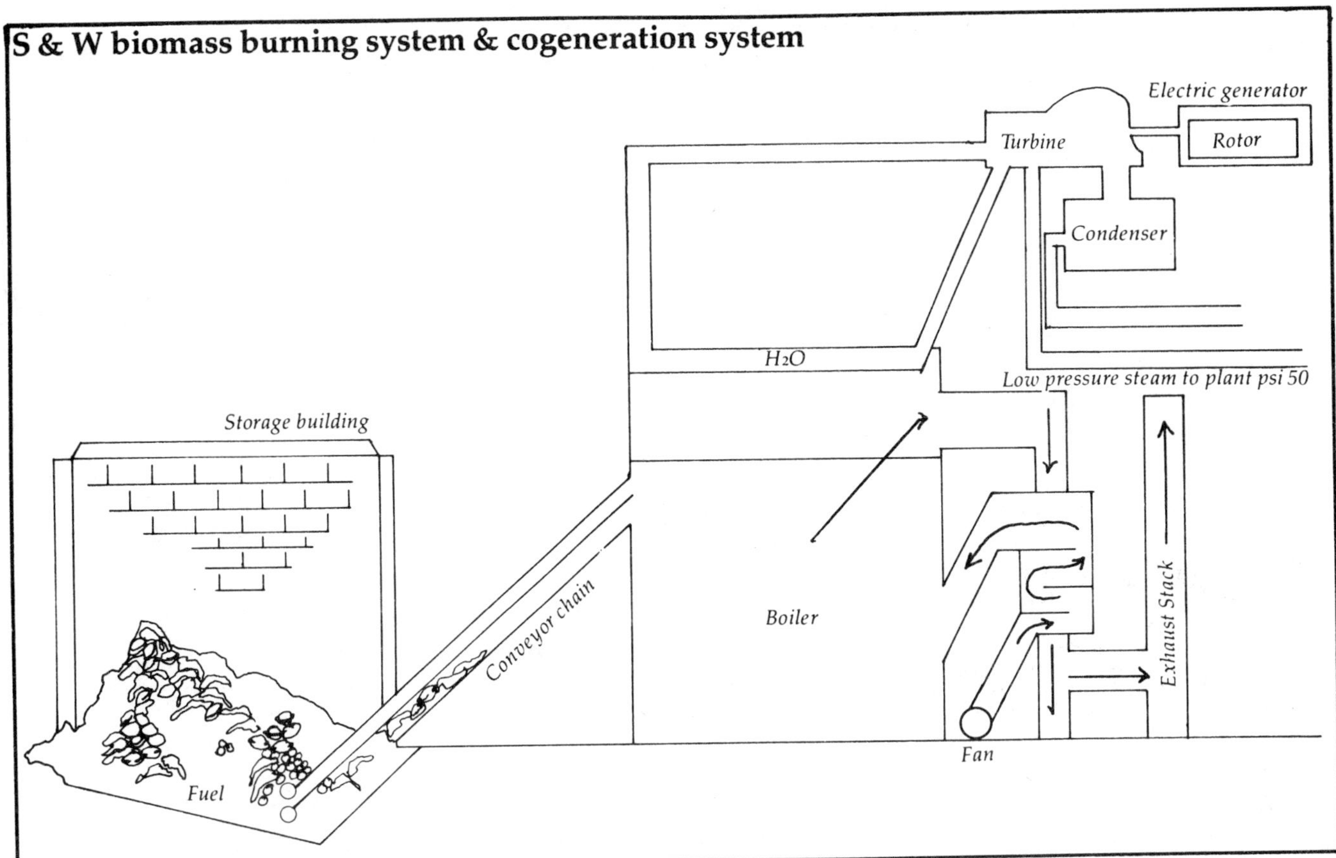

ate can be used as cattle feed, so what we've been doing is to let those materials which have a better use be used for that purpose, such as feed for animals. We'll use something that would have to be burned or dumped in some way. So we want as much as possible to use those fuels which have no other value."

Anti-pollution payoff

Agricultural burning has always created air pollution problems in California's San Joaquin Valley, which is aggravated by the motionless and stagnant air masses over the valley November through January. Wayne Morgan, an official with the Air Pollution Board in Stanislaus County, is optimistic that controlled incineration of agricultural wastes in burners like S&W's may help reduce air pollution problems. However, another biomass generator, operated by a company in Stockton, has been plagued by air emission difficulties. This may be due to its air suspension system of burning—less efficient than pile burning. Morgan also feels that carbon monoxide, sulfur dioxide, and other pollutants will be considerably less for the S&W biomass generator than for a comparable fossil fuel system.

S&W Plant #9 is expected to save 90,000 barrels of oil in a year (or the

equivalent amount of natural gas). Fifteen pounds of peach pits is roughly equivalent, both in size and energy content, to a gallon of gasoline. "Only in the last six months or so," Dr. Thor said, "has biomass energy been recognized as a substitute for petroleum. Because it really is."

S & W Plant #9 is expected to save 90,000 barrels of oil in a year (or the equivalent amount of natural gas). Fifteen pounds of peach pits is roughly equivalent, both in size and energy content, to a gallon of gasoline.

Tri-Valley is also experimenting with ethanol production, though the efficiency of the process is much less than with direct burning. Almost 100 percent of the energy within the biomass fuel is converted to heat and energy with direct burning. With ethanol production, only 30 to 40 percent of the original energy contained in the biomass is recovered.

Byproducts of the direct firing of biomass include potash, which can be used by farmers, and a hard shiny slag, which can be used as aggregate.

According to the US Office of Technology Assessment, wood, plants, and other biomass fuels could supply as much as 20 percent of the US annual energy requirements by the year 2000. Currently, approximately 2.5 percent of our energy needs come from biomass materials.

Awareness appears to be growing about the potential of biomass and cogeneration systems. The Bio-Energy World Congress convened in Atlanta in April of 1980, bringing together 1,700 scientists, businessmen, and policymakers. The leaders in the field at the moment are Sweden, Brazil, and China, as well as the US. Many of the key figures seem to have a healthy sense of caution over possible disruptive effects to the environment if biomass systems are not used wisely, such as overharvesting of forests, or the misuse of food resources.

These are not problems for the San Joaquin Valley's cogenerating canners. Tri-Valley Growers are planning additional cogeneration units at other plants and hope to become a net producer of energy rather than a consumer.

—Lee Purcell

1. The Efficiency of Industrial Economies

In part I of this reader, we looked at the potential of efficiency and renewables in individual applications and energy sectors. To understand the implication of these data for options in energy policy we must look at the flow of energy from well head to final use. This involves defining the various efficiencies in the flow of energy (e.g., from petroleum at the well-head to crude oil delivered at the refinery, to diesel fuel and gasoline at the gas station, to motive power at the rear wheel, and to the final transport service as motive power gets dissipated into the environment through tires and air drag).

One standard of efficiency lies in the laws of thermodynamics. Against these stern criteria, the overall efficiency of industrial economies is dismal: 6 percent for the US, eight for the UK, and 10 percent for West Germany. Almost all "available work" contained in our increasingly precious fossil fuels is dissipated unused. This should be a warning by itself. Of course, 100 percent efficiency, or any figure near there, is not to be expected. But what kind of national efficiency could be achieved with the improved techniques of Part I? Based on an analysis of the Federal Republic of Germany, one would conclude that industrial economies could live on one-fifth their present energy. In fact, the FRG potential probably represents a lower limit; most other coun-tries' present energy efficiency is not as high as Germany's. Because most energy saved with thriftier technologies is much cheaper than other sources of supply, this technical potential also approximates what is economically advantageous.

This potential assures us that we can eventually run even highly industrialized economies with renewables. With this knowledge, energy policy can be dramatically redirected. Until now the official rationale for investing in new forms of future energy supply, including synfuels, nuclear fission, breeder reactors and even nuclear fusion—no matter how expensive, socially damaging and environmentally dangerous they all are—has been that with growing energy demand all new sources of supply are urgently needed.

By contrast, the conservation potential of presently available efficiency technologies gives us a choice between a renewable and nuclear/fossil futures.

Implementing that choice is the subject of Part IV. Meanwhile, we might note that between 1973 and 1980, more than eighty percent of OECD economic growth was "financed" by a combination of technical leak-fixing, structural shifts and behavioral adjust-ments. This encouraging evidence is surprising against the backdrop of grossly inadequate conser-vation policies in those countries.

A Focus on Efficiency

THE TERM "EFFICIENCY" changes meaning as it moves from one realm of thought to another. Engineers calculate it one way, economists a second, and management analysts yet a third, while the general public inherits a notion of efficiency that is a vague amalgam of all of these. In this issue of *Soft Energy Notes*, we report on recent research and development in engine and motor efficiency. To put this material into context, it is important to understand precisely *how* efficiency is defined for each of the different ways it is used and *the limitations* of each definition.

The early scientific definitions of efficiency derive from thermodynamics which employs two distinct measures—one following from the First Law of Thermodynamics, the other a product of the Second Law.

First Law Efficiency =

amount of energy delivered in form and place desired (output)

———————————————————————————————

amount of energy applied to achieve the desired effect (input)

"Energy," in this equation, can take many forms: heat, work, fuel, electricity, the potential energy of water behind a high dam, the kinetic energy of wind, etc.

Second Law Efficiency =

minimum amount of energy needed to perform a function

———————————————————————————————

actual amount of energy used to perform the function

Determining the "minimum amount of energy needed" uses a mode of analysis developed in 1824 by a young French engineer, Sadi Carnot.

Often referred to as Carnot Efficiency, Second Law Efficiency is determined using the following relation, which refers to the accompanying diagram:

Theoretical maximum efficiency $\dfrac{w}{q_2} = \dfrac{t_2 - t_1}{t_2}$

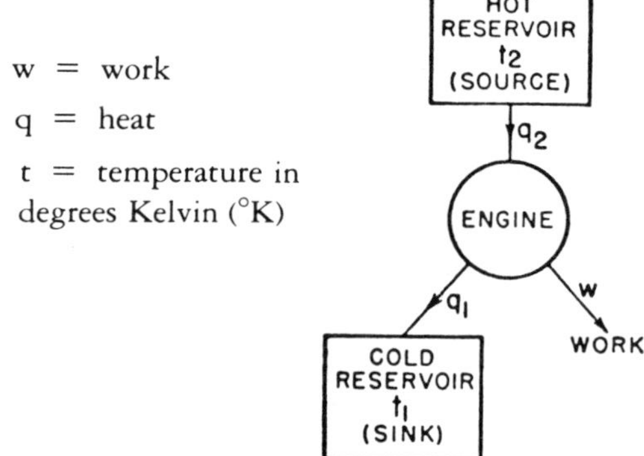

The diagram depicts a *heat engine*, which takes heat (q_2) from a hot reservoir (such as a combustion chamber or boiler) and transforms that heat into useful work (w). "Waste heat" (q_1) is exhausted into a colder environment. In the Second Law equation, q_2 represents the minimum amount of heat required to perform a specified quantity of work. Conversely, w indicates the maximum work that can be extracted from a given amount of heat using an engine that operates between temperatures t_2 and t_1.

With larger temperature differences, more work can be extracted from a given amount of heat, and the engine is more efficient. This principle underlies many common engineering practices, such as the coupling of high temperature-pressure boilers and large cooling systems in electricity generating stations.

In *heat pumps* the same general relations exist, but the direction of action is reversed. In a refrigerator—the most common type of heat pump—work is applied in order to extract heat from a *cold* reservoir (the inside of the refrigerator) and exhaust it into a *warmer* reservoir (the environment). In this situation, w represents the minimum amount of work needed to move a given amount of heat (q_2) from the cold reservoir to the warm reservoir, in a heat pump working between temperatures t_1 and t_2.

A few calculations will illustrate the difference between heat engines and heat pumps. A furnace, for example, is a heat engine which delivers its work (w) as heat. A well-designed home furnace in good operating condition uses, in steady state operation, about 1.25 units of fuel energy to deliver one unit of heat into the ducts.

$$\text{First Law Efficiency} = \frac{1}{1.25} = 0.80 \text{ or } 80\%.$$

Twenty percent of the fuel's energy is being dissipated, or wasted, and not supplied to the heating system. Now consider the furnace's Second Law Efficiency. The minimum amount of work needed, by the Second Law, to deliver a quantity of heat (q_2) into a room at $20°$ C ($293°$ K), when the outside temperature is, say, $0°$ C ($273°$ K), is given by the equation:

$$w = q_2 \cdot \left[\frac{293° - 273°}{293°} \right] = 0.07q_2$$

That is, the minimum amount of work needed to deliver q_2 units of heat into a room is .07 times that amount of heat. The furnace delivers 1 unit of heat while consuming 1.25 units of energy. Its Second Law Efficiency:

$$\frac{\text{minimum energy}}{\text{actual energy}} = \frac{(0.07)(1)}{1.25} = 0.56 \text{ or } 5.6\%$$

While a furnace burns high grade fuel to *produce* low grade heat, a heat pump burns fuel to run a pump which *moves* heat from a low temperature reservoir to a high temperature reservoir. Working between temperatures of $0°$ and $20°$ C, as our example furnace does, a heat pump would use about 0.5 units of energy in fuel to deliver one unit of heat into a room. For such a pump:

$$\text{First Law Efficiency} = \frac{1}{0.5} = 2.0 \text{ or } 200\%$$

$$\text{Second Law Efficiency} = \frac{0.07 \times 1}{0.5} = 0.14 \text{ or } 14\%.$$

The practical conclusion that may be drawn from this relation is that changing devices, from furnace to heat pump, produces much greater increases in efficiency than can possibly be made through improvements in the furnace itself. A furnace can only approach a First Law Efficiency of 100% as an upper limit, while heat pumps range between 150% and 300%. Heat pumps work most effectively when the temperature difference between the two reservoirs is small, and they prove to be most useful in mild climates, or else where seasonally stored heat, or even ground water heat can be used to minimize the difference in temperature between the two reservoirs.

System Boundaries and Efficiency

First and Second Law Efficiencies are calculated for devices or systems with carefully defined boundaries. But in actual operation, these devices are usually components in larger systems. A furnace, for example, is only one part of an entire heating system. The efficiency of the heating system as a whole is determined by taking the ratio of the heat delivered in rooms to the energy put into the system. If a furnace is over-sized for its heating load and if it cycles on and off frequently, then energy is wasted in starting up (when only the furnace is being heated, not the house) and in stopping (when the furnace cools). In the United States, both building codes and engineering conventions encourage over-sizing, and even well designed, well maintained furnaces commonly operate at or below 45% First Law Efficiency, though their bench-test performance may be much higher (see *Notes* 3:1:28). Heat losses elsewhere in the system—heating up and cooling off of ducts, leaks from joints, etc.—cause further reductions in First Law Efficiency.

A highly efficient heating system, however, is not an end in itself, but rather a means of providing a certain level of comfort or amenity in a building. Perhaps a better measure of efficiency would consider the "minimum amount of energy needed to maintain room comfort." This measure would take us beyond the domain of furnaces and ducts into an analysis of the efficiency of the entire structure: its design, the details of construction (e.g., sealing of cracks and joints) and, indeed, even the behavior of its occupants. Using heat in drafty, poorly insulated or little used rooms is inefficient (see *Notes* 2:10-18; 3:1:18-26). Further, if a heating system is designed to maintain a temperature of $20°$ C even during the coldest day of the century, then the furnace will be over-sized for average conditions and run below optimum efficiency the majority of the time. If occupants are willing to wear sweaters on rare occasions, then greater efficiency during routine operations becomes possible. Systems in general are designed for efficiency at specified loads or demands. These demands, however, are set by social factors that lie beyond the present purview of engineering analysis. Yet as the social conditions that affect demand change, then analyses of efficiency should, ideally, change in parallel.

Moving our frame of reference from furnace to house—and more generally from component to system—shows some of the limits of First and Second Law calculations of efficiency. Enlarging the boundaries of a system one step further makes the concept of efficiency even more complex. Consider an electric generating plant in terms of our diagrammatic representation of a heat engine. Waste heat is exhausted into a "low temperature reservoir," which in this case is the environment. The implicit assumption here is that the environment has an infinite capacity to absorb waste heat. But if the reservoir is changed in such a way that energy—including social energy—must be expended to deal with the consequences of waste heat, then the original calculation of power plant efficiency is incomplete. Power plants, like furnaces, are components of larger systems. A focus on the efficiency of one system component not only ignores the costs of external effects, it also tends to overlook large increases in efficiency that may be available elsewhere in the system—once analytical boundaries have been enlarged. This shift of attention brings the question of "load" or "demand" into focus, and the social factors that condition demand become central to the analysis of efficiency.

In engineering analysis, loads or demands for energy are taken, by convention, as exogenous variables. For purposes of *design* they are viewed as given and invariant. For purposes of *planning* they are usually assumed to increase without limit over time. Yet demands are conditioned and formed by any number of time-dependent factors: by prices, by public policies, by events and circumstances. The relations between prices and demands are within the province of economics; while policies and circumstances are likewise important, they receive less systematic attention. Only recently have we begun to re-evaluate the policies that indirectly increase energy demand, for example, by mandating over-size furnaces, and the policies—both foreign and domestic—that have led to a high demand for Persian Gulf oil.

Expectations are now in flux, shaping the private preferences and public policies that will condition future demands for energy. As damage to the environment accumulates, and as demands for energy change, our notions of efficient devices, designs, and systems will necessarily evolve.

—Mark Christensen

Energy Shrinks And GNP Doesn't

PRELIMINARY FIGURES from the US Department of Energy show a marked reduction in total US energy use, from 79.1 quads (83.5 EJ) in 1979 to an estimated 76.1 quads (80.3 EJ) in 1980, despite a year that was substantially (25 percent) colder.

The drop in energy use came with virtually no real change in US gross national product (0.08 *percent* drop, measured in 1972 constant dollars), and provides further evidence that the US is becoming dramatically more efficient in its use of energy.

In fact, the US is using 12 percent less energy today to deliver a dollar's worth of GNP than it did in 1973. In 1973, the US required 60,400 Btu (63.7 MJ) for each unit of constant GNP, while in 1980 the amount fell to 53,200 Btu (56.1 MJ) per 1972 dollar.

The major contributor to the US energy situation was clearly increased efficiency, rather than flat GNP or increased production. At 1979 efficiency levels, the US economy would have used 79 quads (83.4EJ) in 1980 to generate last year's GNP. Production of all types of domestic energy rose 0.90 quads (0.95 EJ), led by a 7 percent climb in coal production. But increased efficiency filled the remaining gap, supplying 2.85 quads (3.0 EJ) in national services.

All sectors of the economy used less energy in 1980 than they did in 1979, but the sharpest drop was clearly in transportation (6.6 percent down). Industrial use fell 1.3 quads (1.4 EJ), and residential/commercial use fell by 1 percent, or 0.25 quads (0.26 EJ).

But the best news of all comes in US oil import levels, and in overall US oil use. Net US imports of oil (some crude oil and coal are exported) fell 28.6 percent from the 1979 level of 16.8 quads (17.7 EJ). National oil use dropped a full 3 quads (3.2 EJ). Because of high imported oil prices, the value of these imports climbed to $71.1 billion, the highest level in history. But at 1973 efficiency levels, the 1980 oil bill would have been an impossible $147 billion, representing a flow of a quarter million dollars per minute out of the nation.

Electricity growth in 1980 was quite small, despite the rugged temperatures. Total US sales climbed a modest 0.98 percent, with a heavy shift from oil to coal fired generation. Coal supplied about 51 percent of primary energy for electricity production (up from 48 percent), petroleum supplied 11 percent (down 19 percent to 7.5 percent of total US energy use), natural gas supplied 15 percent, hydro provided 12 percent, and nuclear energy supplied the remaining 11 percent.

— *Jim Harding*

Reference:

1981 *Monthly Energy Review.* US Department of Energy, February.

2. Country Studies

From international comparative energy analysis we know a good deal about the differences in energy technology and use among various industrial nations. These differences are caused by varying levels of technical efficiency and by variations in economic structure and cultural and geographic characteristics. The technical differences are small compared to the scope of possible improvement, making a detailed investigation of each country desirable. Among other things, an indigenous study will always take better account of the political and institutional climate in which the analysis is going to make its impact.

Western industrialized countries consume the bulk of the world's fuels. Low-energy scenarios for this region set the tone for world developments as a whole. The US, with one-third of the world's energy use, is clearly in a class of its own. The most detailed investigation of US energy prospects so far has been undertaken by the Solar Energy Research Institute, a federally sponsored institution. The SERI study, completed during the final days of the Carter Administration, is a document of the widest political significance, not only for Americans, but for all peoples. It concludes that an energy policy oriented toward least cost (i.e., towards higher efficiency and substantial growth of renewable energy contributions) can eliminate all oil imports by the end of the 1990s. Oil from the politically volatile Middle East constitutes only a third of those imports and could be replaced by other sources long before. It is precisely this oil that the Carter and Reagan administrations have been prepared to go to war for—even at the risk

of nuclear escalation. Rather than putting SERI proposals into practice, the Reagan Administration has suppressed the report and defunded the agency as well as solar and conservation research in general. At the same time, it is vigorously promoting the arms race and foreign policies relying on threat of force. Nowhere is the essence of the hard energy road more apparent than in these policies.

For industrialized countries that are much more dependent on foreign oil than the US, security issues—besides cost advantages—should be an overriding concern. West Germany and Japan can swim free of oil imports, though it may take them longer. The Scandinavian nations are studying a possible solar alliance that would trade Norway's abundant hydropower, Denmark's wind, and Sweden's forests. Caught between oil-rich, hostile neighbors, Israel's solar effort is dramatic and impressive. All in all, the feasible solar contribution by the turn of the century is very substantial in "weatherstripped" economies, contradicting the conventional wisdom of renewables' minor, "supplementary" role.

For the planned economies of Eastern Europe, detailed efficiency/renewables studies are lacking, though Romania is definitely thinking about its solar options. All evidence points to a similar conclusion: there is ample room for improvement. The idea that the Soviet Union will turn into an oil starved giant pushing toward the Middle East is groundless on energy efficiency arguments alone.

A New Prosperity

*I*T IS THE YEAR 2000, *and tankers laden with oil from the Middle East and other OPEC nations have ceased arriving at US ports. Most of the electric power plants under construction two decades previously were never completed, and each year more fall into disuse. Energy consumption has taken a swan dive since the late '70s and an expanding population of Americans is doing literally everything under the sun to find the energy it needs.*

End-use energy demand potentials
(Quads)

One quad per year roughly equals 500,000 bbl/day, or 3×10^6 GJ/day.

Sector	1977			2000 Potential		
	Fuel	Electric	Total	Fuel	Electric	Total
Buildings	13.2	13.4	26.6	5.5	12.3	17.8
Residential	8.8	7.8	16.2	3.8	7.1	10.9
Commercial	4.7	5.6	10.4	1.7	5.5	7.2
Industry	19.8	9.3	29.1	18.7	10.7	29.4
Agriculture	1.3	0.3	1.6	1.4	0.3	1.7
Transportation	19.5	—	19.5	12.6-16.5	*	12.6-16.5
Personal	15.1	—	15.1	6.9-10.5	*	6.9-10.5
Freight	4.3	—	4.3	5.7-6.0	*	5.7- 6.0
TOTALS	53.8	23.0	75.1	38.3-42.2	23.7	62.0-65.9

*An aggressive rail electrification and electric vehicle program could create between 0.75 and 1.15 quad (primary equivalent) demand for electricity in the transportation sector, displacing 0.46-0.76 quad of petroleum demand.

Demand figures are not additive within end-use sectors.

Is this a post-crash depressed America, driven to post-industrial life by unaffordable oil imports, impossible environmental regulations, and endless bickering over resources? Quite the contrary. It is an America that met the challenges of the '80s and emerged victorious: it has cut unemployment rates in half and boosted gross national product eighty percent; it has cut non-renewable fuel use by 40 percent and oil use to a mere trickle by substituting 35 percent reliance on the sun for the "inevitable" expansion in coal, natural gas, and uranium. It is the future suggested as "a least cost" energy scenario by the Solar Energy Research Institute, using today's technologies, modest technical improvements, and common sense to build a "more stable foundation for growth in the American economy by increasing its energy productivity and its use of renewable energy resources."

The SERI study, entitled *A New Prosperity: Building a Sustainable Energy Future*, was commissioned in 1979 by John Sawhill, Deputy Secretary of Energy in the Carter Administration. Instructed to outline the policy requirements for reaching President Carter's proposed 20 percent solar goal by 2000, the study team emerged with even stronger conclusions. US energy consumption would not climb from its current 79 quads/year to 100 or so by 2000. In a "least cost" scenario it would fall to 62-66 quads without the use of any renewable energy technologies (beyond the hydro and biomass we use today). Thirty-five percent of the 62-66 quads could be met with renewable sources. The assumptions do not include a "nightmare of freezing homes, stalled traffic, or massive unemployment," but include a sustained 2.5 percent annual growth in real gross national product, distributed among an increasingly mobile and wealthy population. There will be more for all to share.

Despite increased GNP, energy consumption would fall in all economic sectors, save a slight rise in industry. The increase in the productivity of energy use is accomplished with materials and technologies that are already commercially available, or soon to be. If the assumptions about solar energy use appear farfetched, consider the enormous potential in efficiency improvements, and vice versa. And consider that nearly all the assumptions in efficiency improvements are substantially below the state-of-the-art improvements we try to keep abreast of in these pages. New buildings, notes SERI, can be designed to consume one-fourth the heating and cooling energy of today's structures. (Our reports show cost-effective reductions in residential and commercial buildings to *ten times* the existing level.) Retrofitting existing structures, improved appliances, a better fleet of cars, trucks, and planes, and increased cogeneration are other areas where SERI bends over backwards to be conventional. Its economic growth assumptions parallel those of the Exxon Company and may not be consonant with other resource limitations. All the efficiency improvements predicated by SERI "produce" energy at about half the cost of new conventional sources.

Federal role controversial

SERI cautions, though, that these targets will not be met unless the federal government assumes some of the responsibility for directing the nation's energy purchases—some seven trillion dollars over the next twenty years—into appropriate channels. "It is important to ensure," the report warns, "that the market can fairly compare investments in energy supplies with investments in efficiency and select the most productive use of available capital. The nation cannot afford federal programs which artificially prejudice this choice and encourage inefficient investment."

SERI counsels that investments simply be allowed to flow toward the greatest rate of return, unhampered by government actions. The federal role need be neither costly nor impressive, but can be limited to:
• providing investors with adequate information to make decisions about energy investments;
• desubsidizing energy supply investments, and revising programs that discourage efficiency investments;
• ensuring adequate capital access for demand-reducing improvements;
• maintaining national R & D in energy; and,
• keeping energy investments in line with national concerns for equity, security, and the environment.

In sum, "the government cannot escape responsibility

for national energy policy. Its influence on the economic environment of energy investments will be enormous, whether by design or inadvertance."

Yet even these mild policy initiatives have proven controversial. *Solar Times* reporter Lyndon Stambler describes the Carter Administration as "hesitant" to publish the study, and the Reagan energy team as "stalling" on its release. Denying that the study has been suppressed, acting Assistant Secretary of Conservation and Solar, Frank De George has attempted to discredit SERI's conclusions by labelling its quite conservative assumptions "farfetched." But in spite of the Department of Energy's reluctance to call attention to a responsible study that contradicts central tenets of the Administration's energy policy, SERI has proceeded with the publication of *A New Prosperity*. In addition, the executive summary was entered into the Congressional Record by Democratic Senator Paul Tsongas (Massachusetts), and Brick House Publishing has plans to print the entire report.

The SERI solar/conservation study is both technically and politically significant, and sure to become a standard reference. Highlights of the five principal volumes on residential buildings, commercial buildings, transportation, industry, and the utilities appear on the pages that follow. Until commercial publication, copies may be obtained from the Documents Clerk, Committee on Energy and Commerce, US House of Representatives, Washington DC 20515.

—*Charles Drucker*
Jim Harding

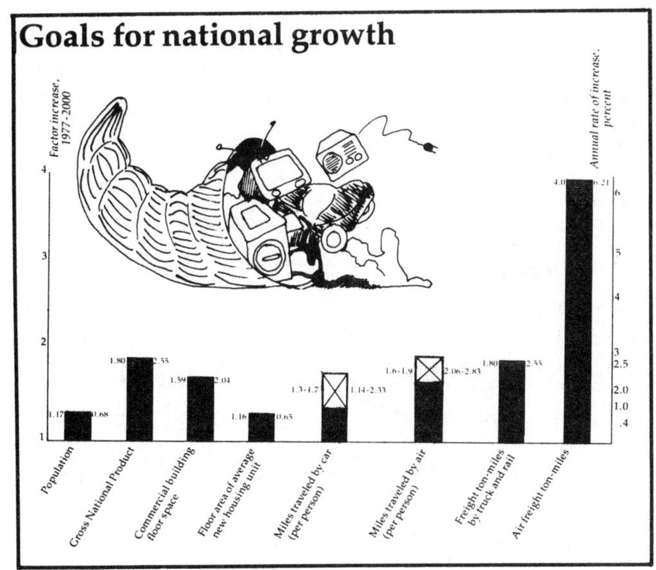

SERI conclusions, the main cost-effective reductions come not from SERI's primary *raison d'être* — solar energy — but improved end-use efficiency — conservation.

Building efficiency potential

Standard US forecasts for the residential and commercial buildings sector by the federal Energy Information Administration project a substantial increase in building energy use by 2000, from 26 to 35.3 quads, equivalent to a 1.5 percent annual growth rate in energy use. Using the same estimate for number of buildings, underlying economic growth, and number of appliances sold as the EIA study, SERI examined the potential for substituting economic efficiency for happenstance. Most of its assumptions in every category of building energy use are far more conservative than others would estimate (its new residential buildings are two-three times more energy using than the current state of the art; its commercial buildings are five-ten times more energy using than a large number of new Canadian designs; etc.).

Nevertheless, the SERI review shows that we have only begun to tap what is possible with removal of market barriers, better pricing schemes, accurate information, and cost-effective regulations. Noting that "five decades of inexpensive energy" have left the United States with buildings that use about 50 percent more energy, despite a milder climate, than equivalent Swedish buildings, SERI notes that US builders are just now beginning to adopt some of the conservation features common in Western Europe, which itself is improving its techniques.

Most building energy is consumed in residences (60 percent), rather than in commercial buildings. (Industrial buildings use tiny amounts of energy for space heat.) Eighty percent of the residential buildings we will have in 2000 are already in place, so the biggest potential savings categories are in retrofits and replacement appliances.

In the residential sector, SERI assumes that most residential investors spend up to the current marginal cost of energy for efficiency measures. (There is a small real

---

Residential Buildings 2000:

Savings Begin At Home

Buildings use one third of all US energy — for heating, cooling, lighting, and running appliances. Because of population and economic growth, federal studies argue that we will need 40 million new buildings over the next twenty years, aside the 60 million buildings (of the current 80 million stock) that will still be standing. But those 100 million buildings in 2000 would use about half the total energy that is used today in a "least cost" energy future, according to the two-volume buildings report of the Solar Energy Research Institute.

US buildings use 26 quads of primary energy, evenly divided between oil and gas for heating, and primary fuels for electricity that is used for cooling, lighting, and appliances. This energy, according to SERI, "is the largest and least expensive source of energy which can be supplied during the next two decades." Consistent with the other

escalation rate on energy costs.) Several different economic standards are used, including three and ten percent real interest rates on efficiency investments, but most are compared to energy prices over a 20-year loan period. These investments are reflected in "supply curves" that identify the cost, priority, and amount of energy (fuel and electricity) that can be saved in the nation's different categories of building. Most of the options fall well under the current cost of fuel oil, the marginal cost of gas, or the marginal cost of electric power. The SERI study is quite aggregated and does not analyze regional differences in any of its economic sectors. These can create profound errors, but it might also be pointed out that the US does not have competent models that perform this function.

This scenario does not assume the use of any novel solar technologies. It assumes only the use of passive solar technology in buildings, a feature that is difficult to differentiate from conservation equipment.

Simple savings: 5 quads/year

Of the projected 21.9 quads that would be used in a "baseline" residential forecast in 2000, SERI finds that simple conservation measures in existing homes (reduced air infiltration, storm windows, ceiling insulation, efficient furnaces, heat pumps, and a small number of active and passive solar features) would save about five quads per year. This would reduce energy use in those uninsulated homes that survive to 2000 by about 85 percent, partly insulated homes by 75 percent, and relatively new homes by about 50 percent. For homes to be built between 1986-1990, it is assumed that they will use a low infiltration BEPS (Building Energy Performance Standards) shell and use an average of 14 million Btu/year. (This efficiency level is about 40 kJ/m²/°C-d, or about ten times less strict than the Saskatchewan conservation houses.) These conservation techniques reduce the space conditioning component of the baseline forecast from 11.03 quads to 4.49 quads. Efficient appliances and water heat use an additional 6.57 quads. This scenario does not assume the use of any novel solar technologies. It assumes only the use of passive solar technology in buildings, a feature that is difficult to differentiate from conservation equipment. Domestic solar water heat, small wind machines, photovoltaics, solar space heat, and other renewables are considered separately. SERI estimates that renewable technologies can cost-effectively supply 3.7-4.45 quads of the remaining residential demand, cutting residential energy demand from conventional sources by 35-40 percent. If only wood fuel, solar water heat, and (mainly passive) solar space heat are considered, solar sources can supply 1.65 quads.

These calculations assume that the cost of solar heating and wind technologies declines by 20-40 percent (constant dollars) by 2000 and that photovoltaic systems reach the DOE cost goals for 1986-2000. The small wind and photovoltaic systems could supply residences with 1.1-1.55 quads by 2000 if these goals are met, argues SERI. About 400,000 American households are located in regions where the output of small wind machines could profitably be sold to the grid, displacing conventional capacity by 2000. Residential photovoltaic systems, selling at a system price of $1.60-$2.20/watt in 1986 and declining to $1.10-$1.30/watt by 2000, could displace 0.3-0.45 quads, depending on system lifetime, ownership (utility or householder), and tax treatment (higher investment tax credits produce a higher photovoltaic level). Most of these photovoltaic systems would be "in new homes where the panels can be used as part of the building shell." If all the south-facing roof area was used for photovoltaic arrays, SERI notes that such houses could be "substantial net producers of electricity in virtually any part of the country."

Appliances can be designed to save considerable amounts of energy, and with the enormous reduction in space conditioning requirements in place, SERI projects that appliances may use 50 percent of residential energy (against 28 percent today). Using the federal Energy Information Administration's projections of appliance lifetimes, SERI estimates that the potential for cost effective investments in improved appliances would cut use by 36 percent. Some types of appliance are particularly notable. Refrigerators, which nationally use the output of fifteen large power plants, can be designed to use three-four times less energy without any change in size or quality. Indeed, Amana's prototype 675 kilowatt-hour/year model comes quite close to SERI's 500 kWh/yr year 2000 goal. (Existing units of this size use 1,800 kWh/yr.) Such models would cost $55-$150 more than existing models, but save more than this over their lifetimes (Nórgaard and Japanese data are considerably more favorable, see *Soft Energy Notes* 1:1:8-9, 4:1:8-9). Freezers, clothes dryers, lights, ranges, televisions, air conditioners and other appliances come under similar scrutiny, with approximately equal levels of conservatism.

SERI recommends a number of programs to ensure such efforts succeed. In research, these include an applied program to determine the most cost-effective combination of efficiency improvements in each building type and climate zone, development of analytical tools for assessing building energy performance and lifecycle costs, and information dissemination. SERI recommends minimum performance standards for appliances, arguing that 65 percent of all appliances are *not* purchased by their end users. On building standards, SERI suggests state-level standards, with federal inducements to ensure development, compliance, and cost effectiveness. A small program of advanced building demonstrations, including the most nearly cost-effective products, could assist developers and homeowners in understanding the meaning of energy efficiency. For existing houses, SERI suggests a strengthening of the Residential Conservation Service (RCS) to expand the skills and functions of energy auditors beyond the current level. Additionally, SERI suggests special programs for low-income homeowners and tenants. Low-income homeowners could be better served, notes SERI, by shifting fuel bill subsidies provided under the Windfall Profits Tax Bill to direct weatherization. Direct weatherization grants of funds or materials could reach a further class of consumer. Rental housing might best be reached with trial programs at mandating modifications on transfer of title.

—*Jim Harding*

Everyone's Business is Saving Energy

COMMERCIAL BUILDINGS *could be constructed to use one-quarter as much energy per square foot as the average building today, reports the SERI study on the commercial sector. Existing stock also shows great potential for savings: energy use here can be cut in half using relatively straightforward retrofit techniques.*

Commercial floor space is expected to increase 64 percent by the year 2000—54 percent of the area will have been built after 1980 and will have an average life expectancy of 45 to 50 years. With the current distribution of floor space among building types changing little in 20 years, energy savings outlined in this report actually *reduce* the sector's total demand for energy by 30 percent. Though standard forecasts peg energy use in commercial buildings at 13.3 quads in 2000—up from 10.3 Q in 1977—SERI estimates a total of 7.26 Q.

"Savings of this magnitude may seem incredible," SERI admits, but not in the face of commercial buildings' heating and cooling systems of the past. Constructed when energy was cheap and reliable controls expensive, buildings often run an air-conditioner and a furnace at the same time, even when outside temperatures are comfortable. "Until recently, engineers calculated the amount of cooling that would be required on the hottest day expected in ten years, installing a chilling unit that was able to meet this peak requirement. Once installed, the chiller was operated at full capacity for most of the year and the space temperature was controlled by reheating the chilled air."

Hardware fixes cut heating & cooling

Saving energy in commercial buildings—office space, hospitals, schools, laundries, churches, warehouses—means different techniques than the housing stock requires. The shells of large commercial buildings tend to be less important than the walls and ceilings of residences: commercial buildings tend to be larger, with less surface area per unit of floor area. For example, SERI notes, doubling the insulation in a Denver office building, which had R-11 walls, would have decreased total energy just one percent, whereas changes in HVAC and lighting systems produced much higher savings. And so, the report stresses equipment for reducing the energy used by the heating, cooling and ventilating equipment. SERI lists: efficient and properly-sized fan motors along with adequately-insulated ventilating systems; heat exchangers as local codes prescribe more outside air per square foot of commercial building than for residences; automatic ventilating systems to flush the building with cool air at night; systems to move heat between a building's various zones of divergent heating and cooling demands; "dead-band" thermostats, which, when retrofitted into each zone of an existing building, have reduced heating and cooling demand 10-30 percent; efficient high pressure sodium vapor lamps to cut lighting energy requirements; and, reflective "solar control" window coatings and overhangs to block heat gains.

Thermal requirements can be virtually eliminated, according to SERI, so remaining energy demand is primarily for electricity. The demand for electricity is considered relatively inelastic with few full substitutes available for cooling and lighting. "These end uses are, however, prime targets for energy conservation." Admittedly far from exhaustive, the Report adds community-energy systems, small-scale cogenerators, experimental heat pump cycles, and solar cooling as other significant technologies which were not included.

Hardware-fixes are the focus of this study; energy savings from building operation and maintenance (O&M) have not been identified. While O&M does offer significant savings, the authors consider it too difficult to implement (measure?) on a widespread scale. "Except for the obvious and easy changes that many building owners have already made, decisions to save energy through the operation and maintenance of buildings involves many small decisions by individuals, the impacts of which are not always obvious to the individual."

Room for error

The diversity of occupancy, function, size, ventilation requirements, and electricity demand among commercial buildings makes totaling the savings potential difficult. Hot dog stands, churches, and hospitals have vastly different energy requirements per square foot, whereas two houses and an apartment building can be treated much more simply. One of the buildings used in this study, for example, used 50 percent of its energy to support office space, 20 percent to supply a retail card store, and 30 percent for a pizza parlor. Hospitals and clinics need more energy per square foot than any other major category—due largely to their use of fresh air which cannot be recirculated—while warehouses use the least energy per square foot.

Information blocks to efficient building design were found to have a significant impact on attaining the commercial building savings potential by 2000, and on the study itself. The authors made it clear they were hampered by a lack of consistent data; contrary anecdotal information in the trades impeded retrofit experiments with practicing engineers and architects.

The Report also emphasizes the built-in market disincentives to conservation. Building owners can easily transfer higher energy costs to their tenants whose energy costs are still a small portion of total operating costs. The tax structure also blocks conservation: the higher the tax rate, the higher the income tax benefits from deductions and so, the lower the incentive for conservation.

Conservation in new building proves cost-effective

Off-the-shelf technologies could push energy savings as high as 65 percent from recent office building design practice with potential life cycle costs lower than typically current, SERI reports. The authors rely on the BEPS (Building Energy Performance Standards) research, begun in 1976 by the AIA/RC (Research Corporation of the American Institute of Architects). The analysis showed that if a sample of 125 various commercial buildings in a range of climates had been built to ASHRAE (American Society of Heating, Refrigeration, and Air-conditioning Engineers) Standard-90—a measure of voluntary energy savings performance—overall energy savings would have ranged from three percent (hospitals) to 41 percent (warehouses). A "Phase-2" redesign study challenged the building designers to improve the energy efficiency without violating clients' requirements or changing cost significantly. Only off-the-shelf technologies could be used; a three-day training session and workbook were offered. Overall savings averaged 40 percent. Phase-3 included just three office buildings by architects with demonstrated skill in designing efficient commercial buildings. Asked to alter the original designs to save energy and give the owner at least a ten percent return on his investment in constant dollars, the designers used careful lighting techniques, state-of-the-art controls, and heat pump systems. Overall energy savings from recent design practice in this phase ranged from 59 to 64 percent.

Seventeen out of 20 cases of the investments were less than 50 cents per ft² and produced estimated energy savings in the range of 20 to 35 percent of the total energy use.

Retrofit experience projects 50% savings

SERI finds significant potential for commercial building retrofits to save energy at well below the marginal cost of producing new conventional fuels. Retrofit conservation estimates are lumped for the sector. "There is no effort to analyze potentials independently by building type or by region, even though such potentials may vary considerably." A 25 percent savings by 1990 in both fuel and electricity and a further 25 percent by 2000 are estimated from SERI's literature search, which included *Energy User News,* Ohio State University research, Minnesota state office building experiments, a private auditing company (EBASCO), and fourteen professionals.

These data indicate that investments are typically low and effective, the report continues. "Seventeen out of 20 cases of the investments were less than 50 cents per square foot and produced estimated energy savings in the range of 20 to 35 percent of the total energy use. Such strategies result in a low cost of conserved energy—about $2/MBtu assuming a ten percent discount rate and 30 year life cycle."

Renewables may save 0.65 quads

SERI stresses the use of daylighting in commercial structures, estimating lighting demand at 50 percent of the energy usage in typical office buildings. Daylighting is estimated to reduce US energy demand at least 0.2-0.3 quads by the 2000, "if aggressively implemented." Experience in North Carolina shows careful daylighting redesign reduced building lighting demands by nearly a factor of two. The Report specifically describes window placement and sensors, which automatically dim lighting fixtures when there is adequate sunlight, as important fixes.

SERI recommends that building designs include fixed lighting only for minimum ambient lighting needs—these can be supplemented with task lighting. As too much lighting can add significantly to cooling loads, the report suggests efficient high pressure sodium vapor lamps over the relatively inefficient fluorescent lights (which convert only 19% energy to visible light). It is estimated that lighting requirements could be reduced to 0.5-1.0 watts per square foot by 1990, down from two watts in 1980 and four watts in typical buildings of 1970.

Hot water demand is not considered large in commercial buildings; SERI estimates 0.1 Q of energy can be contributed by solar water heaters. Photovoltaic systems would contribute 0.1 to 0.25 Q by the year 2000, the authors report.

How efficient?

Since World War II, commercial floor space has almost tripled—from 9.9 to 29.6 billion square feet in 1978. Between 1952 and 1973, the energy used per square foot of building increased nearly 40 percent. While floor space was growing at 4.2 percent annually, energy use increased at a rate 38 percent faster. Rising energy prices turned that trend around: the average building designed in 1976 was more than twice as efficient as its neighbor built in 1972. Still, we have a long way to go, SERI adds: in Sweden, despite a frigid climate, office buildings now use less than half the energy required in the US. And, in Sweden every office must have a window!

To approach commercial buildings' energy savings potential, SERI recommends research and training. A standard method should be developed for rating and monitoring building performance in a way useful to potential renters and purchasers. The US Department of Energy (DOE) should give more seminars to architectural and engineering school faculty (it now gives two per year). SERI emphasizes simple, accessible analytical tools, especially for small architecture and engineering firms. Professionals in the field are advised to form cooperative information programs. Feedback programs, such as feedback billing and metering, should be extended to commercial buildings where their effectiveness may be greater than in the residential sector, according to the authors. SERI prescribes retrofitting federal buildings, and DOE support for state and local initiatives to eliminate restrictive lease provisions, and the development of model lease forms.

Turning to standards, SERI expands the current proposed incentives for states to adopt BEPS in the residential sector to include the design energy budgets proposed for commercial buildings. Finally, the Report cautions that the DOE, ORNL, and ASHRAE programs are not coordinated to produce coherent and consistent documentation on life-cycle cost analysis for new commercial buildings. SERI recommends that existing buildings be included as well.

—*Elyse Axell*

The Policy Role

Accustomed as we are to hearing bad news about the US transportation system, the SERI report is a welcome relief. By the year 2000, it concludes, the total demand for energy in transportation could drop 15 to 35 percent below 1977 levels, despite a significant increase in the services each person receives. From 35 to 45 percent of the remaining transportation energy demand could be derived, on a sustainable basis, from biomass alcohols, dropping oil imports to virtually nil. And, if transportation demands increase less rapidly than SERI assumes, the energy savings could be much greater.

But the SERI report stresses that these changes will not arrive unassisted. The energy requirements of the transportation sector are likely to *increase* over the next two decades in the absence of properly designed policies to foster the "sensible use" of present-day and near-term technologies for improving transport efficiency.

Price decontrol a priority

The policies recommended by SERI include the elimination, or even the reversal, of many historic subsidies, and a general decontrol of the prices for transportation fuels. Until only a few months ago, motor fuels and other petroleum products were kept artificially cheap in the US by controls on domestic crude oil prices that forced some producers to sell at little more than one-third the world market value. These measures were instituted after the first "oil shock" of 1973, when the country awoke to its heavy dependence on foreign oil, and the instability of a transportation system with a nine million barrel a day habit.

More than four-fifths of 1973's domestic petroleum production was consumed just in personal transport, and the sector as a whole required twelve percent more oil than the US brought to the surface that year. This heavy reliance on energy imports made the transportation system sensitive to changes in the price and the availability of foreign supplies. At the first hint of a possible interruption, motorists flocked to neighborhood gas lines, topping off their normally half-empty tanks, thereby helping to create the very scarcities they feared.

Several years, a full-scale recession, and another oil shock later, the situation is not greatly different. Transportation remains the nation's largest single use of oil, consuming over one-half of the petroleum, and one-quarter of the energy budget overall. Price control is one of the reasons why the pace of change has been so sluggish; designed to preserve the *status quo* in the oil industry, it has also preserved the historic inefficiency of energy use in transportation.

Efficiency of Vehicles Used in US Transportation

	1977	2000 No New Programs	2000 With Programs
Automobiles (miles per gallon)	13.9	27.5	34.4-70.8
Light trucks for personal use (mpg)	10.0	20.0	
Commercial aircraft (passenger-miles/gal.)	21.2	30.0	40.0
Medium and heavy trucks (vehicle miles/gal.)	8.3	9.5	10.8

The stated objective of petroleum price controls, in the form of the Domestic Crude Oil Entitlements Program, was to encourage increased domestic production by subsidizing enhanced recovery and accelerated exploratory efforts. Oil from newly discovered wells, and the proceeds from secondary recovery methods, were pegged at higher prices than the oil from wells already producing when the program went into effect. But the Entitlements Program was more than just an investment incentive; it also provided a convenient way to bail out those oil companies which had invested heavily in foreign sources, and who suddenly found it much more expensive to do business. Other oil companies with cheap, domestic reserves were supposed to foot the bill for the combined subsidy/bailout operation through a complex system of inter-company income transfers running into the hundreds of millions of dollars each month.

Ultimately, of course, the public has paid for corporate mistakes and federal misregulation. Petroleum price controls preserved the illusion of a free oil market by removing the post-'73 comparative advantage of a domestically-based producer. But it did so by disguising the marginal cost of oil. This artificially depressed the price of transportation fuels, and failed to encourage more efficient fuel use. The American public never received OPEC's initial price signals, and it took a second oil shock for the market to respond. As late as 1978, more than half the new cars purchased in the US were still powered by gas-guzzling V-8s. Years of opportunity for transportation R & D and retooling had slipped away, and the American ethic of energy inefficiency became even more ingrained. Finally, in 1981, the market scramble for fuel-efficient foreign cars has presented American automakers with the largest annual losses in corporate history. Price controls were supposed to alleviate an energy crisis; instead, they perpetuated it, and turned it into a corporate economic crisis as well.

Estimating future demand

To overcome their present financial troubles, Detroit decisionmakers will have to build cars that more closely match the needs of the driving public in a time of high energy prices. Americans are the world's most mobile people—travelling an average of 11,500 miles each year—and spend more than an hour each day inside a vehicle. One-seventh of individual disposable income, on the average, goes toward travel, with people in upper income groups spending proportionately less, and lower income groups relatively more.

The amount of driving done in the future will depend on several factors: fleet efficiencies, personal incomes, and the price of gasoline. Based on conservative assumptions, SERI calculates that with gasoline prices of $1.50 (1980 $) in the year 2000, per capita miles driven will increase 32 to 52 percent above 1977 levels, depending on then-current fuel economies. Other modes of transport will also increase in volume. Air travel will jump by 55 to 94 percent; bus and rail transit are also assumed to grow in absolute terms, but the relative share of these transit systems is not expected to rise much above their present small percentage of the total.

The US freight transportation bill, like the tab for personal transport, will continue to be higher, on a per capita basis, than it is in either Europe or Japan. As SERI wryly remarks, "shipments of fresh California strawberries to the east coast and tractors from Detroit to Dallas have become a part of the basic fabric of the national economy." Based on recent trends, the study assumes that rail and truck freight (60 percent of the domestic total) will grow with the Gross National Product for the next twenty years. By the end of the century, the freight carried per capita will increase by 53 percent. Rounding out the transportation demand picture, air freight is assumed to

increase at 6.2 percent per year, as it has in the recent past, while water and pipeline shipments are expected to increase only slightly less rapidly (0.5 percent) than the GNP.

Market forces are already moving the US transportation fleet toward greater efficiency, and SERI anticipates significant improvements by the end of the century, even with no policy intervention. But increased efficiency will be overshadowed by heightened demand. With both freight and personal transport up, in 2000, by over 50 percent, the annual energy demand for the sector would rise by 1.0 to 2.5 quads assuming no new programs, representing an increase of five to 13 percent.

Technical improvements to 93 mpg

SERI suggests this increase in consumption is not inevitable. With the successful implementation of present-day and near-term technologies, along with some vehicle downsizing, much lower levels of energy use are possible in 2000, despite greatly increased demand. Key automotive improvements include:
- small-displacement stratified-charge and diesel engines;
- 90-percent efficient continuously variable transmissions;
- lower movement resistance through superior aerodynamics and radial tire installation; and,
- lightweight materials.

A year-2000, four-passenger car incorporating all of these

Energy Use in Transportation *(quads = 1.055 exajoules)*

Transport Sector	1977	2000 No New Policy	2000 New Program Effects	2000 New Program with 50¢/gal. gasoline tax
Personal Transport				
Automobiles and light trucks				
Present technology	10.0	7.5	6.2	5.9
Prototype technology			4.3	4.1
Advanced technology			3.5	3.3
Automobiles and light trucks—fleets				
Present technology	2.9	2.7	2.3	2.1
Prototype technology			1.5	1.4
Advanced technology			1.2	1.1
Commercial Airlines	1.4	1.8	1.6	1.34
General Aviation	0.15	0.37	0.37	0.30
Military Aviation	0.50	0.60	0.60	0.60
Buses	0.13	0.20	0.20	0.20
Rail	0.06	0.09	0.06	0.09
SUBTOTAL	15.14	13.26	7.6-11.4	6.9-10.5
Freight			*(tax effects not considered)*	
Trucks	2.0	3.6-5.4	2.1-2.5	
Rail	0.55	1.0-0.7	1.1-1.0	
Water	1.1	1.5	1.5	
Air	0.09	0.30	0.20	
Pipeline	0.60	0.80	0.80	
SUBTOTAL	4.34	7.2-8.7	5.7-6.0	
TOTAL DEMAND	19.5	20.5-22.0	13.3-17.4	12.6-16.5

changes could meet minimum performance requirements with a 3-cylinder, 27-horsepower engine with a projected fuel economy of 93 mpg. The SERI staff assumed less advanced technology than this, however, in deriving their fuel consumption estimates. Their four-passenger car was assumed to provide 78 miles to the gallon—slightly *less* than Volkswagen has already achieved in prototype turbocharged Rabbit diesels. (Technical details of the automobile efficiency improvements assessed by SERI in this study were described in the first *Soft Energy Notes* transportation issue, volume 3, number 4, pp. 3-5.) Personal transport efficiency may also be increased through modification in passenger aircraft, and air travel management. (Recent developments were summarized in *Soft Energy Notes* 4:2:41-42.)

More efficient freight transport is possible if the trucking industry continues efforts already underway to lower costs through vehicle improvements. Ongoing measures will include turbocharging heavy trucks, more efficient drivetrains, superior radial tires, and increasing the fraction of fuel-saving diesels entering the fleet. (Medium-sized trucks, in particular, are now only about 40 percent diesel.) All told, a 30 percent improvement in average truck efficiency is estimated, and this is a conservative assessment of the savings possible.

Railroads are up to four times more efficient than trucks in intercity freight hauls, but the rail share of freight traffic has consistently declined over the last two decades. Outdated regulations on railroads, and massive federal subsidies for highway use are principally responsible for this energy-inefficient mode switch. (See "Revitalizing the Railroads," *Soft Energy Notes* 4:2:35-38.) To improve freight transport efficiencies, SERI maintains, policy should "ensure that trucks are not inequitably subsidized and that the rail industry is able to raise enough capital." The policy instruments SERI recommends include a broad deregulation of the railroads, and a direct tax on trucks, linked to maximum axle weight and miles travelled, that is sufficient to pay the trucks' fair share of highway costs. Numerous state studies have shown that trucks are disproportionately responsible for pavement deterioration but pay only about half the costs they incur.

Muzzling gas guzzling

US auto manufacturers are required, by the National Energy Act, to produce fleets meeting federally-mandated fuel economy standards. The 1985 target of 27.5 mpg corporate average fuel economy is nearly double the 1973 level. Then the program expires, and it appears unlikely that the present administration will extend it, or in any other way establish automoative efficiency standards. In the absence of mandated fuel economy, the principal policy steps to increase personal transport efficiency are taxes, keyed to fuel economy goals, that will affect consumer choices. One such tax, already proposed, is the "gas guzzler" tax (PL-91-190). This would penalize purchasers of vehicles failing to meet fuel economy standards which grow more stringent each year. In 1986, for example, when the corporate standard would still be 27.5 mpg, a car attaining 21 mpg would carry an additional sales tax of $500; a 12.5 mpg petro-pig would carry the staggering premium of $3,850.

An alternative to a tax on vehicles is a direct tax on fuels, a policy long recommended by Dr. Robert Williams of Princeton's Center for Environmental Studies (see "The $2 Per Gallon Political Opportunity," *Soft Energy Notes* 3:3:20-21). The revenues from these taxes should be

Industry 2000:

Thriving on a Constant Diet

*T*HE US INDUSTRIAL SECTOR *need not suffer a protracted period of decline just because energy costs have increased. By accelerating the implementation of conservation measures, industry could participate in a GNP growth rate of 2.55 percent annually—for a total increase of 73 percent between 1978 and 2000—while keeping to a constant energy diet. Primary energy consumed by industry would creep upward by only one percent over more than two decades, from 29.1 quads to 29.4, yet industrial value added would grow at 1.7 percent per year, and the pounds of material produced by 1.2 percent. These projections imply a drop in industrial energy intensity (the average amount of energy consumed per unit of industrial output) of 26 percent by the year 2000.*

However, the "constant diet" scenario presumes that policy support will free conservation technologies from their present stepchild existence, and move them to their deserved place as the largest economically and socially viable energy resource in the US. In the absence of this support, the growth factors SERI assumes would require primary energy consumption by industry of 34.6 quads in 2000, equivalent to a drop in average energy intensity of only 13 percent. Simply stated, policy measures can just about double the rate at which energy intensity declines.

Contributions to industrial energy supply from direct solar heat and biomass fuels show a similarly strong dependence on policy. If a course of accelerated implemen-

rebated to the consumer to alleviate any economic inequities they might cause. Direct taxes on fuels offer consumers the greatest freedom in their purchasing decisions, and they have administrative advantages as well: they are "the hardest taxes to evade but require virtually no bureaucracy to enforce." Their fault is that their implications are not obvious to the consumer at the frenzied moment of purchase, "when a gleam of chrome may overwhelm reason." Nonetheless, SERI calculates that a modest $0.50/gallon gasoline tax would reduce personal energy consumption by four to six percent, and cut overall energy use in the transportation sector by 0.7 to 1.1 quads, without altering the projected increase in transportation services.

—*Charles Drucker*

tation were to be followed, biomass could supply some 4.8-10.5 quads to industry in 2000, whereas present policy is estimated to yield only 2.4-5.1 quads by then, up from a 1978 contribution of 0.84-0.97. Likewise, direct solar heat could provide 0.5-2.0 quads to industry with an appropriately supportive policy, as opposed to 0.1-1.0 if present policy proceeds. In sum, an impressive 18 to 43 percent of industrial energy needs could be met with renewable energy sources by the year 2000.

Assumptions and Uncertainties

The SERI report bases its assessment of future energy efficiency on presently available technologies known to be cost-effective. Improvements vary considerably over the range of industrial activities and operations: 15 percent in metallurgical coal and chemical feedstocks use; 23 percent in mechanical drives; 26 percent in electrolysis; 35 percent in process heat applications; and 39 percent in the provision of space heating. Clearly, these figures do not reflect the technical potential of economically efficient technologies, but present an estimate of what can be accomplished by 2000. The industrial stock includes much old, inefficient equipment, only some of which can be replaced by 2000; retrofitting what remains will not always be possible, due to technical obstacles and high cost.

Perhaps the most notable conservatism in the demand figures, however, is that *increased cogeneration is not considered*. SERI judged that the looming excess of generating capacity in the utilities sector would make industrial cogeneration uncertain. But if cogeneration were to be introduced in the large, low-temperature, process steam-consuming sectors of industry, up to three quads of primary energy could be saved. This would yield a net *negative* energy growth in industry over the next twenty years.

The SERI report does not give a very detailed description of industrial conservation technologies and their economic potential. Though many detailed studies are cited, and the stress on industrial policy is justified, some further technical details would have made the political argument much stronger. The extreme conservatism of the SERI assessment would be more apparent if the limited efficiency improvements assumed for 2000 were clearly contrasted with their potential in a totally renewed capital stock. Moreover, the energy aspects of an efficient materials policy are not addressed in the report.

To contrast, the solar contributions to industrial energy (biomass and direct solar heat) are reviewed in much greater technical detail. The report's section on biomass is not limited to industrial uses, but covers all applications. The principal organic resource, SERI notes, will be wood, rather than agricultural products or residues, reflecting appropriate concern over soil conservation and food price inflation.

Methanol, according to this evaluation, appears to be the preferred liquid fuel from biomass. The relative maturity of the conversion technologies involved, their economics at present, and the potential to derive methanol from both biomass and coal make it more attractive than the ethanol alternative. Thus SERI concludes that present fuel alcohol programs are insufficient in support and scope, and inappropriately weighted toward ethanol.

Great uncertainties still remain as to the extent to which biomass energy can economically replace fossil fuels by 2000. It is also hard to predict, from SERI's discussion, which of the competing end uses for wood energy—direct combustion and chemical conversion—will predominate.

Nonetheless, the biomass section stands as an excellent review and one of the strong points of the SERI study as a whole.

The estimate of solar process heat contributions is equally clear in its technical and economic assumptions. Industrial process heat applications are broken down by temperature spectrum, and solar fractions are assigned to each, *after* efficiency improvements have been factored in. (The analysis does not, however, consider the fossil-based cogeneration units which would possibly compete with direct solar heat.) The report examines in turn the technical range of present systems, the range of expected costs for solar systems, and their potential contributions under various financial expectations, and climatic, regional and land-use constraints.

Most notable is the strong dependence of the potential for solar process heat on the payback time expected by the investor, and on the future cost of systems ($11-19/ft^2 [$120-200/m^2] for concentrating collectors and $5-10/ft^2 [$55-110/m^2] for flat-plate). The report concludes that R & D programs for solar applications in industry must focus on lowering costs, particularly through the use of much lighter collectors.

Industrial Policy: Go On or Go Broke?

Industrial energy use is concentrated in a few major industrial sectors whose value added and contribution to employment is low compared to their gargantuan appetite for energy. The primary materials industries, notably steel, aluminum, cement, pulp and paper, glass and basic chemicals account for 70 percent of industrial energy use and 25 percent of US energy needs overall. These are the industries most affected by the changed structure of energy prices.

Other factors besides energy, though, enter into the industrial outlook, principally capital and labor costs. US industry is generally capital-short, due to high interest rates and the comparative disadvantage of industrial borrowers vis-à-vis public and utility borrowers. Thus, industry expects very short payback times from energy efficiency investments, on the order of five years or less. This narrows the investments being considered, and energy efficiency improves relatively little as a consequence, leaving industry more vulnerable to the next hike in prices.

So much for industry in general. Now consider the energy-intensive primary industries: in the present phase of US economic development they are no longer the forerunners they once were. Their growth rates are significantly lower, on the average, than either industrial output or GNP and have been so for some time, since the economy is saturated with durable goods, and the economy has found ways to increase GNP with more information rather than materials. Thus, primary materials industries are neither able to generate the large internal profits needed to restructure themselves, nor do they benefit from a "natural" rejuvenation and modernization of their equipment stock through rapid expansion.

This vicious cycle defines the industrial energy problem in the US, and the SERI report appropriately takes this dilemma as a starting point. The policies which the report proposes are designed to improve overall productivity rather than just energy productivity, and to rely as much as possible on the market. These two principles lead to policies which will find capital for industrial investment in energy-efficient technologies; make energy prices reflect the true marginal cost of energy; and improve the climate for industrial research.

Freeing Capital

The three major avenues proposed are:
- reductions in corporate income tax;
- investment in industrial energy productivity by third parties, such as utilities, utility subsidiaries, and energy-service firms; and,
- subsidizing energy conservation and renewables as much as conventional energy supplies.

The latter point includes a proposal for a "scrap and build" incentive for energy-intensive industries, designed to recreate conditions of rapid modernization through turnover of capital stock. The subsidy considered sufficient is 20 percent of new plant costs, or $1.7 billion per year, on the average. This level of subsidy, expressed as dollars per energy saved per year, is close to the credit available for synthetic fuels plants under the Energy Security Act. The necessary capital for the subsidy, SERI argues, could be made available by removing the present subsidies for conventional energy sources.

Future Pricing

Price regulation and federal subsidies to energy suppliers have been and still are distorting the competitiveness of conservation investments. Subsidies to the tune of $6.5 billion are handed out to corporations such as utilities, the nuclear industry, and the oil and gas industry. It is surely consistent with the policies of the present administration to categorize these industries as not among the "truly needy." As a further measure to bring energy prices in line with their full social cost, the SERI report suggests a tax on industrial energy. This tax would reflect, for example, the enormous military expenditure necessary to secure foreign oil supplies. That cost, now the rationale for an unprecedented cutback in services to the country's poor, is not contained in oil prices; nor can the environmental effects of domestic energy production, or the macroeconomic effects of balance-of-trade deficits be adequately reflected by the market price.

Research

Finally, the SERI report points to the need for industrial research, which could be stimulated with increased tax credits for industrial research investments, a liberalized patent policy, and—last, but not least—federally sponsored research facilities.

One industrial problem not addressed by SERI is the possible effect of foreign competition on US basic materials industries. Insofar as production in these industries is highly standardized, does not require a high degree of skill in the work force, and is sensitive in whole-plant technical design, younger industrialized countries with lower labor costs and an unsaturated domestic market (to simulate their growth) will easily gain a comparative advantage. Short of protective measures such as tariffs, this competition may be hard to shake, and could accelerate the restructuring of US industry. This outcome, which would further reduce US industrial energy needs through a declining share of output from energy-intensive industries, might be called "scrap here and buy from elsewhere."
—*Florentin Krause*

Sink or Swim

Electric utilities face a serious challenge over the coming two decades that probably dwarfs their existing problems, notes the utilities volume of the Solar Energy Research Institute report. By implementing a "least-cost" energy strategy in all the major energy-using sectors—residential and commercial buildings, industry, and transportation—SERI forecasts future electric demand remaining virtually constant or declining through 2000.

SERI develops four separate forecasts for electric utilities, ranging from a "business as usual" case to a "cost effective efficiency" scenario that includes industrial cogeneration, and wind and photovoltaic systems. In the business-as-usual case, electricity use grows at two percent per year from its current level. This demand could be covered by a combination of existing powerplants (with

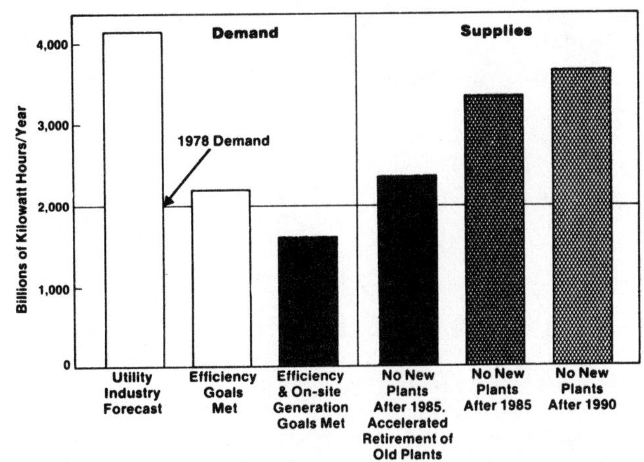

80 percent of oil and gas power retired), units to be completed by 1985, and reasonable growth in industrial generation of electric power.

With the implementation of "cost effective efficiency" efforts, most of which are already underway, demand growth falls to 0.4 percent per year, without assuming much success in active or passive solar heating and water heating, photovoltaics, or small wind machines. In this case, electric energy demand could be covered through 2000 by cancelling all coal and nuclear plants scheduled for post-1985 completion, retiring 80 percent of existing oil and gas capacity, and adding 60-80 gigawatts of cogeneration. This figure is well below the 208 gigawatts estimated to be cost effective (see *Notes* vol. 1, no. 3 and vol. 3, no. 5), and *Business Week's* estimate of 50 new GWe in the 1980s alone.

By meeting most federal goals for passive solar heating, active solar water heating, and daylighting (no federal goal here), the third SERI forecast falls to 0.2 percent per year. The fourth forecast, which assumes growth in industrial cogeneration (60-80 new gigawatts) and successful use of on-site wind and photovoltaic systems, results in declining use of utility powerplants, and a negative growth rate in demand of -1.4 percent per year.

Below industry projections

The SERI forecasts fall well below those of the electric utility industry. The 1980 *Electrical World* forecast assumes a 3.3 percent annual growth rate in demand through 2000, which doesn't sound like a tremendously different estimate from 0.4 percent per year, but represents about 350 new powerplants. As SERI notes, their range of forecasts "should make utility planners reach for a strong drink." Surprisingly, the SERI estimates (in all cases) include "a rather aggressive program for increasing the use of electric rail and electric vehicles, even though the discussion in the chapter on transportation concluded that electric vehicles powered from conventional electric sources are probably not a profitable investment." The forecasts also assume a substantial growth in production of aluminum .

SERI concludes that utilities face a major institutional challenge, through which they can sink or swim. "Perhaps more than any industry, [utilities] have fallen victim to the uncertainties and rapid changes in energy economics witnessed by the past decade." But the report also notes that "the industry's present difficulties are not ephemeral... The trend toward declining demand may be reinforced with an impressive array of technologies designed for increasing efficiency, and for using solar energy."

In practical terms, the SERI estimates pose an enormous challenge to the utility industry. It would involve the cancellation of at least 34,000 megawatts of coal capacity and 41,000 megawatts of nuclear capacity slated for completion after 1985. These figures represent virtually all the plants that are less than 40 percent complete. SERI argues that this strategy is an essential step toward coping with low demand forecasts and strong solar energy prospects. Counting decommissioned nuclear plants, it would leave the nation with only eight more nuclear power plants in operation in 2000 than it has today.

In dealing with new hydroelectric capacity, photovoltaic systems, and large wind machines, SERI assumed severe market tests. It assumed that these systems could only be built by utilities if their combined capital and operating costs were lower than existing depreciated powerplants, and that they would only be built at a rate equal to the need to replace existing generators. Thus, they would be competing with decreasing-cost plants in a shrinking market, not a very optimistic assumption for supporters of these renewable electric technologies. With these strict assumptions, hydro capacity would climb from 2.8 quads per year to 3.4-3.7 quads per year (in terms of primary energy displaced). Wind capacity would displace one to three quads of primary energy by 2000. Photovoltaic systems on buildings could supply a further 0.5-0.8 quads of energy displacement. Such systems as ocean thermal energy conversion, solar thermal electric, and solar ponds would supply very minor amounts of energy, primarily because it is assumed that they cannot compete (except in very special regions) with wind or photovoltaic energy.

The SERI renewable electric estimate ranges from 2.1-5.7 quads of primary energy displaced. Existing hydro plants would contribute a further 3.4-3.7 quads of displaced energy. Fossil and nuclear plants burn 21 quads of energy to supply existing electric energy needs. With a successful program in solar energy technologies, energy conservation, and industrial generation, this could shrink as low as six quads, or about 100 gigawatts of capacity. This is less than half of our existing coal power capacity with no nuclear electric power whatsoever.

With an unsuccessful program in solar energy technologies, industrial cogeneration, and conservation, the requirement would be about fifteen quads, according to SERI, or about 330 gigawatts of conventional capacity. This is about equal to our existing coal and nuclear capacity with no additions at all, and no oil or gas use. The range is a large one, but any way one slices it, this "least cost" energy strategy formulated by a US government laboratory has got to look like a "great pain" energy scenario to most utilities. It's a challenge they'll be tested on, whether they like it or not.

—Jim Harding

US Wastes Three-Fourths of Its Energy, Argues Danish Physicist

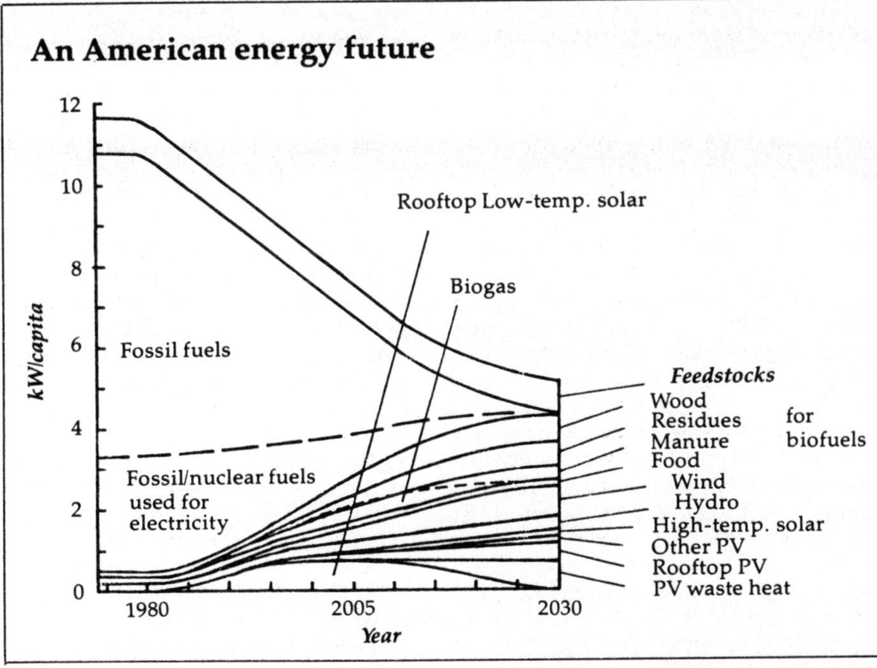

An American energy future

A RECENT SHORT REVIEW of US energy use by Bent Sørensen, a Danish energy specialist, suggests that nearly three quarters of our energy is wasted by inefficient technologies or outmoded practices that perpetuate waste. Sørensen's estimate that 74 percent of US energy is wasted is substantially higher than the proverbial 40 percent estimates of many energy conservationists. Even with added wealth and population growth, Sørensen believes the potential of improved efficiency is so large the US could become "self sufficient in energy before the year 2000, and (use) only renewable energy sources by the year 2030."

Current US energy production, based on 1974 data, totals 11.65 kilowatts per capita. According to Sørensen, the "practical minimum" energy for today's goods and services using known—though not necessarily proven—technologies is 2.98 kW, or 26 percent of what is produced. With substantial growth in goods and services, Sørensen finds that the practical minimum by 2030 would rise to 3.29 kW/capita, but the production of energy would fall by 56 percent to 5.13 kW/capita. More energy is used for goods and services, but less energy is produced. The economy's efficiency climbs from 26 to 64 percent. By 2005, the study midpoint, gross energy would fall to 7.5 kW/capita (about 60 quads with modest population growth). This estimate is similar to that given in the Solar Energy Research Institute's on-going analysis of the potential of solar and conservation. SERI's year 2000 study will be released in October or November.

Sørensen's concept of efficiency differs from those normally used, in that he posits a "minimum practical net energy" to perform a given task. This minimum falls somewhere between today's delivered energy and the constraints of technology and thermodynamics. "The reason for introducing a practical net energy quantity," notes Sørensen, "is that the physical end-use energy is often ill-defined, because the full range of substitution options for achieving the desired task cannot be assessed, or is meaningless because no technologies are known that would allow the physical minimum energy use to be approached. The practical net energy concept does provide a firm ground for planning, since it only uses presently known (that is, conceived as practical, but not necessarily available) technology, and thus leaves the planner with a well-defined question: How soon would it be possible to replace current technology with the presently best conceivable technology?"

The other distinctive feature of Sørensen's review is that it does not depend on sectoral breakdowns of energy use to forecast the future. Typical sectoral breakdowns would be residential, commercial, transport, and industry. Instead, Sørensen uses a finer breakdown of amenities: hot water, space heat, commuting travel, telecommunications, business travel, vacation travel, etc. This approach permits one to consider structural changes in the economy, for example, in the substitution of telecommunications for business travel and in greater vacation travel. As Sørensen comments, "much of the present physical structure is the result of considerations relevant to earlier stages of technology: living and recreational areas should be far separated from the locations of industrial production units, which were 'necessarily' polluting and ugly. The resulting energy use for commuting between home and work is structural. It does not contribute to living standards though it is counted in the Gross National Product."

Sørensen's categories of energy use are closer to those of the economist than the engineer: people and their households; social activities and institutions; distribution and services; agriculture, construction, and manufacturing; and the resource industry. Within each are energy use categories, such as were noted earlier (hot water, commuting travel, distribution of foodstuffs, *et al.*).

In his projection of the future, Sørensen assumes substantial growth in amenities, including an 11 percent increase in per capita floor space, a 250 percent increase in space cooling, and a 250 percent increase in energy available for appliances (reflecting growth in telecommunications, home video equipment, and the like). Sørensen also assumes a doubling in US agricultural production, making exports equal to home turf consumption, a fifty percent jump in recreational and social trips, and a tripling in telecommunications energy. Energy used by government—reflecting life under Jarvis-Gann—diminishes.

Despite the increases in population size and living standards, primary energy requirements of the US economy gradually shrink and renewables pick up an expanding percentage. Currently renewables supply 0.5 kW/capita, including wood, crop wastes, hydropower, and food energy. Many of these sources could be easily doubled, notes Sørensen, including hydropower (from 0.16 to 0.34 kW/capita), and some, like wind energy, could come out of nowhere to become major contributors. Sørensen assumes a per capita contribution of 0.75 kW from wind, well below the 1995 estimate of Lockheed's Ugo Coty (1.09 kW/cap). Solar photovoltaics on half of America's roofs could add 0.41 kW/capita; with waste heat removed from the cells, a further 0.82 kW/capita could be captured. High temperature solar collectors, central photovoltaics, biogas plants, and numerous other smaller options could cover the residual need of a larger, wealthier, albeit more efficient economy, says Sørensen.

The transition years involve some degree of social planning to ensure, for example, that heat exhausted at high temperature by one industry is "cascaded" to a nearby industry with lower temperature needs. This strategy is well known, of course, in petroleum refining, but is virtually unknown in today's industrial parks. "Re-industrialization" for an "era of limits," to marry the political catchwords of the '70s and '80s, is the theory behind Sørensen's path. To take advantage of solar energy's inherent fluctuations, Sørensen reviews storage needs. For the most part, he finds that only a portion of energy produced needs to be stored. Biomass and hydropower contain themselves until a need appears. For solar thermal energy systems, Sørensen assumes a 30 percent storage loss that "could be eliminated by use of chemical phase change storage systems." The latter are not assumed. For electric power from photovoltaics "some of the need for storage is eliminated by using hydropower facilities as backup... However, as the total amount of energy derived from wind alone exceeds the amount of energy derived from hydro towards the end of the period, active storage capable of regenerating electric power must be included in the system. It could be pumped hydro facilities... but it is estimated that other storage systems will be necessary after the turn of the century, such as flywheel or battery storage for short term... and chemical storage for longer terms."

In Sørensen's scenario, food production, biofuels, and biomass for feedstocks supply about 22 quads of energy in 2030, with efficiencies of collection and conversion set at a reasonable 50 percent. Using an assumed population of 290 million Americans, national energy use would be slightly over 44 quads. Wind contributes 6.5 quads, cogenerating rooftop and central photovoltaics another 11.1 quads, with hydro (2.9 Q) and high temperature collectors (1.3Q) picking up the remainder. These estimates are well below the potential figures given in any number of reviews.

The author concludes, "The assumptions underlying the present study do include a number of modifications in the structure of the US distribution, service, and production sectors. However, few actual life-style changes have been assumed, except for trends already becoming apparent. Additional life-style changes are very likely to occur during a period as long as 50 years, and it is my conviction that the present type of planning—relying on high energy efficiency and a large number of individual energy supply systems—does offer the best possibilities for incorporating any modifications that will be required.... It should also be kept in mind that the year 2030 is not an endpoint in the development. New ways of efficiency improvements, as well as new conversion technologies and new life styles, will continue to come into play... a further reduction in gross energy use per capita, a stationary energy use where increased efficiency is used to improve goal satisfaction, or again an increasing use of energy based on extended use of renewable sources."

—*Jim Harding*

Reference:

Sørensen, Bent
  1980 *An American Energy Future.* Niels Bohr Institute, University of Copenhagen, Denmark.

A Soft Energy Path for Germany?

Obstacles Seen to be Political and Institutional Rather than Technical and Economic

PRELIMINARY RESULTS HAVE JUST been released of a study by Dr. Florentin Krause on the potential for the Federal Republic of Germany (FRG) to run on decentralized, renewable solar technologies. Supported by a grant from Friends of the Earth International, Krause suggests that "if a soft energy strategy is feasible for the FRG, it certainly should be feasible for any of the European industrialized countries... Germany is the prototype of a hard nut to crack: small in land area, northern climate, very densely populated, heavily industrialized and motorized, and deeply entrenched in a nuclear program."

The results of Krause's "technical fix" analysis are impressive: Germany has a large potential for improvements in energy efficiency, productive forestry and agriculture, and a large domestic coal reserve to buffer and indirectly finance a transition to renewables. "In fact," notes Krause, "the most serious obstacle to a soft path seems to be neither technical nor economic in nature, but, as in other countries rather political and institutional."

Germany's Energy Use Picture

Most of Germany's energy comes from oil (52 percent), virtually all of it imported. Coal's contribution is 29 percent. Gas supplies 14 percent, nuclear 2 percent, and other sources 3 percent. On the end-use side, 50 percent of German energy use is delivered as low temperature heat, less hot than boiling water; 25 percent is for higher temperature process heat; 18 percent is used for transport and other mechanical work; and 7 percent is used as electricity.

In fact, Germany is "overelectrified," with 15 percent of all end-use consumption used as electricity and over half of it going for low temperature water and space heating applications. Germany's low temperature heat requirements are unusually large, owing to its climate and poor building standards.

But the "Achilles heel of Germany is transport," says Krause. "Here, oil dependence is 95 percent! If oil imports were to be cut off today, Germans would be walking soon, since compared to present gasoline consumption there is no domestic oil extraction worth mentioning." The other big oil consumer in Germany is space heat. Very little oil is used for electricity generation, although the amount has climbed in recent years because of high coal prices. "Today," writes Krause, "government subsidizes coal electricity against that from other fuels in order to keep the mines from closing down in even greater numbers than during the oil glut of the sixties." Were all Germany's coal used for electric generation, 75 percent of today's demand could be met from that source alone.

"In the long run, the FRG could live on solar low temperature heat, liquid fuels from biomass for transport (mostly domestic, minor imports), on electricity from a European wind-hydro grid, and on industrial process heat from domestic coal."

Krause is critical of Germany's nuclear program, since it meets no obvious energy need. "One should expect," he notes, "that introduction of a new source of energy would be geared toward (replacing fossil fuels) for transport and space heat. Light water reactors, however, replace the one fossil fuel that Germany can call its own, i.e. coal, in its predominant application, in electricity generation."

Krause goes on to point out that the German nuclear industry initially argued that high temperature reactors

A Soft Energy Path for Germany?

*The soft path scenario for the Federal Republic of Germany does not assume that
any major changes in society or in people's values and lifestyles are going to occur.*

would meet the high temperature process heat demands of industry and that low temperature applications could be met with a national heating grid connected to a number of nuclear plants. "However, there is no commercial high temperature reactor available, its development would be prohibitively expensive and it would in all probability be much too unreliable to satisfy the strict availability demands in industry." As for the low temperature applications, the idea of a national nuclear heating grid "was abandoned once it became clear that even with present poor insulation in buildings, the initial investment required for such a grandiose scheme would be impossible to recover in most areas of German cities."

"Presently," he continues, "the nuclear industry is trying to make itself indispensible by proposing that space heat be supplied by electric heat pumps, with oil as a winter peak back-up system. This would require some twenty to thirty 1300 megawatt power plants for driving heat pumps, not considering peak load requirements. Traditional electricity markets — household appliances, direct electric water heating, lighting, etc. — are saturating fast. Therefore, any semblance of rationality to the German nuclear program depends on penetrating the space heat market with (nuclear) electricity and on high growth rates in industrial electricity consumption."

Krause shows that the small industrial component of national electric consumption is not increasing, and is most easily met with on-site cogeneration from existing process heat uses. Similarly, space heat in residences, he says, is much cheaper to implement on-site, and when combined with insulation measures can reduce oil imports more effectively than nuclear electricity. Wood is a vastly cheaper back-up fuel for space heating.

Quirks and New Twists in the Analysis

Germany's population is shrinking, its heavy industry is moving south to cheaper labor markets, and its service sector is growing quickly. However, traditional energy analysts with deterministic models of society and energy use don't pick up these trends.

Krause points out that the net reproduction rate in Germany has fallen steadily to 0.65 girls born per German woman, and is continuing to fall. Germany's 1980 population, including foreigners and immigrants, will be smaller than its 1975 population. The Federal Bureau of Statistics plots a number of populations for 2000 and 2030, none of them higher than at present.

The economic structure of the FRG is also changing rapidly. German wage levels, for example, are 20 to 30 percent higher than in the US, four to five times higher than in many southern European and Mediterranean countries. As a result, "all production that requires relatively standardized technology, is not very R & D intensive and has a high share of labor costs is shifting to an increasing extent to developing countries." The growth industries in Germany are thus the chemicals industry, modern steel mill technology, and the development of new machinery, electrical and electronic equipment and measurement technology, and the like. At the same time, traditional consumer goods, steel production, fertilizer production, crude organic chemicals industries, aluminum, and most other primary material industries are emigrating. The energy implications of this sectoral shift are enormous: "German energy growth proponents like to point to the approximate proportionality of industrial and energy growth in the past twenty years. This period was shaped, however, by a process of extensive mechanization and of capital extending investments as well as artificially cheap oil. At the present stage of economic development of the FRG, pursuing further such volumetric growth [e.g. by protective tariffs for uncompetitive weak industries] means destroying our competitiveness on the world market by giving away comparative advantages. For the consumer, it would mean a lowered standard of living. Many studies advocating the need for further energy growth show this contradiction in their reasoning, which always starts from the better life and the more plentiful jobs we should want," says Krause.

Krause calculates that a 50-year halving of the energy-intensive primary materials industries would reduce in-

A Soft Energy Path for Germany?

dustrial energy consumption 35 percent. It would also mean relocating 11 percent of all industrial jobs and losing 14 percent of industrial sales, "while bringing a drastic improvement in environmental quality. In the engineering sector, a 50 percent decline would have twice as strong an effect on sales and jobs while total industrial energy consumption would only be reduced 5 percent."

Some Improvements in Efficiency

It turns out, notes Krause, that the quality of insulation in the German housing stock is "uniformly bad irrespective of age," with the exception of residences built after the mid-seventies, when insulation standards were tightened. A square meter in the average house leaks about 300 kilowatt hours annually with single homes leaking at 400 kWh/m² per year and flats leaking at 220 kWh/m² per year. Krause calculates gradual efficiency improvements in eight different building categories, including historical buildings with unalterable exteriors, to generate an alternative energy forecast for German homes.

Interestingly, solar energy can contribute significantly to well-insulated houses. "To keep the solar and storage system small," avers Krause, "very tight insulation is

The "Achilles Heel" of German energy use is transport.

necessary." A 100 percent solar house in a cool climate with current insulation standards, notes Krause, would require 70 m² of collector and a tank twice the volume of the house itself. For a tightly insulated house, however, 11.5 m² of collector and 30 m³ of water storage would suffice for 100 percent solar space and water heating (30 m³ is the volume of a small basement storage room).

In a house with retrofit insulation, such tightness is not readily achievable — but one cubic meter of firewood burned in the house at 70 percent efficiency is an attractive back-up system. "With all single family homes retrofit with better insulation, only 10 TWh would be needed in a typical winter, that is, 15 percent of available forestry residues and small timber. Family members could fit the gathering and splitting of the firewood into their fitness

program, working off excess beer and cake, or have it delivered to them. No nuclear electric system can deliver 10 TWh of peak demand at anywhere near the price of such a cozy fuel," says Krause.

Automobiles

Though "German car owners have a very intense relationship with their vehicles," and love high speeds, this fact doesn't keep German engineers from improving energy efficiency for a nation of hotfoots. The Volkswagen Rabbit diesel, for example, is 69 percent more energy efficient than the Beetle; while the normal Rabbit is 31 percent more efficient than the unforgettable insect. The next stage diesel Rabbit, however, will be even better: 59 miles per gallon rather than the current 44 and the same percentage efficiency improvements are planned by German manufacturers for all weight classes. Improvements are achieved with further weight reductions, a gradual shift to diesel engines, elimination of idle, aerodynamic improvements, and a variety of other measures. Continuously variable transmissions could bring a 30-50 percent improvement in efficiency, and these are undergoing testing, but manufacturers are quiet about their results.

Despite its high population density and inclement weather, Germany is nearly self-sufficient in agricultural production. There is little surplus land, however, for "energy crops," so wastes from current production are about all that is available for conversion into alternative automobile fuels. Cereal production, potatoes, beets, corn, and forage are large crops in Germany with considerable usable waste. Germany also, in Krause's words, lives "high on the hog," with one pig and three quarters of a cow per family. He concludes that the total energy available for biogas or biofuels use would be equal in energy content to

A Soft Energy Path for Germany?

20 million metric tons of coal (MTC). Another 8 million MTC could be burned directly, or, if necessary, converted to fuel worth 4 million tonnes of coal. "Thus, after an integrated agriculture/forestry/fuel system has been developed, Germany should be able to derive some 25 million MTC of end-use energy from its land. This amount would suffice to power all German cars once tidied up. It would just about cover the long run liquid fuels demand of the whole transport sector. However, even if only part of this potential were to be mobilized for transport energy this would still greatly relieve the dependence of this sector on imported oil."

Germany also has some wind potential along its northern coast. If one 300 megawatt wind plant were located on each 10 square kilometer stretch, all the nation's appropriately electrical energy needs today could be supplied from this source alone. Of course this level of deployment might prove impractical or unacceptable, but the potential is there and a small program based on the German government's favorite windmill (1500 DM/kW installed) would be cheaper than new nuclear reactors and environmentally superior. With the extensive European and Scandinavian grid system, reliability is not a problem, as wind produced in Germany is water not flowing through a dam in Norway and vice versa.

Putting the Uses and Resources Together

Krause's scenario for the Federal Republic of Germany extends to 2030, longer than many energy projections, in order to show the post-2000 potential contributions from renewables. It is a technical fix scenario, "because it does not assume that any major changes in society or in people's values and life styles are going to occur. Rather it relies on conventional forecasts about economic growth and increased levels of consumption. Only technical fixes are introduced to improve energy efficiency in all areas of energy consumption. The rate at which these improvements occur is generally determined by the replacement rates of houses, cars, etc.

> *"The required investment for a soft scenario would be considerably less and it would be faster in freeing the FRG from oil dependence than any other strategy."*

The selected improvements in energy efficiency are only those that are competitive with marginal sources of nuclear electricity or synthetic fuels from coal. "In other words, it describes the paradise of the technocrat who is committed to economic growth, but who also realizes that improved energy efficiency is the only thing that can finance that growth."

Energy Use

Primary Energy Use	1975	348 million MTC
oil	181 (52%)	
coal	101 (29%)	
gas	49 (14%)	
nuclear	7 (2%)	
other	10 (3%)	

End-Use	1975	234 million MTC
industry	84 (30%)	
transport	46 (20%)	
private households	62 (26%)	
commercial, other	42 (18%)	

Electricity Use	1975	319 TWh

Generating Capacity	1976	
public utilities	65 GW	
industrial	15 GW	
railways	1 GW	
total	81 GW*	

* 6.5 GW of this total is nuclear
(one million Metric Tons of coal equals 8.13 TWh)

Occupied, heated (but insulated) residential floor space grows 85 percent per capita by 2030, every house has every significant known appliance, hot water use levels off at a more than adequate 70 liters per person per day, there are no changes in the size of automobiles, and air transport triples. In sum, it is not a future of material austerity. Yet by 2030, energy use is 60 percent its 1975 level. The drop in electricity use is less dramatic, but still notable. "Both reductions occur as GNP and activity levels increase drastically," notes Krause. "This conclusion is the exact opposite of the one reached by other energy studies that project a three or four-fold increase in energy demand to reach the same economic growth goals." The gap, notes Krause, between analyses that begin with a supply goal and those that begin with end-uses, "is even more shattering to conventional wisdom when it is considered that the required investment for a soft scenario would be considerably less and that it would be faster in freeing the FRG from oil dependence than any other strategy.

"In the long run, the FRG could live on solar low temperature heat, liquid fuels from biomass for transport (mostly domestic, minor imports), on electricity from a European wind-hydro grid, and on industrial process heat from domestic coal. The biggest problem might be what else besides oil the FRG could import to balance its exports."

Source: Florentin Krause, *Soft Energy Germany*, work in progress report prepared for International Soft Energy Path Conference, Rome, Italy, May 17-20, 1979.

Jim Harding

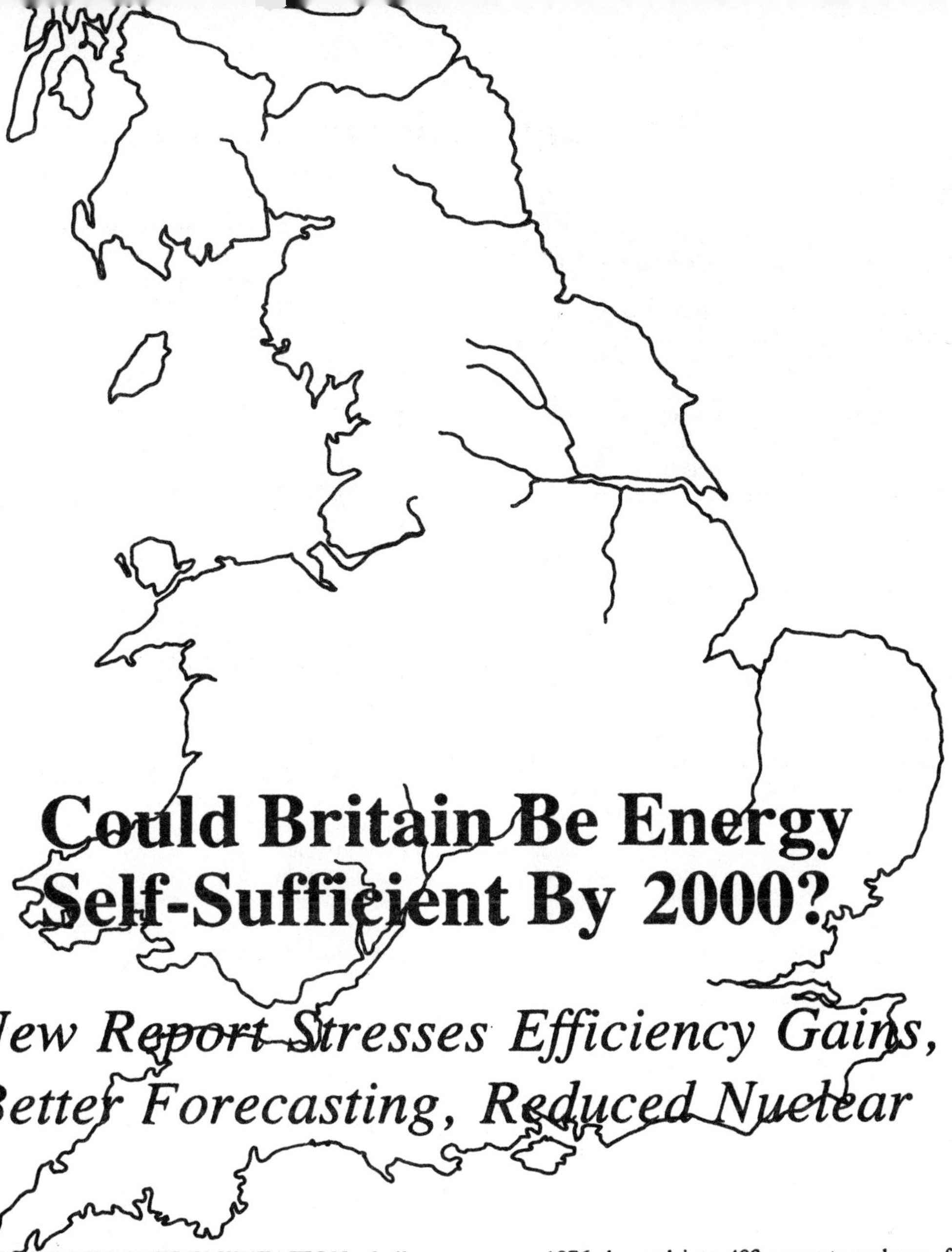

Could Britain Be Energy Self-Sufficient By 2000?

New Report Stresses Efficiency Gains, Better Forecasting, Reduced Nuclear

PESSIMISTIC ABOUT THE IMPLEMENTATION of all ''new'' or ''alternative'' technologies, but hopeful about the chances for improved energy efficiency, a new report by the International Institute for Environment and Development (IIED) proposes that economic and population growth through the end of the century could be easily accomodated despite the virtual. elimination of Britain's nuclear program, and a modest increase in coal use. At the core of the analysis is a deeply persuasive case for careful forecasting of future energy use based on individual end-uses rather than aggregated sums. As a result, improved efficiency, some increased use of coal (less than official plans), and stretched-out use of North Sea oil and gas suffice to make Britain independent of foreign energy sources, without further large commitments to conventional nuclear power or breeder reactors.

The most notable aspect of the Leach study is its careful disaggregation of energy use. Using available studies of British energy consumption, the Leach group was able to break down

The findings of *A Low Energy Strategy for the United Kingdom* by Gerald Leach and four subsidiary authors—Christopher Lewis, Ariane van Buren, Frederic Romig, and Gerald Foley— should hold broadly for other industrial countries as well. It demonstrates systematically, and in detail, how the United Kingdom could have 50 years of prosperous material growth and yet use less primary energy than it does today.

1976 demand into 400 separate end-use, fuel, and appliance categories; apply efficiency estimates for existing and new equipment; then scale each sector and use to an expanded economy and population. This method of forecasting energy use for the long range future is far superior to simple, gross relationships of economic activity and energy use. To drive this point to the readers, the authors note that British GNP has increased by 10 percent over 1973-77 while British energy use has declined 10 percent: hardly a direct correlation, though some might enjoy the possible extrapolations.

Changes

So what are the technical improvements and government policies that could keep coal use below official projections, oil and gas use limited to North Sea reserves, and nuclear power virtually unnecessary? Leach and colleagues assume that energy use per pound of industrial output falls by 22 to 35 percent, depending on the industry considered, over the period 1976-2010. This is an average rate of 0.6-0.9 percent per year and, ''in some cases is slower than past trends towards less energy intensive manufacture.'' Beyond 2010, there are further falls of 10-15 percent. The disaggregation level includes 8 industrial sectors, seven fuels, and 13 end-use purposes.

The domestic sector is fully insulated with ceiling and wall (where possible) insulation over the next thirty years. A small fraction use double glazed windows. New homes use gradually less energy for heating, with the Building Regulations slowly tightened until 1990, when heat losses through the fabric reach 50 percent of 1975 standards. Electric and gas heat pumps gradually acquire 15 percent of the space heat market by 2000; solar has little impact. New efficient appliances use about 50 percent as much energy as today's models, still much higher than current designs under testing at the Danish Technical University.

Fuel consumption per unit of floor area dropped about 20 percent in the institutional and commercial sector between 1960 and 1975. As superior light bulbs and heating controls come into widespread use, savings of 30 to 50 percent should be possible, according to Leach. New buildings are projected to use 40 to 60 percent of the energy used by current floor space, ''following the example of scores of recently constructed buildings.''

Transportation is a sector that turns over rapidly and uses considerable energy. Leach's assumption is that cars and light trucks drop in energy consumption by 34-26 percent by the end of the century, though Britain's Department of Energy projects 40 percent. Heavy trucks reduce their consumption per mile driven by only 10 percent. Electric cars, vans, and buses are assumed to have no impact until 2010.

Leach's scenario also does not depend on vastly increased prices to wring improvements in energy productivity. World oil prices double by 2000 and treble by 2025; other fuel prices roughly follow this trend. All the conservation improvements are attractive to consumers: most, says Leach, are attractive at current fuel prices.

The government policies contemplated in the report include thermal performance standards for buildings (public or private, residential or industrial-commercial), energy performance standards for vehicles, energy efficiency standards for major household appliances, and ''possibly'' legislation to reduce oil use in homes, industry, and public buildings. ''We say 'possibly' because the gradual reduction that we have assumed may well occur naturally,'' notes Leach.

It is Leach's belief that ''none of these measures would . . . affect freedom of choice.. None would go beyond legislative measures that are widely accepted for the common good, such as the Building Regulations, the Clean Air Acts, or the conversion of equipment to take North Sea natural gas . . . '' The group continues: ''All that we have done is to couple . . . political determination with a range of technical opportunities that are now or soon could be available, and forecast the consequences. Many of the changes that we project are likely to occur naturally through technical progress and the turn-over of capital stocks, especially if fuel prices rise substantially. But other changes are predicted on reasonable determined government policies and their implementation in the coming decades.''

Gerald Leach, in a recent *New Scientist* article, describes his group's well-documented findings more fully. (reprinted below by permission)
—*Jim Harding*

A Future With Less Energy

By Gerald Leach

Member of the IIED Energy Project*

Conventional wisdom has it that a country uses more energy as it grows richer. Since governments must assume rising prosperity, their energy forecasts therefore show increasing demands, widening "gaps" for oil and gas, and the need for more nuclear power to fill them. This happens even while everyone is loudly proclaiming the importance of energy conservation.

Two years ago the International Institute for Environment and Development (IIED) Energy Project began to examine this assumption. In a post-1973 era, it seemed likely that some quite moderate technical fixes could couple growth from rising energy use. Wherever we looked in the energy conservation literature, people were reporting actual or potential fuel savings of around 25 to 50 percent from the application of anything from free human ingenuity to investments. Many of these inventions paid for themselves within only one or two years even with today's fuel prices. In some instances such as new house and office designs, it was proving cheaper to halve energy consumption than to build conventionally. Even allowing for institutional inertia, why were government forecasts assuming savings over the next 20 years of only 10 to 20 percent?

"None of these measures would affect freedom of choice. None would go beyond legislative measures that are widely accepted for the common good, such as the Building Regulations, the Clean Air Acts, or the conversion of equipment to take North Sea natural gas . . ."

Even odder was the way people were forecasting energy demand. The game was to take large sectors, such as housing, industry or public services, lump all their types of energy use together, see how this aggregate rose with income or economic output during the pre-1973 era, and then assume that these links hold good for the next 20 or so years. This gives a hypothetical demand which is then knocked down by some percentage owing to conservation effects. This is how the UK Department of Energy forecasts the sectors just mentioned, which account for 67 percent of energy and 92 percent of electricity use.

One oddity is the use of pre-1973 trends, before energy prices soared. Since 1973 many of the links have vanished or gone negative: incomes and output have risen while energy use has stayed level or declined. Why not use these Post-1973 trends as the forecasting basis? Most fundamental is the assumption that energy use is that simple. In industry, for example, energy is used for many different purposes, each with different opportuni-

---

*With Chris Lewis, Frederick Romig, Gerald Foley and Ariane van Buren. The report is called *A Low Energy Strategy for the United Kingdom*, published by Science Reviews and IIED, 10 Percy St., London WIP ODR.

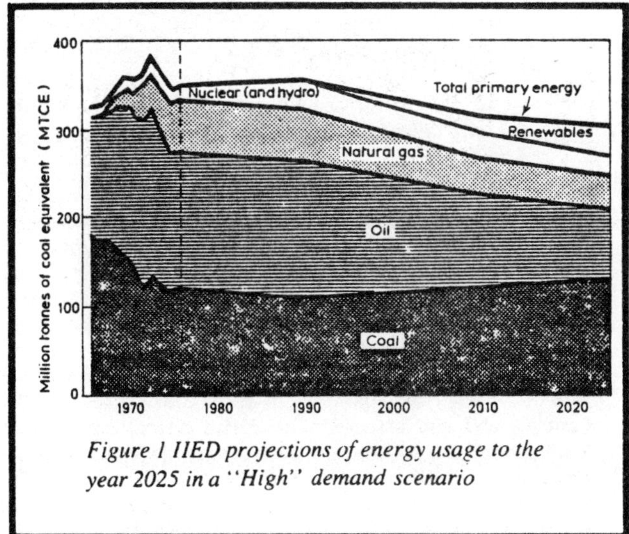

Figure 1 IIED projections of energy usage to the year 2025 in a "High" demand scenario

ties (and constraints) for more effective use. Industries vary enormously in their energy intensity and are likely to grow at different rates. Buildings, in which most energy is consumed, have complex energy interactions which tend to hold down the major use of energy—for heating—as well as obvious limits on heating and lighting levels, however high one's income or economic output.

In short, by its forecasting methods and generally weak assumptions about technical change, might conventional wisdom be greatly overestimating future energy demand—and therefore the need for new energy supplies?

The IIED has published one answer to this question.* It is a detailed analysis of the use of different types of energy in different sectors of the economy in the UK. Its main conclusion is that the UK (and by implication other industrial countries) could have a future of prosperous material growth and yet use less energy than it does today. As Figures 1 and 2 show, if the gross domestic product (GDP) roughly trebles in round terms by 2025, our calculations suggest that a series of simple, known technical fixes could keep energy (and electricity) demand more or less constant from here on. If GDP doubles, oil consumption could be roughly halved, coal output could remain constant. Nuclear output could fall steadily from 1990 (when the stations now under construction or planned are built) and primary energy consumption could be down about 7 percent on today's levels by 2000 and 20 to 25 per cent by 2025.

Obviously this conclusion is only as good as the assumptions that went into it. Thanks to several excellent studies that have appeared only last year, we were able to base our study on a very detailed breakdown of the final use of energy by different fuels, types of appliance and end-use purpose in 1976 which ran to nearly 400 possible categories. For example, the industry scenarios are based on eight separate sectors, each with seven fuels or energy sources including self-generated heat and electricity that provide 13 end-uses including four temperature bands for direct and for indirect process heating. In housing we considered four types (existing and new houses and flats), each with separate components for space and water heating, cooking, lighting, and seven categories of electrical appliances.

A Future With Less Energy

and light vans by 2000. Our figures are 34 percent and 26 percent with only 10 percent for heavy lorries. Electric vehicles are introduced only after 2000 and then at modest rates.

Government's Role

Some degree of government intervention would certainly be required if the energy conservation we envisage were to happen. However, one must remember that massive intervention—and subsidies—are typical on the supply side. In particular, we have assumed that government sets energy consumption targets for new buildings of all kinds, for cars and light vans, and for cookers and major electrical white goods. In all these important categories energy savings of about 50 percent are technically possible—and with the possible exception of road vehicles—at very low cost. But government-led industrial agreement may be needed to secure them. A second policy assumption is that determined efforts are made to provide consumers of all kinds with information about available "best practice" technologies and techniques. Government and professional bodies are rapidly stepping up their efforts on these lines. All that we have done is to couple this growing political determination with a range of technical opportunities and likely growths in energy-using activities, and projected the consequences.

The consequences for energy supply are rather startling. Figures 1 and 2 show that primary energy demand in 2000 is only 330 to 361 million tonnes coal equivalent (mtce) compared with the current Department of Energy forecast of 460 to 570 mtce. Furthermore, energy use declines after 2000—contrary to most expectations. This is due to a slowing growth in some key areas such as the number of households and traffic levels; the long lead times for some major energy saving technologies; and our assumption that renewable energy sources (such as the Sun, wind and waves) and combined heat and power (CHP) schemes do not begin to take off until after 2000. Our assumptions for the contributions in 2000 from renewables and CHP are well below those of the present Department of Energy forecasts. This is not because we think these technologies are unimportant, but merely because we wanted to highlight the enormous savings obtainable from currently available and quite conventional technologies.

Because of this low energy growth, the UK could be almost self-sufficient in oil and gas until 2025. The only significant "gap" occurs for oil if one takes the present central estimate of North Sea reserves. The shortfall opens up to an annual 60 to 80 mtce by 2025, or roughly half recent levels of oil imports. Coal liquefaction, imported biomass (wood, etc.) fuels from the tropical "sun belt," and some imported crude oil could easily plug this hole, which virtually disappears if one uses the upper range of reserve estimates. Coal production rises by 2025 to only 128 to 148 million tonnes a year, or well under the present *Plan for Coal* target of 170 million tonnes by 2000. There is thus plenty of slack.

But the most remarkable policy implications are for electricity and nuclear supplies. We have not been "anti-electric:" there are more electrical devices and cookers in the home, more electrification in industry, more electric trains, and a sizeable fleet of electric road vehicles. However, due to greater efficiencies, demand does not grow over the long-term. Consequently, only 26 to 30 gigawatts (GW) of generating capacity need be built (mostly to replace retired plant) between now and 2000, compared with 83 GW in the present department of Energy forecasts. The savings in capital investment would be of the order of £25-30,000 million, or well over £1000 million a year.

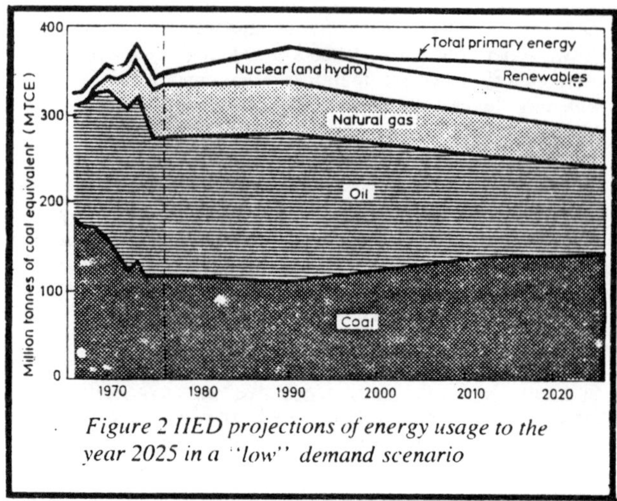

Figure 2 IIED projections of energy usage to the year 2025 in a "low" demand scenario

Nuclear power assumes an almost peripheral role and could easily be abandoned if desired. We have in fact kept the nuclear option open but at a low tick over level. From 1976 to 2000 only 4.5-6.5 GW of nuclear capacity is built apart from plants now under construction and the two advanced gas-cooled reactors planned for Thorness and Heysham. This compares with 30 GW in the Department of Energy's reference forecast. If more were built there would either be a huge surplus of generating capacity, or coal production would have to be reduced to politically implausible levels. The fast breeder reactor is simply not needed for energy supply and could be shelved indefinitely.

The main point of these scenarios is not to say that the future will be like this but to show that it *could* be. An energy future of low risk is possible, with only moderate change. If other countries that currently use a lot of energy followed a similar low energy path, the reduction of international tensions and the drain on world fuel resources could have remarkable effects on global development, equity and self-confidence.

Curiously, five years after the 1973-74 fuel "crisis," such a future now seems increasingly likely. In almost every country energy forecasts are falling as new energy-conscious policies and economies emerge. In Britain the Department of Energy's forecasts for 1990 have fallen at such a rate since 1975 that by 1980 they should be predicting zero energy growth. Actual energy consumption in 1978 is almost certain to be lower than in 1970 despite a rise in GDP of at least 10 percent. In the US six years ago the extreme range of forecasts for 2000 was 124 to 190 quads (10^{15} British Thermal Units). In 1978 the respectable range was 63 (for 2010) to 124 quads: yesterday's ultra-low heresy has become today's record high. Similar sweeping changes in perceptions of the energy future are occurring in Europe.

"A Future with Less Energy" by Gerald Leach first appeared in the January 11, 1979 issue of New Scientist *magazine, London, in The Weekly Review of Science and Technology. New Scientist is published weekly in the UK. Subscription rates are: 20.7 pounds in the UK, 22.9 pounds overseas surface mail, $54 airfreight mail to the US and Canada, and 40.9 pounds airmail anywhere. Write to New Scientist at Kings Reach Tower, Stamford St., London SEI 9LS, UK, or the Schieces, Subscription Department, 2 East 63rd St., NY, NY 10021, USA.*

Sweden Looks to Uranium and Renewable Resources

In *Sweden Beyond Oil—Nuclear Commitments and Solar Options* and a companion piece, *Solar or Nuclear—On the Choice of Energy Futures*, Thomas Johansson, Mans Lönnroth and Peter Steen identify two choices: one future based on uranium and an alternative system relying on wind, solar and hydro power.

The authors make two conclusions: 1) when today's uncertainties are taken into account, it is too early to state that one of the alternatives is more economical than the other; 2) neither choice needs to be a major obstacle to an increased material standard of living.

Sweden

population	8.2 million
area	450,000 km^2
GNP/capita	$6400
annual insolation	800-1000 kWh/m^2
Energy, total	415 TWh
energy/capita	50 MWh
Electricity	80 TWh
Hydro power	57 TWh
Nuclear	12 TWh

But Johansson *et al.* reject the terms of this choice. Their strategy is to preserve both options for the next twenty or thirty years. In light of the nation's current favor for nuclear power, Johansson, Lönnroth and Steen's flexible policy may just be solar's ticket into the 21st century.

Methodology

Johansson, Lönnroth and Steen take a long term approach to energy policy. They develop solar and nuclear schemes for 2015 and then determine what would be required in the short and medium term to implement them. "Most energy policy studies are based on the need to assure supply in the short and medium term," the authors explain. "The longer term development thus tends to be determined more through an incremental series of short term decisions rather than as a result of any overall strategy or vision."

"New technologies will have to compete with old on the old's conditions: taxes and rates, for example, are adjusted to fit existing technical solutions while not necessarily being well suited to any specific new technology; there is also the possibility that important features are overlooked because they do not fit directly into the existing pattern but would do so in a larger context.

"Since we believe such an approach to be inherently biased toward existing technologies and the institutions behind these technologies, we have worked differently."

Both Nuclear Sweden and Solar Sweden use familiar renewable resources: hydro power increases slightly to 65 TWhe/year and biomass (bark and lye) from the paper and pulp industry continues to contribute 35 TWh/y.

Both schemes boast a comfortable lifestyle for a fixed population of 8 million. The production of goods and services doubles, permitting everyone the material standard of living now enjoyed by the top ten percent of the population. The energy needed to supply these goods and services diminishes: industry uses 80% of current demand; the services sector (including transportation) uses half. In the housing sector, the number of flats or single houses increases 40% but retrofits and new building insulation reduce average energy need per unit by 30%. Given these assumptions, the final energy demand rises from the 1975 level of 415 TWh to 547-568 TWh in 2015 (see table).

Nuclear Sweden

The authors envisage sixty-seven 1000 MW nuclear power stations by 2015, providing electricity to industrial, commercial, and domestic sectors. Pumped storage and fuel cells supplement peak load. Six 1000 MW reactors supply district heating by cogeneration to metropolitan areas. Ten additional smaller reactors are for heat production only. The transportation sector relies on methanol or hydrogen used in fuel cells.

Light-water reactors play a major role in the uranium-based energy system. Since known reserves of domestic uranium are expected to provide only thirty years of operation, the authors assume that breeder reactors will be introduced by the turn of the century.

Solar Sweden

Renewable energy sources now contribute about 25% of total supply. Biomass, wind and hydro power, photovoltaics and solar heating fill the rest of the demand in Solar Sweden; the

Sweden's Energy Future	1975 TWh	Production level (compared to 1975)	2015 Specific energy use	TWh
Production of goods	165	+ 100%	-20%	264
Services				
Transportation	75	+ 100%	-50%	75
Others	70	+ 100%	-50%	70
Housing including household electricity	80	+ 40%	-30%	80
	390			489
Conversion losses	25			58-79
TOTAL SUPPLY	415			547-568

biomass contribution dominates. Energy is delivered as electricity, methanol, wood and heated water at appropriate temperatures. Biomass-fueled cogeneration of heat and electricity is the preferred solution for large urban centers. District heating from solar systems with seasonal storage—water and ground stores—heats smaller cities. Process heat is supplied by biomass and some use of waste heat. Hydro power, wind power and solar cells supply electricity.

Energy plantations (willows and poplars are being tested) contribute as much as 280 TWh requiring 2.9 million hectares (1 hectare = 1000 m^2) or 90 MWh/hectare/year. This is 6-7% of Sweden's land, roughly the same area now used for agriculture.

Solar heating contributes 71 TWh/y, most of which is space heating. Wind power accounts for 30 TWh/y using 3700 1-MW units in areas with median wind speeds above 6 m/sec. at 0 or 100 m above sea level. Solar cells—50 m^2/capita—could be used directly for domestic purposes, connected to the grid, or used in the production of methanol. The transportation sector is fueled by methanol, fuel cells or electric batteries.

As with the nuclear scheme, any major commitment to a Solar Sweden will depend on further research. The authors consider the further study of energy plantations and solar heating crucial. Other unknowns include the cost of solar cells, the public acceptance of large scale wind systems, and the environmental effects of energy plantations.

A Flexible Policy

The past 25 years of Swedish energy policy have stressed the nuclear alternative. By 1973, 11 reactors (8.4 GW) were licensed and the utilities planned for 25 GW by 1990, giving Sweden the world's largest per capita nuclear program. Once committed to nuclear plants, electric utilities set rates to protect those investments and, consequently, the authors explain, they dissuade consumers and local authorities from investing in dispersed technologies. Conversely, measures to ensure long term options strengthen the solar industry and threaten utilities' short and medium term stability. To outline a long term energy policy "attuned to freedom of action," the authors list these measures:
- charging local authorities with the production, distribution, and conservation of energy;
- altering the rate structure to stimulate conservation and investment in dispersed and capital intensive supply alternatives;
- encouraging large scale energy users to adapt to supply alternatives;
- creating industry incentives with state-financed R & D of new energy technology;
- ensuring that land is available for reactor sites, biomass plantations, wind energy plants and solar heating stations.

Johansson, Lönnroth and Steen explain: "a flexible policy could be seen as a way to handle and thus reduce overall uncertainty for governments and the public at large, but it has exactly the opposite effect on those groups and industries that already are committed. Their uncertainty is increased by the very same policy. But that can be seen as the price that has to be paid in order to have a democratic freedom of choice."

The feasibility—both technical and economic—of maintaining both solar and nuclear options into the 21st century will be questioned by soft path enthusiasts and industry alike. However, Johansson, Lönnroth and Steen recognize "major uncertainties with both the solar and the nuclear options: technically, environmentally, institutionally and socially." The authors' concern is that Sweden is embarking on a transitional path "without having a clear consensus of where to go." Their proposal is an energy policy for the rest of the century that does not foreclose either of the options.

—*Elyse Axell*

References:

Johansson, Thomas B., Mans Lönnroth and Peter Steen
1979 "Solar or Nuclear—On the Choice of Energy Future." Paper presented at the International Conference on Energy Systems Analysis, Dublin, Ireland, 9th-11th October 1979, 16 pp., work done in the Secretariat for Future Studies, support in part by Swedish Energy Research and Development Commission.

1979 "Sweden Beyond Oil—Nuclear Commitments and Solar Options." The Secretariat for Future Studies, September 5, 30 pp., based on a book with the same title to be published by Pergamon Press.

Steps Toward a Solar Israel

POLITICAL COMMENTATORS describe the State of Israel as a paradox with borders and energy analysts might apply this epigram with equal accuracy. Of all the countries in the Middle East, Israel has the poorest endowment of conventional energy resources. Despite this handicap, its industrial base has become the most highly developed. Hidden within this achievement, according to energy researcher Avram Kalisky, lies a paradox: given local energy costs, Israeli industry should be energy-efficient, yet it is actually one of the most wasteful in the world. And this is not the only riddle within the energy arena. Solar research in Israel got off to an early start in the 1950s, but even with this lead, the country is still in some ways further from a soft energy path strategy than many other technologically advanced nations.

The reasons for these contradictions lie in the general instability of the Middle East and the uncertainty this creates in energy prices, political alliances, and international borders. Long-range plans have little credibility here, since, as Kalisky explains, "in Israel it is a strain to contemplate tomorrow, no

---

"Israeli industry should be energy-efficient, yet it is actually one of the most wasteful in the world."

---

less the day after." If there is any surety in Israel's energy future, it is that solar energy must reduce imported oil dependence in the decades to come. And, Kalisky adds, the technologies that will enable it to do so are already well advanced.

Solar's Early Successes

Solar energy research in Israel began almost three decades ago, with a number of innovative and largely successful projects at the National Physical Laboratory of Israel in Jerusalem. In 1955, Dr. Zvi Tabor and his associates developed a prototype solar collector hot water system. It consisted of two 1.5 m² collectors and a 120 litre tank with electric standby heating in a thermosiphon configuration which has since become extremely popular. Today, more than two hundred Israeli workshops turn out systems developed from this design; the rooftops in Tel Aviv and Jerusalem are cluttered with almost as many collectors as television antennae.

Another Laboratory project was a low maintenance turbo-generator that converted solar energy at low temperature to electricity. While this application proved too expensive, a version of the turbogenerator that runs off conventional fossil fuels sold well as a replacement for diesel generators in remote locations.

Other solar projects, while technically sound, proved to be commercial failures. A solar cooker on a cheap pipe frame foundered, Kalisky comments, because it required a change in customs. Research on solar ponds came to a halt in 1966 when economic studies showed that the energy they produced would be twice as expensive as that from oil. In like fashion, such pioneering solar applications as solar powered irrigation pumps could not compete with the inexpensive oil of the era.

Israel depends on imported oil for 99% of its energy—even with solar collectors providing hot water for 15% of the country's families—and this dependence renders it extremely vulnerable to interruptions in supply. The oil embargo of 1973 revitalized local interest in alternatives to an oil-fired energy economy, and propagated a new generation of solar products and projects. These efforts focus almost exclusively on displacing

Israel Energy Supply Scenario
in Petajoules (10^{15} Joules)

	1975	2000	2025
Oil	301	101	158
Solar Ponds	0	101	131
Other Renewable	3	180	260
TOTALS	304	382	549

the use of oil, and the other components of a soft path—detailed end-use analyses, efficiency improvements, and so on—have received much less attention. This imbalance has led to some curious contrasts in energy use, such as the large number of Israeli households heating water in solar collectors, while using inefficient electric resistance to warm their homes in a country where the only source of electricity is oil.

The responsibility for this paradox rests partly with an economy that offers few incentives for energy efficiency. This is particularly true of Israeli industry, which faces a "supply limited" market in which the consumer absorbs all costs. This leaves manufacturers no motivation to invest in energy savings since they can simply pass their increasing energy costs to the consumer in the form of higher prices. The high cost of credit in Israel turns this into an outright disincentive to invest and so industry there, according to Kalisky, continues to waste the country's precious energy supply.

Promising Projects

Israel's energy problems spin out of the Middle East political process, but their resolution can not wait for the larger regional issues to be settled. Towards this end, government agencies, universities, and private enterprise are all engaged in solar energy research. Kalisky highlights some of the more promising developments:

● Miromit, the world's largest producer of solar collectors, has developed a flat-plate collector seven meters long that could be built right into the roof structure in new buildings.

● Electra is now marketing a low-cost, do-it-yourself collector for swimming pools, using a patented modular plastic reflector, and standard black plastic hose.

● The Israel Institute of Technology (Technion) in Haifa is working on a solar concentrator that heats the transfer liquid to over 400° C with a low cost mirror of two to four meters in diameter.

● Ormat Turbines is adding improvements to the decades-old

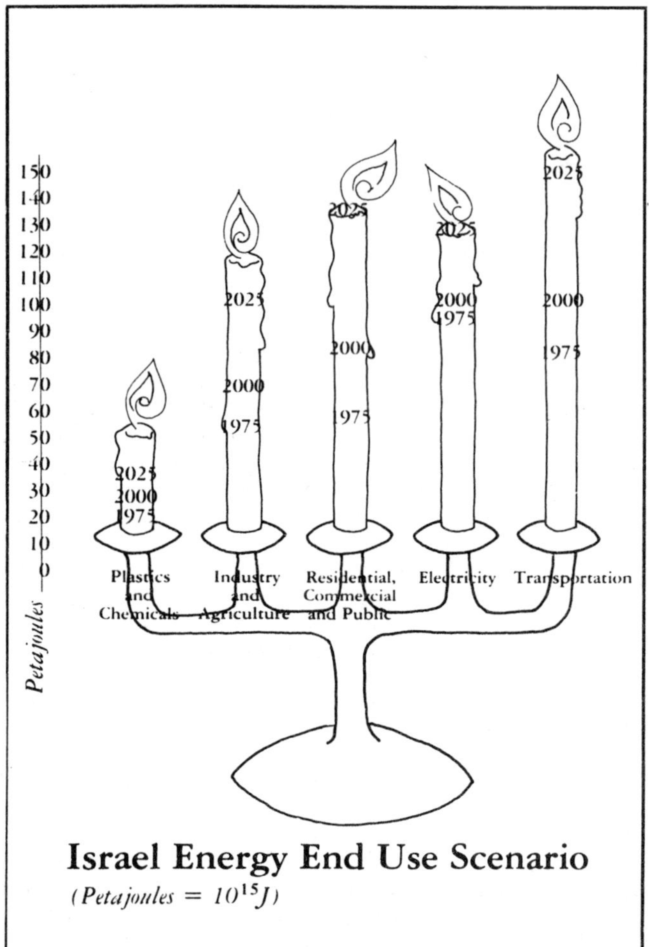

Israel Energy End Use Scenario
(Petajoules = 10^15 J)

electrical turbogenerator design, making it cheaper and more suitable to applications previously considered marginal. This encourages the diffusion of small power sources as an alternative to investment in large central stations.

The low temperature Ormat turbogenerator figures significantly in a unique electric power system based on the solar pond, a technology in which Israeli researchers have pioneered. Conversion of heat into electricity using this system is very inefficient—less than 1%—but construction and maintenance are relatively inexpensive and fuel costs nothing at all, making the solar pond/turbogenerator combination extremely attractive.

Perhaps the most promising research effort is the construction of solar ponds in large lakes and ocean bays (Assaf 1976). Kalisky comments on this technique: "In principle, simply placing a floating barrier about ten meters deep, enclosing several square kilometers, or more, of water isolates a pond. After conditioning the density gradient, heat can be extracted as in a solar pond. (This) may revolutionize future construction of electric power stations." Despite its low net efficiency, the pond/turbogenerator combination still produces 3.5 to 4 MW/km^2. Kalisky calculates that the surface area of the Dead Sea within Israel's borders that is not presently used for chemical production is capable of supplying some 2000 MWe, which is more than the country's present requirements. The net delivered cost is estimated to be only $0.045/kWh.

The high productivity of Israel's farming collective (kibbutz) has also captured the interest of solar energy researchers. Feasibility studies for the conversion of agricultural wastes into energy, feed, and industrial products started in 1974. Results from small biogas test units warranted investment in a large pilot plant of three 100 m^3 digesters, to be constructed at the kibbutz community of Kfar Giladi. Two of the digesters would process the manure from a milking herd of 500 cows and one would ferment other agricultural wastes such as field cuttings. The three together would produce more than 12,000 GJ annually, enough to displace the kibbutz's present consumption of boiler and diesel fuel. Current research in biogas involves a search for productive uses of the digester residues, such as vitamin B-12 extraction, fertilizer applications, and an indirect substitute for fish pond feed by encouraging the growth of edible plants.

A Solar Israel

Energy scenarios are built around assumptions of future resource availability, demand, and consumption patterns, all of which are hard to make for a region as unstable as Israel. Even the borders of this country are likely to change in the years to come, and there is no way to tell whether it will have access to the petroleum supplies controlled by its Arab neighbors. Kalisky contends, though, that certain assumptions are possible, and they suggest that a substantially solar Israel is within reach.

The population, now at 3,100,000 and increasing at a rate of 2.7% per annum, may grow to 5,400,000 by 2000, and to 10,500,000 by 2025. If the new quarters to accommodate this population incorporate energy-conserving designs, half the country's housing stock would be energy-efficient by 2000 and three-quarters by 2025. This trend toward conservation is an essential feature of Kalisky's soft path scenario for Israel. Other key assumptions: the energy now used for pumping irrigation water will not increase, because the available sources, representing about 1.3×10^9 m^3/year are being fully utilized; industry will attempt to overcome the worker shortage by increased investment in capital equipment; machines will replace workers in the future not just in industry, but in agriculture and in the service sector.

The soft path scenario Kalisky constructs reflects a major move toward conservation, but it is dominated by the population increase he projects. In 1975, Israel's energy was supplied by about 7 million metric tons of oil equivalent (304×10^{15} Joules, or Petajoules), almost all in the form of imported oil. Given Kalisky's population and energy demand assumptions, requirements would increase by 2000 to 382 PJ, and by 2025 to 549 PJ.

Soft energy path studies almost invariably conclude that an energy future based largely, if not entirely, upon renewable sources is technically feasible. Only political and social barriers prevent immediate travel down the soft path. But in Israeli society, the move toward renewable, indigenous sources of energy is pervasive and highly visible. The technology is being developed or sold by the people next door, and the motivation to adopt it comes from the people who live just over the hill. This combination of innovative research and political pressure makes an Israeli soft path not just possible, but necessary.

—*Charles Drucker*

References:

Assaf, Gad
1976 "The Dead Sea: Scheme for a Solar Lake."
Solar Energy, Volume 18 No. 4, p. 293.

Kalisky, Avram
1979 *Solar Energy: The Main Soft Energy Path in Israel.*
21 pp. Available from IPSEP for $2.

Soft Paths for Difficult Nations

The Problem of Japan

IF THERE IS A NATION anywhere in the world that has exhausted its potential for domestic energy resources and efficiency improvements and has no options left save nuclear reactors and massive fuel imports, that nation must be Japan. Its back to the wall, Japan imports over 99.7 percent of its oil, three quarters of that from the mideast, 86 percent of its natural gas, 76 percent of its coal, and all its uranium. Japan's spot oil purchases from Iran have inspired America's ire.

Japan's nuclear policy is similarly controversial: by shipping wastes from 22 reactors to France and Britain for reprocessing (at great cost to remove the appearance of an unresolved problem), it has almost singlehandedly made reprocessing plants in those nations commercially feasible—with all the attendant nuclear proliferation risks. This is a policy based on scant resources, entrepreneurial zeal, and panic.

In this precarious environment, the idea of a non-nuclear energy strategy that would phase out nuclear reactors and foreign oil and increase national security was simply a matter of thinking the unthinkable in a nation that thinks together. Nevertheless, a self-sufficient "soft energy strategy" for Japan is well underway, and a brief English summary by Dr. Haruki Tsuchiya, a physicist with the Research Institute for Systems Technology, shows the unthinkable is not only possible, but well within Japan's reach.

Tsuchiya's study is actually only one of a number of studies now underway on aspects of energy conservation and renewable energy resources. One would expect that a nation with half the US's population and a fifth its energy consumption might have exhausted the potential for energy efficiency improvements. But this assumption, it turns out, is not so. Using detailed analyses compiled for each economic sector by the respected Institute for Energy Economics (reviewed in *Notes* volume 3, number 1) and available data on the potential for renewable resources, Tsuchiya shows that Japan's energy use could be cut 38 percent by 2010 without reducing standards of living.

Oil and gas currently provide 77 percent of Japan's primary energy (11.3 of 14.7 quads; 11.9 of 15.5 EJ). Tsuchiya's study aims to phase out fossil fuel dependence over thirty years—half the current oil and gas use would be eliminated with efficiency improvements, the other half with renewable technologies (mainly hydropower, wind energy, solar heat, and biomass). By contrast, the plan prepared by the government energy agency's Advisory Committee on Energy recommends a massive (seven to eight-fold) increase in nuclear generation, and a doubling of imported coal and liquefied natural gas supplies by 1995. But the plan—"to lower... dependence on imported oil through even further energy conservation and by increasing alternative energy resources"—falls short of its language. Oil imports rise 13 percent by 1995.

The Tsuchiya study assumes that Japan's economic activity and standard of living will not change substantially over the next thirty years, a calculational simplification that Tsuchiya argues is balanced by very conservative estimates of energy savings potential. Refinements in the data are expected shortly.

Energy Efficiency

In the home and in commercial buildings, tasklighting, better light bulb designs, improved appliances, computer control of energy use in large buildings, solar hot water heaters and passive space heat design, and improved efficiency in cooking offer substantial savings potential. Task lighting and better bulbs, says Tsuchiya, can reduce lighting energy requirements 44 percent and reduce air conditioning requirements at the same time. He adds that Japan's appliance manufacturers are today producing home appliances (vacuum cleaners, washing machines, and refrigerators) that use 40-60 percent of the energy used by these appliances only a few years ago.

One of Japan's fastest growing industries—slated to grow by 50 percent in the next five years—is electronics. But even in the electronics field, energy efficiency is well within reach. Tsuchiya estimates that miniaturization and new silicon chip designs will allow the electronics industry to halve its energy consumption despite a 50 percent boost in business.

Japan's iron and steel industry is among the world's most efficient. The US steel industry, for example, uses 38 percent more energy per ton of crude steel than Japan's industry does. Electricity and heat are typically recovered in Japan, Sweden, and West Germany. Tsuchiya does not allow, however, for some of the further potential efficiency improvements underway in Sweden (see *Notes,* volume 1, number 2) that would reduce energy use further. Tsuchiya adds that it is difficult to forecast the future of this industry; steel is an important current source of foreign exchange, much of which pays for imported energy that might not be needed in a self-sufficiency scenario. If imports of energy are reduced, energy-intensive industries might wisely be reduced in preference to high technology, low energy use industries like electronics.

Tsuchiya finds that electricity use in many industries is higher than necessary. Various factories are substituting small electric motors closely matched to actual mechanical needs in preference to the older oversized electric motors. "Motors consume about 30 to 80 percent of their rated energy use while idling," notes Tsuchiya, adding that such techniques as "load management, peak shifting, and use of flywheels with clutches save about 40 percent of the electricity (in a typical industrial application)."

Automobiles account for 12 percent of the primary energy used in Japan, and ships, railroads, and planes require an additional 2 percent. The average mileage of Japanese cars is about 23 miles per gallon (9.8 km/l), but Tsuchiya notes that auto manufacturers are producing cars capable of 44.5 mpg (18.9 km/l) in "ten-mode driving tests." Tsuchiya assumes only a 35 percent improvement in performance by 2010, a 30 percent improvement in airplane efficiency (versus the 50-100 percent improvement in the new generation of planes), and a shift of some transportation to light rail (7-10 times more efficient than private autos).

Japan's Energy Supply Matrix

Unit: 10^{13} kcal (1 million tonnes of oil equivalent)

	Solar Heat		Solar			Biomass			Geothermal		Hydro-electric	Wind			Hydrogen	Wave power	Domestic coal
	Low temperature	Moderate temperature	Electricity	Thermal heat	Photovoltaics	Solid fuel	Liquid fuel	Gas fuel	Electricity	Heat	Electricity	Electricity	Thermal conversion	Mechanical works	Hydrogen	Wave power	Domestic coal
Lighting											4.41	1.40				0.01	
Mechanical work in home					0.90						1.72	0.39		0.01		0.01	
Mechanical work in commercial					0.40						3.01	0.60					
Electronics					0.20						1.04	0.10					
Heating & Cooling	8.95			0.80		0.10							1.50				
Hot water	6.67																
Cooking								5.85									
Iron industry electric											8.15	3.00					
Iron industry heat																	4.46
Industry mechanical works			1.20		0.80				0.89		4.03	3.11		0.20		2.48	
Electrochemistry											4.51						
Process heat	10.61	9.13		0.40		0.40				0.80			1.50				
Industry heat	0.42	0.84				0.42	1.20	1.30							0.80		10.94
Automobile							21.81								0.56		
Aircraft							1.04										
Ship							3.28										
Railways											3.50	0.60					
Agriculture mechanical works									0.01		0.14	0.02		0.10			
Agriculture heat	6.17									0.10							
Elec. distribution											1.22	0.68					
Biomass Conversion	13.79																
TOTAL	46.61	9.97	1.20	1.20	2.30	0.92	27.33	7.15	0.90	0.90	31.73	9.90	3.00	0.31	1.36	2.50	15.40

from Haruki Tsuchiya, "Soft Paths Planning for Japan"

Renewables' Potential

Renewable resources in Japan are poorly documented and only recently investigated. "After the oil crisis in 1973," notes Tsuchiya, "government began to shift their main energy supply sources from oil to coal and nuclear power..." Little thought was given to the potential for renewables or efficiency improvements. MITI (the Ministry for International Trade and Industry) spends only about $20 million per year for solar energy sources. Wave energy is under advanced development at the Japan Marine Science and Technology Center and a 2 megawatt floating facility is under construction. Biomass research is sorely underfunded by the fisheries and agriculture departments.

Nevertheless, says Tsuchiya, a substantial potential exists. In the 1950's, Japan had 2 million solar collectors (California has about 75,000 today) for heating hot water. Many were victims of low oil prices, but Japan still produces over 170,000 solar heaters annually. Several hundred solar homes exist in the nation. In Tsuchiya's soft path, solar heating would supply home space and water heating, low temperature process heat for agriculture, and some medium temperature "preheat" for industry and biomass conversion plants. These applications, 82 percent at low temperatures and 18 percent at medium temperatures, would supply 2.25 Q (2.37 EJ) of the 9.1 Q (9.6 EJ) needed in 2010.

Geothermal energy is also substantially underdeveloped, according to Tsuchiya. A number of studies peg the potential at 20,000 megawatts, but Japan has only developed 170 megawatts for electric generation. Tsuchiya relies on a very modest increase in geothermal energy (500 megawatts), citing potential environmental problems as a reason to limit development.

Biomass potential is substantial in Japan: 67 percent of Japan's land area, according to Dr. Tsuchiya, is forested! Agricultural waste currently totals 12 million tons per year of dry organic material, wood waste in the lumber industry contributes 11.2 million tons, municipal waste could supply a further 10 million tons, and used wood materials, wastes from livestock, sludge from the paper, food, factory, and sewage industries, and forest slash, add a further 33.7 million tons for a total biomass potential of 67 million tons. Tsuchiya believes that special energy crops could double or triple this quantity, but only a 28 percent increase would be necessary to supply liquid and solid biomass fuels to transport and industrial sectors.

Hydroelectric power supplies the bulk of Japan's current indigenous energy supply — 23,000 megawatts. The nation's potential is much larger, 56,000 megawatts. Thirty six percent of the energy in new hydro plants would come from small facilities, less than 10 megawatts in size. Solar electricity from thermal plants and photovoltaic cells would contribute small amounts, as their technical status is less assured than for some of the other cited sources (0.14 Q, about 9.8 GWe). Domestic coal use for industrial heat and ore reduction remains constant at 0.61 Q (0.64 EJ).

Wind turbines are widely used in Tsuchiya's projections. A total of 23,400 megawatts are deployed, 15,500 of which are used for electricity generation, 4,000 are used to produce hydrogen by electrolysis, 360 megawatts are used for mechanical work, such as pumping, and 3,530 megawatts are used directly to supply heat. A small amount, 3,900 megawatts, is projected to come from wave power, using about 95 miles of coastline in offshore buoys.

Further studies by the Research Institute for Systems Technology are underway on the economics of each of the energy conservation efforts recommended in the Tsuchiya study, on the comparative costs — including government subsidies — of soft and hard energy paths, and probable shifts in Japan's industrial structure over the coming years.

—Jim Harding

References:
Tsuchiya, Haruki
1980 "A Soft Path Plan for Japan." Research Institute for Systems Technology, 403, Taiyo Bldg. 3-3-3, Misakicho, Chiyoda-ku, Tokyo 101, Japan.

Committee for Energy Policy Promotion
1980 *Japan and the Oil Problem.* No. 11 Mori Bldg., 6-4, 2-chome, Toranomon, Minato-ku, Tokyo, Japan.

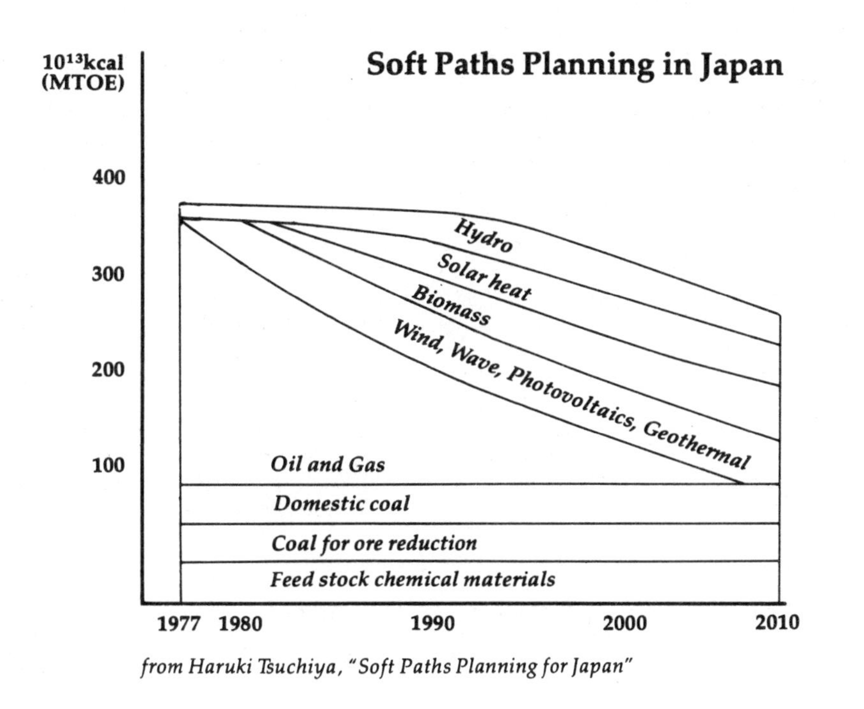

from Haruki Tsuchiya, "Soft Paths Planning for Japan"

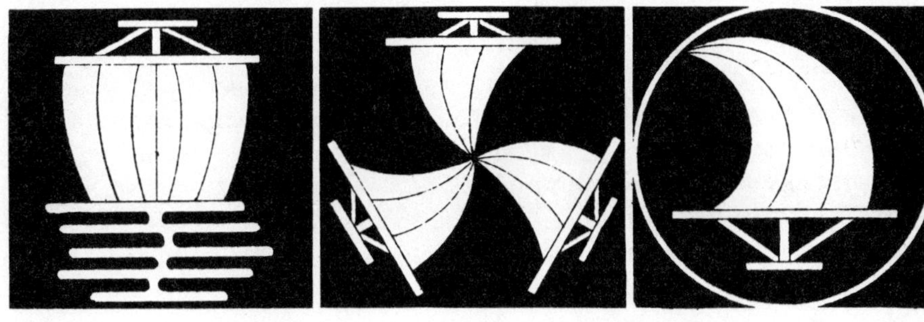

Oil & Water Mix Well
In Norway

T HE VITAL SIGNS of Norway's energy system all point to hard path symptoms: 53 percent of total energy demand supplied by oil; yearly electricity use now nearing 19,000 kWh/capita—almost twice that of the US; half of space-heating needs provided by electricity, and that fraction will rise; a constant population, but an individual material standard official projections expect to double between 1976 and 2015.

The effect on primary energy demand could be a 47 percent increase overall, according to Stellan Atterkvist and Thomas B. Johansson in *Solar Norway*, a study funded by the Norwegian Department of the Environment. Even so, they report, Norway could have an energy system based entirely on renewable energy flows by 2015.

Norway's ace is hydropower. Nearly half the nation's total energy, and all of its electricity, is supplied by hydro, with a vast potential for expansion. Oil reserves are also substantial: enough to last 100 years, even with a nine-fold increase in production. But Atterkvist and Johansson propose it may be economically attractive to look for domestic oil substitutes, giving Norway an advantage in a world economy of rising oil prices.

"The final mix of the technologies is an open question."

Supply

In 1976, Norway's hydroelectric production, at 68 GJ/cap, stood as the world's largest. Though the economic hydropower potential is twice the 1979 production level of 306 PJ, the authors of *Solar Norway* assume an increase by 2015 of only about two-thirds, to 504 PJ/a. Of the total increase, 40 PJ/a derives from improvements on older installations, and the remainder from the construction of some low-head stations. Industrial electricity demand, currently 58 percent of the total, grows, in this scenario, to 82 percent.

In all cases, the technical limits "have not been pushed."

Solar Norway includes a mix of other renewable sources; it does not depend on the commercial development of any one of them as each component contributes a relatively small share to total supply, and can be replaced by another. In all cases, the authors state, the technical limits "have not been pushed." The renewable energy system will have a small growth rate—relative to historic rates in the energy sector—of 3.2 percent per year (this is an average, as most technologies are scheduled to come on-line during the 1990s or later). Hydropower expansion, however, would continue at a rate of 1.4 percent/a. "The final mix of the technologies," Atterkvist and Johansson explain, "is therefore an open question," but could include these components:

● 1600 windpower stations of 4 MW each will contribute 57 PJ/a—a small fraction of the potential, as median

wind speeds along the Atlantic coast reach nine to ten m/s at the standard height of 100 m.

● With an average power density of 23 kW/m of wave front, or 2000 PJ/a theoretically available along the coast, the wavepower potential is also large. This scenario includes a 67 km

station with 20 percent conversion efficiency, contributing 10 PJ/a.

● Again, although the theoretical potential is large, photovoltaics would contribute only 10 PJ/a—a corresponding installation would cover 3.5 percent of the total urban land area.

● Industrial cogeneration will make an even smaller contribution to electricity supply.

● Biomass for methanol-blended gasoline or pure methanol fuel will be produced from 0.4 million hectares of energy plantations to contribute 120 PJ/a from an average yield of 190 GJ/ha (methanol production yield is assumed to be 50 percent of the energy content of the biomass). These 400,000 hectares correspond to 1.3 percent of Norway's land area or 6.2 percent of commercial forest land—not considered large by the authors. They add that it may be attractive to use electricity through hydrogen or in battery storage or to keep oil in the transportation system for a longer time to reduce Norway's biomass needs.

● Solar heating with seasonal stor-

age will meet less than half the space heating demand.

Demand

In *Solar Norway*'s every sector, Atterkvist and Johansson claim to make economically and technically

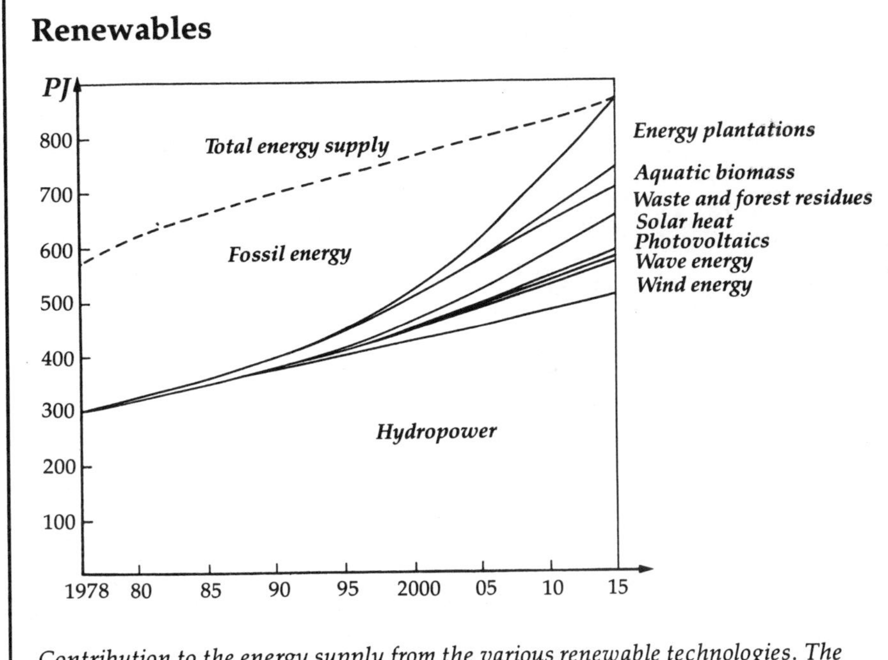

Contribution to the energy supply from the various renewable technologies. The dashed line shows the energy demand.

conservative estimates of efficiency potential. "Today," they state, "it is technically and economically possible to cut the energy demand in buildings to half of the level assumed here. Automobiles with better fuel economy are already on the road." Likewise, the potential for increased energy efficiency in industry is large as the industrial capital stock will be renewed "between one and two times" by 2015.

Costs

The economics of this system are coupled to rising hydropower costs, Atterkvist and Johansson explain. While the cost of electricity in 1976 was 0.7 US cents/kWh for large industrial consumers and 1.8¢/kWh for residences, electricity from new hydropower today costs 2.3¢/kWh at an interest rate of nine percent. This cost will increase with new, less favorably located installations, and the distribution system will add a further 50 percent cost increase. The energy supply system's total gross production value would be 6.2 billion dollars in 2015, three times the 1976

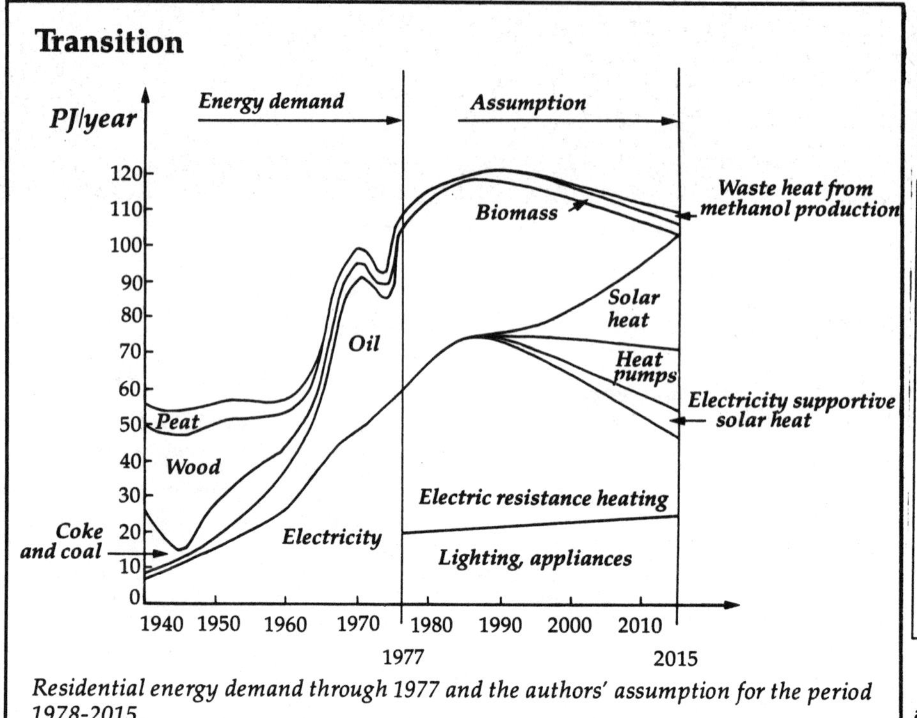

Residential energy demand through 1977 and the authors' assumption for the period 1978-2015.

value, given high estimates of wind-power costs (4¢/kWh, 1976$), wave-power (6¢/kWh), photovoltaics (13¢/kWh), solar heat (6.3¢/kWh), and biomass (1.4¢/kWh) with an annuity of twelve percent. "This indicates," the authors project, "the economy from an overall point of view, would not be in conflict with the assumed doubling of the production of goods and services."

Atterkvist and Johansson conclude with a summary of *Solar Norway's* environmental costs, citing the benefits of avoiding sulfur and carbon dioxide emissions. And, if developing the additional 110 PJ/a hydropower is unattractive, it can be replaced with some mix of other components.

Using official demand projections, *Solar Norway* remains really a "worst case" scenario. Moreover, Atterkvist and Johansson have used renewable sources at a fraction of their potential. "A less materialistic development, or a more efficient use of energy," the authors explain, "would lead to a lower energy demand. But, if a renewable energy supply system can be designed to meet this energy demand, it can also be designd to meet

Norway

Area	324,000 km²
Population	4.04 million
Area per capita	0.080 km²/cap
Average yearly insolation	220-350 kJ/cm²
Automobiles/1000 persons	290 (1976)
GNP per capita	$7,767
Primary energy use	622 PJ
Total consumption	579 PJ
Energy use per capita	154 GJ/cap
Electricity use per capita	18,769/cap

General data for 1977.

any lower level of energy demand. The assumption made here thus separates the discussion of future lifestyles from the discussion of energy supply."

—*Elyse Axell*

Reference:

Atterkvist, Stellan and Thomas B. Johansson

1980 *Solar Norway.* Supported by the Department of the Environment, Oslo, Norway, 7 pp. This article, submitted to "Energy Systems and Policy," is based on the book, *Sol-Norge* (in Norwegian) by Atterkvist and Johansson, Universitetsforlaget, Oslo, 1980, 115 pp.

Romania
Plans for Diversity

*R*OMANIA MOVES TOWARD *achieving energy self-reliance within the next decade. Coupled to this will be an improved standard of living: the nation aims at a per capita income of US $3,000–$3,500 for a projected population of 25 million.*

The past several decades have seen considerable progress toward this goal: from 1950 to 1979, energy production multiplied by a factor of eight and the median wage jumped fourfold. Agricultural production has increased by 350 percent and since 1938, industrial capability increased 42 times.

This degree of improvement in the Romanian standard of living will require significant amounts of energy, much of which will go to satisfy heightened industrial demand. In 1978, industry was both Romania's largest energy consumer, and the country's principal contributor to Gross National Product, accounting for slightly more than 60 percent of the total energy demand, and slightly less than this proportion of GNP. But Romania faces serious energy problems in its effort to increase industrial output—even to maintain production at its present level.

Foreign oil dependency is a prime concern, as it is to many other countries, industrialized and developing alike. Petroleum provides nearly half of the country's energy, yet oil reserves appear near the limits of primary recovery. Domestic production has been on the decline since 1975, and economic growth has required a vast expansion in coal production, along with the importation of almost as much oil as the country produces.

Romanian government energy planners recognize that dependence on energy imports must be reduced and eventually eliminated, if the country is to achieve its social and economic goals. They have formulated a program of research and develop-ment that will lead Romania toward energy self-sufficiency over the course of the next decade.

A diversity of resources and technologies dominates Romania's energy program. Diversity, however, is no novel feature of Romanian energy planning; it represents historic policy. After World War II, Romania resisted the temptation to become a wasteful, petroleum monoculture in the era of cheap oil. Natural gas was also developed as a major power source, and numerous energy conserving features were incorporated into the rapidly growing urban and industrial infrastructure: well-insulated brick buildings, extensive district heating, large-scale mass transportation systems, and cogeneration facilities. The society's attitude is one of energy austerity—an important asset that should help it to develop in the face of a limited resource base.

Conservation efforts will remain an important part of the Romanian energy use picture, and planners assign new and renewable sources of energy—particularly solar radiation—a significant role. At the same time, however, conventional energy technologies will be pushed to their limits in order to reduce the level of petroleum imports. This will mean more intensive exploration for fossil fuels, more efficient extraction methods, and a reliance upon nuclear power to provide an increasing share of electrical generation.

Fossil Fuel Exploration And Extraction

The national Five Year Plan that outlines energy research and development for the period 1981-1986 calls for investments in geological research 50 to 60 percent greater than in the preceding five years. Modern techniques for exploration, including teledetection, will search for deep oil and gas reserves, as well as offshore sites in the Black Sea. Even with the additional sources identified by more intensive exploration, oil and gas production will be limited in order to maintain geological reserves at adequate levels. Extraction in 1985 will accordingly be held to 12.5 million

---

In Romania, as elsewhere, efficiency improvements are the cheapest new source of energy.

---

tonnes of crude oil and 26.5 billion cubic meters of methane gas, both figures representing some 10 percent *below* 1978 production. To make the most of reserves, enhanced recovery techniques will be introduced, raising the recovery factor from 31.5 percent in 1979 to 37 percent in 1981, and to as high as 40 percent in the 1990s.

The essential aim, with respect to these hydrocarbons, is to save them for use as petrochemical feedstocks. This involves restricting their combustive consumption, and at the same time developing new types of fuels, including synthetic fuels, for transportation.

The Move to Coal

Coal will meet much of the increased demand for electricity, as oil and gas gradually move out of thermal power production. Forty percent of all electricity in 1980 will be derived from coals and bituminous shales, rising to

55 percent or more by 1985. All new conventionally-fueled power stations will be designed for coal, and plants now operating on gas and oil will shift to solid fuels.

The rise in coal extraction will be striking; 1978 production, at a little over 31 million tonnes, will be dwarfed by the 85-88 million tonnes expected for 1985. Lignite production, which is assigned principally to energy generation, will reach 75 million tonnes by 1985, a growth of more than 300 percent in only seven years. Most of the new coal will be taken from large, open pit or strip mines. The rest of the 1985 coal production will consist of bituminous shales, and two large thermal power plants are to be commissioned specifically for their combustion.

Nuclear Prospects

No nuclear plants are presently in operation in Romania, but two are already on order: one a 440 MWe VVER

The society's attitude is one of energy austerity—an important asset that should help it to develop in the face of a limited resource base.

---

reactor manufactured in the USSR. Installed nuclear capacity in 1985 should be 660 MWe of a total 22,000 MWe (3 percent), for a yearly national electricity production of 80-90 billion kWh. By 1990, however, the nuclear fraction will rise to about 14 percent (3,960 of 28,000 MWe). At the turn of the century, nuclear plants will amount to 10,560 MWe of installed capacity.

One objective of the Romanian nuclear program is to reduce the fraction of electricity provided by coal and shales; peaking at 55 percent of electric power in 1985, the share of solid fuels will drop to 44 percent by 2000.

Hydropower Growth

The diversity of the Romanian energy program is evident in its approach to hydropower development. With 30 percent of the national hydropower potential harvested in 1980, there is a determined commitment to use this resource fully by the end of the century; 45 percent by 1985 and 65 percent in 1990 are interim targets. This growth is both absolute *and* relative; hydropower's 18 percent share of electricity production will rise to 20 percent in 1985 and 24 percent two decades after that.

To exploit the country's hydro potential, power stations of varying sizes will be needed: large generators on the Danube and inland rivers, and micro-hydro facilities widespread over the landscape.

The Renewable Component

Advanced exploration and extraction methods for oil and gas, coal conversion, damming major rivers, and an aggressive nuclear program, taken all together, make Romania's energy strategy look very much like the capital-intensive, environmentally hazardous energy path that has lead many another country into debt and debility. But these energy sources and technologies tell only half the story.

The Romanian government has included within its program the goal of 20 percent of total energy production from new sources and technologies by 2000, the same solar fraction that the US government calls for by that time. The sources Romania includes in this rubric are geothermal, solar radiation, wind and waves, biomass, urban wastes, and "trapped" coal reserves (unapproachable by conventional mining methods but potentially gasifiable underground). The new energy technologies envisaged include heat pumps to recover waste heat, fuel-saving techniques for thermal engines, gaseous and liquid synfuels from coal, hydrogen energy, and electrochemical sources and storage.

Using only existing, first-generation technologies such as solar flat plate collectors, new energy sources could provide Romania with the equivalent of more than 20 million tonnes of coal each year. To guide the deployment of these technologies, the National Council for Science and Technology developed an initial inventory of first-generation technology solar applications, the results of which were incorporated into the 1981-1985 energy plan. In the next two years, more than 400 energy consumers will be adopting solar heat, geothermal water, and biogas, in order to reduce or eliminate their use of hydrocarbons. The 1985 anticipated fuel savings will be distributed among several demand sectors: 18 percent of the savings in industry (washing, preheating, drying, etc.); 19 percent in agriculture, primarily food processing; 22 percent in transportation; and about 41 percent in domestic consumption, mostly hot water and space heating.

Initially, the new sources of energy will be used for low temperature heat; in the last decade of the century, however, it is expected that high temperature needs will be met as well. Fuels from biomass conversion, industrial waste heat recovery, and large-scale solar concentrator systems (such as central receivers) will all contribute to this area of energy end-use.

End-Use Efficiency

A central tenet of the Romanian energy plan is that "total energy

Romania

Population—1980	22.3 million
Total Land Area—1977	23,750,000 hectares
Total Commercial Energy Consumption—1978	657.6 Billion kWh/yr
Total Commercial Energy Consumption Per Capita—1978	713.6 kWh per capita/yr
Fuelwood and Charcoal Production—1978	7.9 Billion kWh/yr
Renewable Energy Potentials	
Solar Insolation	47,500 Billion kWh/yr
Biomass Energy From Residues	200.4 Billion kWh/yr
Wind Energy Near Ground Level	305.1 Billion kWh/yr

from "World Energy Data Sheet, "World Resources Inventory, 3510 Market St., Philadelphia, PA 19104 USA.

consumption can rise in the decades to come only if new sources of energy are identified and technological measures applied to economically utilize them. The chief source and decisive conditions of the further steady development of the national energy and of the whole people's welfare is a substantial increase of the economic efficiency in the utilization of fuel and electric energy." In Romania, as elsewhere, efficiency improvements are the cheapest new source of energy.

A "basic target," then, is to achieve a rapid decrease in energy consumption per unit of GNP. By 1990 this is expected to drop by 40 percent, and a further 20 percent in the century's last decade. In part, this higher energy productivity will stem from increased industrial materials recycling. Re-use goals for 1985 are: 40 percent of the iron, 41 percent of the copper, 50 percent of the processed polyethylene products, 52 percent of the pulp and waste paper, and 90 percent of the glass. But the chief line of action is "to lower end use energy consumption by a rigorous reappraisal of the consumers and allocation of energy in the strictly necessary amount, form, and quality."

Romania's President, Nicolae Ceausescu, emphasizes the national commitment to conservation: "Taking into account the international economic pattern, the impact of the world energy and raw materials crisis, a central task of the next Five Year Plan is the radical improvement of the structure of our industry, the firmer direction of the efforts towards the development of the low-energy consuming branches and sectors... It is only obvious that we will have to mainly develop those low-energy consuming industries yielding high value products, embodying complex labour, superior technical skill, and to continuously diminish production involving large consumption of raw materials and energy, as well as to eliminate outdated technologies requiring large consumption of fuel and energy."

Conservation, in the broadest sense, is a matter of Romanian national policy, and the criterion of fuel saving governs the country's philosophy of development. But efficient end-use and renewable sources of energy go hand in hand here with intensified development of conventional energy systems, including an accelerated nuclear program. At the heart of Romanian energy planning is a call for flexibility: for structural changes in the country's economy and industry, and for a diversity of energy sources and technologies to rely upon in the future.

*—Charles Drucker
and Adrian V. Gheorghe*

Reference:

Gheorghe, Adrian V.
1980 "Energy Today and Tomorrow—a Romanian Point of View." Manuscript prepared for IPSEP, 25 pp., available for $2.50. Additional materials may be obtained from Adrian Gheorghe, Bucharest Polytechnical Institute, Faculty of Power Engineering, Bucharest, Romania.

3. Global Outlook

We all live in one and the same greenhouse, and it's getting hotter. That is one reason to worry about global energy use. Early projections put long-term world energy needs as high as six to eight times the present eight Terawatt-years we use per year. These projections assumed that the entire world would use as much energy per person (6kW) as the advanced industrial countries use today. More recently a group of experts at the International Institute of Applied Systems Analysis (IIASA) concluded that a figure of three kilowatts per capita would be a minimal requirement, essentially halving previous presumptions. Yet even this scenario would involve twice today's fossil fuel consumption, and would, in essence, make survival of world climate as we know it a matter of rare luck.

When aspirations for greater economic welfare and likely population developments are blended with high efficiency, a world suddenly becomes conceivable that is supplied with energy from renewable sources alone and lives, on average, as well as Western Europe does today. The inequities of such a future seem more manageable than the inequities we face today.

IIASA's Fantasy Forecast

by Florentin Krause

Consider the following scenario for the next half-century: World primary energy demand increases from the 1975 level of 8.2 terawatts (8.2 billion kilowatts) per year to 22.4 terawatts.

NUCLEAR ENERGY use expands as fast as the capacities of the nuclear industries allow, leading to an increase of more than forty-fold (to 5.2 terawatts). Global fossil fuel consumption doubles from 7.5 terawatts in 1975 to 15 terawatts. Oil consumption exceeds 1975 levels by 40 percent, drawing upon costly and environmentally damaging unconventional resources. Coal consumption triples, with more than half converted to liquid fuels in hyge synthetic fuel plants. Natural gas consumption doubles.

The world oil trade still resembles the 1975 volume with an affluent ten percent of the world's population competing with the poorest 45 pecent over available exports. The world trade in coal and nuclear fuel expands to unprecedented levels.

Atmospheric carbon dioxide (CO_2) rises steeply, increasing the risks of irreversible climatic changes.

Primary energy use per capita in North America remains twenty times higher in 2030 than among the still-undeveloped 3.5 billion people of Southeast Asia and Africa.

To many energy analysts, this scenario is an unthinking mix of unbalanced growth and increased risk. Yet a prestigious group of researchers from the International Institute for Applied Systems Analysis (IIASA), in Austria believes it to be a "low" scenario. It provides,they maintain, only the minimum amount and the optimum mix of energy necessary for social stability and world economic growth.

As outlined in their report, "Energy in a Finite World," the IIASA authors, let by the West German head of nuclear research, Wolf Hafele, see their energy strategy for the next five decades as a crucial step toward a "sustainable" energy future based on nuclear breeder technology.

IIASA, sponsored by several national academies of science, drew upon the talents of 140 consultants from many nations over the seven-year duration of this project. Yet, the resulting report has drawn more informed criticism from the scientific community than any other energy study.

A flawed technical analysis.

Ironically, the report claims rigor and consistency precisely in those areas where it fails most decisively: engineering and economics.

Of the 80 published research documents related to the report, not a single one contains a thorough and comprehensive assessment of the state of the art in efficient energy use. IIASA assumes, for example, that the US automobile in 2030 will consume 6.8 liters per 100 km—the average for cars imported into the US today. Similarly, they state that US houses will need 60 percent of the heating energy they need now. Simple weatherization can lower heating needs 35-50 percent, while more extensive retrofits can cut demand by 90 percent, and new houses can be built with virtually zero space heating needs—all cost effectively.

IIASA's professed energy strategy of least economic cost should have assumed efficiency improvements far beyond those they actually employed. But they failed to compare the costs of efficiency efforts with those of increased supply, for reasons that appear to be more ideological than technical.

This lack of analytical rigor leads to unlikely scenarios. For example, energy/gross domestic product (GDP) ratios drop about as slowly with rising real energy prices as they did in the past with declining prices.

Cost comparisons among technologies are undertaken on the supply side. However, the emphasis on nuclear energy is justified with cost data that appear asymmetrically selected—nuclear energy is made to look better, and renewables worse, than their actual records.

Total primary energy demand is inflated by assuming a larger share of electricity than physically necessary or economically advantageous to the final consumer, who must pay the costs of conversion losses.

The IIASA study correctly emphasizes that time is of the essence, but chooses as principal carriers for its stategy two technologies which have proven, even with subsidies, to be logistical lame ducks: nuclear reactors and synfuel plants. Technical efficiency improvements, on the other

"Are any of these for me?" By Richard Willson.

hand, could move very rapidly because they can piggyback on the renewal and repair of capital stock.

Environmental blind spots.

At the outset, the IIASA group declares all environmental and safety problems solvable. Thus, nuclear, fossil, and renewable sources can be treated as equivalent, and considered on the basis of their perceived economic potentials alone.

The hazards of radioactive pollution from the nuclear fuel cycle, only recently shown to be more accident prone than expected, are simply ignored. The climatic risk posed by the IIASA "low" fossil fuel consumption is recognized without proposing the logical solution to this dilemma: reduced fossil burning by means of conservation and substitution with renewables. Instead, the report meekly suggests that levels of CO_2 be monitored, and hopes that the world might sneak by this problem.

A western bias.

The model of global development used by the IIASA authors is implicit in their economic growth assumptions.

Continued growth in the rich nations will lead poor nations down the path of western urbanization and industrialization. This approach ignores the alternative concepts of ecodevelopment, rural development, and a new international economic order, along with their implications for future energy service demand.

Other analysts found that the energy required for food, transport, and materials depends strongly on whether economic activity is inequitably centralized in large cities, or localized in smaller communities, where most Third World people live today.

The IIASA report assumes increases of 85-255 percent in per capita GDP for North America. It does not, however, explain how average US consumers are supposed to use more energy services, or why they should want to, given the current widespread dissatisfaction with the consumption ethic.

Further, a recent analysis of West German GDP showed that 19 percent goes to fight crime, environmental pollution, and illness. An increase in GDP and energy services may actually mean a decrease in welfare.

Freedom to waste money.

Conservation and efficiency efforts that might reduce future global energy needs below the IIASA minimum are labeled attacks on present lifestyles and on the principle of free choice. Yet it is questionable whether cars and houses that waste both fuel and money bring freedom or constraint. Moreover, the security systems required by the nuclear fuel cycle come at a heavy cost in civil liberties.

The inequities in global energy use and resource distribution have led to a potentially explosive competition over access to oil resources. The IIASA scenario suggests a dramatic increase in problems of access and distribution. A nuclear economy on top of this situation is likely to increase weapons proliferation, already a cause for military confrontation.

Rather than searching for an energy strategy which takes these real-world issues into account, the report blithely assumes an unprecedented future degree of international harmony and cooperation. Hard technologists may then make their favorite technical schemes into a reality.

Perhaps the most regrettable aspect of the IIASA study is than an opportunity to develop a constructive energy strategy, in a spirit of true international cooperation and stewardship for mankind and earth has been wasted.

Climatic Changes Can Be Avoided

CO₂– A Pseudo-Problem?

Carbon dioxide (CO₂) build-up in the atmosphere from large-scale burning of fossil fuels could lead to climatic changes with drastic social, agricultural, and economic effects worldwide. To delay or wholly avoid these hazards, fossil fuel use should be reduced through a combination of efficiency improvements and renewable energy sources.

THE WORLD'S POPULATION of 4.3 billion people uses energy at a rate of about nine terawatts (nine billion kilowatts) per year, including non-commercial sources. Roughly eight of these nine terawatts are derived from burning fossil fuels, releasing into the atmosphere carbon that has not been airborne for eons. These emissions disturb the delicately balanced engine or global climate.

Roughly half of the CO₂ from fossil fuel burning is retained in the atmosphere. Each terawatt-year of heat generated by burning fossil fuels thus increases the atmospheric concentration of CO₂ by 0.16-0.18 ppm, depending on the mix of fuels involved. Climatologists now generally agree that rising CO₂ levels tend to warm the earth's climate. Temperatures could climb high enough to melt polar ice caps, inundate the fertile lowlands of many densely populated coastal regions, reduce crop yields by 20 percent in breadbaskets dependent on cooler climates, disturb rain patterns, and shift fisheries through changes in ocean currents.

Preindustrial Levels Surpassed

Before the spinning wheels of industry created a demand for fossil fuels, atmospheric carbon dioxide stood at 280 ppm; by the end of 1981, the level had risen to 339 ppm. And if the known fossil fuel resources were to be burned in their entirety, CO₂ levels would jump to between four and eight times the preindustrial concentration.

> Reductions in fossil consumption, achieved now, buy a disproportionately large amount of time. Because our present level of carbon release is so high, we are, in effect, sitting on a high-speed train racing toward an abyss. The challenge is to slow this train.

Two aspects of the CO₂ increase make a solution particularly difficult. To begin with, the minimum carbon dioxide level of climatic concern is unknown and may never be accurately determined. Certainly a doubling of CO₂ to about 550 ppm would bring with it a substantial temperature rise of several degrees centigrade; this much warming would, in all likelihood, prove severely disruptive. The concentrations up to that level constitute a grey zone of considerable uncertainty.

Secondly, and more ominously, carbon dioxide levels cannot be reduced by human means. The decline of CO₂ by natural processes would be very slow, in the worst case taking many centuries. In all likelihood, the decline of human civilization would be faster.

The prudent policy would be to remain as close as possible to present levels of atmospheric CO₂, while we find out more about climatic thresholds. Unfortunately, at historical rates of increase in fossil fuel use—three to four percent per year—we are racing away from present conditions, and could reach a doubled CO₂ level early in the next century.

But, if we were to add fossil carbon to the atmosphere at a constant, rather than an increasing, rate (eight terawatt-years/year burned), the same terrain of high climatic risk would not be reached until about 100 years later. And if we could reduce fossil fuel consumption to two-thirds the present rate by the year 2000, and to only an eighth (one terawatt) by 2025, we would gain about 1,000 years. Finally, if a way could be found to eliminate fossil fuel consumption entirely, at a rate of 0.2 terawatts/year over the next forty years (equivalent to an immediate crash program of extreme proportions, and not very realistic), atmospheric carbon would remain within ten percent of present levels. Table 1 gives the dates by which various CO₂ levels would be reached for different fossil fuel use trajectories.

Reductions in fossil consumption, achieved now, buy a disproportionately large amount of time. Because our present level of carbon release is so high, we are, in effect, sitting on a high-speed train racing toward an abyss. The challenge is to slow this train to the speed of a horse-drawn carriage, without forcing a virtual standstill in economic development, particularly in the Third World.

Energy Policy and Climate

Last year, the federal government of West Germany commissioned a report

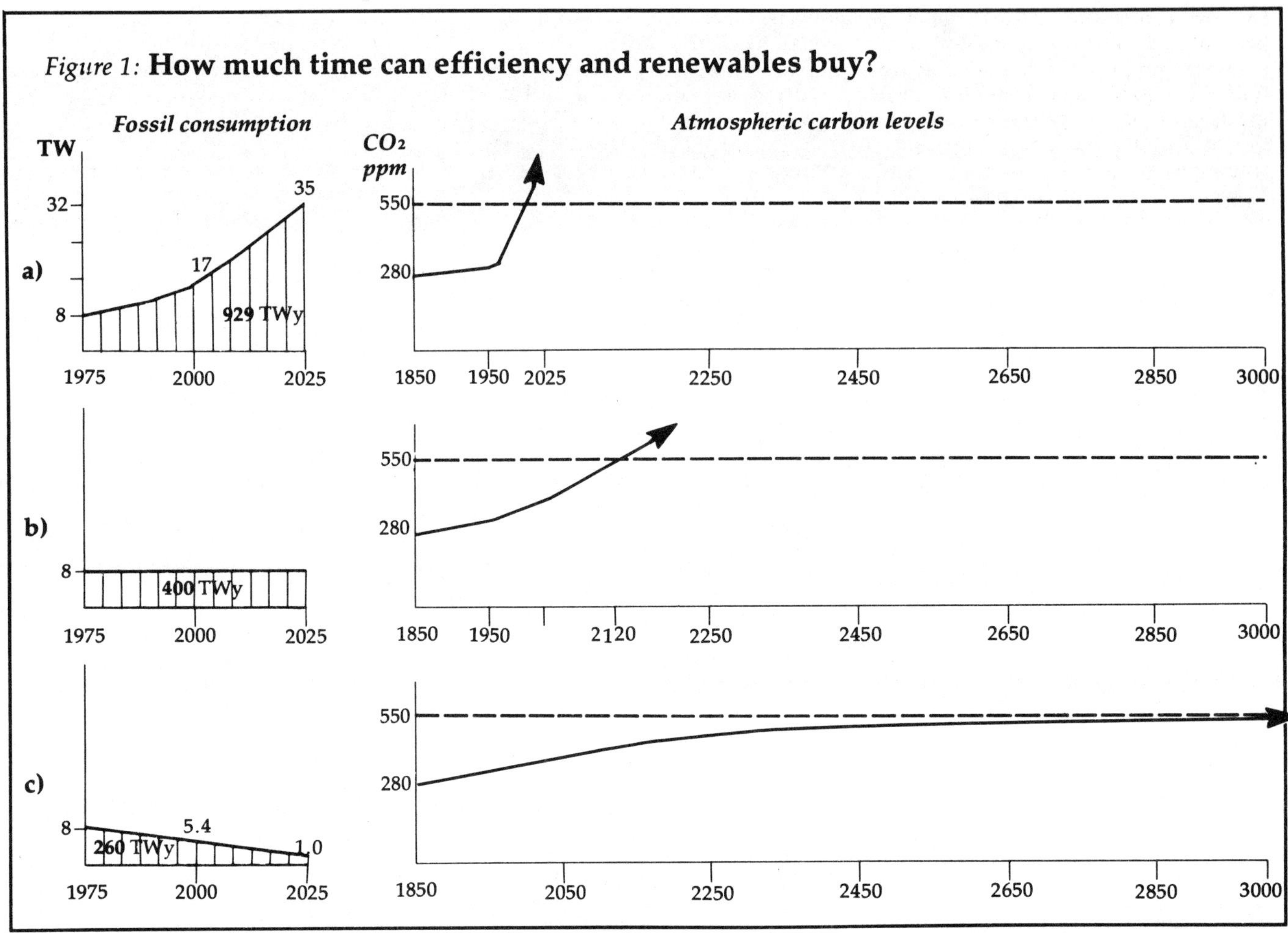

which combined the analysis of energy and climate in the development of viable policy. Written by Amory B. Lovins, Wilfred Bach, and this author, the report concluded:

• The CO₂ problem we presently face results from inefficient energy policies, notably those of industrial nations.

• It is possible over the next few decades to stabilize, and then gradually reduce to near zero, the global rate of fossil fuel burning with available energy-efficient and renewable-energy technologies.

• World economic development would not have to be stunted as a consequence. On the contrary, efficiency and renewable sources offer the most economical and logically expedient avenue for growth.

• A highly prosperous world of eight billion people, if it were to use energy in a way that saves money, would still consume somewhat less energy than at present. Global energy demand in the long run could drop as low as four terawatts, even if the world's Gross Domestic Product were to climb to about five times its 1975 world level.

• The potential of economically attractive renewable energy sources is sufficient to satisfy all or most of this demand both globally and regionally within a few decades. Readers of *Soft Energy Notes* are by now fairly familiar with the technical aspects of this argument. However, the methodology for estimating future world energy demand used in the report deserves special mention.

World economic growth was disaggregated according to the seven regions identified by the International Institute for Applied Systems Analysis (IIASA) in their recently-issued global energy forecast (see "IIASA's Fantasy Forecast," *Soft Energy Notes* 3:5:139-140). The same growth rates were assumed as in the IIASA "low" scenario. The increases in Gross Domestic Product were translated into energy demand figures by employing two multipliers. One multiplier corrected for the changing energy *service* intensities of developing economies (where infrastructure is being built up) and of advanced countries (where services and information gain increasing importance).

The second multiplier corrected for technical efficiency improvements over time. For the year 2000, the multiplier is derived from the best national-level efficiency studies of various industrial countries, where for each case, rates of introduction have been considered in detail. As a conservatism, developing countries were not assigned higher rates of introduction, even though they are not as restricted by old capital stock. For the long run, a technical potential for efficiency improvements was established (see "Efficiency Improvements: The Largest Energy Source," *Soft Energy Notes* 4:2:55-57).

West German Affluence Worldwide

One industrial economy, the Federal Republic of Germany (FRG) was taken as a prototype: a detailed sectoral analysis served as the basis for an aggregate national coefficient. This figure indicates the technical and economically-attractive potential for improved energy thrift in the long run, after all capital stock has been renewed. Since the savings in all sectors of the FRG economy are of comparable size, this coefficient is not very sensitive to variations of thermodynamic end-use structure. Furthermore, the FRG has very favorable characteristics for extrapolation: efficiency of its present capital stock is relatively high, primary energy and embedded energy trade is in good balance and industrial production has a relatively high share in total GDP. Thus, a prototypical treatment of the FRG will tend to lead to conservative estimates of possible efficiency gains in other countries.

—*Florentin Krause*

References:

Bach, Wilfred

1981 "Fossil Fuel Resources and their Impacts on Environment and Climate," *International Journal of Hydrogen Energy*, 6:185-201. Pergamon, UK.

IIASA

1981 *Energy in a Finite World: A Global Systems Analysis*. Ballinger, Cambridge, MA.

Krause, Florentin

1981 *An Efficiency- and Development-oriented Approach to World Energy Prospects*. IPSEP, San Francisco.

Lovins, A. B., H. Lovins, F. Krause, W. Bach

1982 *Energy Strategy for Low Climatic Risk*. Brickhouse Publishing, Andover, MA.

Scandinavia Shares The Solar Wealth

EACH OF THE NORDIC COUNTRIES is differently endowed with renewable resources. Norway and Sweden have substantial hydropower potential, Finland and Sweden have huge forest resources, while the largest wind resources are in Denmark and Norway.

All of the Nordic countries have solar energy potential, but the seasonal variations increase toward the North. This diversity points to the possible advantage of considering the region as a whole in order to construct an optimal long-range energy plan. A combined system may have a higher degree of supply security and may offer less expensive solutions, along with the possibility of larger markets for the energy technologies involved. The resulting dependency among the Nordic countries would only be a natural extension of the close economic and cultural ties they now share.

Four Scenarios

The first report from a joint Nordic project explores these possibilities in the form of a set of energy scenarios for the year 2030. The project tries to break new ground in the style of international cooperation. Rather than seeking to reach a consensus among the participants from the individual countries, each nation has been asked to furnish its own picture of society in 2030. Two values for the total gross national product in 2030 were prescribed — 20 percent or 100 percent higher than today — each to be combined with two scenarios for the 2030

energy intensity (energy input for a given product or service). The two scenarios for energy intensity are 0.7 and 0.4 of today's value: the higher figure corresponding to the best presently economic technology, and the lower to the best currently known technology believed to become economical before year 2030.

Each country then specified the distribution of economic activities by sector to arrive at the prescribed overall GNP figures. In this way they constructed scenarios for each of the four combinations of GNP growth and energy intensity. The scenarios turned out to be very different from what would have emerged if the same model had been used for all the countries.

Next, each country provided estimates of the maximum use of renewable energy it considered compatible with environmental and social concerns, together with a second, lower bid aimed at minimizing unwanted effects. Again, the results showed that different environmental and social criteria had been employed in each country.

The final step involved an attempt to match the energy demand scenarios with the proposed use of renewable energy resources. For the combination of high GNP growth with modest im-

provement of energy intensity, only the maximum use of renewable resources could furnish the requisite supply. In the other three scenarios, lower level use of renewable resources was possible, provided that energy could freely cross national boundaries. Only in the scenario involving low GNP growth combined with strongly decreased energy intensity was there any hope for local, single-nation solutions to the supply problem.

Second Project Phase

The Nordic study group is now seeking financial support for a second project phase, in which a more detailed investigation of the scenarios will be conducted. This will include an economic and social evaluation of each scenario with its assumed introduction of new technologies, both for the conversion of renewable energy, and for achieving the prescribed change in energy intensity. (Decrease in energy intensity is equivalent to an improvement in the efficiency of energy use, either through improved conversion efficiency, or by introducing new paths for providing the same end result.)

Further, the second phase aims to describe the transition from the present energy system to each of the 2030 scenarios, with particular emphasis on areas where cooperation among the Nordic countries will be of advantage. As the Finnish participant joined the study quite recently, the four scenarios have not been completed for Finland, so this country's data do not appear in the figures of the now completed report. In the continuation of the project, Denmark, Finland, Norway, and Sweden will be dealt with on an equal footing. It is expected that detailed models will describe the time sequences of energy transfers among the countries, and assess the need

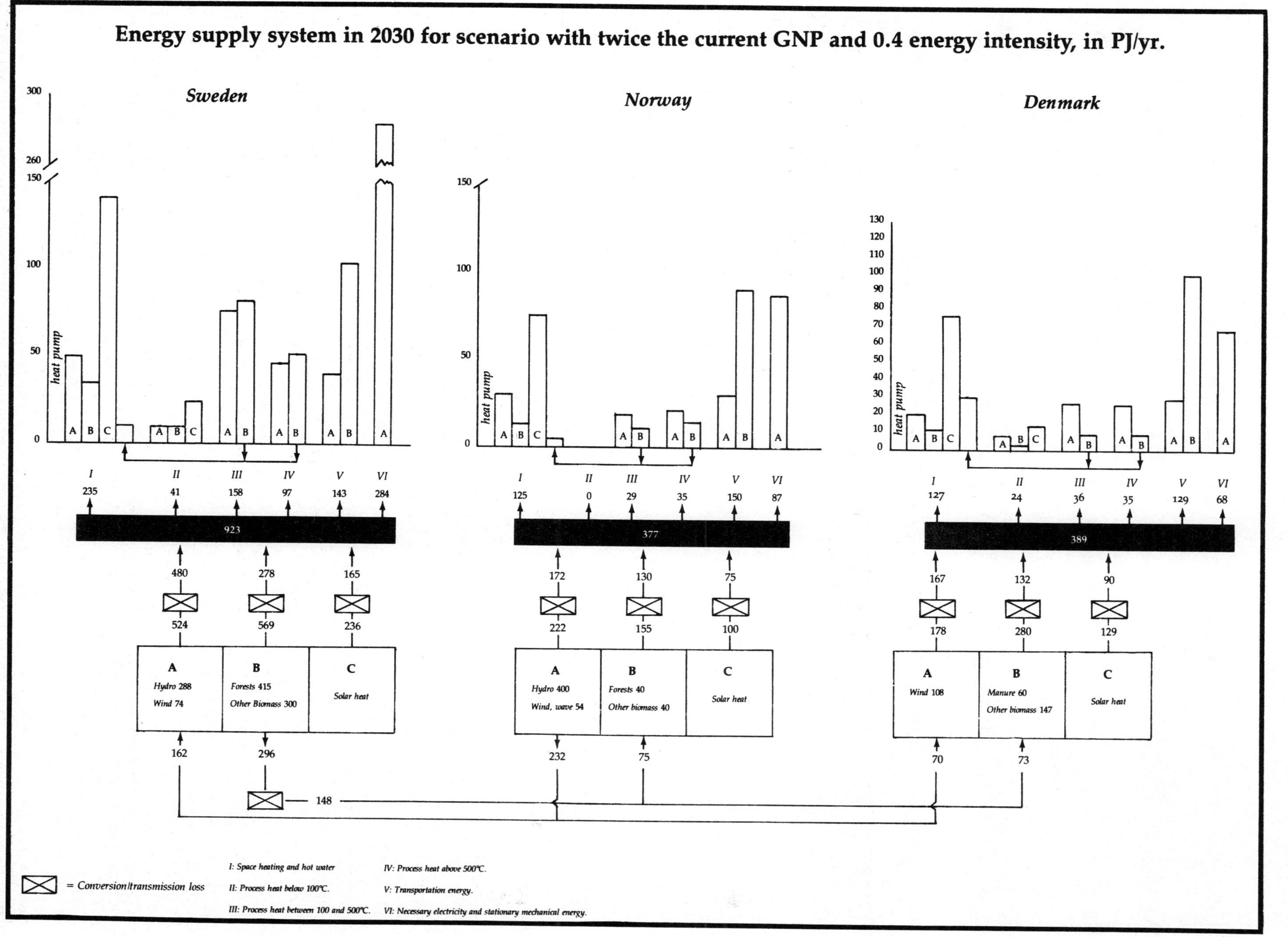

Energy supply system in 2030 for scenario with twice the current GNP and 0.4 energy intensity, in PJ/yr.
Sweden
Norway
Denmark
heat pump
I: Space heating and hot water
II: Process heat below 100°C.
III: Process heat between 100 and 500°C.
IV: Process heat above 500°C.
V: Transportation energy.
VI: Necessary electricity and stationary mechanical energy.
= Conversion/transmission loss
Hydro 288 Wind 74
Forests 415 Other Biomass 300
Solar heat
Hydro 400 Wind, wave 54
Forests 40 Other biomass 40
Solar heat
Wind 108
Manure 60 Other biomass 147
Solar heat

Ratio of per capita delivered energy in 2030 and 1980

GNP ratio:

Energy intensity ratio:		**1.2**		**2.0**	
0.7	*DENMARK:*	0.81	*DENMARK:*	1.16	
	NORWAY:	0.81	*NORWAY:*	1.39	
	SWEDEN:	0.85	*SWEDEN:*	1.29	
0.4	*DENMARK:*	0.54	*DENMARK:*	0.78	
	NORWAY:	0.46	*NORWAY:*	0.79	
	SWEDEN:	0.52	*SWEDEN:*	0.81	

The present energy use in the three countries (energy delivered to the customer) is 2200 PJ or 4.03 kW/cap.

for storage of heat and mechanical energy. (The existing hydropower reservoirs can probably take care of all necessary mechanical energy storage.) In addition, the environmental impacts of each scenario will be assessed.

The experience of the Nordic study group has been that the pluralistic approach is a success. The initial sessions soon revealed that it would not be possible to define common criteria for the detailed energy scenarios that would please all participants. One alternative was to include all the views in the set of scenarios, but this approach would have produced so many of them as to be unmanageable. Instead, the overall model prescriptions left enough room for each participating country to incorporate its preferred model of development.

It is also true that the several Nordic countries, and certainly the diverse interest groups within each of them, would select different year 2030 scenarios from among the four provided as the most probable or desirable. In future work, the group will discuss the basis for such expectations, and examine, if possible, the causes underlying these differences in judgment.

— *Bent Sørensen*

Reference:
1981 "Nordisk Energisamarbete — mögligheter och begränsninger i ett långsiktigt perspektiv." Report from Phase I of a project on the Nordic Energy System, February. Participants: Niels Enrum, Niels Meyer, Jørgen Nørgaard, Sigurd Lauge Petersen, Thorkild Saxe, and Bent Sørensen (Denmark); Kyosti Pulliainen (Finland); Paul Hofseth, Vidar Myhrer, and Elin Tyse (Norway); Thomas Johansson and Jaquette Lyttkens (Sweden).

1. Energy and Social Development

Soft energy path concepts do not deserve the name unless they lend themselves to a socially just and equitable future. There are serious and widening inequities within industrialized nations, but the most glaring ones exist between the industrialized "north" and the poor nations, and within these poor nations themselves. The energy situation is an expression of those gaps. Much of the energy-using economy in the Third World has yet to be built. If questions of social development can be postponed in affluent countries in favor of technical fixes to the existing energy infrastructure, the inverse is evident in developing countries. For whose benefit and what kind of development will energy be used? Any proposal must answer that question.

There are two crises of energy in the Third World: that of the commercial, modernized sector and that of the rural sector, where most people live. Third world energy policies, and economic policies in general, largely focus on the highly oil-dependent, modernized sector. Growth and development in that sector, so the rationale goes, will lay the groundwork for tackling poverty in the countryside later. There can be no doubt that expansion of domestic industries is vital to the survival of Third World economies. However, the lopsided emphasis on the urban sector leads to a form of self-cannibalization: soaring oil costs must increasingly be paid with money borrowed from western banks and international institutions. In order to be credit-worthy, national elites are commercializing ever larger portions of their coun-

tries' resources, dislocating the rural populations. These, in turn, cornered by their poverty, are forced to deforest the countyside in ecologically fragile regions. The vicious cycle of trickle-down development thus closes in on itself.

Against this background, introducing new energy technologies is no isolated technocratic maneuver, though, as rural electrification efforts demonstrate, western development aid has naively hoped to quell Third World political unrest by just those means. Well-meaning advocates of more efficient fuelwood use must also guard against simplistic proposals. Inequity at the rural village level can be an insurmountable inhibitor, even when national governments try to bring rural development along. Nothing illustrates the connection between rural social organization and technology better than the relative failure of biogas digesters in India and their relative success in the People's Republic of China.

Recognizing the interdependence of equity, social development, conservation of ecological resources and of energy headaches, Third World analysts have recently come up with promising strategies. A.K. Reddy demonstrates just how the Indian oil crisis might be unraveled by such an integrated approach. Here again, social reform is an indispensable part of the solution. Any power that pursues a policy aimed at repressing Third World movements for social reform, is therefore choosing to be part of the problem rather than part of the solution.

Distribution

Renewable energy sources will help Brazil become independent of foreign oil imports, report José Goldemberg and Hartmut Krugmann. But to keep energy demand from skyrocketing, they counsel, development must include a more equitable distribution of resources.

BRAZIL, with a per capital Gross National Product of about US $1,500, is an upper middle income developing country. Its energy problems stem from a social and economic pattern common in Latin America and throughout the developing world.

Industrialization is relatively high, with the industrial sector accounting for more than half of the GNP. Urbanization has brought roughly half of the population to the cities. But, Brazil has a dual economy, with a distinct cleavage between the dynamic, modern, industrialized sector, and the stagnant, traditional, subsistence agricultural sector. The distribution of income is horrendously skewed, and worsening; the richest five percent of the population earns as much as the poorest 60-80 percent, and the incomes of the rich grow faster than those of the poor. Finally, imported petroleum supplies at least half of primary energy consumption.

In developing countries with these characteristics, developed urban centers may be visualized as prosperous islands in a sea of poverty. Surrounded by belts of slums that lead to vast, poverty-stricken rural areas, the wealthy urban cores contain most of the modern industries, and house most of the people who benefit from modernization. Their lifestyles, copied from the industrialized West, include many energy-expensive habits, most notably the use of the automobile. Indeed, access to a car is one way of defining, or identifying, members of the urban elite. In Brazil, 10-20 percent of the population drive, though far fewer are really affluent, as many of those with cars use them at the expense of essentials such as decent housing.

The society's elite is as powerful politically as it is economically dominant. It is thus able to prevent the social ascent of the masses, keeping their demands under control with police or army force, if necessary.

Energy is only one of the actors in this drama, but the petroleum crisis has shown just how important its role actually is. Soaring oil bills have created severe balance-of-payments problems, pushing some developing countries to the brink of bankruptcy. Continued economic growth has been threatened, and with it whatever small benefits may have trickled down or spilled over to the poor. Furthermore, worsening economic conditions reduce the financial and political margin for social reform. On the other hand, the petroleum crisis has a stronger direct effect upon the modern sector and the social elite than it does upon the poor. Perhaps this common cause will help to unite rich and poor in an effort to increase energy self-sufficiency and overall self-reliance.

Energy Independence and the Rural Poor

In Brazil, the movement toward energy independence has already begun to bear fruit. Biomass conversion, mainly sugar cane into ethanol, has increased significantly, which can effectively replace gasoline and other petroleum products. A second major effort derives ethanol from wood and other cellulosic materials. Unfortunately, the Brazilian alcohol program in its current form appears to exacerbate, rather than mitigate, social problems associated with land distribution, subsistence food production, and migrant labor.

Even though biomass is particularly suited for decentralized use in rural areas, the Brazilian program is geared toward cities, where most petroleum products are consumed. For biomass to substitute for gasoline and effectively displace oil, the plant-based fuels will have to be shipped to these centers for centralized use, much the way food has been.

The move away from the "petroleum civilization" to a "biomass based civilization" might be a way for Brazil and other developing countries to reduce drastically their dependence upon industrialized countries without completely dismantling capitalist structures. However, significant improvement in the situation of the needy, while employing currently available resources, could also result from a sociopolitical change toward more egalitarian conditions. The energy requirements of satisfying basic human needs are smaller than current per capita world energy consumption, and less than twice the current per capita energy consumption in the developing world. (The exact magnitude of basic energy needs, of course, depends on how these needs are defined.) Policies directly or indirectly fostering redistribution of income are bread and butter for the poor in the Third World. Important moves toward equitable distribution of wealth have already been taken by a few developing countries, including China, Cuba, Sri Lanka, and Taiwan.

Poor people spend a larger fraction of their income, directly and indirectly (through purchase of goods and services), on energy than do the rich. This raises the question whether redistribution of income would not increase total energy demand, aggravating the problems caused by the petroleum crisis. Recent studies focusing on the State of São Paulo in Brazil indicate that within the income range found in poorer countries energy consumption rises almost linearly with income.

These studies also find that for a given income (monetary plus non-monetary), energy consumption does not appear to vary appreciably with locale. Urban slums, metropolitan centers, and rural areas differ little with respect to per capita primary energy consumption because greater energy use—principally electricity—with modern amenities is balanced by more efficient energy end use, particularly cooking with bottled gas rather than wood. This finding should not undermine the importance of improving rural living conditions and keeping people in rural areas; urban growth brings about a number of problems not directly related to energy.

Two main concerns emerge from the foregoing considerations. First, indigenous fuels—especially renewables—must be substituted for petroleum products in developing countries now heavily dependent on imported petroleum. This will ensure continued economic growth and increase energy self-sufficiency. The economies of developing countries are dynamic; most investments in infrastructure and industry lie ahead, and not behind, as in industrialized nations. This flexibility will be an asset in the shift away from petroleum.

The second concern is that the extreme contrasts in well-being that currently prevail throughout the developing world must be reduced to give the poor a chance. This goal will not be easily achieved, since economic and political power are usually in the same hands.

—José Goldemberg and Hartmut Krugmann
Institute of Physics
University of São Paulo
São Paulo, Brazil

Diversity

Vaclav Smil, Canadian geographer and China energy analyst, takes a hard look at the soft path. He cautions against large-scale biomass conversion and opts for a mix of conservation, small-scale hydro, coal and oil shale. Like Goldemberg and Krugmann, he finds poor countries need huge increases in average energy consumption.

ANY SERIOUS CONTEMPLATION of developmental energetics should start from a simple premise: unlike the industrialized nations, the poor countries need huge increases in average consumption of energy. Western countries, the Soviet bloc, and Japan can take advantage of multifarious conservation opportunities, technological innovations, and pricing adjustments to sustain their (more or less) affluent societies with considerably less energy than they waste today.

Opportunities for energy conservation undeniably exist in poor countries as well. There are rich rewards from replacing antiquated technologies, from more perceptive building design (frequently derived from ingenious traditional structures), and from better conversion, beginning with that ubiquitous and exceedingly inefficient device, the simple cooking stove, or its even more primitive open-air counterparts.

But even if all the holes in the extant thermal sieves were plugged up—a most unlikely assumption—impressively large, unserviced needs would remain at all levels of consumption.

Households need energy for family cooking and washing, for food and feed processing, and for heating. This last requirement is all too often ignored in the erroneous belief that "most Third World countries are fortunately located in sun-drenched tropics." In reality, the tropics are far from sun-drenched, and, more importantly, at least 750 million people in developing countries live at latitudes, or altitudes, where seasonal heating is unavoidable or highly desirable.

At the village level, energy is needed to pump water for irrigation and drainage, and to run small manufactures. Cities consume energy to house the urban populations, to educate and transport them, and to employ these rapidly expanding concentrations of mostly young people and children. Rapid urbanization throughout the developing world, producing high residential densities, is a fundamental consideration conveniently overlooked by the uncritical advocates of decentralized solutions, who cater just to small, subsistence villages.

Finally, all the larger, populous, poor nations—those with over 50 million inhabitants account for nearly three-quarters of the developing world—need substantial quantities of extremely concentrated power to build up or to greatly expand the large modern industries critical for their economic survival: iron and steel, synthetic fertilizers, cement, and electricity. Without these inputs there is little prospect for the sorely needed increases in crop yields, emplacement of efficient transportation links, housing construction, or expansion of labor-intensive industrial capacities.

A Non-Exclusivist Approach

To satisfy these disparate power requirements, no energy source and no commercially viable and environmentally acceptable conversion technology should be excluded *a priori*. Some richer nations may indulge their preferences and choices, but the leaders of poor countries would be both negligent and irrational to limit their energy strategies to any predetermined notion.

Certain small-scale, renewable technologies—above all, hydro, biogas, and some direct solar conversion—are already fitting very well into the household and village niches in many poorer nations. But a critical examination of any of these systems shows that there are clear and often unsurpassable limits—environmental, social, and technical—to their generalized, easy, and rapid diffusion. There are also limits to their ability to fill the needs for more concentrated energy flows very quickly, if they can fill them at all.

One of the seemingly simple, and current fashionable, approaches to harnessing more concentrated fuels from solar energy on a larger scale is through biomass. Materials and methods range from the cassava, which is Amory Lovins' favorite, to corn gasohol, to fast-growing tropical silvicultures. However, our still meager understanding of the planet's primary productivity, combined with our already alarming knowledge of the deterioration and often irreversible impoverishment of most climax ecosystems, counsel extreme caution.

Taking advantage of all practicable approaches should include such often neglected options as the local uses of low quality fossil fuels like coals and oil shales. Frequently, these are readily accessible, they can be extracted with a minimum of equipment, they are quite concentrated, and they store well. Even though their combustion in small units is neither highly efficient nor particularly clean, they can be burned in households and in many local industries at lower cost and with less environmental damage than the wood, crop residues, and refined fuels currently used. Industrially submarginal coal deposits are very widely distributed throughout the world, and very much like small-scale hydro can make a substantial local difference in some poor nations.

The only technology I would not insist on including among energy conversions for the developing world is nuclear generation, mainly because of its potential military misuse. Most certainly not because I would consider it "hard" and beyond redemption. I do not think, for example, that biomass energy crops are "soft" and adorable, and I perceive the whole "hard/soft" dichotomy to be artificial and all too simplistically contrived. All those in the direst need would only benefit from a catholic, rather than from an exclusivist, embrace of energy sources and technologies.

—Vaclav Smil
Department of Geography
University of Manitoba
Winnipeg, Canada

References:
Smil, Vaclav

1979 "Energy flows in the developing world." *American Scientist*, vol. 67, no. 5, pp. 552-531 (September-October).

1979 "Renewable energies: How much and how renewable?" *Bulletin of the Atomic Scientist*, vol. 35, no. 10, pp. 12-19 (December).

1979 "Energy strategies for the developing world." *Energy International*, vol. 16, no. 12, pp. 27-29 (December).

Smil, Vaclav and W. E. Knowland, eds.

1980 *Energy in the Developing World*. Oxford: Oxford University Press.

Dr. Smil is presently working on a book entitled *Biomass Energies: A System Approach* for Plenum Energy Series. He invites readers of *Soft Energy Notes* to send him their publications on biomass so that references to their work may be included in his volume. Materials should be sent to Vaclav Smil, Department of Geography, University of Manitoba, Winnipeg, Canada R3T 2N2.

Population

RAPID POPULATION growth makes the global energy problem doubly difficult to solve. In fifty years, the earth will support twice the present number of people, according to most projections; and, if each individual consumes energy at only twice the present per capita rate, total demand will quadruple. Most of the population growth, and the consequent demand increase, will occur in the Third World. Faced with staggering capital and environmental costs, energy planners must find new ways to balance supply and demand.

Decentralized energy systems may prove an effective remedy. They encourage an overall pattern of decentralized economic development, de-emphasizing the role of the cities. This would stem the tide of rural-to-urban migration, which is a major contributor to net population growth. While increasing the productivity of agricultural labor, decentralized energy development would generally strengthen the agricultural sector, and remove some of the incentives for farm families to have many children. Investments in decentralized energy systems thus have a two-fold return: they increase supply while they help to slow demand.

On Third World Shoulders

THE WORLD'S POPULATION is more than 4.1 billion and rising. By 2030 it may reach eight billion, or nearly double its present level, with the plateau not even then in sight. This growth, however, will not be uniformly distributed, since populations in developing nations are increasing much more rapidly than in those already industrialized. The annual rate of natural increase (crude birth rate minus crude death rate) for the Western countries rests at 0.6 percent, possibly the lowest it has ever been, for any stretch of time, since before the adoption of agriculture in the late Neolithic. Developing populations, in contrast, increase at the historically unprecedented rate of two percent annually, for a population doubling time of only 35 years. This discrepancy in growth rates means that most of the population increase in the next half century will occur in the developing world, which by many measures, appears already overpopulated.

The lion's share of future energy demand will similarly fall to the less developed countries. With larger populations, they will require ever more energy to meet their basic needs. On top of this growth, per capita energy consumption in the relatively poor regions will have to increase if there is to be any improvement in the quality of life. Global commercial energy use is now about 8.2 terawatt-years/year (245 Quads), or roughly 2 kilowatt-years per person per year (42,000 kilocalories per capita per day). This average figure hides enormous discrepancies: actual consumption in developing countries is less than half the world average, while individuals in the US and the other industrialized nations consume up to eleven times the allotment of their less wealthy neighbors.

A more equitable distribution of resources between and within countries, according to José Goldemberg and Hartmut Krugmann, would satisfy the basic human needs of the current population without increasing world energy consumption (see *Soft Energy Notes* 3:4:18-19). However, the political and economic changes involved in such a massive resource redistribution are unlikely within the next half century. Moreover, the expected doubling of world population in that period makes additional energy use almost inevitable, even with some more equitable pattern of allocation. More conventional estimates of global energy demand for 2030 call for three to four times present levels. The Austrian-based International Institute for Applied Systems Analysis, for example, projects 22 to 36 terawatt-years/year by that date.

The developed countries, already approaching their limits of both population growth and per capita energy use, will capture only a small part of the projected increase in energy consumption. The global energy problem of meeting future energy demand is thus one that falls principally on Third World shoulders, and an expanding population base will be the largest single contributor to demand escalation.

Development planners typically accept population growth as a given—an unfortunate factor that heightens demand, but is beyond their control. Their best efforts go, instead, to increasing supply, and when they think of demand reduction at all, they do so in terms of efficiency improvements which lower per capita consumption. Putting the brake on resource demand through lower population growth rates is outside their perceived scope of work.

This view is almost certainly short-sighted. Accumulating evidence indicates that development and population trends are, in fact, closely linked. The effects of development on the labor market reach deep into the structure of rural family life, and influence decisions about marriage, child-bearing and migration. Development policies and projects can actually *encourage* population growth, and make the overall problem worse by increasing demand more rapidly than they expand supply. Or, more optimistically, those projects that increase supply while checking population growth can be doubly beneficial.

Global Demographics

A FIRST STEP in population-conscious development planning is understanding the factors responsible for growth. The explosive increases of the past several decades are principally due to a rapid, worldwide decline in mortality. Between 1950 and 1980, life expectancy (at birth) rose from 65 years in developed countries, and 42.5 years in developing countries, to 72 and 56 years, respectively. There is no simple explanation to account for this change, and demographers look to a combination of more medical personnel, extended transportation systems, superior nutrition, and the widespread use of antibiotics and DDT.

Most of the greater life expectancy results from lower infant mortality, and the infant's improved chance of surviving into adulthood helps to account for a second global demographic trend: the decline in fertility. Over the past 30 years, the crude birth rate dropped from 23 per thousand to 16 in industrialized areas, and from 42 to 33 in developing areas. Contrary to Malthusian population notions, better health and nutrition have *lowered* birth rates, rather than increased them. The reason: even without modern contraception, people have some measure of control over their reproduction, which they exercise to achieve socially and economically optimal family size. With more of their infants surviving, parents op⁺ for fewer births.

The family size that parents deem optimal varies inversely with income: the more well-to-do the family, all other things being equal, the smaller the optimal family unit. For poor people, children provide a source of labor, of income, and a kind of social security, particularly in the traditional agricultural sector. In rural India, for example, birth control programs failed because they did nothing to make smaller families economically advantageous. Mahmood Mamdani points out that birth control, rather, "contradicted the vital interests of the majority of the villagers. To practice contraception would have meant to willfully court disaster."

A weak rural economy thus provides incentives for large farm families, and prevents fertility rates from dropping more rapidly. A third global demographic trend—rural-to-urban migration—is both symptom of this weakness, and contributor to it. The rural labor market, cycling seasonally from shortage to surplus, cannot match the opportunities provided by the expanding urban centers in developing countries. The flood of migrants is expected to swell the urban fraction from 29 percent of the Third World's population at present to 41 percent by 2000. Mostly members of the 20-35 age group, the migrants are the cream of the labor crop; their departure leaves rural villages critically labor-short during peak agricultural periods, especially harvest. Agricultural wage labor rates double, or even triple at these times, but then fall back to levels below those found in the developing cities. Farm families consequently rely more heavily on their children as substitutes for agricultural labor that is expensive, or unavailable.

The rural regions of developing countries, then, serve as the perpetual wellsprings of population increase. Yet the agricultural sector is not sufficiently developed to retain all the individuals it generates. The uncertain and relatively unrewarding rural economy propels workers to the expanding urban centers, leaving the reproductive capacity of the farm population to make up for the loss of their labor services. Over the next several decades, agricultural underdevelopment and the consequent rural-to-urban migration, will be a principal force behind sustained high rates of population increase in the Third World.

Decentralized Development

A DEVELOPMENT POLICY that bolsters the agricultural economy could check urban migration, and lead to greater fertility declines than generally forecasted. The crucial element is energy development, because the geography of energy supply has a powerful effect upon regional development.

Industry and commerce flourish where energy is inexpensive and readily available. Energy development that follows conventional, centralized lines encourages urbanization and weakens the agricultural sector by concentrating investment capital, physical plants, labor opportunities, and economic activity in cities and a few resource regions. The additional energy supply from centralized systems is available principally to primary industries and urban consumers, creating jobs and amenities the rural areas can not offer. In addition, the commercial energy that reaches the countryside is priced beyond the purse of small farmers, and benefits large agribusinesses. Their higher productivity and access to capital often push the small farmer into an even more marginal niche, encouraging rural outmigration.

Highly centralized energy systems can have this effect even when they are based upon renewable resources. As José Goldemberg comments on the Brazilian alcohol fuel program: "Large sugarcane plantations are being established in regions where many small farms existed . . . This has had the very negative social consequence of forcing the exodus of small farmers and labourers in the fields to small cities when it is difficult to get jobs."

Decentralized energy systems, however, would have multiple benefits. Microhydro installations, biogas digesters, and solar-powered irrigation pumps directly aid the rural poor, whose plight has demonstrably worsened under the centralized development regime. The modest energy inputs provided by these technologies would raise crop yields per unit area (among other improvements), and increase the food supply without displacing indigenous populations. Agricultural labor productivity would also climb, reducing the farm labor shortage, and with it the incentive for large families. The stress on rural, rather than purely centralized and urban development would further stabilize the agricultural economy and restrain migration to the cities.

The choice between centralized and decentralized energy development goes beyond a ranking of nuclear, coal, oil, alcohol, biogas, minihydro, and other options by their cost per unit energy delivered. The social consequences of development policies can actually *undermine* the original development objective. Energy development that promotes population growth by weakening the agricultural sector, fuels demand more rapidly than it boosts supply. Local-level energy development, while it may make a more modest contribution to total supply, helps to check demand by slowing population growth. If increased per capita energy availability is the goal, decentralized systems are surely the better investment.

—*Charles Drucker*

References:

Goldemberg, José
1980 "A Centralized 'Soft' Energy Path." Preprint IFUSP/P-233, Instituto de Fisica, Universidade de São Paulo, Caixa Postal-20.516, Cidade Universitaria, São Paulo, Brasil.

Goldemberg, José, and Hartmut Krugmann
1980 "The Energy Costs of Satisfying Basic Needs." Instituto de Fisica (address above).

Mahler, Halfdan
1980 "People." *Scientific American* (Special issue on Economic Development), September, vol. 243, no. 3, 66-77.

Mamdani, Mahmood
1972 *The Myth of Population Control.* New York: Monthly Review Press.

Mauldin, W. Parker
1980 "Population Trends and Prospects." *Science*, vol. 209, 4 July, 148-157.

The Energy Crisis: A Matter of Access

Supply and Demand—the familiar tug-of-war over prices—may well be replaced by Supply and Access as a way of thinking about international energy problems. The essential constraints on the energy-poor regions of the world, according to Hartmut Krugmann and José Goldemberg of the University of São Paulo, Brazil, are access and availability.

Hartmut Krugmann, José Goldemberg, and Amulya K. N. Reddy estimate the energy cost of satisfying basic human needs at less than current per capita world energy consumption. Their reports suggest that the energy crisis is really a problem of unequal access and uneven distribution.

It is an "inescapable conclusion," Krugmann reports, "that the energy problems of less developed countries (LDCs) could be solved at today's world energy consumption level by redistributing resources more equitably between and within countries.

"It is political and economic constraints that prevent redistribution from taking place. The 'energy crisis' of LDCs is thus intimately linked to inequities in the access to energy sources, not to their physical lack."

Krugmann sees the rate of energy consumption as a barometer of a society's standard of living. He notes that LDCs average 18,000 kcal/capita/day—more than six times less than their industrialized neighbors. But even greater disparities in energy consumption lie within and among the less developed countries themselves. Data from 1975 show Brazil's energy appetite to be *eleven* times the 2,300 kcal/ca/day of Bangladesh. And within Brazil, demand in the southeastern region of São Paulo, at 28,000 kcal/ca/day, is *fourteen* times that of the rural northeast.

If the energy crisis is really an access crisis, Krugmann suggests that the solution may be found without increasing the present world average per capita consumption of 42,500 kcal/day. He estimates the energy cost of satisfying basic human needs at below this level, and he cites three estimates which hold similar conclusions: the Haffner model, the Palmedo index based on Morris and Liser's Physical Quality of Life Index, and the Bariloche Foundation Latin American World Model.

Krugmann and Goldemberg find that on the basis of the Bariloche Model, the cost of satisfying basic energy needs

Haffner's Minimum Energy Budget for a satisfactory life in the USA		
Need	*Energy Requirement*	*Percentage*
Food	6,200 kcal/day	20
Housing	6,200	20
Clothing	2,065	6.7
Transportation	4,130	13.3
Leisure	12,400	40
Total	30,995	100
Present Total	*243,000*	*784*

from Reddy, page D-48

in LDCs is approximately 30×10^3 kcal/ca/day which is comparable to the assessments by Haffner (31×10^3 kcal/ca/day) and Palmedo *et al.* ($23 - 28 \times 10^3$).

Though lower than the present world average, even these estimates of basic energy needs may be high, according to Amulya Reddy of the Indian Institute of Science, Bangalore, India. In "Energy Options for the Third World," a report to the Salzburg Conference for a Non-Nuclear Future—April 1979, he undercuts Krugmann's figures by more than 50%. Noting that the typical Indian village now consumes 5,250 kcal/ca/day, Reddy asserts that 14,000 will satisfy basic needs.

He begins with the Haffner estimate of 31×10^3 kcal/ca/day, and deducts 10%, corresponding to the space heating component of Haffner's number, "because most third world countries are fortunately located in sun-drenched tropics." Reddy then cuts the remaining 28,000 in half, to reflect decentralized production, needs-oriented product mixes, and energy-saving technology. These low estimates suggest, says Krugmann, that resource redistribution between and within nations could solve the energy problems of developing countries at today's world energy consumption level.

Reddy and Krugmann approach the redistribution of energy supply by redefining development. Krugmann explains, "there has been a tendency on the part of LDC governments and development agencies in developed countries to equate development with high economic growth rates. According to this philosophy of maximizing GNP, investments have been made primarily in projects promising fast returns, and production has focused primarily on goods yielding large profits."

As an alternative to this growth option, Reddy propses that the energy sector promote development. He cites, as a model, the UN definition of development:
- the satisfaction of basic human needs (material and non-material), starting with the needs of the neediest, in order to achieve a reduction of inequalities between and within countries;
- endogenous self-reliance through social participation and control; and
- harmony with the environment.

Reddy sums up: "whereas it is mainly the elites of developing countries who benefit from the growth option, it is the people, and in particular the poorest section, who stand to gain by the development-oriented energy option." Rural electrification in India, rarely a success story according to Reddy, serves as his example. Even though 25 percent of the Indian population lives in villages of less than 500 people, only 11% of those villages will be electrified in the next decade. And once they are electrified, they will receive only about 100 kWhe per day, which is equivalent to 400–500 kWht per day, or about 10 percent of current energy consumption. In fact, this 10 percent is consumed by the rich who constitute approximately one-tenth of the rural population.

Reddy finds the energy budgets of third world villages inconsistent with development and he seeks village-scale production of energy from renewable sources. The starting point, he says, is improved productivity of human and animal labor and increased efficiency of non-commercial energy use. He argues that continued reliance on human labor would only be putting back the clock of history if its arduousness, drudgery and low labor productivity are also preserved. Instead, Reddy calls for new technologies: anaerobic fermentation of wastes to biogas, energy forests, solar and wind systems, micro-hydro plants, and mechanical aids such as levers, pulleys, cranks, and pedal power.

"One point is clear," concludes Reddy, and Krugmann would agree, "if third world villages are to be the peripherals to growth, their futures, as well as those of the people in them, are bleak, but if they are made the core of development, their futures are bright."

—Elyse Axell

References:

Krugmann, Hartmut and José Goldemberg

1980 *The Energy Cost of Satisfying Basic Needs,* Instituto de Fisica, Universidade de São Paulo, C.P. 20.516, São Paulo, Brazil, 22 pages.

Reddy, Amulya Kumar N.

1977 "Energy Options for the Third World," paper prepared for Earthscan press briefing seminar in the Hague, April 18, 1977 and included in papers from the Salzburg Conference for a Non-Nuclear Future—April 29, May 1, 1977, 30 pages.

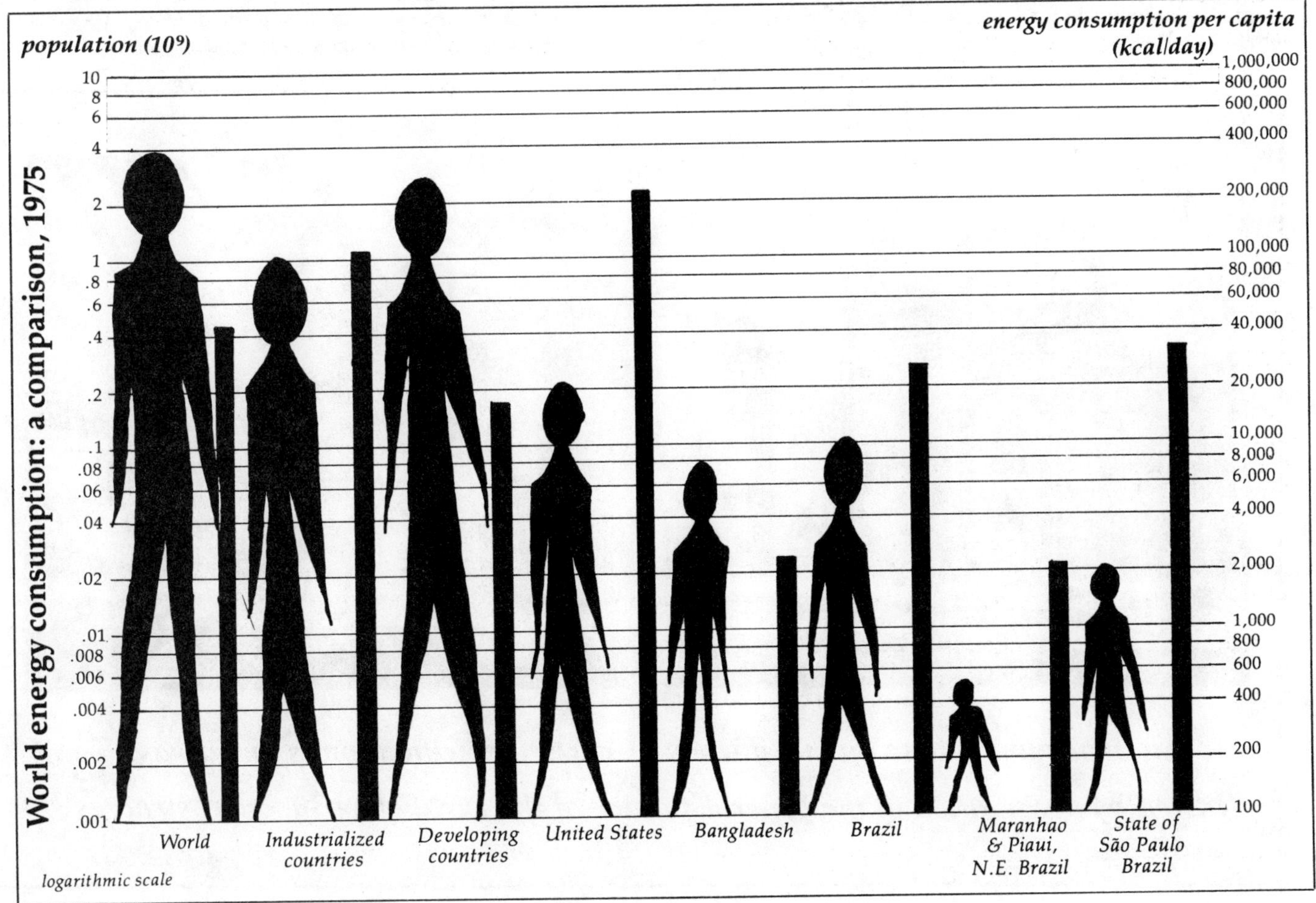

Rural Electrification: Towards a Crisis in Confidence

A Variety of Reports Conclude that Rural Electrification is not only Inefficient but Provides Substantial Benefits to Relatively Few

A GROWING CONCENSUS that electrical energy system use, in rural areas has proven inefficient and beneficial to relatively few persons has prompted many energy policy makers and Third World researchers to reevaluate the future implementation and development of such large-scale projects. Materials from the US Agency for International Development (AID), Development Alternatives Inc., Massachusetts Institute of Technology (MIT), Pacific Studies Center, and the Papua New Guinea government among others, indicate the inadequacy of existing evaluations of rural electrification programs and the need for more thorough and wide-ranging critiques.

"Rural electrification," says MIT Energy Laboratory consultant Douglas V. Smith, "continues to be a catch phrase of considerable emotional appeal among politicians, generals and power engineers." Referring to rural electrification efforts in Bangladesh, Smith explains that, "there is no evidence of the development-inducing effects of rural electrification although there have been monitored projects since 1963. In any event, rural electrification in practice means electrification of town centers, lights and fans for the influential, and volumes of talk about energizing tubewells," he adds. "A few merchants and small industries are able to receive the subsidized electricity as well but there is no evidence that their expansion or location decisions are influenced by the presence of electricity."

Methodological Critique

Reports on AID rural electrification programs also point to the need for changes in present development plans. Analysts hired by the agency's Bureau for Program and Policy Evaluation have noted that existing rural electrification studies are methodologically flawed. Meanwhile, the Washington-based Development Alternatives Inc., has outlined problems with AID's general approach to supplying energy to rural areas. In its report on AID's rural electrification subcontractor—the (US) National

"An important debate on the efficacy of rural electrification is underway. . . .
fueled by observation of the general failure of the rural electrification strategy."

Rural Electrification: Towards a Crisis in Confidence

Rural Electrification Cooperative Association (NRECA)—Development Alternatives Inc., states, "everything written by NRECA and by AID on the subject of rural electrification reflects the belief that rural electrification is a universally and immediately desireable goal. That rural electrification is not only a desireable goal but also a necessary vehicle of development is taken as axiomatic. This belief creates a relationship between AID and NRECA in which AID funds are used by NRECA to promote the desire for, and the presence of electric service in areas outside capital cities."

After reviewing the existing studies of rural electrification, Development Alternatives Inc, also has noted that, "the more objective and more thorough the data collection and analysis techniques, the fewer benefits can be attributed to rural electrification." The problem, it says, with partial or unconvincing impact assessment studies "is that the results may be taken as gospel by those promoting rural electrification and used to convince other countries to undertake programs modelled on a 'proven' success story." AID funding for rural electrification has already reached the $800 million mark, with $40 million going to the Philippines in 1978 alone.

AID Justifications

AID studies have not yet been able to clearly delineate the productive benefits of rural electrification according to Judith Tendler, AID consultant on rural infrastructure. Tendler describes the Agency's reports as "somewhat forced attempts to 'squeeze' New Directions[1] justifications out of rural electrification projects, trying to smooth over the fact that household electricity will be used by the better-off.

"Either the poorest of the poor are excluded" she says, "or their gain is limited to the substitution of electricity for other fuels in lighting." It needs to be shown, she adds, "that this gain is greater than those to be had from the development of non-

"When given a choice of these alternatives, in the face of the real costs and benefits of each, it is debatable whether rural electrification would rank high on a list of village peoples priorities."

household uses of electricity or through investment in other rural services like water supply."

AID justifications of rural electrification assume but do not demonstrate that electricity is cheaper than kerosene, and reduces the ecological damage caused by wood-fuel cooking, Tendler also points out. Though the substitution of electricity for wood as a source of household energy is said to help prevent deforestation, AID studies actually show, that even those poor who hook up to the systems continue to use wood for cooking and ironing. This suggests that electricity is not competitive with wood—at least for the poorest—and does not therefore lead to the alleged conservation benefit. "In reality, then," Tendler concludes, "not much is being achieved by rural electrification in the fight against deforestation, and the 'conservation benefit' is hardly worth mentioning."

Documentation Needed

The effects of rural electrification on food productivity and new business growth, likewise, have not been well-documented. Ralph Lukens' 1978 AID environmental assessment of the model Phillippines program (at Misamic Oriental) points out that rural electrification has not resulted in a "significant increase in the number fo electric pump irrigation systems" nor did it have "much impact on the establishment of new businesses except for some small establishments and no impact on the growth of large industries."

In Nicaragua, reports Development Alternatives Inc., "only on the largest, most heavily capitalized farms and ranches in the three cooperative areas was electricity used for production or processing purposes." The group also indicates that there was no evidence that electricity was a factor in starting these activities, or that it proved cheaper.

Rural Electrification: Towards a Crisis in Confidence

Political Considerations

A variety of energy researchers view political factors as instrumental in determining the present course of rural electrification development. In a 1977 memoranda, AID official, Hal Datta notes that, "political considerations [of rural electrification] are sometimes given greater priority than socio-economic [considerations]." This fact is difficult to express, he says, in the basic socio-economic technical survey documents which are usually prepared for the agency.

Commenting upon local government backing of rural electrification systems, a 1973 University of Florida report to AID on projects in Costa Rica and Colombia, indicates the role which rural electrification assumes in political controversy. "Is the particular area under consideration one in which low standards of living are creating political instability, unrest, and turmoil?" the report asks rhetorically. "If so, a social benefits project might help to calm such unrest and lead to renewed confidence in the government, avoiding costly and damaging repercussions throughout the country and internationally."

Rural electrification, assert Walden Bello and Peter Hayes in a report prepared for the California-based Pacific Study Center, has not increased rural industry and employment because of lack of demand for mass consumer goods and inequitable income distribution. "Until change in patterns of land-ownership and political power arms them with purchasing power, the link between electricity and rural industry will remain utopian," they conclude. However, Bello and Hayes point out, that the very process of rural electrification may increase perceived class inequalities and contribute to class antagonism. Similarly, a study of rural electrification in Colombia by Augusto Torres et

> *"This is not to say that generation of electricity is invariably a suspect form of investment in development, but that a great deal of careful thinking must be applied to each request for electricity in rural areas to see whether the provision of electricity is indeed the best solution, both economically and ecologically, in relation to satisfying that particular demand, and to fulfilling the development goals of the country."*

al, notes that rural electrification, "became the focal issue for exacerbation of long-standing rivalries between hostile factions or individuals . . . and . . . tended to exacerbate poor social relations between classes." The study also explains that, "the poorer people made it explicitly clear to observers that they could not install and use electricity for lack of funds." In one town, for example, it was reported that, "the wealthier townspeople refused to keep the generator in operation (although

Table 1 Household Profile of the "Typical Beneficiary of Rural Electrification" in the Philippines, 1977.

Percent of Total Households Using Rural Electricity	Characteristic
72	Have a combined household annual income above $533/y, that is, is in the top 46 percent of households by income level
82	Have household heads educated above grade 4
56	Have more than 7 simple household items
70	Earn income from occupations other than farming
79	Live in solid housing
69	Own their dwelling
79	Live closer than 2 kilometers from a provincial road

Source: Based on National Electrification Administration 1978 report as derived by Bello and Hayes, 1979.

they apparently could afford to do so) because they did not want the poorer people, who had had the electricity installed, but who could not pay for it, to benefit from the electricity at their expense. Hostility between the two classes of people was thus increased, rather than diminished by the presence of the generator," the Colombian study concludes.

Questioning International Energy Aid

Though total energy aid to Third World countries climbed to $21 billion between 1970-1977, a thorough reevaluation of rural electrification and of urban-industrial central electrification strategies, has yet to emerge. To date, policy discussions have questioned the utility of electrification systems until their benefits can be well-established. As the 1978 energy policy paper issued by the Papua New Guinean government argued:

"Proponents of rural electrification infer general improvements in the standard of living, but they are unable to point to specific

Rural Electrification: Towards a Crisis in Confidence

benefits that only electricity can provide, or can provide in a socially more effective way than other alternatives. Yet it is clear that investment in rural electrification is investment forgone in such areas as improved roads, water supply, schooling and health services.

"When given a choice of these alternatives, in the face of the real costs and benefits of each, it is debatable whether rural electrification would rank high on a list of village peoples priorities.

"Moreover, the research we have conducted into the distribution of existing energy forms in Papua New Guinea villages informs us that a relative few, the wealthier villagers, benefit most, and these families will certainly be the first, and perhaps the only benefactors of rural electrification. Because rural electrification would have to be heavily subsidised in PNG it could well turn

into a subsidy of only the well off. Rural electrification is likely to be a much more effective instrument creating rural elites than would be the case with investment in schools, roads and other community facilities we have cited. Rural Electrification could quite easily, then, become a Government subsidised mechanism to contravene the Governments own development objective of improving and maintaining social equity.

"This is not to say that generation of electricity is invariably a suspect form of investment in development, but that a great deal of careful thinking must be applied to each request for electricity in rural areas to see whether the provision of electricity is indeed the best solution, both economically and ecologically, in relation to satisfying that particular demand, and to fulfilling the development goals of the country."

Source:

W. Bello and P. Hayes, "Nuclear Energy and Underdevelopment" 1977, *Pacific Studies* Pacific Studies Center, 867 West Dama St., Mountain View, California 94941. USA (60 cents plus postage); H. Datta, US Department of State (AID) memo, 22 June 1977, Washington DC, USA; J.M. Davis *et al*, "Rural Electrification: An Evaluation of Effects on Economic and Social Changes in Costa Rica and Colombia," Center for Tropical Agriculture and Center for Latin American Studies, University of Florida, Gainesville, Florida 32611, USA, report to AID no. AID/csd 3594, August 1973; Development Alternatives Inc., "An Evaluation of the Program Performance of the International Program Division of the National Rural Electric Cooperative Association (NRECA)", report to AID no. AID/otr-C-1383, 28 January 1977, 1823 Jefferson Place, N.W., Washington DC 20036, USA; Donovan, Hamester and Rattien Inc., "Review of Literature, Conferences and Programs Concerning Energy Assistance To Less Developed Countries," Report to Brookhaven National Laboratory, Appendix C, Volume III, 30 December 1977, 1055 Thomas Jefferson St, N.W., Suite 414, Washington DC 20007, USA; R. Lukens, Environmental Assessment of the Rural Electrification Project: Philippines," (mimeo) US AID, Interagency Committee for Ecological Studies (Manila). April 1978 (available from Susan Weintraub, AID Information Center, Department of State, Washington DC 20523 USA; F.C. Madigan *et al*, "An Evaluative Study of the Misamis Oriental Rural Electric Service Cooperative Inc (MORESCO)," report to AID by the Research Institute for Mindanao Culture, Xavier University, Philippines, March 1976; P.D. Almario, "Lumber Firm in Kolambugan Answers Criticism," *The Philippine Times* (Chicago), 23-29 December 1978, p. 6; National Electrification Administration (Philippines), "Nationwide Survey on the Socioeconomic Impact of Rural Electrification," Manila 1978, available from Martha Brady US AID, 1680 Roxas Boulevard, Manila, The Philippines; Policy and Planning Division, Department of Minerals and Energy, "Energy in Papua New Guineas Future,", 1978 from Ken Newcombe, P.O. Box 2352, Konedobu, Papua New Guinea; D.V. Smith, "Small Scale Energy Activities in India and Bangladesh", MIT Energy Laboratory Working Paper no. MM/D1018, 31 August 1977, MIT, Cambridge, Boston, Massachusetts, USA; J. Tendler, "Rural Infrastructure: Roads and Electrification", (mimeo), 1979 from Robert Berg, PPC/E, UA AID, Department of State, Washington DC, 20523, USA; A. Torres *et al*, "Social and Behavioural Impacts of a Technological Change in Colombian Villages," report to AID no. AID/csd/775, April 1968, copies from Lily Griner at cost of xerox ($8.70) at the American Institutes for Research, 1055 Thomas Jefferson St., NW, Washington DC 20007. USA.

—*Peter Hayes*

[1] *New Directions* is the Congressionally mandated reorientation of AID's programs towards a "participation" and equity ("poorest first") strategy. Conceived of in 1971, the Agency is still struggling to implement the mandate, especially in the controversial "rural infrastructure" field.

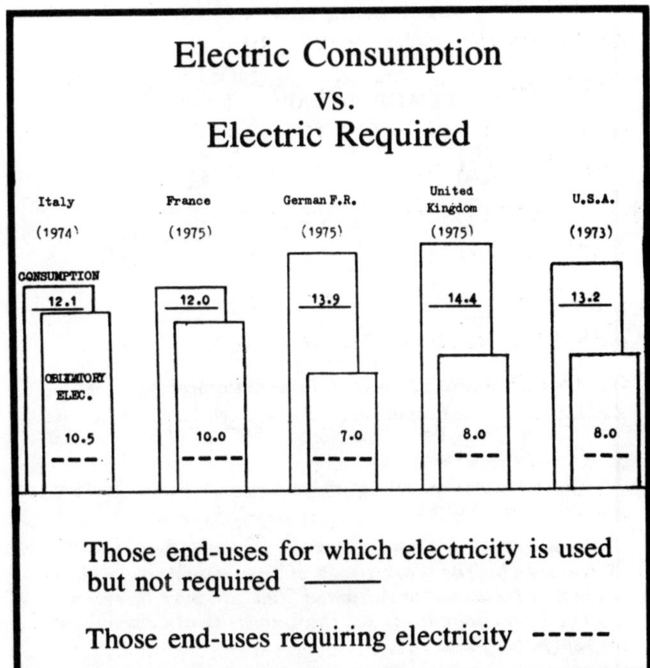

Energy Policy For Development

Agriculture Only One of Many Priorities

"Philippine Rice Harvest" photo by Charles Drucker

WHILE energy studies and policy in industrialized countries aim at maintaining or improving quality of life, the research focus in developing countries must be on basic needs. The use of energy "must increase considerably in the farms and villages of the developing world if rising populations are to be fed adequately, let alone better," begins a recent report

---

Energy Consumption per capita in Denmark for heating and cooking

YEAR	GCAL PER YEAR[1]	MOST IMPORTANT FUEL TYPE
about 1500	7-15[2]	firewood and peat
1800	7[3]	firewood and peat
1900	3	coal
1950	7	coal and oil
1975	17	oil

---

[1]Gcal is the energy content in the food which an adult consumes one year. It is eventually given off to the surroundings as heat.

[2]Estimate on the basis of information in Troels-Lund: Dagligt Liv i Norden (Daily Life in the North), Vol. 1 p. 19 ff, Gyldendal, Copenhagen, 1929.

[3]Chr. Olufsen. Danmarks Braendselsvaesen (Denmark's Fuel Supply), Copenhagen 1811.

From Energy in Denmark 1990: a case study, *Report no. 7— a survey by The Work Group of The International Federation of Institutes for Advanced Study c/o Sven Bjornholm, The Niels Bohr Institute, The University of Copenhagen, September* 1976.

---

by U.K. researcher Gerald Leach. "Rural energy deprivation of *all* kinds—for cooking, lighting, village industry, food processing and transport as much as for agricultural tasks—is an insistent and potentially explosive fact of the Third and Fourth Worlds."

In a report originally presented at the International Conference on "Agricultural Production: Research and Development for the 1980's," Gerald Leach surveys current research on the use of energy to increase farm yields. This study asserts that energy used in agriculture cannot be treated separately from energy used in other sectors. "We cannot focus only on the agricultural dimensions of energy use. The best solutions for agricultural energy problems (and the highest priorities for energy policy) usually lie outside the boundaries of 'energy and agriculture'." Other developing sectors, Leach points out, compete with agriculture for energy resources, and global fuel prospects place serious constraints upon the overall process of growth.

Energy Use in Developing Countries

Leach outlines five significant features of energy use in developing countries. First he notes the global inequity in commercial fuel use: "About 1.5 billion people live in countries where per capita consumption is less than 7 GJ or 250 kg coal equivalent per year and a further 1.1 billion lie between this and the 20 GJ or 750 kg of coal level. Per capita consumption in the USA is close to 11 tonnes of coal equivalent and in Western Europe 5 tonnes." Related to this inequity is the intra-national imbalance in fuel use. Urban-industrial centers of developing countries use commercial fuels as industrialized countries do—for transport, industry, building heating and cooling; rural areas use virtually no commercial fuel. In these areas the dominant fuels are "non-commercial": wood, charcoal, animal wastes and crop residues. Leach estimates that "although dependence varies greatly by country, about 2.4 billion people in Africa and the Far East depend on non-commercial fuels for more than half of all energy supplies."

"A second significant feature of developing countries," he explains, "is their abnormal dependence on oil." In developing countries 63% of total commercial energy needs come from oil, compared to 51% in industrialized countries. Moreover, growth in oil consumption is nearly 6% per annum in non-OPEC developing nations, while only 3% for North America, 2% in Western Europe, and 4% worldwide.

A third feature of the rural Third World is the almost exclusive and inefficient use of locally produced organic fuels—mostly firewood. In many countries, non-commercial fuel use exceeds 90%. Leach quotes a 1978 estimate for total rural fuel use in Kenya: "94.7% wood, almost exclusively for cooking; 3.1% kerosene for lighting; 1.7% indirect energy for fertilizers; 0.1% for tractor fuels; and 0.4% for rural industries and transport." Fuel used for cooking exceeds that in developed countries, not only as a percentage of the total, but also in actual per capita consumption. "Wood use in the Sahel ranged from 0.5 to 1.0 cubic metres per person per year," reports Leach, "or roughly 7 to 14 GJ—the kind of range often quoted for other countries and regions. This is much greater than the fuel use for cooking in developed countries (typically 1.5 to 2.0 GJ/person/year) because of the deplorably low efficiencies of most village cooking stoves." Citing the 1979 energy use survey for the 357-person village of Pura, India, Leach notes total fuel consumption of 10 GJ/person/year, or one-thirtieth of US levels. Leach contends that "because of the low efficiencies with which this energy is used, final or 'useful' energy consumption was probably no more than 1 percent of the US level." (See SEN VII, "Energy Use in Pura Village, India.")

The fourth significant feature of energy use in developing countries is the agricultural sector's relatively small share, even in comparatively developed rural areas. The Pura study itemizes energy use as 90.6% domestic activities, 4.4% industrial purposes, 2.8% in agriculture and 2.2% for lighting. Even in the major cereal growing region of India, agriculture's share is small: a study by India's National Council for Applied Research indicates that in the six Northern states, roughly 14% of total commercial and non-commercial fuels is for agricultural activities. Diesel oil, accounting for 4.3% of the total, is used in irrigation pumps, trucks and tractors, while energy for nitrogen fertilizers comes to approximately 10% of the total.

The final feature is the relatively rare use of commercial energy in rural areas, where it is usually more expensive than in wealthier and more developed regions. "Owing to high transport or transmission costs, fuel and electricity prices in most developing countries and rural areas are several times higher than in the USA or Europe," states Leach. "The costs of buying and maintaining equipment that runs on these fuels also prevents their use for most people."

Supply Options and Problems

The traditional alternative to commercial energy is fuelwood, but this resource has increased in scarcity and cost to

World Fuel Consumption
10^9 GJ, 1973

	Commercial Fuels	Fuel Wood	Agricultural Waste	Total	Fuelwood + Agricultural Waste as percent of total
Africa[1]	1.9	3.3	0.6	5.8	68%
Mid East	3.1	0.1	0.3	3.5	13%
Far East[2]	7.1	4.1	2.8	14.0	50%
Latin America	8.8	2.8	0.9	12.5	30%
West Europe	45.2	0.5	0.2	45.9	2%
Cpl Europe[3]	51.0	1.2	0.7	52.9	4%
North America	78.4	0.2	0	78.6	—
World[4]	227.1	14.3	4.8	246.2	8%

[1] Excluding South Africa
[2] Excluding Japan
[3] USSR and East Europe
[4] Including Japan, South Africa, Australia etc.

From J.K. Parikh, "Energy Use for Subsistence and Prospects for Development," Energy 3, pp. 613-637.

Note: converted from million tonnes coal equivalent.

the point where it now competes with food production for land. "To the extent that the problem is not solved," Leach points out, "in many countries substitute fuels for cooking may take first claim over additional fuels for agriculture." In the Sahel region, firewood consumption is 16 million m³ or 12 million tonnes per year—0.6 m³ per person. With this rate of consumption, Leach estimates that 150-300,000 hectare/year must be planted in forests to meet needs in the year 2000, assuming non-irrigated forest yields of 3-6 m³/hectare. While this is about 50 times greater than present forest planting efforts, the cost of oil is sufficiently high to warrant serious consideration of such massive planting schemes. Leach judges the present cost of replacing wood with oil at 22% of the region's G.D.P. He infers that "by the year 2000, when the population now dependent on fuelwood is expected to double, the costs of oil substitutes on any substantial scale are likely to be 'utterly unbearable'."

Meanwhile, the urban Third World continues to depend heavily on imported oil and Leach foresees no easy transition away from this petroleum-based commercial energy diet. Conservation in countries already developed, however, offers a vast energy resource for those now developing. "Since the developed countries use 83% of world [oil] supplies, one needs only moderate oil conservation policies in them to move the lifetime of oil resources into the 60 to 100 year bracket *and* allow a large rise in consumption by developing countries." He illustrates: "if in 2000 oil use in developed countries were 15% below present levels and in developing countries increases until then by 5.5% a year (giving a 3.1-fold increase by 2000), world consumption would increase by only 1% a year over 1976-2000. At the end of the century, remaining oil resources would be 190 billion tonnes, giving a further 60 year life at the 2000 consumption level of around 3.1 billion tonnes."

Another important option for developing countries, contends Leach, is to develop their own energy resources. According to 1979 UN estimates, only one-fifth of the oil in developing countries has been identified. Citing 1978 World Bank figures, Leach projects that non-OPEC developing countries will increase oil production from 3.7 million barrels a day in 1976 to 8.3 million barrels by 1985. Thirty or forty countries producing no petroleum now have the potential to do so quite economically, with average investment costs of $3 to $6/bbl. (compared to imported oil at above $24/bbl.). Leach notes that even these investment requirements are still very large—in some countries as much as one third of all new investments.

Although coal reserves are unevenly distributed, the World Bank forecasts output growth rates in Third World countries increasing from 1.5%/year in the early 1970's to about 5.6%/year by 1985 (apart from China's huge reserves). The largest gains are expected in Colombia, India, Mexico, Mozambique, and Viet Nam. Coal imports may be another important new energy source. Coal could replace oil for power generation, nitrogen fertilizer production and a wide range of petrochemical products. Coal-based smokeless fuels could be used for many industrial and domestic heating tasks, including cooking. Other potential coal uses involve liquefaction for portable fuels, acetylene gas production for lighting, and a powdered form for mobile and stationary machines.

Rural Energy Consumption in Northern India

	QUANTITY	10^{15}J	PERCENT
Coal and coke	0.862 Mega tonnes	21.0	2.2%
Electricity	2.889 Terawatt hrs.	10.5 (a)	1.0
Kerosene	535 M litres	19.3	2.0
Diesel oil	1105 M litres	41.1	4.3
Petrol	21 M litres	.8	0.1
COMMERCIAL FUELS TOTAL		92.7	9.6
Firewood	18.799 Mt	370.4	38.8
Dung cake	24.486 Mt	251.0	26.3
Vegetable waste	15.876 Mt	233.0	24.4
Charcoal	0.265 Mt	7.5	0.8
NON-COMMERCIAL FUELS TOTAL		861.9	90.3
TOTAL FUELS		954.6	99.9
Additional Sources:			
Nitrogen (fertilizer)	1.263 Mt	13.8	
Manpower	8.533 M days	66.6 (b)	
Animal draft power	3.060 M days	235.9 (b)	

(a) Excluding energy losses in generation: i.e., at 1 kWh=3.6 MJ instead of approximately 11 MJ.
(b) Food input basis using rather high rates of 1863 kcal/work day and 18400 kcal/animal day.

Natural gas production is not a likely candidate for domestic consumption due to the lack of pipeline distribution networks and the high costs of other transport modes. Leach does mention its potential as a nitrogen fertilizer feedstock, a low cost/high value export for OPEC producers.

The nuclear option is an even less likely prospect for developing countries. "The enormous investment costs, long construction times and the impossibility of absorbing single packages of about 600 MW capacity into small grid systems are the main technical reasons," explains Leach. "Only 10 developing countries have grid systems large enough to take a conventional nuclear system." Equally important are the political problems of relying on "high technology" and outside expertise. "These worries are tuned to the increasing technological 'internalization' of developing countries, especially those where meeting basic human needs by selective, appropriate technologies is gaining a high priority."

Without negating rural electrification's potential importance, Leach notes that opinions differ on how this might be achieved: "creeping electrification" from the center, or "scattered electrification" based on small-scale, localized sources. "Whether it should come from a 'central' grid, from local fossil-fueled generators, or from renewable sources such as mini-hydro sets, wind generators or solar devices is a central and extremely complex issue of rural energy policies," he explains. While electricity could improve food production and enhance village welfare and incomes, the high costs of extending electricity grids to rural areas is a major disadvantage. According to Leach, "costs are exorbitant so that electricity prices are high, producing low demand." Citing a 1979 paper by T.S. Tushak, Leach notes that "in Kenya, electricity costs range from 28 Kenyan cents/kWh in large urban areas to 94 cents in rural areas already connected to the grid and 180 cents for connecting isolated villages. In one case, connecting a village of 43 consumers gave a cost of 600 cents/kWh." In India and Egypt it is current policy not to connect villages of less than 1,000 people unless they are very close to transmission lines. But hooking villages up to the grid is not always prohibitively expensive, "except perhaps for the most isolated communities."

Conservation can be important in urban-industrial sectors even as total energy use increases. Leach suggests "there is much to learn from the evolving energy conscious technologies of the North as well as a chance of avoiding some of the developed countries' mistakes." Cookstove efficiency improvements are "the most important single priority" in rural areas—two and three-fold gains are possible through simple redesign. (See SEN VIII, "Fuel Efficient Stoves for Rural Households.")

Specific Energy Inputs to Agriculture

Leach examines the direct inputs to farming—fertilizers, chemical sprays, irrigation and mechanization—and concludes that high energy cost does not impose a major constraint upon food production. Central to his conclusion is the assumption that energy inputs can be self-limiting. Leach uses US corn production from 1945 to 1975 as an example of this process. Primary energy input increased from 19 to 30 GJ/hectare/year, with energy for fertilizers up from near zero to 11 GJ. In the same period, yields soared from 2 to 5 tonnes/hectare. Thus five tonnes of corn are today (1975) produced with 17.5 fewer gigajoules and 60 percent less land than in 1945. Nationwide, 20 million hectares were released. It is also worth noting that labor input decreased five-fold, but the figures do not specify how much of this may be attributed to mechanization and how much to increased yields due to fertilizer use.

For the developing countries, Leach finds "there is overwhelming evidence that higher chemical inputs are the driving force of higher outputs." Nitrogen fertilizers are the most significant of these chemical inputs as they are extremely energy-intensive, requiring roughly 1.8 tonnes of oil equivalent for every tonne of nitrogen. But Leach points out that energy prices are only a secondary concern in determining the prices of fertilizers. High capital costs of new fertilizer plants, infrastructure, financing costs, and conditions in labor and world markets play the determining roles. Indeed, he asserts, "the effect of higher feedstock prices has mostly been swamped by these non-energy factors." Alternative nitrogen sources, such as coal, biogas and small electric arc units, may help to lower fertilizer costs in the future.

Feedstock prices similarly do not constrain production of synthetic herbicides, insecticides and fungicides, even though these agrichemicals, like fertilizers, are energy-intensive. Leach stresses, though, that synthetic pest and weed controls have hidden costs. They represent a drain on energy and transport, they cause a loss of foreign exchange, and the ecological impact of their use is considerable.

Irrigation, too, is extremely energy intensive, but the energy inputs involved are less significant cost components than construction and maintenance.

Leach concludes his study with a review of alternative energy sources: fuel cells, solar cells, solar engines, wind power, solar drying and refrigeration, and biomass technologies. He generally finds great potential in these alternatives, but also indicates the need for further research and development to reduce costs and to match systems to local conditions, practices and tastes. Leach's major concern, though, is that alternative technologies are most likely to be developed and marketed by industrialized nations, perpetuating the technological dependence that presently afflicts developing countries.

—*Elyse Axell*
Cissy Wallace

References:

Leach, Gerald
    1979 "Report of the Energy Resources Working Group," International Conference on 'Agricultural Production: Research and Development Strategies' for the 1980s." Bonn, West Germany, 8-12 October, 1979. 86 pp., double spaced xerox available from IPSEP for $8.

Tanzania National Scientific Research Council
    1978 *Workshop on Solar Energy for the Villages of Tanzania.* Dar-es-Salam.

Tuschak, T.S.
    1979 "Energy Policy Planning in Developing Countries—Problems and Challengers." Paper delivered to forum on Third World Energy Strategies and the Role of Industrialised Countries, Royal Institution, London, UK.

Soft Path to Dependency?

Decentralized energy systems that use local and renewable resources do not necessarily break Third World countries from their dependent past. Self-reliance in energy, as in other technologies, requires an investment in indigenous capabilities and expertise.

L ARGE, CENTRALIZED ENERGY SYSTEMS, based on fossil fuels or nuclear power, draw developing economies into long-term relationships with external suppliers of equipment, fuel, and know-how. Renewable energy technologies, however, are both simpler and smaller in scale, putting them within the reach of Third World production and distribution capabilities. With the right financing and institutional support, they could rapidly provide a major portion of a developing country's energy needs. Further, they could contribute to local self-reliance by making at least this segment of the economy wholly independent of foreign assistance and influence.

But decades—even centuries—of dependency relations between Western countries and their one-time colonial possessions are not so easily put aside. As H.K. Hoffman, of the University of Sussex Science Policy Research Unit, and others have noted, the strength of the industrialized economies is so overwhelming that it perpetuates the historically unequal relationships between advanced and developing nations, even at the level of village energy technologies (Hoffman 1979; Makhijani 1978; Rao 1980).

Dependency means a reciprocal but *unbalanced* connection between regions or nations. Since colonial times, the developed countries have viewed the Third World as an investment and market opportunity, and present-day multinational corporations consider developing states in much the same light. The Third World, for its part, has traditionally looked to the First—and now the multinationals—as a source of capital and technical innovation. The dependency relation works both ways, but the primary advantage rests with the stronger, more advanced economy, which can dictate the terms of trade.

Industrial and economic development, Hoffman points out, fail to liberate Third World countries; worse, they draw the bonds of dependency even tighter. He cites as examples the urban transportation and energy systems which developing nations have acquired from overseas

suppliers. The conventional fossil fuel technologies they are based on actually promote a double dependency: upon the oil producers and refiners for energy supplies, and upon the large Western corporations for equipment, spare parts, and technical support. The financial burden imposed is enormous. Not only do oil imports create a huge trade deficit, but between 15 and 20% of *all* capital investment in the Third World is dedicated to increasing electricity supply (Makhijani 1978). This level of financial commitment is beneficial to the multinational manufacturers of power generating equipment, who now send one-third of their exports to developing countries, and, according to Hoffman, hope to expand this market as their sales to the industrialized nations decline.

Rural energy systems, many believe, could begin to undo the dependency knot. The urban and commercial sectors of developing countries are heavily committed to conventional energy systems, but 88% of all the rural areas in the Third World have not been electrified. Hoffman predicts that much of the future increase in energy demand will come from the countryside, but it is unlikely to be met in the conventional, expensive way. Renewable resource technologies are much more suitable to the energy needs of the rural Third World. They are scattered and decentralized, much like the rural population; they are matched in scale to an agricultural community's requirements, such as irrigation pumping, or the threshing and hulling of grains; capital costs do not exceed the scope of peasant incomes; and the devices can generally be serviced and repaired by those who use them, with only limited outside technical help.

E VEN SO, Amulya K.N. Reddy, the noted Indian electrochemist who has shifted his work to pioneering rural technologies, points out that one cannot develop village scale technology from a research laboratory in Cambridge, Massachusetts. Locals must be involved at every step. "This is considered the obvious thing to do for the urban architect," Reddy explains, but the needs and preferences of the rural poor are rarely taken into account. "Then they don't like what we have done and we say they are stupid."

Inspired by the vision of sturdy self-reliance, research programs investigating renewable energy sources and technologies were initiated in many Third World countries. Unfortunately, these efforts have been poorly co-ordinated, and less effective than hoped in reducing energy dependency. Hoffman suggests that there has been much duplication of effort, and that the energy systems produced have not always been appropriate to the needs of rural villages. The limited research funds available to developing nations have been used inefficiently.

At the same time, the industrialized countries have mounted an effort to develop alternative sources of energy that is massive and highly productive in comparison (however inadequate it may seem to domestic critics). The Western R & D juggernaut now threatens to overrun the smaller, less well funded programs in the Third World, and if it succeeds in establishing an initial market lead, the prospect of reduced dependency through alternative energy technology may vanish.

Ultimately the renewable energy technologies will assume a major supply role in developed countries, but for the present they are not generally considered to be cost-competitive. The suppliers, realizing that large initial purchases can lower production costs, are looking for their first major markets overseas, often, Hoffman claims, with the assistance of governmental and international agencies. Rural areas in developing nations appear attractive, since the cost of providing energy via conventional systems is so high there. Successful market penetration, though, would forestall the development of indigenous energy expertise.

E VEN WHERE LOCAL EFFORTS at research and development have met with some success, emerging industries have been stifled by government policy. In India, the Lok Dal government (which acceded to power in mid-1979 after the collapse of the Janata administration) was convinced that "the country would be better served by imported technology rather than by indigenous expertise." (Rao 1980). Solar irrigation pumps were purchased from a US multinational supplier instead of from a government-owned concern which had "convincingly demonstrated its ability to develop the type of solar pumps that the... government wanted to import. Against such a disturbing backdrop, apprehension grew that foreign exploitation in the strategic field of energy might also eventually stifle indigenous technology in other areas." (ibid)

It is still possible for soft energy technologies to contribute to local self-reliance and reverse the course of increasing Third World energy dependency. Hoffman identifies three components to a technology policy for meeting this objective. "Firstly, and most obviously, is the need to generate the relevant information relating to energy use in rural areas and the economic and social viability of various technical options... Secondly, there must be some effort to control the activities of Western firms and of the aid agencies promoting the interests of these firms..." In support of this view, Reddy asserts, "The cultural hangover of colonialism is such that if there is an institutional collaboration, any credit for achievement will always go to the Western institution." This undermines local confidence, and Reddy insists that "the growth of confidence that you can tackle your own problems is the crux of development."

The third component in Hoffman's strategy for energy self-reliance is educational and social planning to "generate the technological capabilities needed to meet the needs of a long-term rural energy policy." This strategic response involves some costly, short-run trade-offs. "Investing in the creation of local technological capabilities rather than importing foreign energy systems... may mean that fewer rural people will have access to adequate energy supplies in the short-term." But, however high the price of energy self-reliance, the costs of continued dependency are certain to be even greater.

—Charles Drucker

References:

Hoffman, H.K.
1979 "Alternative Energy Technologies and Third World Rural Energy Needs: A Case of Emerging Technological Dependence." To appear in the July, 1980 issue of *Development and Change*.

Makhijani, Arjun
1978 "Economics and Sociology of Alternate Energy Sources." United Nations Environment Programme seminar paper FP/0404-78-04 (1092).

Rao, Radhakrishna
1980 "When alternatives are inappropriate." *New Scientist*, 3 April 1980, pp. 28-30.

Reddy, Amulya K.N.
1980 Quoted in "Pioneering Rural Technology in India," by Constance Holden, in *Science*, Vol. 207, p. 159, 11 January 1980.

photo by Charles Drucker

2. Technologies for Third World Energy Needs

If the pitfalls of a technocratic approach to Third World energy problems are great, so is the usefulness of soft energy for those countries. Cooking and lighting are the principal household needs, and both are now done inefficiently. A careful application of fluid combustion theory to Third World cooking stoves could do wonders, but, as K. Krishna Prasad points out, a technology consists of more than just hardware. For example, introducing a new stove may require that a group of local craftsmen exist to disseminate it within the local culture. "Inefficient" open fires can also be valued in local religion, or serve as insect repellents. Biogas units can be very effective in preserving forests and supplying fertilizer byproducts, but there are similar hitches to be overcome.

Less problematic are wholly new appropriate technologies. Microhydro electricity generators, unlike large hydro projects, are within the reach of local production capabilities; New Guinea is our case in point. Local communities are liable to see more direct benefits from such small projects than just a high voltage transmission line overhead. Similarly, photovoltaic pumping would be very attractive for irrigation; why not make the same sunshine that evaporates the vital fluid replenish it?

One aspect that can only help the spread of renewable technologies is that they have been used in simpler form for centuries in many Third World cultures. Passive solar architecture is not an invention of twentieth century New Mexico, while wind-cooled buildings may sound like a far-out concept in The Whole Earth Catalog, but are old hat in Iran and Pakistan.

Household Energy Use In Kenya

*I*N NAIROBI, *the capital city of Kenya, all-electric homes, complete with the full array of modern appliances, stand within view of tiny, ramshackle structures whose inhabitants burn charcoal in stoves called* jikos. *Not far from the city, rural people continue to gather firewood for their energy needs, as their ancestors have done for generations. As in many developing countries, household energy use in Kenya includes extremes of modernity and tradition.*

The energy used in Kenya's more than two million households accounted for but a small share of the 85 x 10^{15} Joules (85 Petajoules, or PJ) of commercial energy consumed nationally in 1979. Of this quantity, over 90 percent derives from imported oil, with less than four PJ of petroleum products reaching the residential sector directly. Homes use an additional 0.8 PJ of electricity—20 percent of the country-wide generation.

When all forms of energy are considered, however, the picture changes dramatically. The vast majority of Kenya's population still uses traditional energy sources—wood and charcoal—to meet household energy needs. Despite increasing urbanization, most Kenyans still live in rural areas where commercial fuels have made few inroads, so that at least two-thirds of all the energy consumed in Kenya, in all forms, lies within the residential sector.

The pre-eminence of wood-based energy stands out as a principal finding of the Rural/Urban Household Energy Consumption Survey, conducted over 1978-79 by the Kenyan Central Bureau of Statistics. All in all, wood supplies over 95 percent of residential primary energy demand. Kenya's resource distribution and population growth rate raise questions whether the resource base will be adequate to meet anticipated demand increases. Now growing at the world's highest rate of more than four percent per year, Kenya's population could soar to two-and-a-half times today's by the end of the century. With commercial fuels increasing in cost, wood fuel demand could escalate to an unsustainable level—which some analysts claim has already been reached —provoking fears of environmental deterioration. But damage may be avoided through the careful husbanding of indigenous resources, and an increased efficiency in their use.

Electricity

In Kenya, the grid reaches a small percentage of all households. Even in Nairobi, only six percent of the households are electrified, according to East African Power & Light (EAPL), the national utility. Connections there increased rapidly, about seven percent per year, from 1972-78, at which time they accounted for 60 percent of total residential sales. Rural electrification proceeds at a much slower pace.

The all-electric homes of the urban upper class claim a large share of total consumption. While lower class households may use 120 kWh/year for lighting and little else, executive houses frequently consume 100 times as much electricity or more. The Survey revealed that 70 percent of the electricity used in the households it polled (upper-income households not included) was used for lighting, the rest for cooking. When all electric customers are considered, the pattern changes; EAPL data show that 20 percent of total residential consumption goes to heat water, an end use that scarcely showed up in the Survey results.

Forces of change

Population changes, rising incomes, and declining fuel-wood availability interact to influence future patterns of household energy consumption in Kenya. While rapid population growth drives up the absolute demand for household energy services, new settlement patterns are shaping the nature of this demand. Continuing urbanization in both large cities and provincial towns brings with it large increases in the demand for charcoal. More transport-the growing residential energy sector suggests that attention must also be paid to reducing the demand for fuelwood. This is possible through some combination of fuel switching and increased efficiency of consumption.

As in the industrialized countries, fuel switching depends on the relative prices of competing fuels, and consumers' ability to afford the capital and fuel costs of the switch. Kenyan household energy use today thus varies markedly with income level. The lowest-income households rely exclu-

---

Fuelwood

Between ten and twelve million tonnes of wood (145-175 PJ) are consumed by Kenyan households each year, much of this in the form of charcoal. (An additional one or two million tonnes go to industry and the rural services sector.) The Rural/Urban Survey found an average annual consumption for a wood-using household of almost five tonnes (some 70 GJ), or 650-750 kg per person. A recent study in the rural Machakos District arrived at a figure even 30 percent higher. In rural areas, fuelwood is usually collected free, and consumption depends largely on availability. For the nation as a whole, over 80 percent of fuelwood is used for cooking, with most of the rest providing heat. Distinguishing between cooking and heating uses of fuelwood is sometimes difficult, since the survey was made during late fall and winter, when people would sit around the fire as food was prepared.

Fossil fuels: kerosene and gas

Total residential consumption of kerosene and LPG (liquefied petroleum gas) can be gauged from oil company records, but the number and type of users is uncertain. In rural areas, villages use kerosene extensively, often purchasing a month's supply at a time at local trading centers. The urban poor use kerosene, which they refer to as paraffin, principally for lighting. But in Nairobi, 75 percent is used for cooking, suggesting consumption by lower middle-class people since kerosene cooking is less common among the very poor. As income rises, kerosene's use tends to diminish. LPG is used mostly by middle-income households for cooking and domestic hot water, but its overall role in the residential energy picture is relatively small.

---

able than wood, charcoal can be brought in from regions where wood is more plentiful. Charcoal distribution stations in Nairobi—as common a sight in the capital as in the other cities—are linked to production centers up to 250 kilometers distant, which supply urban energy needs at the same time as they generate local income and employment. The shift to charcoal, as rural people migrate to cities, portends a tremendous increase in the demand for wood, since households using the processed fuel indirectly consume twice as much of the primary resource as those who simply cook with wood. The base for this resource, however, is not great.

Only three percent of Kenya's land area is forested, with one-fifth of that classified as open woodland. Demand for wood already exceeded annual yield in 1979, according to one estimate, and even highly optimistic forecasts anticipate this occurring by the mid-1990s. Changing patterns of land ownership exacerbate the problem; the formerly collective family or village woodlots have largely moved to individual tenure, and many people who once collected fuelwood freely must now pay for their principal household energy source. In some areas, the shortage is already being felt. Nearly one-third of the households surveyed in the rural Machakos districts said that fuelwood had become scarce, and another 48 percent called it very scarce. In many rural villages, people are burning grass, sisal leaves, and cow dung, because fuelwood is either unavailable or unaffordable.

Household fuel switching

Although government forestation efforts are underway, and village woodlots are being encouraged, the sheer size of sively on charcoal and kerosene; electricity for lighting and gas cooking appear among those slightly better off; gas is the dominant cooking fuel of the middle-income group, while electricity is used for other appliances; in upper-income homes, the switch from primary fuels to electricity is virtually complete.

Despite Kenyans' increasing use of commercial forms of energy, and the higher efficiency of their use, economics work against this substitute for fuelwood. If half of the households now using fuelwood (at ten percent efficiency) were to change to kerosene stoves (at 40 percent efficiency), the annual residential kerosene demand would quintuple to almost 500 million liters. Moreover, the average household, now using four or five GJ of fuelwood for cooking each month would have to pay $90 per year for the equivalent services from kerosene, even at today's subsidized prices. Such a cost is prohibitive, since half of the households in Kenya have annual incomes of less than $400 (1974).

A more likely fuel switch from fuelwood is to charcoal, which will further stress forest resources if present inefficiencies in production and consumption are not overcome. Unlike fuelwood gathering, charcoal manufacture involves the felling of whole trees, with a considerable woody mass cut down but not used. Production methods need not be held to the present ratio of useful wood to charcoal. A device called the Cusab kiln can make charcoal from the abundant shrubs and bushes of the Kenyan rangelands. An improved version of the Cusab, designed by a Nairobi firm, reportedly produces a tonne of charcoal from only three tonnes of woody material. Though the Cusab kiln's cost—$500 to 800—places it beyond the reach of most rural farmers, its

Charcoal

Charcoal is the favored fuel of low-income urban dwellers, and one of the consequences of Kenya's high rural-to-urban migration is an increase in its use. This fuel is, in a sense, the poor person's electricity: more convenient and efficient to use than less-processed alternatives, but highly consumptive of resources not seen by the end user. Most kilns require nine tonnes of wood for each tonne of charcoal they produce: although charcoal has twice the energy content by weight of wood, and is used perhaps twice as efficiently (20 percent *vs.* ten), a household using charcoal still claims more than two times as much primary wood resource as one using fuelwood.

The Rural/Urban Survey pegs total annual residential charcoal consumption at 540,000 tonnes (16 PJ), requiring 2.5-5.0 million tonnes of wood. The upper end of this range (probably more realistic) represents 40 percent of total estimated fuelwood consumption. The national average in charcoal-consuming households was 660 kg, about 19 GJ. This is well below the annual energy use of fuelwood-consumers, reflecting the greater efficiency of charcoal use, and the fact that it is applied to fewer tasks.

---

high yields make it attractive for cooperative endeavors or small companies.

More efficient end-use devices could also enhance charcoal as a household energy source. Substituting clay stoves for traditional metal ones reduces charcoal consumption by 40 percent, and the cost of the more efficient device—about one dollar—is the same as for the wasteful one. Improved *jiko* stoves can, reportedly, achieve similar efficiencies. Introducing such stoves into Kenya would reduce charcoal use per household, but would likely increase the number of users. Unless more efficient production techniques are encouraged simultaneously, the fuelwood situation could deteriorate even further.

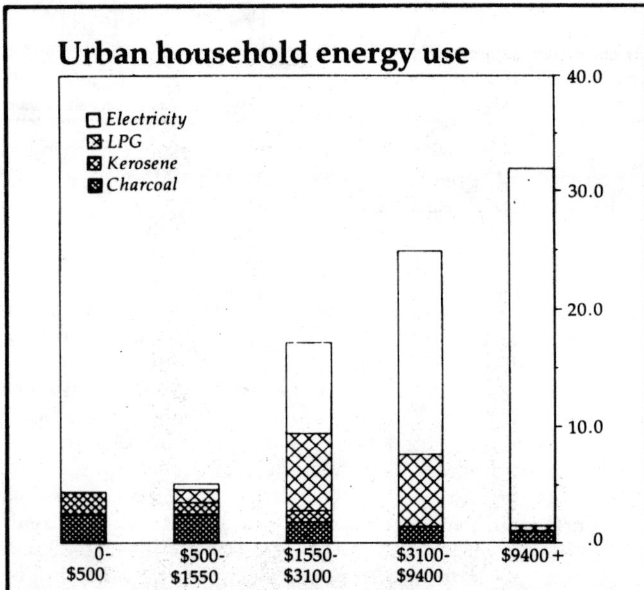

Basic urban household energy demand by income group (annual), Nairobi, 1978. Basic demand refers to "useful" energy delivered to the task. Incomes are from 1970.

A dual strategy, aimed at improving both production and use of charcoal, has several advantages over continued (or increased) subsidies for kerosene. If the combined efficiency of charcoal kilns and stoves could be increased by a factor of four (as demonstrated here), the energy demands of a rural household could be met with half as much primary fuelwood as is used at present. The fuel cost of switching to charcoal would be about $50 per household per year, which compares favorably with the present cost of kerosene. And the clay stove is much less expensive than one that uses kerosene.

If half the households now using fuelwood were to switch to charcoal, some 660 million kg would be needed. Twenty-two hundred Cusab kilns, each producing 800 kg per day would have to be constructed, at a capital cost of one to two million dollars. In contrast, the increased refinery capacity to make an equivalent switch to kerosene would cost about $40 million, based on World Bank estimates. Instead of importing expensive oil to refine subsidized kerosene, Kenya would do well to consider the efficient use of indigenous, renewable resources.

–Steve Meyers

Steve Meyers is with the Energy Analysis Program at Lawrence Berkeley Laboratory, California.

References:

Kokwaro, J.O.
1979 "Indigenous and Introduced Common Firewood and Charcoal Plants of Kenya." University of Nairobi; presented to the International Workshop on Energy and Environment in East Africa, May 1979.

McGranahan, C., S. Chubb, R. Nathans, and O. Mbeche
1979 "Patterns of Urban Household Energy Use in Developing Countries: The Case of Nairobi," draft. State University of New York at Stony Brook and the University of Nairobi.

Mott, F.L., and S.H. Mott
1980 "Kenya's Record Population Growth: A Dilemma of Development." *Population Bulletin*, vol. 35, no. 3, October.

Mung'ala, P.M.
1978 "Estimation of Present Consumption and Future Demand for Wood Fuel in Machakos District of Kenya," thesis. University of Dar es Salaam, Morogoro, Tanzania.

Openshaw, K.
1979 "A Comparison of Metal and Clay Charcoal Cooking Stoves." University of Dar es Salaam, Morogoro, Tanzania.

Schipper, L., J. Hollander, M. Milukas, J. Alcamo, and S. Noll
1981 *Energy Conservation in Kenya: Progress, Potentials, Problems*. Berkeley, CA: Lawrence Berkeley Laboratory.

Western, D. and J. Ssemakula
1979 "The Present and Future Patterns of Consumption and Production of Wood Energy in Kenya," draft. Presented at the International Workshop on Energy and Development, Nairobi, May 1979.

The Combustion Connection

*I*N TESTS *conducted by Eindhoven University, the Nomad stove achieved an efficiency rating of 40 percent. Specifically designed to boil water rapidly using little fuel, this stove is used by nomadic communities such as those living on the outskirts of the Sahara region. Recent studies of traditional fuels in less developed countries show the Nomad stove to be 20 percent more fuel efficient than most other wood stoves, and 35 percent more than open hearths.*

In many parts of the world, as much as 90 percent of all the wood cut is used solely for cooking and heating. The Food and Agriculture Organization estimates that the annual wood fuel consumption in developing countries is about 1.3 billion m³. Firewood, still the primary energy source for roughly 1.5 billion people in developing countries, is rapidly being depleted by a growing population. Half a billion people lacking wood meet their needs by burning animal dung and crop residues which are badly needed to replenish the impoverished soil.

The cooking efficiency of most Third World stoves is less than 10 percent. Little more than pots balanced on rocks, covered buckets, or scooped-out clay forms, these stoves are the culprits in deforestation and erosion. However, stoves with high heat transfer efficiencies, coupled with reforestation efforts could help alleviate these problems by reducing family fuel consumption while multiplying fuel supplies.

Combustion dynamics

To realize the full thermal value of firewood, its combustible gases must be burned completely. Incomplete combustion is the greatest energy offender in a wood-burning stove. Yet simple changes in design and cooking procedures can enhance a stove's performance.

For optimal efficiency, the stove should have a small base and wide top (like an inverted cone) which provides ample combustion space. The combustion chamber, or firebox, can vary in size depending on the use of the stove, but should be large enough to contain one or two cooking pots. To derive the maximum benefit from the heat generated, the pots should be lowered directly into the firebox, thereby presenting the greatest possible heating surface. To further efficiency, heat can be kept close to the cooking pots by using baffles which obstruct the flow of gases.

Ideally, combustion should take place within a closed firebox using multiple air intakes to regulate oxygen flow. Pre-heating the air supply to the firebox is important, as

cool air can impede proper combustion. The primary air flow, entering via an opening in the ash box door below the grate, is heated as it passes through hot embers. Secondary air, which enters the stove through an opening in the firebox door, can be preheated by passing first through a small metal box riveted to the door. Dampers situated at the entrances to the stove and chimney can also help regulate air input and gas evacuation. A disadvantage of this system however, is that ambient air is admitted directly into the firebox, cooling the gases.

The function of the chimney is twofold: it allows the escape of gases, and promotes a draught which facilitates combustion. For a chimney to work well, it must be perpendicular to the stove, and rise at least half a meter above the roof. The optimal diameter will depend on the chimney's height and the firebox's thermal capacity. The chimney should be topped with a cowl to keep rain out and increase the draught regardless of wind direction.

Alternative fuels

In areas where wood shortages are most acute, and kerosene, gas, and electricity are not options, cookstoves are fired with alternative fuels—rural waste, bark, straw,

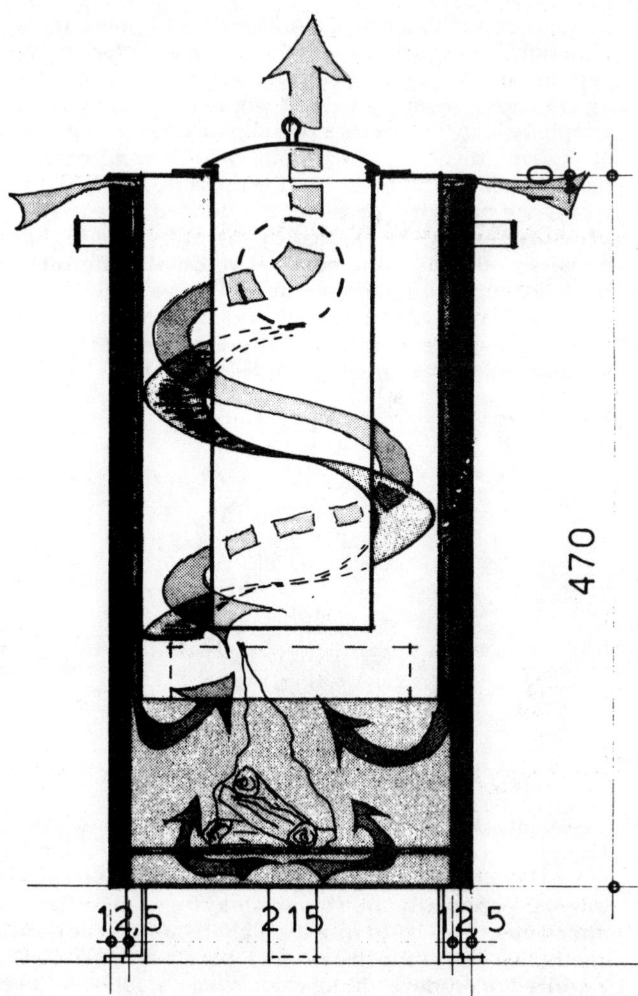

Cross-section of Nomad stove, from "Modern Stoves for All" by Waclaw Micuta.

dry leaves, and municipal garbage. These normally fast-burning materials can be used for cooking purposes if they are processed into bundles or faggots. Once the dry waste has been pressed, the air supply is diminished and combustion is slowed. Bundles of sticks, hay, leaves, etc. can be simply tied together, but their heat value increases if they are pressed in devices such as those illustrated below. Efficiency can be further enhanced if each bundle or faggot contains a piece of wood at its center.

Bundle pressed with rope and stick, from "Modern Stoves for All" by W. Micuta.

Another method which works well is to press dry waste material into compact briquettes. Several different presses capable of exerting a pressure of approximately 1,000 kg/cm² have recently been designed in Switzerland. A simple hand press exerts a pressure of over 500 kg or about 10 kg/cm² on a pressure plate which compresses the briquette. To further increase the pressure, the lever which works the pressure plate can be extended. The higher the pressure that can be exerted by the press, the higher the density and heat value per unit volume of the resulting briquettes. Designed by the Bellerive Foundation of Geneva, a non-governmental organization interested in problems of development, these hand presses were satisfactorily tested in Nairobi, February 1981.

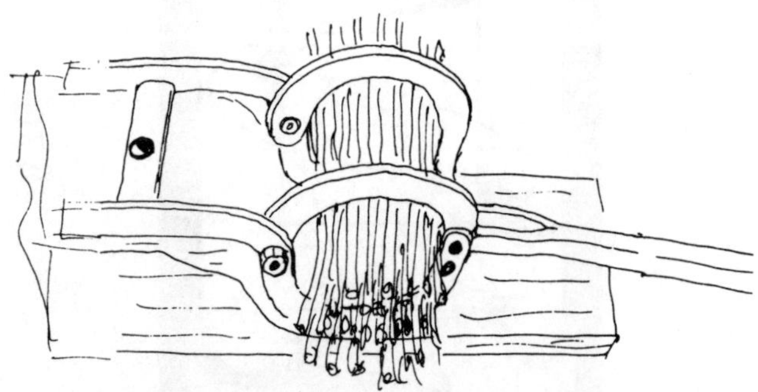

Small metal press, from "Modern Stoves for All" by W. Micuta.

Waclaw Micuta, energy consultant to the Bellerive Foundation and author of *Modern Cookstoves for All* makes it clear that charcoal as fuel for cooking is a wasteful use of energy especially in regions having acute firewood shortages. The kilns in which charcoal is produced are inefficient and as much as eight kg of air-dried firewood is required to produce one kg of charcoal. "It follows," claims Micuta, "that it is possible to extract four to five times more heat from any given volume of firewood if, instead of being transformed into charcoal, it is burned directly as fuel." (If the heat value of charcoal is 30,000 kJ/kg, one kg of wood yields about 3,750 kJ when converted to charcoal. Since the heat value of air-dried wood is 16,000 kJ/kg, the efficiency of charcoal production is thus only about 23 percent.

The Polish stove

Eleven stoves of varying sizes and materials are reviewed by Micuta. Among them is the Polish stove which runs well on wood, wood-waste, and other combustible materials.

Cylindrical in shape, the Polish stove has double walls, the inner structure consisting of an open-cone firebox. Holes are drilled to provide for the entry of air, passing between the body and the inner structure before entering the firebox as pre-heated air supply.

In April 1981, tests conducted on a windy day in St. Lucia, West Indies demonstrated the stove's fuel efficiency (according to the formula of the University of Eindhoven) to average 50 percent. Generally, performance characteristics can be measured by the amount of fuel required to bring a measured amount of water to a boil and maintain the boiling for a specified time. In this case, .34 kg of mixed wood waste brought 4.0 kg of water to a boil in eleven minutes, and continued to boil and simmer for an additional 31 minutes with no addition of fuel. In comparison, a similar volume of water in the same pot took 28 minutes to boil on a modern gas stove.

Despite considerable efforts to improve the cookstove, open fires continue to be the most common method of cooking in many developing countries. This has been attributed to limited local materials and the unsatisfactory performance of many stoves. If modern cooking methods are to be adopted in developing countries, they must give

Ethiopian woman tends inefficient, open-hearth cookstove.

satisfactory results and they must be within the means of the poor rural communities for which they are intended. Moreover, they must suit the fuel-using and cooking styles of the local culture, and this is often the greatest challenge to the stove's designer.

—Katy Slichter

References:

Eindhoven University

1981 "Some studies on open fires, shielded fires and heavy stoves," a Report from the Woodburning Stove Group, Departments of Applied Physics and Mechanical Engineering, Eindhoven University of Technology and Division of Technology for Society, TNO, Apeldoorn, The Netherlands. A highly technical explanation of combustion dynamics, along with efficiency ratings of open fires and stoves, with suggestions for improving the fuel economy of wood burning stoves.

Knowland, Bill and Carol Ulinski

1979 "Traditional Fuels: Present Data, Past Experience, and Possible Strategies." Prepared for the Agency of International Development.

Micuta, Waclaw

1981 *Modern Stoves for All.* Available from the Bellerive Foundation, Case Postale 6, 1211 Geneva 3, Switzerland. Price: US$ 10 Europe; US$ 12 elsewhere.

Sodha, Dr. M. S. and Dr. Rajendra Prasad

1981 "Efficient Chulhas: Household wood cookstoves," *Science for Villages*, January 1981, no. 39, pp. 9-12.

VITA News

1981 "The Search for Better Stoves" *VITA NEWS*, October 1981, p. 4. Subscription rate: $15 per year. Address: 3706 Rhode Island Ave., Mt. Rainier, MD 20712.

Kenya's Lesson In Conservation

*H*OUSEHOLD ENERGY CONSUMPTION *in Kenya is based on wood, and no other source is likely, over the next few decades, to supplant it. But if consumption continues at present rates, the country will suffer a shortfall in the year 2000 of 15.4 million tonnes, due primarily to increased demand for charcoal as urbanization accelerates.*

Several strategies for conservation and supply enhancement have been advocated, but an eighteen-month study by the Beijer Institute Fuelwood Project on Kenyan energy resources suggests that these solutions will exacerbate, rather than ease, the energy problem.

Two conservation strategies are frequently proposed as solutions to reducing fuelwood consumption: improved stove design and improved pyrolitic conversion processes for charcoal manufacture (see *Soft Energy Notes* 4:4:114-115, "Household energy use in Kenya," and article in the present issue). Both strategies have limited impact because both disregard the social setting into which they are introduced—the peasant mode of production and consumption.

Stove design

Naive rules of thumb abound in the literature on improved cooking stoves, which concentrates on the combustion-related technical parameters of stove design. But, as T. Harris notes in a recent *Ambio* article, "...it is important to remember that, while fuel efficiency may be the principal interest of the designer, it can only be achieved if the more diverse values of the user are first satisfied" (Harris, 1981).

In peasant production and consumption systems, the traditional open fire (which can be as efficient as some stoves, when carefully managed), has several fuel sources and multiple end uses. Many stove designs are fuel-specific, thus limiting source material, often because of the size of the fuel load required by redesigned combustion chambers. Yet peasant and semi-urban families may not even possess the elementary tools—heavy cutting knives, or *pangas*—to cut fuel to appropriate sizes. Such stove designs do not satisfy the values of the end user.

More importantly, current design efforts focus on only one end use: cooking. Traditional open fires, however, serve several functions, including rapid boiling, lighting, space heating, reducing insect populations (which both controls disease vectors and preserves thatch), drying or flavoring of food stored above the hearth, and providing a social focus. Designs that seek principally to improve the efficiency of simmering do so at the expense of other end uses. Peasants consequently require additional end-use devices, which raise the real costs, in both monetary and labor terms, of energy procurement.

Is improved stove design, then, really more socially efficient? Should the emphasis in the household consumption sector perhaps be on accelerating movement "up the energy ladder," promoting fuels such as kerosene and gas, rather than reinforcing the poverty-ridden *status quo*? The growing sideshow of technical appliances absorbs considerable amounts of capital and effort, but tends not to reflect on these structural questions. In our view, improved stoves are most likely to be valuable in urban areas, where *jikos* are already widely employed. There, the main constraint is cost: above US $4, a stove is beyond the financial reach of the average Kenyan.

More efficient charcoal production

Peasant pyrolitic conversion technology is relatively inefficient, at about eight or ten percent, and modern production methods could increase yields threefold. But such techniques involve centralized, frequently capital-intensive systems, more likely to be controlled by the urban, industrial middle class, than by the peasants themselves. At present, charcoal manufacture is the single most important source of peasant off-farm income, and its disappearance would threaten the entire peasant production system. The loss of this income would increase rural-to-urban migration, exacerbating problems of underdevelopment and further distorting the energy system.

Designs that improve the efficiency of simmering do so at the expense of other end uses. Additional end-use devices are needed, which raise the monetary and labor costs of energy procurement.

There are few examples of peasant production cooperatives that could serve as models for community control of pyrolitic production processes. Moreover, the projects now scheduled for the large state forests—promoted by transnational energy corporations—focus on charcoal for industrial process heat, rather than urban household consumption.

Current strategies for wood conservation are, in sum, inadequate because they do not consider the social context of energy use. Steps to improve stove and charcoal conversion efficiencies, taken alone, may actually increase the burden on poor peasants.

Supply enhancement

If demand management strategies call for careful reconsideration, supply enhancement strategies require total revision. The Beijer Project's survey of natural wood supplies in Kenya showed that most wood is presently extracted from high-potential agricultural lands and fringe areas. The rangelands, which cover 87 percent of Kenya's land surface, support 600 million tonnes of wood, with a sustainable production of about 14 million tonnes. These low yields, and the high transport costs from the rangelands, preclude large-scale extraction. Higher demand for fuelwood thus stresses agricultural areas, which must also supply food and export crops.

The supply of wood is only temporarily sufficient, and is likely to dwindle over the next decade. The expansion of the arable land frontier through clearing has produced an artificial surplus; and selective clearing of forest stock disguises the actual rate of forest depletion. By cutting principally the young trees, fuelwood gatherers are consuming the future stock of wood capital, and creating forests that are already "dead on their feet."

The limited potential of Kenya's natural woodstocks emphasizes the need for planned afforestation efforts, but several constraints must be overcome for them to succeed.

The jiko stove

Administrative Constraints. The government's forestry administration concentrates on watershed management and maintenance of surveyed forest areas. It responds principally to the pressures of industrial wood consumers (for feedstock and building materials), who account for less than 20 percent of the wood demand. Eighty percent is used for energy, and until the administration changes its priorities, there will be little progress in energy afforestation (O'Keefe, Weiner and Wisner, 1981).

Social Constraints. Recommendations for planned community woodlots on, or near, arable land, do not consider several key social constraints. Firstly, land in Kenya's rural areas has been privatized as capitalist relations of production there consolidate. Communal relations and "community spirit" have been severely eroded. It is impractical to suggest that energy supply policies should seek to function through relations of production that have already been destroyed.

Secondly, community-based projects most frequently are ill-disguised attempts by development agencies to extract free labor from the peasantry, generally womens' labor. Even during the best of times, there is a labor constraint in the peasant production system, and the demands of energy production would compete with agricultural needs. Women traditionally carry the heaviest burden of agricultural and domestic work; now, as children are often away all day at school, and men are going to urban centers seeking wage labor, the work load has increased even further. To design community-based forestry projects which assume free labor donations, especially from women, is to misunderstand the nature of the peasant economy.

Thirdly, the East African landscape does not display a village settlement pattern. In the densely populated areas, individual tenure, with the household settled on the land, is the established order. Such an ownership pattern makes it difficult to envisage a community (*i.e.*, village) woodlot,

as exists in Asia. Moreover, the distance to be travelled to collect wood from "community" projects is frequently too great.

Finally, and most importantly, many decentralized schemes for fuelwood production in Kenya will fail because not all the population has access to land. Increasing privatization of land, coupled with a rapidly growing population, will mean less access in the future. This, in turn, will mean longer periods of time to collect an ever scarcer resource. But this scarcity is a direct result of the changing social order, and not a decline in biomass productivity.

Production Constraints. The most fundamental constraint is that *forests are unimportant for wood production*. The key issue is how to establish trees outside the forest, where most fuelwood is gathered. There are two major areas of intervention: improved agroforestry within farms where, at present, much of the land is underutilized, especially along pathways and hedges; improved pollarding techniques with the design of simple tools for non-destructive removal of secondary vegetation on "within-farm" trees. The Beijer Project is still experimenting with ways to increase agroforestry yields without increasing energy inputs.

Development Constraints. Current forestry, and particularly nursery practices, have no adequate extension staff. It will be necessary to decrease the number of large-scale nurseries, because distribution from them is difficult, particularly in the rainy season when the seedlings are required for transplanting. For a successful plant diffusion program, more small-scale nurseries will have to double as seed orchards. The seeds and seedlings must be carefully selected for energy production; the current practice of stocking nurseries with ornamental and fruit trees might fulfill the desires of the still-resident settler community, but it does nothing for the peasant trying to maintain fuel supplies.

Conclusion

If the administrative, social, production, and developmental constraints can be overcome, then a viable energy-supply enhancement policy might be possible. If not, we will merely retain the current situation, in which most young seedlings are gathered spontaneously by children, but, unfortunately, not at a rate sufficient to meet the growing demand for wood. The poor planting rate, and the limited conservation opportunities, may force a move "up the energy ladder," regardless of what the Kenyans themselves desire.

—*Don Shakow*
Cara Seiderman
Philip O'Keefe

The authors of this article are presently working with the Beijer Fuelwood Project at the Beijer Institute, Nairobi, Kenya. A full report of their findings, *A Survey of Natural Wood Supplies in Kenya and an Assessment of the Ecological Impact of its Usage* is available from the Institute.

References:
Harris, T.
1981 "Designing Fuel Efficient Wood Burning Stoves." *Ambio,* Vol. X, No. 5.

O'Keefe, P., D. Weiner, and B. Wisner
1981 "The Tail that Wagged the Dog: A Precautionary Tale of Forestry Planning in Kenya." In L. Buck, ed., *Proceedings of the Kenya National Seminar on Agroforestry, 12-22 November, 1980.* ICRAF University of Nairobi, Kenya.

Biogas Development in Nepal

Affluence from Effluents

In the Himalayan kingdom of Nepal, the fuelwood crisis is acute. Cooking fuel demand rises in lockstep with population, while supplies grow increasingly scarce. Despite ambitious and expensive efforts at reforestation, trees are being cut several times more rapidly than they can regenerate. The need for an inexpensive, renewable alternative to firewood for cooking is clear, and numerous development projects have focused on biogas from human and animal wastes.

If the people of Nepal continue to consume fuelwood as they do at present, all of the commercial forests—twelve percent of the total—will be gone in little more than a decade, according to H. C. Rieger of the Swiss Association for Technical Assistance. The loss of this much forest will mean not only fuel shortages and high prices, but severe environmental damage. Habitat destruction reduces wildlife populations, including endangered species, more drastically than poaching. The geologically young soil structure of the Himalayas is easily eroded; unforested hillsides, washed away by the torrential monsoon rains, may become unproductive and unusable as either farm or forest, perhaps for centuries. And forests, once gone, are hard to re-establish; livestock thrive on tree seedlings, and overgrazing hinders reforestation projects.

Preventing this damage and preserving the forest resource is only possible if a cost-effective replacement is found for firewood. Biogas energy is frequently suggested as a promising candidate, capable of meeting the cooking and illumination needs of developing populations. In Nepal, biogas projects using primarily cow and buffalo dung have met with moderate success. But community biogas systems, which work with human wastes, have encountered both technical and social problems.

Three Digester Designs

More than 1,000 "gobar gas" plants (gobar is Hindi for cow dung) have been built in Nepal, and the demand for digesters currently outpaces the supply of materials and technicians trained in their installation. The most successful design is the simple KVIC (Khadi & Village Industries Commission) pit-type digester, originally developed in India. A ten-foot deep well is lined with bricks or rocks, then plastered with cement to form a water-tight container. The digester operates on a slurry of cow dung and water mixed in equal proportions, and capped with a floating gas holder. Pipes on top of the holder, or inside the digester, feed the gas to household lanterns or cookstoves, with the weight of the gas holder supplying the required gas pressure.

Each day, fresh manure slurry recharges the digester, forcing digested effluent out the other side. Manure and effluent discharge are equally valuable as fertilizers, but many farmers complain that the liquid is more difficult to

Building the KVIC pit-type gas digester in Nepal.

transport—especially if their fields are upslope from their digester.

A gobar gas plant typically supplies enough cooking gas for as many people as there are livestock providing manure. Containing only a few moving parts—the gas holder, a few gate valves and stopcocks—the device is easily maintained by the owner. The gas holder, fabricated of mild sheet steel, is the most complicated component; it is now manufactured in at least five locations in Nepal, and marketed through a combination of government agencies and private corporations.

Digester designs more complicated than the Indian variety have not experienced the same success. One farmer briefly experimented with a "Korean" model, where the manure is digested in an enclosed tank, and the gas collects nearby in oil drums floating in a water bath. The oil drums, unfortunately, proved small, expensive, and easily corroded, while the cement roof of the digester was never perfectly gas-tight.

Another fixed-roof design has been extremely successful in the People's Republic of China, where more than seven million plants were in operation by mid-1978 (see *Notes* December 1979, pp. 77-81). The Social Science Division of Nepal's Agriculture Department constructed its own prototype version of this digester, hoping that it might find equal popularity in Nepal. But, like the Korean type, the Chinese digester incorporates high gas pressures, and depends upon them to move the slurry in and out of the fermentation chamber. Traditional Nepalese materials and construction techniques were not up to air-tight roofs, so that gas escaped from the prototype, and material movement was inconsistent. Later versions, however, avoided this problem by incorporating an acrylic plastic emulsion paint into the cement mixture, and ten modified Chinese-style digesters are now operating in central Nepal.

The advantages of the simple, Indian biogas digester are clear to many Nepalese farmers, but the investment capital required is often out of reach. The typical gobar plant costs about $400, equivalent to nearly a year's wages for a skilled worker. The Agricultural Development Bank offers an attractive loan program, which has bolstered gobar gas plant construction, but the program is understaffed. Many subsistence farmers, moreover, find that the digester saves them the labor of collecting fuelwood, but provides virtually no financial return. Working largely outside the modern, cash economy, they are unable to service the debt of plant construction.

Wealthy farmers are the principal beneficiaries of the Nepalese biogas digester program, as is the case with many of the innovative technologies introduced in developing rural areas. But the benefits may yet diffuse to the other segments of the agricultural population. Preservation of Nepal's forests—a community and a national resource—can only benefit the entire citizenry. Widespread adoption of biogas and other non-wood-based energy technologies would relieve the pressure on the country's shrinking forests. This, in turn, would deflate the price of traditional fuels, benefitting the poorer villagers who cannot afford the alternatives.

Small-scale biogas digesters in Nepal, working mainly on animal manure, tend to benefit only individual, wealthy families. Community-based systems, in contrast, promise a more equitable distribution of the energy produced, and of the burdens of human waste management. Two such projects in Nepal, however, illustrate the problems typical in these community development efforts.

Community Digester Projects

An 18-stall women's community latrine/gas digester system was installed in 1977 in an urban suburb of Kathmandu. Part of a US Agency for International Development "Study of Energy Needs in the Food System," the system was designed to generate gas while mitigating a serious sanitation problem. Considerable attention was given to possible hitches: all influent pipes were straight, so they could be easily cleaned when

Lined with brick and plastered with cement, the ten-foot pit will be water-tight.   Photos by Broughton Coburn.

clogged; all surfaces near the compound were cemented over to prevent rocks and other foreign matter from entering and plugging the system; all water taps were self-closing, to control the water/waste mixture; stalls were private, but holes between them allowed the women to talk freely, as the community sanitation area was traditionally a social space.

For both cultural and technical reasons, the gas from this community digester was less valuable than anticipated. The Hindus of the village considered food cooked on any sewage byproduct to be ritually polluted, and could therefore use it only for cooking animal food or for lighting. The gas that was produced had too high a concentration of CO_2 to burn continuously, probably a result of the acidic chemistry of the digesting sewage. Apparently, the digester was loaded too quickly, and contained insufficient material of high carbon content for proper digestion. Future community systems might well include an intermediate mixing or holding tank, rather than direct feed, to control the loading rate.

As a local sanitation facility, the system operated well for over a year, mostly through the efforts of a janitor whose monthly wage was paid out of the AID project budget.

Capped with a floating gas holder, the digester operates on a slurry of cow dung and water.

beyond technical feasibility entering into the design process.

Research efforts to optimize digester net energy production will probably be less effective, in the long run, than simple cost reductions. Field personnel need to be trained in digester financing and construction. Planners must also learn from the successes and failures of previous Nepalese projects, if biogas is ever to help preserve the Himalayan forests.

—*Broughton Coburn*

Broughton Coburn spent four years with the Peace Corps, two of them as co-principal investigator for an AID-funded project, "Study of Energy Needs in the Food System," in Nepal. He is now employed by a Seattle engineering firm, surveying the biogas potential of dairy farms.

When his salary was discontinued, the city government was unable to supply funds. So, community residents arranged with the ward representative to levy a small tax (about four cents per month for each patron) to pay the janitor directly. This procedure, however, worked only for two months, and when the janitor quit, water supply problems and lack of maintenance made the latrine unusable.

Another community biogas project, in the village of Dobare in southeast Nepal, faced a different set of problems. The $3,000 digester was constructed there in 1978 under the supervision of the Agriculture Department and a Peace Corps volunteer. Designed to produce 500 ft³ (14 m³) of biogas per day, with an energy value of 300,000 Btu (75,600 kcal), the plant could serve the cooking and lighting needs of five households. Most of the labor of plant construction and pipe installation was donated by the five beneficiary families, who also agreed to supply it with their daily wastes.

One year after the plant's completion, a visitor from the Butwal Technical Institute found only two of the mantle lamps in working order. (Corrosive elements in the gas and inadvertent improper usage make mantle burnout common.) Two of the five households were using neither gas lamps nor stoves, and had withdrawn from the cooperative arrangement. Waste collection and gas distribution systems had not worked smoothly. And many Nepalese feel that the burner does not emit a sufficient flame, though it is in fact hot enough for most cooking. It is the smoke they miss: firewood smoke cures and protects the thatch of their homes and its structural members.

For biogas to reach the level where it can significantly reduce fuelwood consumption, foreign aid and government assistance programs must expand. But they must also be carefully planned and executed, with many factors

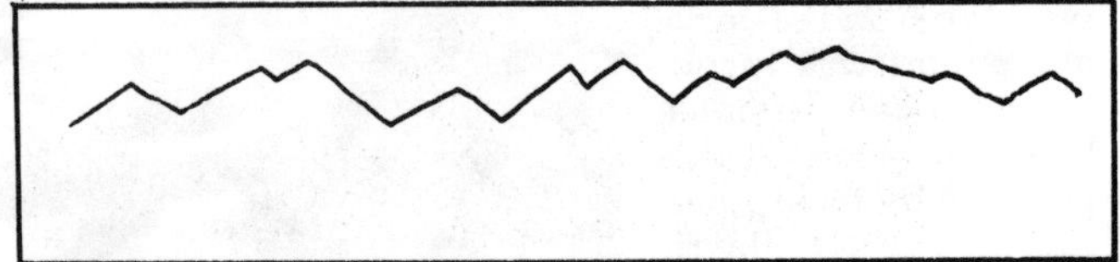

References:

Bulmer, Andrew
1980 "A Survey of Three Community Biogas Plants in Nepal." Unpublished report, Development and Consulting Services, Butwal, Nepal.

Eckholm, Erik P.
1976 *Losing Ground*. W. W. Norton & Co., Inc., New York, New York.

Karki, A. B., and B. A. Coburn
1977 "The Prospect of Biogas as One of the Sources of Energy in Nepal." In *Proceedings of the Tenth World Energy Conference*, Istanbul, Turkey.

Rieger, H. C.
1976 "Floods and Drought: The Himalayas and the Ganges Plain as an Ecological System." From *Mountain Environments and Development*, Swiss Association for Technical Assistance, Kathmandu.

Skrinde, R. T.
1978 "Review of International Biogas Programs." Unpublished paper, Olympic Associates Co., Seattle, Washington.

van Buren, Ariane
1979 *A Chinese Biogas Manual*. Intermediate Technology Publications Ltd, London.

TOOLS for the SOFT PATH

photos by Carollee Pelos

Welcoming the Wind

From zephyr to tempest, the wind has many moods. Most architecture shelters us from it. But in some regions of the world, intriguing forms of traditional architecture welcome the wind.

The wind blows hard in western Afghanistan. Sweeping south from the Qizil-Qum steppes of inner Asia, a torrid gale known locally as the "wind of 120 days" and "the wind which kills cows" occurs almost daily from June through September. Often reaching 100 miles an hour, this wind blows hardest in Sistan, a dry, hot region crossed by the southern border between Afghanistan and Iran.

Sistanis were the first people to focus the wind's wild energy into work. During or before the 9th century, turning a scourge into a benefit, they invented the windmill. It is virtually certain that the sails of these initial mills rotated horizontally, thus bypassing the problem of translating vertical into horizontal torque.

The windmill is a relatively simple machine. But once it inspired admiration shading into wonder. Of Sistan, "the land of wind and sand," a 10th century chronicler writes, "there is no place on earth where people make more use of the wind." A 13th century commentator even compares the Sistanis' power over the wind to that of early wind-master Solomon, the first person in legend to travel—with his entire court—on a flying carpet.

Superseded by mills powered by diesel motors, working horizontal windmills have now almost disappeared. But the one pictured here, erected over twenty years ago, still grinds wheat. Built of sun-dried mud-brick, the mill stands two stories high. The shaft rises through a domed roof and, above, carries six vertical sails made of mats of grass. The upper story is open on one-and-a-half sides. The open half-side faces the wind, the closed half shielding the returning sails. When the wind is particularly fierce, additional mats can be placed in the wind-slot to keep the sails' speed down, so the flour will not burn and blacken.

In central Iran in the late 19th century, a local ruler tried to import the horizontal windmill. At considerable expense, he built a huge one for power to raise water. The mill never pumped a quart. Local winds were far too light to turn the mill's sails. This mismatch of regional wind and vernacular wind-device is rare if not unique. The rule is a quite exquisite pairing, best shown in traditional architectural forms using the wind not for energy but to cool.

The strength of a region's wind governs the form of vernacular "air-conditioning," including the size of openings both to receive the wind and to vent it. Since in and near Sistan the wind is powerful, wind-scoops are small, and small ventilators are needed only over kitchens. Upper Egypt's prevailing summer wind is mild, so ventilators—which help to draw the breeze through the building—yawn at least as large as the quite large wind-catchers (Fathy 1973).

In Iran's central plateau and along the Persian Gulf, torrid temperatures and light breezes have inspired thermal ingenuity. The area's wind-towers—tall to reduce the admission of dust—are topped by slender openings up to thirty-five feet high. Cleverly, the giant slots—connected to separate passages within the tower's shaft—alternately function as intake *and* outlet. Facing all four quarters, they can harvest the slightest fickle breeze. After sundown, should the wind die, the heat stored in the tower's mud-brick warms the air which—because it is less dense than the

---

In a portion of Sind's flat desert, an onshore south-west wind blows from April through June. Dry, steady, and cool, it is a blessing. The area's wind-catchers all face this wind, turning their backs to the winter wind, which, blowing from the opposite quarter, is not needed indoors when the weather is cool.

---

cooler surrounding night air—rises. This creates a draft which pulls a refreshing night breeze through doorways and windows (Bahadori 1978).

In southern Pakistan's Sind region, the prevailing wind is moderate, so wind-openings are large and ventilators infrequent. In a typical village hundreds of wind-catchers reach like giant periscopes high into the sky to clear the "wind-shadows" of neighboring roof-lines.

Their design is elegantly simple. Most consist of just three flat planes and a post. Two vertical surfaces at right angles form a base. A third, one of its corners tucked into the angle, tilts forward at about 45 degrees and deflects the wind down into the house. A post anchors the deflecting plane.

In a portion of Sind's flat desert, an onshore south-west wind blows from April through June. Dry, steady, and cool,

it is a blessing. Its season coincides with the area's most brutal heat, when the daily mean maximum temperature climbs to 43°C (105°F). The area's wind-catchers all face this wind, turning their backs to the winter wind, which, blowing from the opposite quarter, is not needed indoors when the weather is cool.

Like that of most vernacular architecture, the history of lower Sind's wind-catchers is obscure. Local tradition states only that their use is "very old." The first written Western account is from an English traveller who, visiting the city of Hyderabad in 1815, observed that wind-catchers appeared on every house "from the governor's palace to the lowest hovel." Other travellers to Hyderabad mention wind-catchers' popularity in the 1840's. Strangely, a general view of Hyderabad published in the 1870's shows not a one, while a photo taken some fifty years later from the same spot shows hundreds.

The large town of Thatta still boasts wind-catchers on the great majority of its houses. The plan of most is the standard, free-standing square. But there are interesting variants. Sometimes wind-catchers do not rise from the roof but are incorporated into a building's highest story. Many are double or "quadruple," the east-west axis two or four times the one north-south. In a few cases, the ratio can be as much as a splendid seven to one. Below this type extends a long, narrow "wind-room" whose entire ceiling is a wind-catcher.

A twenty year old well-to-do house in Thatta boasts a wind-room 28 feet long. On its lee wall, four windows and two doors can be opened in any combination to regulate the flow of air into two bedrooms, and through them out to an open court and its wide colonnade. In the colonnade's four walls stand twenty-one doors with transoms. A total of eleven wind-catchers and two ventilators cool the entire house. Kitchen and toilets—and their odors—are to the lee of all bedrooms.

Every wind-catcher has a metal grate to prevent entry into the house via the roof. Except for the very large ones, all also have trap doors. Propping these at different angles governs the amount of air admitted. In winter they are shut. In multi-storied houses, an internal vertical shaft conveys the breeze to rooms on lower floors.

Wind-catcher technology runs the gamut from traditional to industrial. In many, a base of sun-dried mud-brick supports a deflecting plane of wood, the whole sheathed with a layer of mud mixed with dung and straw. In more modern ones the base's bricks will be kiln-dried, its sheath plaster, and the deflecting plane corrugated steel supported by a wooden frame. The most modern ones are entirely concrete.

One type of traditional air-conditioning is used from northern India through Iran. A suitable local bush is either woven into a loose, thick mat or packed into a frame. It is placed outside a window or door and drenched, continuously or repeatedly with water. A breeze evaporating the water can easily cool a room 15 degrees. In Indian cities, an electric fan may replace the breeze. The entire appliance, a clever wedding of traditional and modern technology, is called a "desert cooler."

Architecture to catch a cooling wind is very old. Ancient Egyptian houses featured triangular wind-devices. At least 1200 years ago wind-scoops caught the sea-breeze on buildings in Peru. As rising fuel costs make industrial air-conditioning prohibitively expensive, age-old ingenuity may come to the rescue (Fitch 1972). Meteorologically appropriate areas of the world will do well to adopt specific wind-devices—ecologically sound forms of air-conditioning at once cheap, efficient, and often beautiful.

—*Jean-Louis Bourgeois*
Carollee Pelos

References:

Bahadori, Mehdi N.
1978 "Passive Cooling System in Iranian Architecture." *Scientific American* 238:2 (February), pp. 144-154.

Fathy, Hassan
1980 Architecture for the Poor. Chicago: University of Chicago Press.

Ferdinand, K.
1963 "The Horizontal Windmills of Western Afghanistan." *Folk*, vol. 5, 71-89.

Fitch, James Marston
1972 *American Building: The Environmental Forces that Shape it*. Boston: Houghton Mifflin. Drawing on vernacular examples, Fitch makes an eloquent plea for matching modern architecture to climate.

Jackson, P. and A. Coles
1975 "Bastakia Wind Power Houses." *Architectural Review*, vol. 158 (July), 51-53.

This article is reprinted courtesy of Natural History

Micro-Hydro Designed for Papua, New Guinea

L ow cost electricity for developing villages need not come from large-scale centralized power systems. In Papua New Guinea, micro-hydro power is both economically and technically appropriate.

Many developing countries have abundant hydrological resources that could provide electricity to rural communities isolated from central power grids. But commercially available hydroelectric systems are expensive; one of the cheapest, a Chinese 5 kW unit, costs $5,000 (US) for just the turbine and generator—more than most small villages can afford. Conventional units are also complex and therefore unsuited for use in remote locations where spare parts and technical expertise are scarce.

Low-cost, rugged micro-hydro systems, however, can be fabricated from domestically available materials. The Appropriate Technology Development Unit (ATDU) in Papua New Guinea (PNG) has shown how. Although tailored to PNG, their design is flexible and can be adapted for other countries.

The Design

To be appropriate for village use, a generating set has to meet several conditions:

- *Low cost.* Lack of capital is a major constraint on village development projects. The hydroelectric set has to be less expensive than conventional gasoline and diesel driven generators to be within reach of cash-poor villages.
- *Maximum use of local materials and skills.* Local self-reliance keeps money in the community, promotes local employment, and keeps community interest high.
- *Simple construction.* The villagers should be able to install the system with minimal technical guidance.

- *Simple, rugged design.* Villagers should be able to operate and maintain the system with minimum training or outside assistance.

Relatively small hydro sets are therefore best suited for remote villages: they not only lend themselves to the greatest degree of simplification, but they also allow villagers to begin electrification with the least capital outlay. The ATDU design, rated at 5 kW, is at the lower end of what is considered micro-hydro.

The system prototype consists of four basic components: a penstock made of polyvinyl chloride pipe, a Pelton runner, a commercial 5 kW alternator, and a concrete base. The penstock leads water down to the generating set, and through two small nozzles; the two high pressure jets then turn a runner, which drives the alternator to generate electricity.

All of the system parts were purchased in Lae, PNG. The Pelton runner presented the only construction difficulty. Individually cast buckets had to be bolted on to a steel disk, increasing the amount of labor. In the future, the runners will be cast as a single unit at a local foundry (another ATDU project). Assembly of the generator set is straightforward and requires only basic tools. On-site installation and calibration, meant to be done by the villagers, is even easier.

Maintenance requirements are minimal. Other than regular inspection and occasional replacement of the belts, the only necessary upkeep is replacement of the alternator brushes, runner shaft bearings and alternator bearings after every 5000 hours of operation. With the bearings and brushes costing $60 and the belts $9, maintenance costs are low.

The Pelton runner may not have been the best choice because it requires a relatively high head of about 80 m. The long penstock needed to provide the head adds substantially to the system's overall cost, especially at sites with gradual slopes. Depending on the site conditions and the cost of PVC pipe, low head designs may be more

---

Total Installed Cost of the Micro-Hydro Set (US dollars)

Hardware (alternator, belts and pulleys, concrete, runner materials, misc.) and Labor	$ 840
30% Mark-up	252
Penstock @ $220/kW	1,100
TOTAL COST	$2,192
COST/kW	$ 438

---

economic. They require, however, larger volumes of water, and this means greater expense for bigger dams to tap larger streams. Taking this trade-off into account, the ATDU design seems well suited to the steep terrain of PNG.

One problem with simple micro-hydro sets is runner speed control. The two common governing methods—adjusting the rate of water flow to the runner to match the load, and keeping the flow constant while using electric means to ensure that all the available power is used—add too much cost and are unnecessarily sophisticated. But some means of keeping the voltage within broad limits and protecting the system from no-load situations is needed. With the runner speed unregulated, the generator speed, and thus the supply voltage, could easily increase by more than 60 percent between full load and non-load, risking serious damage to the system.

The ATDU has experimented with several low-technology methods. One village, for instance, keeps the hydroelectric set operating at full power while maintaining a constant load by switching equivalent loads in and out of the system: when they turn the village lights out at night, a water heater is turned on. In systems with much more power available than usually needed, a large permanent load can be connected to provide some useful service like crop drying, water heating or refrigeration. When a small device like a water pump, a small electric tool, or a light is switched on, the load change will have only a minor effect on the voltage.

To protect the runner and alternator from an open circuit in the distribution system, one village inserted a relay in the main line. When an open circuit occurred, the relay switched in an emergency load capable of dissipating the power. Over the course of its first year of operation this device saved the generating set three times.

System Cost

The hardware and labor used in fabricating the ATDU generating set cost about $840. Allowing a 30 percent mark-up for a local manufacturer, this means that the system could be marketed for about $1,092. This does not include the cost of labor for installation or the necessary penstock. Installation costs are presumed to be minimal because local labor should be able to perform all the necessary tasks. For this reason, the hydroelectric generating system makes an ideal self-help project.

Penstock costs vary by site. The amount of PVC pipe required depends on both the penstock gradient and the pipe diameter used. With a vertical drop (the ideal case) and a 80 mm diameter pipe, for instance, the Pelton runner only needs a 70 m gross head (70 m of pipe) to attain rated

Capital Costs (5 kW sets)	
Petrol	$145/kW
Micro-hydro	$438/kW
Diesel	$580/kW

power. But, with a 14° incline, it requires a 90 m gross head (the difference is due to friction between the water and pipe), and that takes about 360 m of pipe.

For a 19° penstock gradient—typical in PNG—about 250 m of 80 mm pipe would be needed to provide a 83 m head. At current prices this penstock would cost about $1,100 or $220/kW (compared to $45/kW in the ideal case of a vertical drop). Adding this to the price of the generator set, the cost of the system installed at a typical PNG site totals $2,192.

Cost Comparison

Conventional gasoline and diesel driven generators are generally considered more cost-effective at providing power to isolated rural areas than small hydro. The common argument against small hydro is its high capital cost. A comparison of the cost of the ATDU unit with the

comparable alternatives available in Lae suggests that the argument should be re-examined. The micro-hydro capital cost of $438/kW is less than the $580/kW outlay for diesel. The gasoline driven generator, with a capital cost of $145/kW has a considerable initial advantage. But with the current PNG gasoline price of $.43/liter, the higher operating costs make up the difference in less than a year. Assuming only 4 hours of daily operation, the annual gasoline expense would be more than $580.

The micro-hydro system has the additional advantage of a durable, simple design which makes regular maintenance easy and major repairs unlikely. In contrast, internal combustion engines operating in remote areas tend to wear rapidly because of rugged conditions and inadequate maintenance.

The ATDU project demonstrates, therefore, that where suitable sites are available, locally fabricated micro-hydroelectric sets are better than conventional means of providing power to remote villages. Because many Third World countries—Nepal, Turkey, Pakistan, Philippines, to name a few—have an abundance of rivers and streams, decentralized micro-hydro systems could contribute to development worldwide. Perhaps this small-scale and self-reliant approach to rural electrification would be more successful than past large-scale efforts.

—*John Fore*

References:

Alward, Ron, Sherry Eisenbart and John Volkman.

1979 (January) *Micro-hydro Power: Reviewing an Old Concept*. This report, prepared for the US Department of Energy (doc. #DOE/ET/01752-1), describes the basics of micro-hydro technology and provides useful design information. It is available from National Technical Information Service, US Department of Commerce, 5285 Port Royal Road, Springfield, Virginia 22161 for $5.25.

Inversin, Allen R.

1980 (June) *A Pelton Micro-hydro Prototype Design*. A report prepared for the Appropriate Technology Development Unit, available from the ATDU, PO Box 793, Lae, Papua New Guinea.

Photovoltaic-Powered Water Pumping: Giving Small Farmers a Lift

"A TREE is watered at its roots, not its branches," the Bengali proverb says, but when no water is available, the tree is left to wither. A shortage of irrigation water in some of the Third World's most heavily populated areas hampers millions of small farmers in their efforts to increase production.

A pumping technology matched in scale to the needs of the small farmers may now be available. Paradoxically, it is based on the epitome of the high-tech, Space Age device, the solar cell. Small, photovoltaic-powered pumping systems offer simple, low-maintenance operation and even at today's prices for solar arrays, they are very nearly cost-effective for many applications.

Irrigation can just about double annual food production by permitting a second, dry season crop and by protecting the main harvest from drought. In the delta regions of Asia and Africa, water is available year-round from wells, canals and ponds with lifts of five meters or less. Unfortunately, all the methods available for low-head water lifting have disadvantages. Many simple devices use human muscle power, but this work is physically demanding and diverts labor effort from other agricultural tasks. Water buffalo and other draft animals provide modest amounts of water at low cost if free grazing land is available, but farmers with less than two hectares rarely have enough suitable land.

Diesel and gasoline powered pumping systems support many medium to large-scale water supply programs throughout South and Southeast Asia. But in the size range of one-third to one horsepower needed by small farmers, they are presently uneconomic and becoming more so with the rising cost of fuels. Cheaper to operate are pumping systems drawing electricity from the grid, but a distribution network capable of reaching small motors on millions of tiny plots would be massive and costly. In addition, efforts to organize many small farmers into cooperatives for effective use of conventional pumpsets have come up against both technical and social obstacles. Consequently, the benefits of mechanical irrigation lie just beyond the grasp of small landholders, but not their wealthy neighbors, and so the gap between the two classes widens.

The Match to Energy Demand

Photovoltaics are particularly well-suited to the energy demands of irrigation. The output of the solar array coincides precisely with varying pumping needs because plants use water only in proportion to the amount of sunlight they receive. Both transpiration and electrical generation peak on clear, bright days, drop when it is overcast, and fall to zero at night. Expensive battery storage is rendered unnecessary. Moreover, the photovoltaic pumping system is quiet, pollution-free, incurs no perpetual fuel costs, and promises high reliability over a long service life.

The ratio of costs to benefits for any water pumping system depends on the value of the irrigation water to the farmer. Smith and Allison (1978) report that 10,000 m^3 annually of irrigation increases production in rice cultivating areas by at least 2.5 tonnes/hectare, adding $250 to yearly income for each hectare the farmer owns. The value of the irrigation water thus falls in the range of 2-3 cents/m^3. Further analysis by Tabors (1978), assuming a pumping system efficiency of 50% and a system life of 15 years, shows that irrigation from surface sources with lifts of about 1.5 m is marginally attractive at solar array prices of $10/peak Watt, providing water at 2.9 cents/m^3. With array prices of $4/Wp, the cost of surface irrigation drops to 1.3 cents/m^3. At the lower array cost, ground water pumping with lifts of 4.5 m is also cost-effective at 2.4 cents/m^3. These figures assume fairly low balance of system costs, but if the 1986 US Department of Energy goal for solar cells of 50 cents/Wp (1975 $) is met, then photovoltaic-powered pumping will be a bargain. Units sized for extremely small

Lifting water, Portugal

Photograph from World Enough: Rethinking the Future, by Margaret Mead and Ken Heyman

landholdings of 1/2 hectare or less will provide water costing 50% more than from the larger units, but even this should be economically feasible.

The major drawback of photovoltaic water pumping is high initial capital cost. The system may well be cost-effective for subsistence farmers, but this alone does not make it affordable. At array prices of $10/Wp, a typical surface source pumping system would cost well over $2,000, a sum virtually impossible for Third World farmers to finance. Small landholders find almost any capital investment a problem, since their savings potential is so low. Bulk purchases, financed through rural credit facilities assisted by international funding agencies, would help to drive down the unit cost. Under United Nations auspices, test and demonstration programs are already underway in India, Mali, the Philippines, and the Sudan, and their success could generate further capital assistance.

The market potential for low-lift, PV-powered pumps is enormous, as is the difference they would make to regional food production. Smith and Allison estimate that some 50 million small farms in developing countries might benefit from PV-powered irrigation, representing over 12,000 MWp from solar cells. This is a far cry from the 1 MWp generated by all the photovoltaics sold in 1978. If, by the year 2000, 10 million units were in operation, they could increase food production by some 25 million tonnes, or by about 50% of present yields in those areas.

Photovoltaics *vs.* Conventional Pumping

The short lifts and low power requirements of micro-irrigation create a niche in which photovoltaic pumping is nearly cost-competitive. As fuel costs rise, and the price of solar arrays declines, developing countries will look to photovoltaics for drinking water and for irrigation on a larger scale. Diesel engines have served these needs for decades, but compared to photovoltaics they are unreliable and have a thirst for fuel that grows more expensive almost daily. Diesels are questionable investments for oil-poor countries where small increases in the bill for imported fuel mean a huge jump in foreign debt. Even existing systems face an uncertain economic future.

At array costs of $2.50/Wp, photovoltaics match the economy of large diesels. An analysis by the Lewis Research Center of the National Aeronautics and Space Administration, examining installations under 15,000 kWh/year, shows that PV systems at $3.25/Wp—the Department of Energy's 1981 cost goal—also compete with diesel systems. The analysis assumes a fuel cost of $2/gallon, and a conservative 7% annual rate of fuel escalation.

Potable water supply systems powered by photovoltaics are now marketed by several manufacturers. Demonstration sites, funded by international development agencies, range in size from 0.6 to 1.8 kWp; they provide 10-50 liters of water per capita per day in villages with populations as high as 2,000.

Widespread use of PV systems, mass production, and lowered unit costs appear just around the corner. By 1986, it is estimated that foreign sales of photovoltaics will be as high as 130 MWp, and pumping systems will account for 55% of these exports. It is highly unusual for a sophisticated, modern technology to find its *first* major market in rural, developing areas, but this is apparently the course on which photovoltaics are set. The potential market in the Third World is large enough to initiate a cycle of mass production and lowered costs and, curiously, help to make solar arrays competitive in the industrialized nations.

But the perennial capital shortage in developing areas keeps this round of mutual benefits from beginning. Few systems are likely to reach the small farmers who need them most without active programs for low-interest loans, supported at both national and international levels. In the absence of adequate financing targeted at the small, subsistence-level farmers, the benefits of photovoltaic technology may never be realized. As Douglas Smith points out, "The mere fact that photovoltaics are smaller in scale than their competitors does not guarantee that they will be accessible to the debt-bound villager. No villager can sit down and by the sweat of his or her brow manufacture a solar cell. If access to capital, irrigation water, etc., were restricted prior to the arrival of photovoltaics, it is not likely to be less restricted after their arrival. In fact, the relative social and economic power of the already more well-to-do farmer is likely to increase if they alone gain access to this source of electric power."

—*Steve Meyers*

References:

Pacific Northwest Laboratory

1979 *Export Potential for Photovoltaic Systems,* 180 pp. Available from E. Smith, Office of Conservation and Solar Applications, US Department of Energy, Washington DC.

Rosenblum, Louis, *et al.*

1978 "Photovoltaic Water Pumping Applications: Assessment of the Near-Term Market," 18 pp. Available from NASA Lewis Research Center, Cleveland, Ohio 44135.

Smith, Douglas V., and Allison, Steven V.

1978 "Micro Irrigation With Photovoltaics," 22 pp. Available from the Energy Laboratory, Massachusetts Institute of Technology, Cambridge, Massachusetts 02139.

Tabors, Richard D.

1978 "The Economics of Water Lifting for Small Scale Irrigation in the Third World: Traditional and Photovoltaic Technologies," 17 pp. Available from MIT Energy Laboratory (see above).

Daily Energy and Peak Power Requirements for Micro-Pumping Installations				
Plot Size	Surface Water[1]		Groundwater[2]	
(Hectares)	Non-Rice Cereal	Rice	Non-Rice Cereal	Rice
(watt/hour/day)[3]				
0.25	80	160	240	480
0.50	160	320	480	960
0.75	240	480	720	1,440
1.00	320	640	960	1,920
(peak watts)[4]				
0.25	16.7	33.3	50	100
0.50	33.3	66.7	100	200
0.75	50.0	100.0	150	300
1.00	66.7	133.3	200	400

1. Lift 1.5 m.
2. Lift 4.5 m.
3. At 50% pumping system efficiency.
4. At 4.8 watt-hours/peak watt/day.

Energy requirements for water lifting in heavily populated river valleys and deltas of developing countries. Smith & Allison (1978).

3. Country Studies

Despite economic underdevelopment, some Third World countries are making rapid progress in their quest for solar futures. The People's Republic of China is the uncontested world leader in bringing biogas technology to rural villages. The Chinese constructed five to seven million family units in the last twenty years, while other Asian countries are still struggling to get beyond the range of tens of thousands of units.

Brazil is another success. Its pioneering fuel-alcohol program has influenced western research and development efforts and helped promote similar schemes in the industrialized world. Some doubt does remain about the long-term ecological viability of the Brazilian program, and it may be questioned whether fuels for private automobiles are an appropriate priority for Brazil's socio-economic development.

For other countries in the Third World, a lot of hard and constructive thinking is being done about the alternatives to the current energy dilemma. India has produced some of the most innovative proposals on energy strategies and is blessed with a group of socially aware and untiring analysts; so is Malaysia. These data-rich studies are all the more impressive when seen against the dire lack of research resources, trained professional talents and—all too often—academic freedom in Third World countries.

India's Energy Alternatives

WHEN AMERICA'S OIL import bill surged last year to 27 percent of its total foreign exchange earnings, economists shivered over the prospect of higher inflation and unemployment, and lower economic growth. When India's oil bill gnawed through 85 percent of this year's export earnings, its economists quaked. Caught in a desperate squeeze between firewood energy shortages for the rural poor and catastrophic oil bills, many developing nations have grown to see the energy problem as the main threat to their survival.

In a paper delivered last August to the UN Conference on New and Renewable Energy Sources, the Indian physicist Amulya Reddy outlined a strategy for India that could avert the squeeze. Combining an ambitious program of rural electrification to substitute for kerosene lighting, biogas to substitute for wood fuel, and energy efficient transport to reduce oil use, Reddy suggests that India could vastly improve the life of the poor, substantially reduce oil imports, and move increasingly to renewable energy resources.

Eighty percent of India's oil is used for transport (60%) and home cooking and lighting (20%). An additional 14 percent is used in diesel-fuelled agricultural pumps. The lion's share consists of "middle distillates" (kerosene and diesel), which are price-controlled, and sell for about half or a third the price of gasoline or jet fuel. Such controls are common in many oil-importing developing nations, and the Reagan Administration cites the removal of controls as a key prerequisite to approving foreign aid or creation of a World Bank energy affiliate. While it is true that many oil companies would not wish to sell their products in a price-controlled market, the removal of price controls, as Reddy explains, is a complicated matter.

Virtually all the kerosene in India is used for lighting and cooking in the household sector, with 80 percent of the nation's 116 million households wholly dependent on kerosene for lighting. With diminished supplies of cooking fuelwood, and no alternatives for lighting, kerosene price increases would leave the poor with no option short of the voting booth and worse options beyond it.

The transportation sector has the makings of a market, but petroleum price controls have diverted freight from railroads to more energy-gobbling trucks. Twenty years ago, notes Reddy, 84 percent of India's freight travelled by rail, but this has shrunk to 64 percent today. About one-third of India's oil is consumed in diesel trucks, despite (roughly) a six-fold penalty in energy efficiency. Trucks were traditionally used for short hauls, and Reddy calculates that the average haul length is three to four times longer than the 100-130 kilometers that would make sense after a 50 percent increase in fuel prices (see Table 1). At oil prices above $30 per barrel, Reddy calculates that 95 percent of truck freight

travelling beyond 130 km would shift to rail, reducing transport oil consumption to a level "within the capability of indigenous production."

Table 1: **1977-78 originating traffic, average lead and break-even distance**

Commodity	Originat ing traffic	Average lead (kms)		Break-even distance*
	×10⁶tonnes	Rail	Road	(kms)
COAL	69.26	691	408	106
FOODGRAINS	23.65	1278	277	130
IRON ORE	17.23	529	96	106
IRON & STEEL	15.21	1101	371	128
CEMENT	15.20	717	286	116
FERTILIZERS	11.21	1039	267	107

*Based on 50% increase over 1979 diesel prices
SOURCE: national transport policy committee report, 1980
Average lead distances >> break-even distances diesel subsidy!

The difficulty with this approach, notes Reddy, is that trucks would simply shift from diesel fuel to kerosene, as they have done in the past when the two fuels diverged in price. Legal strictures against such conversions would be difficult to administer, if not impossible. Trucks would simply switch from high priced diesel to cheaper kerosene and supplies to the poor would dry up with nothing gained. The answer to the apparent dilemma, suggests Reddy, lies in providing alternatives for household lighting and cooking. Only with such a program in place can petroleum product prices be increased to a level that would increase the energy efficiency of the transportation sector.

The first prong of Reddy's alternative strategy is a vigorous program of rural electrification so that electric lights, 200 times more energy efficient than kerosene, can be provided to all homes. India's present rural electrification program, Reddy notes, is alarmingly inadequate: with less than one-sixth of the nation's power sent to rural areas, only 44 percent of the country's half million plus villages receive electricity. Even in these villages, only 14 percent of the homes are electrified; some 87 percent of rural electricity is used for agricultural pumping rather than lighting. Thus, affluent farmers rather than the kerosene-dependent poor, Reddy says, receive the benefits of the program.

Reddy's longtime specialty, apart from chemical physics, has been biogas facilities for the rural poor. Citing India's high population density and limited land area, Reddy suggests that the Brazilian ethanol program would probably be unsuitable in India. Planting of sugar cane or cassava for fermentation to ethanol would "lead to serious competition with food production," according to Reddy. But India's ample supply of cattle dung (let alone other wastes), if converted to biogas in anaerobic digesters, would meet *all* the cooking energy needs of *every* rural Indian household.

Figure 11: **Two-pronged strategy for resolving oil crisis**

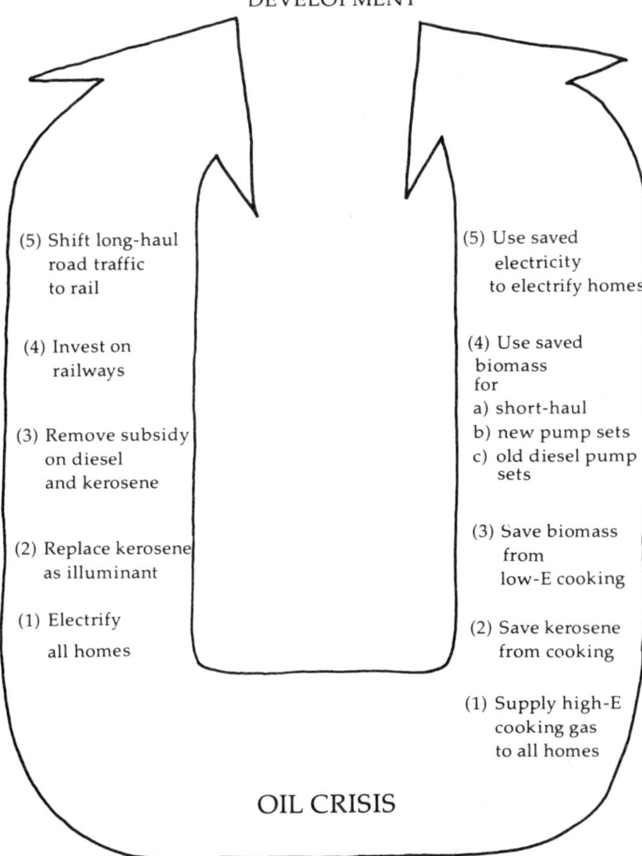

Urban homes, now 90 percent dependent on firewood and kerosene for cooking, could shift to piped mixtures of sewage, coal, and producer gases, Reddy notes. By providing gaseous cooking fuels to all homes, Reddy estimates that the 33 percent of India's kerosene now used for cooking could be phased out, along with 130 million tonnes of increasingly scarce firewood.

When there is no further need to depend on firewood for cooking, then, Reddy comments, it can be used to reduce oil use. About 60 percent of this firewood converted to producer gas and methanol could replace all current use of diesel fuel in trucks and buses.

Another 33 percent of existing firewood use could be used to fully substitute for the diesel fuel used in 2.7 million existing agricultural pumps (roughly the 14 percent

of national oil use cited earlier), and the 7.4 million new pumps projected for the year 2000. About 118 million tonnes of firewood, compared with 130 million tonnes used today, would be required to fuel all diesel trucks and buses, plus existing and projected agricultural pumpsets. Though Reddy does not estimate a sustainable yield from forests, he notes: "Once the pressure on trees as a source of cooking fuel decreases, conditions for growing energy forests will be dramatically improved leading to a much greater availability of firewood."

With the substitution of biogas-fuelled agricultural pumps for projected electric units, Reddy estimates savings of 17 terawatt-hours of electricity, equivalent to the output of about four large power plants. The electricity requirement for lighting all Indian homes is 21 TWh, leaving a net requirement of four TWh, about the output of a 700-900 megawatt power plant, which could come "from decentralized generation ... such as biogas generating sets, microhydroelectric plants, and energy forests."

As important as Reddy's detailed and optimistic conclusions are his guidelines for conducting energy studies. As Reddy writes, "if development is viewed as a process of satisfying basic needs, starting with the needs of the neediest, of strengthening self-reliance and of ensuring harmony with the environment, it turns out that this strategy for resolving the oil crisis also promotes genuine development of the country."

—Jim Harding

Figure 13: **Comparison between present energy consumption and proposed strategy**

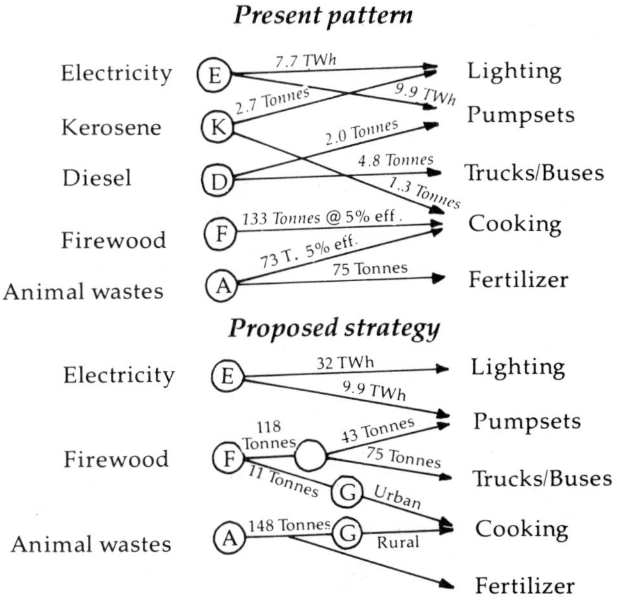

Soft Energy in China

People's Republic Develops Both Centralized and Small-scale Energy Sources

CHINA IS THE WORLD'S most interesting example of self-reliant development. For the last several decades, China has undertaken a prodigious effort to develop its indigenous energy resources. Those most commonly cited are biogas and small hydro stations, and rightly so, but the bulk of Chinese energy production has come from vast expansion of offshore oil production, coal mines, and large hydroelectric stations.

It is important to point out when talking about Chinese energy development that it involves not just one path, but many. Large-scale, highly centralized power facilities complement the ''appropriate,'' ''renewable,'' ''decentralized'' sources that are now widespread in rural China, providing an important lesson for the rest of the world. The People's Republic could not afford to make investments in large industrial equipment without such far-sighted rural development. A reasonably self-reliant rural population, educated and not starving was a prerequisite to self-propelled development.

Beginning in the late 1950s, in a frenetic and unplanned way, China began to stress the development of small-scale hydro plants and fermentation of human and animal waste into fertilizer and biogas. The program has now come of age. A recent report by Vaclav Smil concludes that the ''Chinese have become the world's leading practitioners of frugal energetics and their achievements have been notable in two areas directly growing from their ancient experience in handling and recycling animal and human wastes and building admirable waterworks: Chinese successes in rural biogas generation and small-scale hydroelectricity production will remain unmatched for a long time to come.''

During the Great Leap Forward, in 1958, Chairman Mao beseeched peasants to build millions of biogas digesters immediately, but the program fizzled and many digesters were soon abandoned. A second effort began in 1970-1972 in the Mianyang and Zhongjiang counties of Sichuan province, where several hundred digesters were installed.

By the spring of 1974, Sichaun province had 30,000 digesters, and by fall, 120,000. In June 1975, the number had reached 410,000, and by the end of 1977, Sichuan had 4.3 million digesters providing 20 million villagers with biogas for fuel and lighting and fertilizer for food production. The nationwide total reached 7 million in August 1978, 15 times as many as in 1975.

By contrast, India had 15,000 digesters in operation in 1977, or one digester per 10,000 rural families. China has 300 digesters per 10,000 rural families.

Like all biogas plants, the Chinese units rely on microbiological digestion of plant and animal wastes in an oxygen-free atmosphere. A high grade gaseous fuel and ammonia and phosphorous rich fertilizer are produced. The Chinese strongly claim that digesters improve village health, as schistosoma cells are destroyed after about a week of fermentation and thus do not appear in the fertilizer.

Most plants in China range from 10-20 cubic meters in capacity. Unlike India, where plans for biogas plants are predominantly for village-scale units, Chinese plants are family size. In both cases, the reasoning behind the choice was social and political, rather than technical. In India, individual biogas plants would increase the gap between the rich and poor by turning dung — now used by the poor as a cooking fuel — into a commodity that rich owners of biogas plants would buy. The poor would end up with

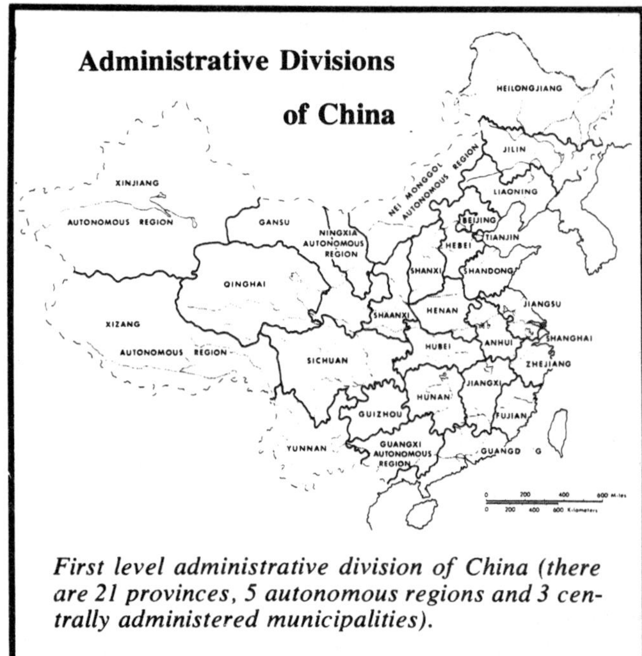

First level administrative division of China (there are 21 provinces, 5 autonomous regions and 3 centrally administered municipalities).

neither biogas nor cooking fuel. Most current Indian efforts, as a result, are for communal projects.

A typical family digester costs about 30-40 yuan in the Sichuan and Hunan provinces of China. Cheap, locally available materials are used to build the structure — mortared rocks, bricks, 100-200 kilograms of cement, a one-meter long PVC or steel pipe (for extracting gas), and about 100 kilograms of lime are about all it takes. Bamboo or plastic pipes are used to transport biogas to each house. A simple pressure gauge is used to measure leaks, estimate the quantity of biogas present in the tank, and prevent a pressure build-up that could crack the chamber.

A latrine and pigpen (China has 300 million hogs) are usually directly attached to the intake chamber. Other fermentable materials — grasses and straw — are dropped into the intake chamber; from there they move to the main fermentation chamber, which is held under pressure by a rigid wood cover. Ample pressure for distribution of biogas to the home is achieved by forcing up water columns in the intake and outlet areas against outside air pressure. The gas can be piped 50 feet.

Family-scale units have many advantages; they have

spurred economic health in China, reinforced a sense of self-reliance, contributed to environmental protection, and consequently promoted a relative political stability. The fuelwood crisis that looms in most of the Third World may well be avoided in China. In the summer of 1978, at the nation's second national biogas conference, Chinese planners proposed a target of 20 million units in use by 1980 and 70 million in use by 1985. The latter total would imply that two-thirds of China's rural households would use biogas for heating and lighting by 1985.

Decentralized Hydropower

The expansion of China's small hydroelectric capacity parallels the biogas experience. In 1949, China had two large power stations built during Japan's brief occupation and 50 small plants, with a total capacity of 5.6 megawatts. It was not until 1967 that the program began to self propel.

Nationwide figures show the total number of small hydro stations growing from the initial 50 to 6,000 in 1965, 35,000 in 1971, 50,000 in 1973, more than 60,000 by September 1975, and nearly 87,000 by the end of 1978.

Thrift, use of local resources, and even local financing are hallmarks of the program. Most of the stations have generating heads of less than 10 meters, with the average station producing about 57 kilowatts of power. The generating equipment varies from simple wood propellers to rather sophisticated exported Francis turbine sets of 0.418-28 kilowatts in size. The exported units larger than 3 kilowatts can be directly coupled to on-site machinery.

The smallest Chinese hydro generators of sophisticated design are 400 watts in size, for use on creeks and springs. By the mid-1970s, small hydro units contributed to a third of China's total hydro generation which, in turn, represents about 30 percent of the nation's total power production. But, as Vaclav Smil points out: "The importance of small hydros goes beyond the fact that they have been producing roughly every tenth kilowatt-hour in China: in most cases these stations have been the first source of power for the villages and have served as the foundation of the rudimentary electrification of China's countryside."

Irrigation and drainage pumps, food processing equipment, and power for local small industries have taken precedence over household use beyond lighting.

Small hydro plants are not without environmental disadvantages and social risks, however. Reservoirs silt up, requiring prompt reconstruction; droughts occur; and the complete emptying of the small reservoirs that often back up hydro stations can bring local food production to a complete halt. But, for the most part, small hydro plants have permitted local electrification in small amounts, without major national capital investments or foreign exchange. By contrast, most rural Third World electrification programs have been dismal failures, expensive and ineffective. Douglas Smith of the Massachusetts Institute of Technology Energy Laboratory points out, "Rural electrification continues to be a catch phrase of considerable emotional appeal among politicians, generals, and power engineers." Reviewing the case of Bangladesh, Smith adds, "There is no evidence of the development-inducing effects of rural electrification although there have been monitored projects since 1963. In any event, rural electrification in practice means electrification of town centers, lights and fans for the influential, and volumes of talk about energizing tubewells."

This is not so in China, because of a synergism between good cooperative planning, development of local resources by local people, and, in Ivan Illich's phrase, "convivial tools." In fact, the Chinese example makes one wonder whether technology assistance from developed nations can really contribute anything to Third World nations. There is considerable evidence to suggest that aid programs are Trojan horses, brought in on a pretense of cooperation to leave only after extracting dear victory. The Chinese biogas program is an instructive example of self reliant development in energy and agriculture and a people, neither hampered nor "assisted" by the schizophrenic efforts of the developed world.

Source: Vaclav Smil, *Renewable Small-Scale Energetics in China,* work in progress report prepared for International Soft Energy Path Conference, Rome, Italy, May 17-20, 1979.
Jim Harding

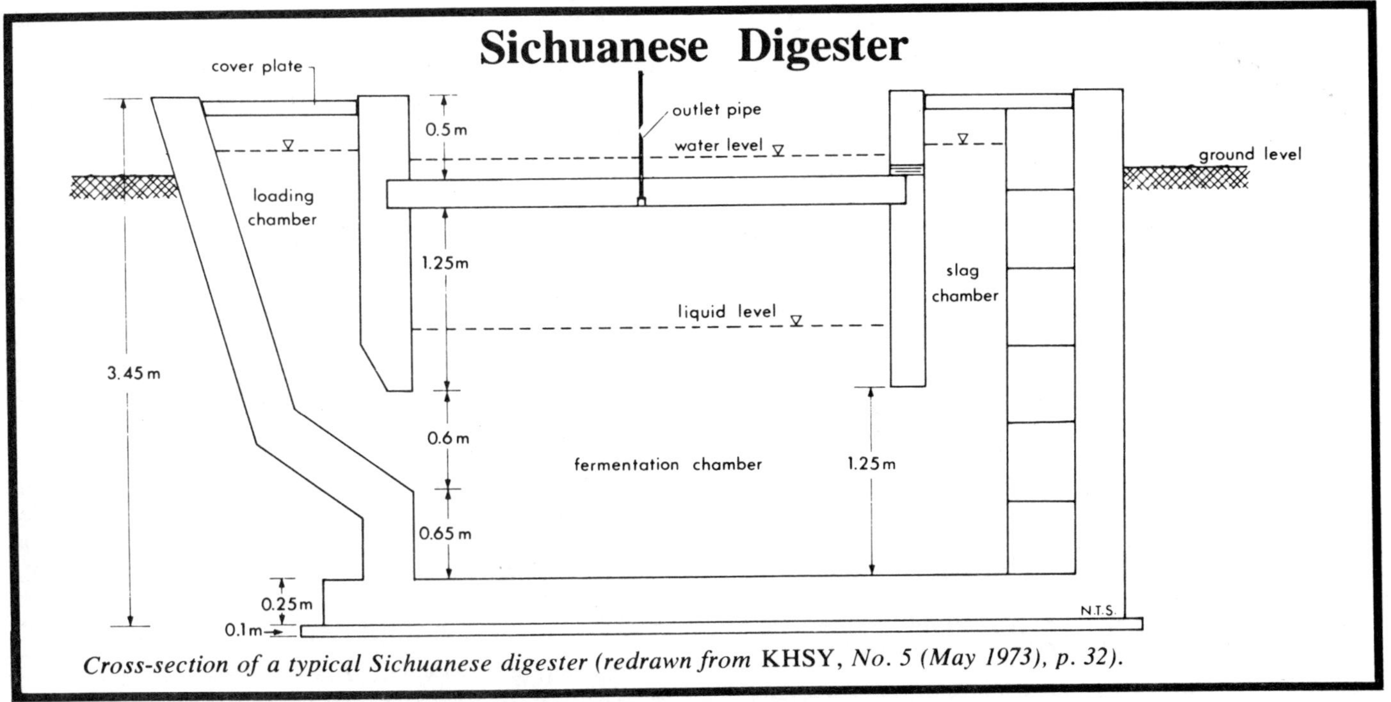

Cross-section of a typical Sichuanese digester (redrawn from KHSY, No. 5 (May 1973), p. 32).

Malaysia's Soft Energy Path

by Gurmit Singh

*M*ALAYSIA IS A SMALL COUNTRY of 13 million
people located on the southernmost tip of the
*Asian land mass. It is one of the comparatively
richer members of the Third World, with a large
natural resource export base, being the world's
largest producer of natural rubber, palm oil, and tin.
It is also currently a net exporter of oil: 1979's net
oil exports were worth 2.28 billion Malaysian
dollars (US$ = M$ 2.1). However, this oil surplus is
expected to become a shortfall by 1990 unless soft
path options are taken.*

Energy statistics

As in many other developing countries, there is an
appreciable discrepancy between official energy statistics
and projections, and those estimated by soft energy
workers. This difference arises because official figures
derive from the large-scale commodity trade, and ignore
the rural, domestic, and renewable sectors. Energy
projections vary for a number of reasons: official projections based total demand on GDP growth; they target
energy use in the form of electricity because they equate
this with modernity; and they ignore the improved
efficiencies possible when sources are matched to end
uses.

The following tables describe the current energy
situation in Malaysia, according to government statistics,
and include projections of future consumption and supply.
It will be apparent that the official figures contain
inconsistent elements, and should not be taken at face
value.

Energy consumption by source *(barrels/day oil equivalent)*

	1970	1975	1977	1978[*]
Oil	63,310	98,168	110,055	110,000
Hydro	1,392	1,184	1,023	2,500
Others, incl. wood and charcoal	7,727	5,468	13,411	36,000
Total Demand	72,429	105,308	124,489	148,500
% oil	87.4	93.7	95.2	74.1

[*] *(Not official statistics)*

Energy demand projections 1980-2000

Year	Gasoline & Diesel (bbl. day oil equiv.)	Total Demand	Avg. annual growth rate
1980	72,400	175,000	8.6%
1990	138,000	400,000	5.6%
2000	—	700,000	

Energy consumption by sector *(in percent)*

	1970	1973	1975
Agriculture	9.1	8.1	8.2
Mining	14.0	11.2	10.3
Manufacturing	12.2	10.2	11.0
Construction	1.5	2.2	2.1
Electricity	17.7	22.0	24.3
Transport	10.0	10.5	9.7
Wholesale/retail	7.5	6.8	5.3
Dwellings	0.9	1.1	1.0
Public Admin.	4.3	3.8	3.7
Other services	0.5	0.4	0.5
Private consumers	22.4	23.2	23.8

Annual electricity generation by plant type

	Hydro%	Thermal, oil-fired %	Thermal, gas-fired %	Total (GWh)
1970	33	67	—	3,388
1975	18	82	—	5,444
1980	15.6	84.4	—	8,716
1985	23.1	64.1	12.8	9,413
1990	15.7	61.9	12.8	26,740
1995	11.6	44.1	44.3	42,860

Annual rate of increase = 10.21%, 1970-80.

Households supplied with electricity

		Peninsular Malaysia		Sabah		Sarawak	
	Rural	345,602	32%	7,480	7.5%	4,459	6.3%
1970	Urban	387,097	77	17,605	90.0	18,823	—
	Total	682,693	43	25,086	21.1	23,282	—
	Rural	515,000	41	23,170	18.4	12,026	10.6
1975	Urban	329,754	67	22,321	90.0	22,826	—
	Total	835,754	48	45,492	30.2	34,922	—
	Rural	790,000	55	42,850	29.3	29,100	19.3
1980	Urban	564,000	85	32,350	90.0	27,500	—
	Total	1,354,000	64	75,200	41.3	56,600	—
	Rural	1,090,000	65	69,777	38.4	46,100	25.8
1985	Urban	714,000	91	44,433	90.0	31,000	—
	Total	1,804,000	73	114,210	49.5	77,100	—

Soft energy options

The most optimistic estimates place Malaysia's fossil fuel reserves at 2.8×10^9 barrels of oil, 15×10^{12} cubic feet of natural gas, and 600 million tons of low grade lignite coal. There is some possibility of significant uranium deposits, but no reliable data on the size of the reserves.

The following soft energy options have been identified to date:

Biomass

The following (1978) potential was calculated by Kien Keong Wong:

Source	Quantity (10^6 tonnes)	Energy value (10^6 GJ)
Rubber wood	7.0	88
Timber waste	0.5	6
Palm oil		
shell	1.4	27
fibre	0.9	13
bunches	0.7	10
effluent	5.8	5
Domestic & animal wastes		7
Rice wastes	1.2	15
Sugar cane wastes	0.9	11
Solid municipal wastes	1.8	11

Of the above, 50 percent of the first five items is already being utilized in one form or another. There is a double motivation to develop the energy potential of the other biomass sources, as these materials cause serious pollution problems.

Hydropower

The energy potential with conventional, large hydro installations is 16,100 GWh/annum for Peninsular Malaysia, 65,700 GWh/annum for Sarawak, and 42,000 for Sabah. The largest single plant is envisaged for Balui in Sarawak, with a rated potential of 2,830 MW. The snag in all these large-scale schemes is that the majority of the hydro potential is located on the island of Borneo, while the greatest power demand is on the Peninsula. Power transmission of this magnitude via undersea cable represents a serious technical barrier.

More than 100 mini-hydro projects throughout the country are now under investigation by three international engineering firms recently appointed by the Malaysian government. The National Electricity Board is currently implementing twelve of these on the Peninsula.

Solar

Malaysia receives every day an average of 4.5 kWh/m² of solar radiation, but no reliable data indicate how much of this is now being utilised by either commercial or traditional sectors. It has been estimated, though, that if 0.1 percent of the land area would be used to capture solar energy at an average efficiency of one percent, the net available energy would be 15 GWh.

Solar technology applications are, unfortunately, not extensive. A handful of water heaters has been installed in middle-class homes, a few prototype crop dryers are in the fields, and some photovoltaic work is in progress in the universities. Multinationals have promoted a few radio/TV and telephone applications in remote locations.

Of the above, 50 percent of the first five items is already being utilised in one form or another. There is a double motivation to develop the energy potential of the other biomass sources, as these materials cause serious pollution problems.

Wind and wave energy: Hardly any work has been done to evaluate the potential of both these sources in Malaysia, and many planners have apparently dismissed them offhand.

Geothermal: A preliminary study is underway in Sabah, on the northern tip of Borneo, but the findings have still to be publicly released.

Energy conservation: The Malaysia Government recently launched an Energy Conservation Campaign, concentrating on radio and TV. Its main targets are electricity and petroleum consumption. The campaign's second phase envisages as yet undescribed fiscal, regulatory, and legislative measures. The Ministry of Energy also proposes energy conservation incentives to induce capital investment but the start date for the program is not yet set.

A step in the right direction was recently taken when the electricity tariffs were restructured. The rate per unit now increases with higher levels of consumption. For example in the Peninsula, households now pay 20 ¢/kWh for the first 100 units, 23¢ for the next 900, and 26¢ for each additional unit. Industrial consumers are charged 23¢/kWh for the first 200,000 units, and 25¢ for each additional.

Soft path obstacles

- *Complacency.* Even after price increases for oil and electricity and the National Energy Conservation Campaign, there is little evidence of any conscious energy cutbacks. Private motor car sales are still increasing, and air conditioning for offices and middle class homes proceeds unchecked.
- *Obsession with progress and modernity.* Most Malaysians seem obsessed with these two concepts, and link them to higher energy consumption—particularly electricity. This view is reinforced by the government, which employs electricity use as a quality of life indicator. Meanwhile, multinationals plug the advantages of electric washers and

newspapers proclaim that air conditioning is no longer a luxury but a necessity.

• *International economic pressures.* Malaysia is pressured into spending its considerable export revenues on energy-intensive equipment and technologies. The nuclear industry has been knocking at our door for several years, while others are trying to launch us on paths that will exhaust our energy resources in processing raw materials, for ourselves or for others.

• *Misguided elites.* Politicians, businessmen, engineers, and scientists aspire to emulate and catch up with their peers in the First World. Hence, soft energy is seen in the same light as appropriate technology: something second-hand, if not worse.

• *Raised expectations.* The general public's demand for goods and services seems to be realisable only through large energy and resource use.

• *Foreign domination of soft energy technologies.* In all the options discussed so far, multinationals and foreign consultants are swamping local initiatives.

• *Lack of funds and local effort.* Total funds allocated to date for non-hydro soft energy work is probably not more than $1 million (Malaysian).

• *Absence of national energy needs inventory.* Soft energy options have not been adequately identified, since no exhaustive study of sources and energy matching possibilities has been undertaken.

• *Exclusion of public participation.* A sad feature of Malaysian life is the exclusion of the public from the energy debate, save at a few token seminars and workshops. All that the Environmental Protection Society of Malaysia has achieved over the last seven years is sporadic newspaper coverage, and input to the draft Malaysian Social Report.

The soft energy path in Malaysia is not an easy and straight one. It has many pitfalls and is likely to take many turns before we can become self-reliant, and the poorest sector of the population is assured of adequate quality of life. Malaysia cannot travel alone on this path, especially if the countries of the First World do not themselves undertake a similar journey.

Gurmit Singh's article on the Malaysia soft energy path was originally presented at the Second International Conference on Soft Energy Paths, held 16-19 January 1981 in Rome.

Brazil Moves Toward Energy Self-Sufficiency

"AS LONG as people crowd into cities," writes Brazilian energy researcher José Goldemberg, "a hard energy path is unavoidable." Small, rural communities may be able to satisfy their energy needs with local agricultural wastes and small-scale hydropower, but large urban populations, he claims, call for centralization of supply. This need not, however, stop Brazil from becoming energy self-sufficient by 1990, and in *Soft Energy Paths: The Case of Brazil,* Goldemberg examines the route that South America's largest country is taking to eliminate its dependence on foreign energy imports.

Since 1950, Brazil's urban population has soared from 40 to 60% of the national total, accompanied by a steady increase in the country's reliance on energy sources that are environmentally, economically, and socially damaging. Soft energy technologies can affect only a part of this process: they can help to alleviate problems associated with energy supply, but by themselves, they cannot reverse the course of rural-urban migration. Brazil is unlikely to shift entirely to soft technologies, even though the technical potential to do so is there. "Once adopted," Goldemberg explains, "a hard path technology is hard to abandon because a great deal more than just technology is involved."

Brazil is now developing renewable energy sources which contain components of *both* soft and hard paths. The current move toward hydropower and biomass, for example, is less environmentally destructive than the use of fossil fuels, but the technologies are not always small in scale and decentralized.

Goldemberg notes that "although there is an important role for minihydro plants in Brazil, it is clear that the largest share of the power will come from large stations, which are destructive to the environment." Producing large volumes of alcohol from biomass also requires an industrial infrastructure, which presents many hard path environmental problems. This energy system requires, further, that vast areas of land be diverted from food crops into energy production, bringing the interests of Brazil's energy-hungry affluent elite into conflict with her largely subsistence population.

Other aspects of biomass production in Brazil, however, seem to move the country toward soft path goals. The heightened

> *"Alcohol from biomass diverts land from food crops, bringing the energy-hungry elite into conflict with those who are simply hungry."*

demand for agricultural material will increase the number of job opportunities in rural areas, and this economic shift could reduce the flood of urban migration, encouraging a more decentralized form of society.

Brazil's Energy Profile

A growing urban population, the large number of private vehicles, and an energy-intensive industrial base combined, between 1940 and 1975, to increase Brazil's dependence on imported petroleum. Primary energy consumption grew at a rate of 7% per year (doubling roughly every ten years), reaching 4 quads by 1975. The contribution of petroleum rose from 9.6% in 1940 to 45% thirty-five years later, while the share provided by biomass dropped from 80 to 30%. Hydropower, accounting for 20% of Brazil's primary energy, and coal, with a 5% share, round out the country's 1975 energy mix.

Only 14% of Brazil's enormous hydroelectric potential (estimated at 119,000 MW) is now being used, and government programs call for an average yearly increase in generating capacity of 11% over the next two decades. Only conventional sites—that is, those larger than 10 MW—are to be developed. Included will be Amazon River tributaries in Northern Brazil which pose transmission problems since these sources are 2,000 km away from the large population centers of Sao Paulo and Rio de Janeiro in the Southeast. High voltage, direct current transmission lines are now being introduced and it has been suggested that new energy intensive industries be located near the hydropower sources in the North.

Brazil's minihydro potential is unknown but large. Goldemberg sees Amazon Valley river bank villages as textbook examples of places where minihydro would be appropriate. "There are hundreds of villages," he explains, "scattered in the immense valley where power is produced by diesel sets which have to be supplied by boats or airplanes because the terrain rules out ground transportation." This system is expensive and unreliable, and requires regular maintenance of the diesel generators. Unfortunately, the government's present hydropower program slights smaller, unconventional sites, and there is only one pioneering effort, in Aripuana village, where minihydro generators are being installed.

The abundant potential for hydroelectricity in Brazil should, according to Goldemberg, obviate the need for a large nuclear program. Brazil has nonetheless agreed to purchase eight 1,300 MW reactors from Germany, but Goldemberg comments that "the implementation of this deal is behind schedule and one has serious hopes it will be curtailed in the near future."

Reducing Oil Imports

While hydropower can displace petroleum in the generation of electricity, biomass liquid fuels can reduce the oil imports required for transportation (see *Notes,* July 1979, *Brazilian National Alcohol Fuels Program*). The first phase of the biomass program, commenced in 1975, involves a 1 to 5 blend of ethanol to gasoline, which present-day engines can use without modification. In the second phase, engines will be altered to accept 100% ethanol, and by 1982, major car manufacturers will install ethanol-modified engines in 15-20% of the automobiles they produce.

Since alcohol fuel burns cleaner than gasoline in well-tuned, modern engines, air quality in the Sao Paulo area, with over a million cars, is expected to improve. At the distilleries, though, an alcohol by-product called *vinhoto* is being dumped into nearby streams, heavily polluting them. But *vinhoto,* Goldemberg notes, "can be used as a fertilizer and returned to the fields, or dried up and used as a protein source."

Critics within Brazil are concerned about the use of sugarcane and prime agricultural land for energy crops when much of the

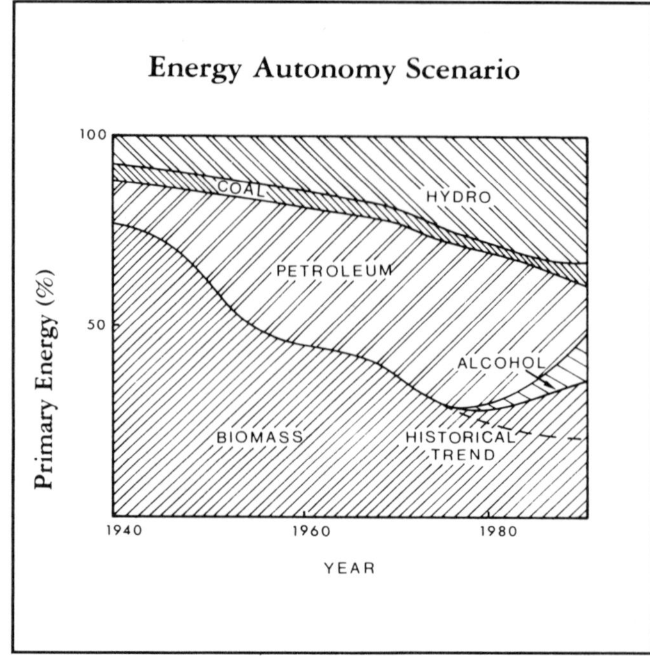

population hovers at or below the subsistence level. They argue that trading food for fuel—much of which is dissipated in unnecessary urban auto traffic—commits an injustice against the nation's poor. Sensitive to this argument, the government is experimenting with cassava (manioc) grown on marginal land as a fuel feedstock, and methanol from trees may provide an even more benign solution.

The wave of government enthusiasm for renewable energy sources has extended to the recycling of urban waste for methane gas and/or power generation. And, Goldemberg adds that the bagasse that remains after the production of ethanol can also be used as an energy source.

Energy Autonomy

Goldemberg reports that Brazil's current efforts to develop renewable, indigenous energy supplies could result in complete energy self-sufficiency by 1990. This scenario incorporates the following assumptions:
- 7.3% increase per year in overall energy consumption;
- 10% annual increase in hydroelectric generation;
- constant proportion of energy supply provided by coal;
- displacement of petroleum by biomass fuels.

The transition, Goldemberg asserts, would require a heavy capital outlay, much of which is already committed to hydropower development and to foreign oil purchases. These two energy systems presently abosrb 2% of GNP, and Goldemberg calculates that 4% of GNP must be invested in indigenous energy production for Brazil to achieve self-sufficiency. No reduction in overall consumption will be necessary, only the shift to renewable sources. But, as Goldemberg points out, this in itself will not put Brazil on the soft path. The concentrated, Southeastern urban centers will continue to be served by centralized, large-scale, but renewable energy systems.

—Cissy Wallace

Reference:
Goldemberg, José
1979 *Soft Energy Paths: The Case of Brazil.* Prepared for IPSEP, 15 pp. Available from IPSEP for $2.

1. Barriers and Impediments

One of the most contested aspects of soft energy path proposals has been their claim to economic cost-effectiveness. After years of debate, a consensus is emerging that the technical efficiency improvements described in Part I are cost-effective by virtually any criterion applied to hard energy technologies. Renewable energy technologies are generally competitive with new oil, gas, coal, or nuclear energy. Had we resolved that debate ten years earlier, Roger Sant tells us, each of us would now be saving several hundred energy dollars per year while enjoying the same services.

What is holding us back? Obviously, the market won't automatically lead us in a cheap direction. It is rigged with subsidies to conventional and nuclear sources of energy. These subsidies make conventional energy appear cheaper by more than a third and hide an estimated half of the cost of nuclear electricity in the US.

Back in 1970, utilities misprojected 1980 electric demand by nearly 50 percent and built powerplants at costs 3, 4 or 5 times higher than their estimates. We pay for that market inefficiency, and that's not the only one. One serious problem with improving the efficiency of residential and commercial buildings is the landlord-tenant relationship, but there are other examples of such potential gains limited by "split incentives." Homebuilders strive for low first cost; homeowners pay the big utility bill.

If the direct economic losses of misdirected energy policies are staggering, indirect costs can be just as damaging. Soft path investments create significantly more, and more equitably dispersed, jobs than hard road developments, with their boom town effects. As the economy ails and unemployment plagues many communities, the social inequity of centralized energy investment becomes increasingly evident.

The lesson of all this is that energy policies have never merely relied on market forces, and such reliance would be foolish now. As a first priority, industrial societies must establish explicit and sensible goals for their future energy development. For setting these goals, economic cost, ecological sustainability and social vulnerability questions (i.e., equity and security) must all be brought into a balance. In achieving these goals, all instruments of implementation, from setting standards for manufacturers to subsidizing market mechanisms, have a place.

Is the Soft Path Cheap, Too?

A RECENT report by the conservative Carnegie-Mellon Energy Productivity Institute concludes that if it weren't for government interference in energy pricing and regulation we would be using 24 percent less energy than we do, and paying almost $200 less per year each for energy services. Titled "The Least-Cost Energy Strategy," the report concludes that the major waste is in centrally generated electricity, where we could be using 43 percent less if cost-effective conservation and industrial power generation had been implemented properly.

In fact, the report suggests that our problems in 1978 could have been very much less serious than they were. A 1978 economy that was "optimally designed" for 1978 energy prices would have used, according to the analysis, 28 percent less oil than was actually used, 34 percent less coal, and 43 percent less centrally generated electricity. "Theoretically," writes the lead with all the heat, light, and mechanical motion—the end results of energy use—that they require, or even want, at the least possible cost."

To encourage competition between energy sources and conservation techniques, the Sant report urges deregulation of the oil, gas, and electric industries, except for pipelines and transmission systems, which could be "common carriers." In particular, the report cites four cases of "over-regulation" that worked to the detriment of a "least-cost" conservation based strategy.

"Heavy government research and development emphasis has been focused on increasing utility supplied electricity, despite the fact that in straight cost competition electricity consumption would have decreased.

"Regulations have forced industrial conversion to coal, as a means of reducing oil imports, despite the fact that at 1978 fuel prices our analysis shows that it was not in the industrial users' economic interest to do so.

"The most cost-effective opportunity to reduce oil consumption is in the building sector, yet government programs have focused on the transportation and industrial sectors.

"Controlled, artificially low prices for natural gas have created excess demand; this has resulted in over-use in the building sector because of assured supply to high priority customers and under-use in the industrial sector because the supply was not available. The Natural Gas Policy Act has increased that distortion by passing on the price of higher-cost gas to industry only, thereby encouraging continued over-use in buildings."

And, finally, "The present popularity of a massive government program to produce synthetic fuels well beyond the testing of commercial feasibility is a further example of the propensity of legislators to enact measures that work against achieving the least-cost objective."

The 50-page summary does not do justice to the detailed calculations of cost-effectiveness in each of the principal energy using sectors: buildings, transportation, and industry. Nearly author Roger Sant, "a least-cost strategy applied during the 10 to 12 years prior to 1978 could have reduced the cost of energy services by roughly 17 percent in 1978 with no curtailment of services."

Over-regulation and under-pricing are at the root of our energy problems, according to the group. "We have concluded," writes Sant, "the implementation of a least-cost strategy would utilize much of the traditional free market system. . . .Our conclusion about the value of market forces in the operation of a least-cost strategy arises from the evidence of our analysis that, in regard to most of our energy problems, a freely competitive environment would work, primarily because of the numerous, diverse, and actually or potentially competitive technologies that can be brought into play in the energy economy."

Formerly a vice-president of Saga Foods, Sant joined the Federal Energy Administration in 1973 as Assistant Administrator for Conservation, to the consternation of many environmentalists. Sant later resigned over the unwillingness of the Ford Administration to push conservation efforts.

The Sant report stresses the concept of energy services in language that could be Amory Lovins's: "The resolution of the energy problem might be achieved through what can be called the 'least-cost strategy,' a strategy that concentrates on the ultimate goal of providing industrial and individual consumers

one-fourth of the savings in their hypothetical least-cost 1978 economy come from improvements in residential buildings. A seventh of the savings come from automobile weight reduction and power-train improvement. And a tenth comes from better air conditioning equipment in large commercial buildings. All the other factors: 15 or 20 small economic adjustments to higher prices yield the total savings. Conservation is not monopolistic, but small and diverse, though additive.

Interestingly, the Sant report carefully estimates costs for each improvement. The total capital cost of the package of improvements, all of which increase national productivity, is calculated at $364 billion. This appears to be a considerable sum, but the amount of energy saved is also considerable, nearly a fourth of total US energy supply (19.3 quads). The average cost of these energy savings, then, is less than $12 per equivalent barrel of oil. Had the investments been made, the Sant group calculates that over $400 billion would have been saved in unnecessary power plants and petroleum imports.

The report concludes on an upbeat note: "A wave of optimism—and commitment—is beginning to emerge from many quarters: these changes are possible, desirable, and necessary. Perspectives have and will continue to change rapidly. When coupled with ingenuity, new technology, and improved management, these changes can be powerful enough to master the energy problem. In fact, seen in this perspective, the problem is transformed into an opportunity—increased employment, new markets, and enhanced environment, a more secure energy future, and most important, less onerous levels of energy service costs. We are definitely not stuck with our old attitudes about energy and energy conservation. Our analysis to date shows that we can move to higher levels of productivity through a more competitive, consumer-oriented energy policy."

It is useful to see energy conservation presented in a way that appeals to a new constituency (in this case, *laissez-faire* economist Milton Friedman and friends) and the Sant report does that. But the overall impression, that one can create a "least-cost" result merely through deregulation and higher prices, is deceptive. Cogeneration can be stimulated by deregulating the generation of electric power, or by setting rates for purchased power that are equal to the cost of new supplies. (California's Public Utilities Commission is urging cogeneration rates at 90 percent of new supply cost.) Energy-efficient automobiles take considerable time to develop, and tight mileage standards have been an essential component. Standards and regulations may force concerted risk-taking in a society that has become fearful about long-range investments.

There are equity questions as well. High energy prices don't encourage renters to insulate their apartments, they just make them poorer. And in sectors where consumer demand outpaces supply, as in new homes, people take what they can get and builders build what builders want. The marginal cost of new home improvements can be as low as $1,200 to save two-thirds of the heating cost, with a probable payback of several years, and efficient houses sell but there aren't enough of them.

If something already cost-effective at today's prices is not being done, will doubling the cost of energy make it happen? Possibly, but doubtful.

—*Jim Harding*

Reference:
Sant, Roger W., *et al.*
  1979 *The Least-Cost Energy Strategy.* The Energy Productivity Center, 1925 N. Lynn St. #1200, Arlington, Virginia 22209. $5.

US Nuclear Subsidies Top $37 Billion

TAXPAYERS HAVE SUBSIDIZED the development of commercial nuclear energy to the tune of more than $37 billion (1979 dollars), according to a draft report recently released by the Energy Information Administration, of the US Department of Energy. Without these subsidies, the EIA estimates the cost of nuclear electricity would be 66 to 100 percent higher. Continuing federal subsidies signal a government commitment to nuclear technology, the report comments, reducing the uncertainty that would otherwise be shouldered by private enterprise.

The EIA analysis is the first in a series of reports on how government actions "differentially affect energy markets," and on the criteria for subsidizing new energy technologies. The present report reviews "the rationale for selected Federal subsidies to nuclear power, including subsidies to reactor design and sales, subsidies to uranium producers, subsidies provided via government pricing of enrichment services, and subsidies provided via government control of waste management." While many of the original R & D costs of nuclear development are sunk, the EIA report strongly suggests that the annual rate for subsidies is growing, exceeding $3 billion in 1979 (all cost figures are in 1979 constant dollars). Amortized over the entire production period, the nuclear R & D subsidy calculated by EIA is equal to $28-42/barrel of oil equivalent in delivered energy.

The EIA estimate is taken from a number of sources, including a 1978 report by Battelle Pacific Northwest Laboratories that placed a subsidy pricetag on nuclear energy of $15.3-17.1 billion. EIA has updated the data through 1979 and added several subsidy categories not found in the earlier study. Even so, the evaluation is incomplete. It does not consider such important subsidies as the tax treatment of nuclear plants, or the limitation of accident liability as provided by the Price-Anderson Act.

Nor does the review of nuclear energy subsidies include research and development for military purposes, though many facilities, such as the gaseous diffusion uranium enrichment plant in Oak Ridge, Tennessee, have served both masters. The naval reactor program similarly "contributed directly to designs later adopted for civilian commercial use." In such cases, "only those (military nuclear activities) which contributed most clearly to the development of civilian reactor technology are considered to have provided subsidies."

Subsidy History

The Atomic Energy Commission laid the foundation for federal subsidies in 1951 with its Industrial Participation Program, designed to encourage the involvement of private industry in nuclear power generation. Early studies concluded that the barriers to industrial involvement in nuclear development were higher—in capital requirements and technical uncertainty—than private industry could surmount. In 1954, the Atomic Energy Act was amended to allow private ownership of nuclear facilities, and to provide AEC services and access to technical materials. This was of little help, however, and it was not until 1955 — when the AEC instituted its Power Reactor Development Program (PRDP) — that commercial interest truly surfaced. Under this program, and AEC performed R & D on request without cost to private firms, provided lump sum R & D grants, and waived nuclear fuel charges to utilities with power reactors. An ancillary program permitted the AEC to buy, build, and operate a power reactor for a utility company. The company would find a site and own the conventional part of the power plant, buying steam from the AEC, and selling electricity at the normal rate.

In the PRDP's second round, the AEC instituted programs to develop small reactors for publicly owned utilities. In a third, the AEC further shouldered development costs for utility companies, waived fuel costs, loaned heavy water, and offered R & D grants. Three light water reactors were built during this phase of the PRDP, laying the technical groundwork for further commercial use. In the final stage of the program, which began in 1962, only light water reactors were eligible for support. The AEC limited its direct financing role to ten percent of R & D, and waiver of fuel charges. Under this program, the Connecticut Yankee plant at Haddam Neck was completed.

At this point, subsidies for further development of commercial reactors came from reactor vendors in the form of "turnkey" contracts. Companies such as General Electric or Westinghouse would contract for and construct an operable reactor for a fixed fee, assuming the risks of technical failure or cost overrun. Hundreds of millions of dollars in overruns were assumed by private firms competing for turnkey contracts; these subsidies are, of course, not considered in the EIA analysis. Subsequently, the AEC's role shifted to R & D for advanced nuclear breeder reactors.

Breakdown of Subsidy Components

The principal direct federal subsidy for commercial nuclear power has been the research and development effort, totaling $23.7 billion between 1950 and 1979, according to EIA's estimate.

Foreign reactor sales are another type of federal subsidy. Under the Atoms for Peace program, research reactors and equipment were sent to 27 countries between 1953-1962 at a total cost of $29.1 million. The Agency for International Development (AID) provided a $72 million loan to the government of India on extremely favorable terms, to help build the Tarapur nuclear reactor. The AEC also waived or deferred fuel costs for US-designed reactors to be built in

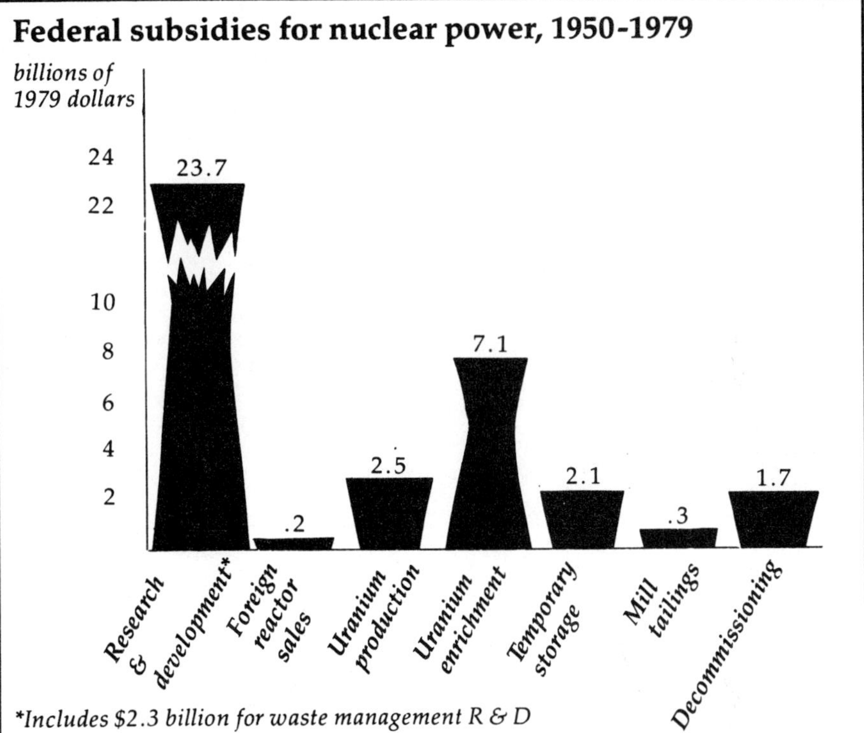

Europe. Research grants, training programs, and maintaining the Export-Import Bank program — with $6.5 billion to date in loans and guarantees — constitute additional subsidies. EIA estimates the total cost of the quantifiable subsidies for foreign reactor sales at $237.4 million.

Uranium production subsidies have cost the taxpayer $2.5 billion, according to EIA. Beginning in 1948, the AEC offered a ten-year guaranteed price for uranium yellowcake, a $10,000 bonus for high-grade discoveries, provision of access to public lands, payment of development allowances and transport costs, and a guaranteed price for the vanadium that is often a co-product of uranium mining. By 1955, the US was the world's largest uranium producer. By the mid 1960s, the program was phased out, with uranium prices fully decontrolled by 1970. The current federal role in uranium development is limited to an intensive exploration program, the National Uranium Resource Evaluation (NURE). Aerial assessments, stream sediment surveys, and hydrogeochemical analyses are conducted under this program. The principal purpose has been to create a uranium mining industry, reduce financial risks, and support uranium prices at a level capable of meeting vast increases in demand. An embargo on importing foreign uranium also supported domestic mining, though EIA makes no attempt to estimate this cost to consumers.

Uranium enrichment is also heavily subsidized. The three government-owned enrichment operations are, according to the EIA review, almost entirely available for civilian uranium production. The prices charged, consequently, do no include a rate of return, taxes, insurance, or eventual enrichment plant decommissioning. Power is the principal cost component of the enrichment process, and the EIA report notes that the "government buys power at a price well below the price that a commercial enrichment facility could expect to pay." A third enrichment subsidy takes the form of plant depreciation. In contrast to other AEC and DOE facilities, the enrichment plants have been depreciated over a 50 year period, rather than the 23 years typical for such installations. All depreciation prior to 1960 is charged against military uses, with subsequent depreciation to civilian programs. Together, the enrichment subsidies total $7.1 billion.

Subsidies for waste management total $6.5 billion to date, according to EIA. Direct research and development efforts make up $2.3 billion of this sum (and are accounted by EIA in the

> Economist Duane Chapman of Cornell points out that nuclear plants, in fact, have a negative income tax status under federal law, though they are owned by firms considered to be in a 48 percent corporate tax bracket.

$23.7 billion R & D total). Clean-up of mill tailings from abandoned private uranium mines comes to about $300 million. Temporary storage of used reactor fuel represents a further $2.1 billion subsidy, given the EIA assumption that the spent fuel handling cost will be $352/kilogram of used fuel. Decommissioning costs are estimated by assuming expenditures of $110 million per reactor. With 22 percent of the average life of existing reactors "used up," the value of the funds that should have been collected to pay for decommissioning is $1.7 billion.

Additional Tax Benefits

The EIA report makes no attempt to calculate the cost of the tax benefits that utilities enjoy when they purchase nuclear reactors, yet some believe this to be one of the largest taxpayer contributions to the nuclear industry. Economist Duane Chapman of Cornell, in a 1979 report to the California Energy Commission, found that state and federal tax subsidies for a new nuclear plant amount to over $200 million per year (in 1988 dollars). Also in 1988 dollars, he finds that a nuclear plant with power generating costs of 6.8 cents/kilowatt hour actually costs 10.65 cents/kWh (or 3.79 cents/kWh more) when subsidies from taxpayers are included. Chapman only considers a model plant to be complete in 1988, and he does not calculate the current cost of subsidizing construction or operation of facilities.

The principal sources of nuclear tax subsidies are the federal investment tax credit for nuclear plants and the short tax lifetime of these installations. Doubling declining balance depreciation techniques, paired with a 16-year tax life, make for a

facility with little tax obligation. Chapman points out that nuclear plants, in fact, have a negative income tax status under federal law, though they are owned by firms considered to be in a 48 percent corporate tax bracket. A very rough estimate of tax subsidies for nuclear plants could be calculated by using Chapman's percentage of tax subsidies to direct costs (56 percent) and applying this to facilities in operation and under construction. This method points to an annual taxpayer subsidy to utilities with nuclear plants that probably exceeds $15 billion per year, substantially higher than the $3 billion that EIA estimates for R & D activities.

Lessons from History

The record of government involvement in the nuclear industry provided by the EIA report is instructive. No industry has been more carefully shepherded through the difficult process of maturation than the nuclear industry. Its development was never subjected to a thoroughgoing "benefit-cost" analysis. Demonstration facilities were built and decommissioned and built over again. Generous grants were given to private industry so that companies could learn, profit, and aspire to a role in commercial nuclear generation.

For twenty-five years the wisdom of this commitment was unquestioned. But nuclear power has failed to live up to its expectations, and proponents of large government programs in synthetic fuels or solar energy should take note. Subsidies appear vital, at first glance, because they reduce the risk or uncertainty of innovations. Nuclear power, however, clearly received this benefit to excess. The industry now is so immune from risk that reactor operators carelessly drop light bulbs behind control panels; reactor operators sue the government for failing to regulate them stiffly; and the public is forced to assume the risks of any catastrophic accident. Nuclear power has been over-protected to death.

—Jim Harding

References:

Chapman, Duane
  1979 "Nuclear Economics: Taxation, Fuel Cost and Decommissioning." Cornell University, October.

Energy Information Administration
  1980 "Federal Subsidies to Nuclear Power: Reactor Design and the Fuel Cycle." Analysis Report of the US Department of Energy, March.

Electricity Forecasting: Round to the Nearest Trillion

ELECTRICITY FORECASTING has traditionally been a simple art. Some have described it as fitting a ruler and pencil to semi-log paper and drawing a straight line. For some time, the performance of the forecasters was irrelevant: demand growth on electric grids was a constant 7 percent. In the last eight years, roughly since the time of the oil embargo, electricity demand has been utterly unpredictable. Utilities that must plan 10 to 15 years ahead find their plans shattered and resurrected on the basis of one August's peak power demand growth.

In 1972, the Chase Manhattan Bank's energy economics division predicted, without a trace of equivocation, that demand on electric grids in 1980 would exceed 3.2 trillion kWh. It will be lucky to top 2.2 trillion kWh; the difference is equivalent to more than 200 absolutely unneeded power plants that would have cost, in today's dollars, nearly a third of a trillion dollars!

In 1975, the Energy Research and Development Administration's assistant administrator for planning and analysis, Roger LeGassie, prepared four wide-ranging forecasts for electricity growth through 2000. The "low case," or estimate, stressed "stringent conservation," and assumed electricity growth at 5.8 percent per year. In the high estimate, electricity returned to its historic 7 percent rate. LeGassie's forecast for 1980, even in the lowest case, would leave us with more than 80 plants that we wouldn't have needed. The nation's *excess generating capacity* (defined as those units not necessary to meet a prudent 15 percent reserve capacity at time of system peak) is already equal to 60 large plants.

As proof of his argument, LeGassie appended an industry forecast even worse than his own. This one was a collaborative effort by the IEEE Energy Forecast Working Group, representing General Electric, Westinghouse, Chase Manhattan Bank, four of the nation's largest utilities, and Ebasco, a construction management firm. Citing eleven different forecasts and "industry judgment," the group forecast 2.8 trillion kWh for 1980, an overestimate about equal to 150 power plants that could not—not to say, should not—have been built.

In 1979, the National Electricity Reliability Council (NERC), a nationwide consortium of utility planners, predicted 7 percent demand growth for the summer. Peak demand was flat—zero percent, and NERC, after guessing wrong for too many consecutive years, threatened to stop forecasting entirely. The NERC 1978 forecast for the amount of oil required in 1979 to generate electricity was off by 23 percent. A worse forecasting record could not have been compiled by an unregulated private company, if only for the simple reason that it could not have survived eight years of catastrophically erroneous financial planning.

Calculating Conservation

Utility forecasting techniques are mightily unprepared for the energy future of unsteady prices, fluctuating demand, energy standards, conservation action programs, improved appliances, and new industrial processes designed to save energy. Their forecasting techniques, with a very few exceptions, are based on gross economic data, compiled by sector. In most cases, utility companies collect data in an even more "aggregated" fashion—"large light and power" and "small light and power." This gives them no ability to calculate the impacts of particular programs, for example, the California appliance efficiency standards or the federal Residential Conservation Service.

In the early '70s, when conservation was considered a patriotic duty, utilities used a simple one-year adjustment (often 10 percent, more out of respect for round numbers than circumspect analysis) in their demand models. Hence 1973's demand was below that of 1972, but people would soon return to their old energy use habits. The following years, from 1974 to 1980, proved this theory grossly wrong.

An alternative to gross economic forecasting is end-use forecasting, a technique that depends on careful collection of data on individual energy uses. A forecast is then built up from minute details on the number of types of appliances or electrically heated homes. This technique enables a forecaster to change the energy-use characteristics of, for example, a particular refrigerator if energy standards change. Data on the number of solar

---

Utilities still wait until August to see when they should next postpone new plant construction to. California has an end-use demand forecasting model; few other states do.

---

heated homes under construction, on the impacts of residential and commercial conservation efforts, and on proposed standards of energy use can be fed into a planning model to generate a future growth rate. This effort of data collection is expensive—about half a million dollars were spent developing data in the commercial sector in California—but its payoff can be considerable.

Despite the apparent desirability of end-use data gathering and forecasting, it is happening at a snail's pace. Utilities still wait until August to see when they should next postpone new plant construction to. California has an end-use demand forecasting model; few other states do. And California's model is only end-use specific in the residential sector. It works for energy sales; not so well for peak. The

Drawing by Dana Fradon; © 1979 The New Yorker Magazine, Inc.

more difficult to disaggregate commercial sector—schools, hospitals, office buildings, small factories—is vastly harder to evaluate.

California's end-use forecasts have also uncovered substantial anomalies in utility forecasts. When the state Energy Commission released its end-use based residential model, the state demand rate dropped from its former 6-7 percent range projected by utilities to 4.5-5.0 percent per year. Later refinements in the data caused it to drop further to about 3.5 percent. Though the commission had not done exhaustive end-use analyses of the commercial sector, the residential end-use model suggested that for the utility forecasts to be correct, electricity use for lighting and air conditioning the average square foot of commercial floorspace would need to double in the years ahead. Recently enacted state standards and data on construction industry trends suggested that it would fall by half. It was assumed electricity use in existing buildings would remain constant, but audit programs by the utilities showed 10-30 percent savings easily achievable at minimal cost (0.3 ¢/kWh). With commercial sector conservation programs and standards for lighting and air conditioning new buildings, the forecast falls to 1-2 percent. Industrial cogeneration with firms generating a portion of their needs and supplying surpluses to the utility industry plus improved household appliances could cause the forecast to fall below zero, but the impacts are impossible to judge and programs difficult to choose among unless the individual energy end-use needs are apparent, a process that takes time and cooperation from the industry.

Models—Mock Power Soup

Heritage has also posed great problems on the supply modelling side. At one time the choice for utilities had to be made among oil fired plants, nuclear plants, and coal plants. These units were judged by their costs and reliability at the generator terminal, but many assumptions in this process were grossly distorted. Nuclear plants were projected to run at 80 percent capacity factors and cost $300-700 per kilowatt of capacity. Their capacity factors have been 55-60 percent and their costs have topped $2,000 per kilowatt of capacity. Choices between new power plants have become a matter of guesswork, rather than engi-

neering cost estimation. Of course some utilities still choose power plants on the basis of a fine cost difference, but inflation rates, soaring oil costs, construction delays, demand projection reevaluations, and environmental issues make a mockery of the fine detail.

Despite the overall uncertainty, there are details and habits of thought that are important for planning around the future. Utilities make choices among individual central-station, base-load power plants. They choose the most marginally attractive source. But in a world of energy conservation options, industrial power generation, wind turbines, and a host of modular sized future options, this planning technique is grossly inadequate. Utilities often use the term "baseload" to describe the kinds of plants they want to build, but this term is ill-defined. To some, "baseload" means reliable, to others it means "low operating costs," to still others it means "big." But what it re-

---

Current federal regulations require utilities to "buy back" power from small generators for what they'd pay to generate it themselves. Most rate schedules are set far too low.

---

ally defines is the plant a utility has decided to build. There is no reason that a "baseload" plant must be any one or all three of its supposed defining characteristics. What is important is that the system a plant is inserted in, be cheap and reliable. And the system cannot stop at individual plants. It must include the capital and marginal operating costs of transmission systems, distribution systems, and reliability. *No existing utility model* for adding to system capacity can choose among end-use and central station options to add new capacity at a fixed level of reliability and minimum cost.

This failing of utility methods to select the cheapest new source among many is a principal problem in planning an efficient energy future. Let us assume that we need to add reserves to a utility system to meet a growing load. Reliability planning is done by using the individual probabilities that certain components of the "bulk grid" will fail: individual plants and large transmission lines. Thus new plants

TOOLS for the SOFT PATH

are added to create a bulk grid failure risk of a certain level, say one day in ten years. Yet reliability at the consuming end bears little resemblance to this goal. Nearly ninety percent of the lost electricity in a typical electric grid—ninety percent of the system failures—is in the distribution system, not in the "bulk grid." The result is that a much larger number of power plants is built than is in fact "demanded" by consumers, and that investments in better substations, redundant lines, and stiffened control systems are given short shrift. John Peschon of Systems Control, Inc. cites a study completed several years ago on end-use reliability: "We examined a system, which we think is typical, and found that end users would see equal reliability from a one kilowatt battery in the distribution system as they would see from 2.5 kilowatts of central station generation at peak."

This astounding example suggests that one could afford to spend much, much more on the 1 kilowatt battery (per unit of capacity) than on the 2.5 kilowatt central station plant, yet no utility expansion model would produce this answer because costs and reliability are only compared at the bulk grid rather than end-user level.

Current federal regulations require utilities to "buy back" power from small generators for what they'd pay to generate it themselves. Most rate schedules are set far too low. In California, for example, buy back rates are set at 4 to 5 cents/kWh and revised annually as oil costs climb. This isn't the way utilities plan capacity additions. When a new plant is proposed, its *life cycle* (30 year) costs are compared against the *life cycle* costs of the competition, and the cheapest alternative for the ratepayer is selected. In many cases, this alternative will cost more initially, but save fuel costs later (coal or nuclear versus oil). When a utility enters into a contract with a windpower company, the long term equation should hold: a ten-year contract should earn revenues equal to the current value of costs displaced. Even without considering many of the fac-

tors noted earlier, this provision alone would double the value (8 to 10 cents/ kWh) of cogenerated, wind generated, or hydro generated power.

Over the past decades we have built institutions, methodologies, economic subsidies, habits, and inertia into energy planning. Conventional sources are blindly favored at virtually every step, whether in intricate models or tax codes. Some would have us throw regulation to the wind and let utilities fend for themselves—perhaps a useful experiment in a state unlike California, where utilities out-lobby energy conservationists by 2½ orders of magnitude. What makes more sense to us is force and co-option in the transition: force to ensure utilities and their regulatory commissions to live up to their "least cost energy service" charter; co-option to ensure that what makes sense in the market and the larger environment doesn't lose out in the discussion.

—Jim Harding

Breaking the Rental Barrier

DESPITE THE RISING COST of energy, few commercial buildings now under construction include solar systems or incorporate energy-efficient principles of design. Even fewer existing structures are being retrofitted to conserve energy, although such modifications are known to be cost-effective. The owners of these buildings care little about energy efficiency, and have no strong incentive to improve energy performance, because they generally don't pay the fuel and utility bills: their tenants do.

The businesses that lease commercial space, in turn, have no more reason to invest in solar or conservation improvements than do the building's owners. The anticipated payback period is usually longer than the unexpired period of the tenant's lease, so there is no guarantee of recouping the original investment, much less profiting from the capital outlay. Few commercial buildings are individually metered, so there is no way to determine how much energy any given tenant is saving, or how to adjust their share of the total. Moreover, there are legal restrictions on the structural modifications that commercial tenants may make on the premises they occupy.

The split between occupancy and ownership stands in the way of commercial building efficiency. In a recent issue of the *Solar Law Reporter*, Counihan and Nemtzow write, "tenants control much of the pattern of energy use, while landlords establish the efficiency of consumption. Both parties therefore affect the quality of total use, but only one party pays the bills." Their comments, though directed at the rental housing market, apply equally well to the commercial sector, where an even larger proportion of the floor space in use is leased, rather than owned. In both cases, overcoming the rental barrier will require creative combinations of new financial incentives, regulatory changes, and cooperative arrangements among renters, owners, utilities, and government.

Building in inefficiency

Between 1945 and 1978, commercial floor space expanded from 9.9 to 29.6 billion square feet, with new office buildings leading the advance. The architects of the commercial districts designed highrises in the "International" style: the high-tech tower whose uniform facade consists mostly of inoperable windows. Intended to insulate its human occupants from the natural environment, the modern commercial structure has energy inefficiency built in.

By the year 2000, according to Oak Ridge National Laboratory projections, commercial floor space is likely to increase 64 percent, to 51 billion square feet. Much of this additional footage will be just as inefficient as today's commercial space unless the way of allocating energy costs among owners and tenants changes.

The exact percentage of commercial floor space that is tenant occupied cannot be determined from available data. However, a recently-released Department of Energy report notes that 35 percent of all buildings in the commercial sector are entirely occupied by tenants; of the remaining buildings, at least some portion of each is occupied by the owner. Regardless of the fraction of rented space, though, nearly all large commercial buildings are master-metered for electricity, and have single heating and cooling facilities. This means that commercial tenants rarely, if ever, pay their own energy bills directly. Electricity and gas or fuel oil costs are commonly paid by the owner, and included in the charge per square foot stipulated by the lease (which may contain a pass-through provision or an escalation factor to cover energy price increases. Alternatively, the landlord may divide energy bills among all the tenants, prorating by the fraction of the total area each occupies.

This system for allocating energy costs fails to encourage thrift, when it does not actually reward waste. With energy bills held constant by the lease, or distributed among a large number of users, a tenant winds up paying about the same amount, regardless of how much or how little is consumed. The energy-conservers enjoy but a small fraction of their savings, while the excesses of the wasteful are subsidized by their neighbors. The consequences of any one tenant's energy consumption habits are unknown and unfelt.

Building owners are discouraged from solar or conservation for different reasons. Lease provisions may allow them to pass energy operating costs on to tenants, but not the investment cost of energy-saving building modifications. Moreover, many commercial buildings are owned as tax shelters, a status which discourages further capitalization, and provides little incentive to reduce deductible operating costs.

A meter of your own

In residential rentals, the savings with separate metering can run as high as 26 percent, because tenants know, and pay for, their own levels of use. Individual meters in the commercial sector are less common, found principally in smaller buildings, where there is only one tenant, or just a few. Changing over to separate metering, according to Counihan and Nemtzow, can alter consumption habits, and reduce energy use somewhat, but this movement toward thrift is outweighed by the "catastrophic effect on the investment behavior of landlords." Separate metering removes any incentive building owners may have had to invest in energy conservation, or even maintain existing equipment, by freeing them from the financial responsibility of increasing fuel bills. Paradoxically, with separate metering, "the total energy use of a building may increase," despite tenant efforts to conserve.

Tenants responsible for their own energy bills would realize immediate savings from efficiency or solar improvements, but they are unlikely to remain in the same premises long enough to recover the full value of their investment. Most commercial leases run only three to five years, while payback periods for major energy savers such as solar collectors and insulation are typically longer than that. The major benefit is the landlord's, not the tenant's.

Federal tax credits for solar and efficiency investments can effectively shorten the payback period, but the present program leaves the commercial sector out in the cold. *Energy Forum in New England* notes the discrepancy:

"a . . . homeowner who installs storm windows, caulking or insulation is eligible to receive a 15 percent tax credit for a maximum of $300. But the owner of a restaurant, clothing store or movie theater, who makes the exact same improvement to increase the building's efficiency, will not qualify for a comparable incentive. Similarly, a large industry that installs a computerized energy management system to control the energy used in manufacturing processes will receive a ten percent investment tax credit, in addition to the regular ten percent credit. An advertising firm, that installs a similar device to control heating, cooling, and lighting, will not receive the additional ten percent investment tax credit."

Without substantial credits to drastically shorten payback on major improvements, the only measures commercial renters can economically take are low-cost, rapid-payback alterations such as high-efficiency light bulbs, rewiring to permit task lighting, weatherstripping, and the purchase of energy-saving appliances, which tenants can take with them when they move.

First steps through third parties

Removing these inconsistencies, and giving commercial building investments the same federal tax status as their residential and industrial equivalents, would make energy-saving measures more attractive to both owners and renters. Additional credits could be offered by state governments, further reducing effective payback period. For example, the 20 percent tax credit for energy investments, available to owners of residential buildings in Rhode Island, could be extended to leased commercial structures. (For the program to be effective, however, the upper credit limit of $5,000 over five years would have to be increased.) States can also mandate minimum standards for energy efficiency in rented commercial space, as Wisconsin did in 1979 for rental housing.

Tax incentives and regulatory mechanisms will be ineffective if they are directed at commercial tenants, whose transience limits their concern for the structure they temporarily inhabit. Building owners are the proper targets for such measures. Their investment incentives, though, could be reduced by the trend toward separate metering that many states are now following in an effort to promote tenant conservation. One provision of PURPA (see article, this issue) requires state public utilities commissions to consider prohibiting or restricting master metering of electricity in new buildings. But a recent report from the Solar Energy Research Institute strongly cautions that master metering in rental housing could have adverse effects on energy use (Levine & Raab, 1980). A similar trend in commercial rentals could be equally damaging to conservaton goals.

Other ways of breaking the rental barrier involve a nongovernment third party who can bring the split incentives back together, by absorbing most of the up-front capital costs of energy improvements. Utilities or energy management firms could, for example, lease solar equipment to building owners who cannot pass higher energy costs on to tenants, and lower their operating costs without requiring large capital outlays. Separately-metered renters of commercial building space could similarly lease equipment (with the owner's permission), with the lease transferable to the next tenant.

A more complex arrangement is needed if a building is master-metered, and the owner passes energy costs on to tenants under some fractional area rule. To avoid the waste of atomistic consumption, tenant groups could be organized to promote collective savings through cooperative leasing. The formation of such a group would also increase awareness of energy consumption habits, and bring group pressure to bear upon the most wasteful tenants.

Structural improvements to commercial buildings such as insulation and furnace modification can also be arranged through third-party intervention. No-interest or low-interest loan programs could make these conservation measures economical for owners, individual renters, or tenant collectives, depending upon metering and whether energy costs are passed on. Because building types, owner-tenant arrangements, and local energy needs are so varied, the key to reducing energy use in rented commercial space is a flexible approach that recognizes the inadequacy of single solutions.

—Charles Drucker

References:

Counihan, Richard, and David Nemtzow
1981 "Energy Conservation and the Rental Housing Market." *Solar Law Reporter* volume 2, number 6, March/April, 1103,1131.

Energy Forum in New England
1981 "Energy Efficiency Increase in Commercial Buildings." Volume 2, number 2, Summer, 1-3.

Levine, Alice, and Jonathan Raab
1981 "Solar Energy, Conservation, and Rental Housing." Solar Energy Research Institute document RR-744-901.

Tax Credits For Landlords

OWNERS OF COMMERCIAL and multi-family residential buildings will enjoy additional tax incentives to invest in renewables and conservation, if a bill recently introduced by Senator Paul Tsongas becomes law. Co-sponsored by Tsongas and fellow Massachusetts Democrat, Senator Edward Kennedy, the bill would make rental housing and commercial buildings eligible for tax credits. The July, 1981 *Solar Times* reports that "the bill would provide a 20 percent tax credit to commercial facilities for conservation measures; increase from $4,000 to $6,000 the allowable credit for renewable energy systems; increase homeowner credit for conservation improvements from $300 to $450; and extend energy tax credits for business to December 1985, instead of the scheduled 1982 expiration."

The Shadow On Solar Jobs

IN A PERIOD of rapidly declining employment opportunities, the solar market offers a ray of hope. The new products to be designed, manufactured, and installed could create many new jobs, with semi- and unskilled workers filling the majority of the employment slots. However, the solar industries' first few years of experience suggest major discrepancies between expected and actual labor demand, casting a shadow on the promise of solar jobs. The reasons for these discrepancies, and what might be done to reduce them, are the subjects of Solar Energy & Jobs, *a report from the Citizens' Energy Project (CEP), by Ken Bossong.*

"Nuclear Destroys, Solar Employs"

The familiar pro-solar bumper sticker points to one of the strongest arguments of the solar proponents. Conventional energy systems, like most capital-intensive industries, tend to replace human labor with machinery; creating additional jobs through centralized energy technologies involves enormous investments. Solar strategies and energy conservation, in contrast, are labor-intensive, and per dollar of energy investment produce anywhere from two to ten times the new employment opportunities. Professional slots would open in such fields as solar engineering, architecture, law, real estate and appraisal, sales, and consumer protection. Solar energy systems would also call for tradespeople and craftworkers including carpenters, masons, electricians, plumbers, sheet metal workers, air conditioning and heating technicians, glazers, crane operators and teamsters. Individuals with lower skill levels would also find work in the solar field.

Estimates of the new employment solar industries might create vary. One conservative calculation by the former Federal Energy Administration projected that under a Project Independence development scenario, up to 4 million person-years of employment would be needed by the year 2000. This averages to 200,000 positions—more than enough to put all the unemployed autoworkers back on the payroll.

The CEP report stresses, however, that it is both naive and facile to assume that these jobs will automatically create themselves in the solar future. Moreover, there is no guarantee that the new job opportunities will reach the unemployed and unskilled, low-income and minority groups who inhabit the inner cities. The solar movement, Bossong claims, has unfortunately "closed its eyes to this fact.... If no action is taken and solar commercialization is allowed to continue on its current path, the promise of solar jobs will be frustrated."

The Policy Role

Federal policies influence the distribution of solar employment through the programs and incentives that encourage new energy technologies. The present federal solar program, "whether by design or by accident," promotes installations in middle and upper income, suburban single family homes, "at the expense of inner-city, lower-income, and multi-family building installations." The cornerstone of the government's program is a system of tax credits, which is of value only to those whose incomes are high enough to tax; "unemployed persons or persons living near the poverty line pay no taxes against which a solar tax credit may be charged." Low-interest loan programs through private banks and utilities are similarly of little use to low-income families. Most of the solar installations appear in suburban, residential neighborhoods, too far away for most inner-city, low-income workers.

Federal standards and information

programs also favor the "more expensive and highly-engineered active solar technologies" rather than less costly alternatives. Breadbox water heaters, attached solar greenhouses and other passive designs are passed over in favor of solar consumer products that only middle and upper income families can afford. High-tech solar affects employment two ways: it discourages installations appropriate to low income neighborhoods, creating fewer inner-city solar jobs; and, it raises the training level for competent design and installation. As such, it generates fewer semi-skilled and unskilled positions, professionalizing the solar industry.

These factors push solar businesses and solar jobs out to the suburbs, where their impact on inner-city employment is slight. Of the 100,000 solar installations in 1979, less than one percent were federally-funded low income residential projects, even though 20 percent of US housing qualifies as low income. The jobs for installing and maintaining this solar equipment will naturally fall to members of the local community. As Bossong points out, "it's not likely that semi-skilled, inner-city workers would be imported into the suburbs to install the highly-engineered and more sophisticated solar systems that, to date, have characterized single-family installations."

Sharing the Solar Employment

If low-income groups are to share in the solar job opportunities, some difficult political decisions must be made now. Innovative financing, job training, and community organization programs are needed, but they are unlikely to be implemented without opposition.

The Citizens' Energy Project recommends federal financial incentives to promote solar technologies for dwellings that are now beyond the reach of market incentives, such as rental or low-income units. CEP allows this is likely to mean diverting some research and development funds from high-technology systems to low-cost and do-it-yourself devices, especially passive retrofit systems. But with one-fifth of the US population below the poverty level, the Project points out, "at least this percentage of the federal solar budget should be directed towards this audience."

In addition, CEP outlines job training programs that provide inner-city workers with solar skills. Although federal and state agencies recognize the need for solar job training programs, and have already initiated a number of courses, these are generally not aimed at unskilled and unemployed inner-city residents.

A significant exception is the SUEDE (Solar Utilization, Economic Development, Employment) Program, which provided solar training

Industry	Capital Investment/ Employee
Petroleum	$108,000
Public Utilities	105,000
Chemicals	41,000
Primary Metals	31,000
Stone, Clay, Glass	24,000
All Manufacturing (average)	19,500
Food & Kindred Products	18,000
Textile Mill Production	11,000
Wholesale and Retail Trade	11,000
Services	9,500
Apparel & Other Fabricated Industry	5,000

for the unemployed, and then conducted actual installations on the homes of low-income individuals. Unfortunately, federal inertia, lack of funding, and a dearth of jobs for graduates have reduced the impact of pioneering programs such as SUEDE, and led to their cancellation.

Opposition to more extensive job training efforts for the urban poor may well come from the labor unions. Sensitive to the overall job shortage, they are anxious to protect their own ranks. Unions have actively discouraged outside training and apprenticeship programs which might lead to non-union employment in the solar trades. At the same time they are unwilling to take on many new apprentices, as there is not enough work to employ their current membership.

A compromise suggested by some unions is to use CETA funds to support training programs for their union members. But by limiting the more skilled jobs to present union members, this concept would prevent low-income and semi-skilled workers from entering the training programs and moving into the higher-paying solar trades. A more effective compromise would direct CETA funds to subsidize union training for the economically disadvantaged who are not presently members. This training "should focus on solar retrofit skills and the application of solar/conservation technologies in multi-family living units."

Finally, CEP maintains that if these projects' jobs potential is to be realized, community organization and participation in solar projects is essential. "Inner-city communities will benefit from solar jobs," Bossong says, "if they get organized politically at an early date and if projects are small-scale and widespread, eventually retrofitting the community, neighborhood by neighborhood." The national solar commercialization effort should stress decentralized, small-scale and community-based solar technologies, rather than the big business projects that now swallow most of the solar budget. The "cornerstone of solar development programs" should be energy conservation, passive designs, and biomass, small hydro and wind systems. "A phased-in program of mandating this type of solar activity at the community level," he concludes, "would both accelerate the pace of solar commercialization as well as provide more certainty in the solar job market."

—*Liza Gimbel and Charles Drucker*

Reference:
Bossong, Ken
   1980 *Solar Energy And Jobs.* Report Series no. 49, 11 pp. Available for $1.25 from the Citizens' Energy Project, 1110 Sixth Street NW, Suite 300, Washington DC 20001.

2. New Actors and Approaches

Most US energy supplies flow through the utility industry on their way to pilot light and socket. It is only logical to implement the soft path through new approaches for old actors.

The principle here is to convert an enterprise based on selling energy to one geared towards providing energy services, whether this is achieved with more efficiency or with diverse sources of new supply. Since investments in conservation and solar technology by consumers are far cheaper than new billion-dollar power stations, utilities should be financing those investments in order to keep the rates down.

Likewise, the electric grid could be redefined as a public road system, in which all electricity generators compete on fair terms, whether they are central stations, wind turbines, cogeneration units or new, decentralized producers.

To many analysts, these schemes, when first proposed, seemed far-fetched and impractical. Indeed, many utilities have fought such proposals tooth and nail and are still putting up strong resistence. At the same time, two factors mitigate against them:

● the failure of nuclear power plant investments means that to get onto more solid financial footing, utilities must turn to conservation and alternative supplies.

● rate payers, angered over the escalation of utility bills, are turning to the regulatory process to reassert public control over utility policies. The result has been a surprising upsurge of new institutional arrangements. Utilities representing some 40 percent of total industry sales in the US, are now offering conservation financing, ranging from weather stripping and attic insulation loans for individuals to rewards for peak electricity load shaving by communities. Special municipal utilities are introducing solar equipment, and federal law requires all utilities to offer energy auditing services.

With the 1978 regulatory act (PURPA), utilities must buy back surplus electricity from decentralized producers at avoided cost. This rule introduces marginal cost comparison as an investment criterion and opens the grid to many new technologies: domestic total energy systems, industrial cogeneration, microhydro sets and wind turbines. Small scale generation also opens the way for efficient cogeneration in urban district heating systems.

Utilities are by no means the only center of new activities. Increasingly, local governments are realizing their stakes in energy matters and are establishing their own energy offices. Communities see they are not unlike small nations that export capital and jobs in order to pay for their energy imports. Such insights have loosed a wave of local energy initiatives and led to the rediscovery of domestic energy sources, including that of long-forgotten garbage, while ride sharing and other means of reducing commuter traffic are losing their old stigmas.

Urban systems, by sharing walls and transport, can have advantages in energy efficiency, but incur heavy indirect energy costs. Cornbelt farmers, meanwhile, are making up for rural disadvantages with something the city often lacks: cooperation.

"Drilling for Oil and Gas in Our Buildings"

Conservation Program Could Save 50% of US Space Heating Needs

A RECENT REPORT by Bob Williams of Princeton's Center for Energy and Environmental Studies concludes that the United States could find the equivalent energy of 2.5 million barrels of oil per day by "drilling for oil and gas in our buildings."

Williams proposes a large effort to capture 50 percent of the energy now used in heating residential and commercial buildings, as an alternative to the more draconian measures — mandatory thermostat setbacks, gas rationing, huge subsidy programs, further speed reductions — that might be needed to stave off a serious oil crisis in the mid to late 1980s. Two and a half million barrels per day is equal to two-thirds of our present dependence on Arab and Iranian oil. It is also about four times the net output of all the nuclear power reactors presently operating in the US.

The cost of this effort, Williams calculates, approximately equals oil at US $14 per barrel. In other words, each $14 spent on energy efficiency measures frees up another barrel of oil for another use. By comparison, the Department of Energy assumes that 4 percent of our present space heat use will be saved by efficiency improvements, but their criterion for judging what makes sense is quite different: DOE assumes that 7 percent of the nation's buildings will annually be "audited" for space heat use and that 75 percent of these audited buildings will be "retrofitted" with insulation and other measures.

Williams believes a much larger effort is cost-justified and, with proper institutional arrangements, can be mounted. In their extensive retrofits of "well-insulated" townhouses in Twin Bridges, N.J. (SEN III), the Princeton energy group found that insulation, caulking and weatherstripping, window improvements, and furnace modifications could reduce the energy needs of a typical home by 50 to 75 percent.

In another carefully documented effort, the National Bureau of Standards installed the more conventional storm windows, wall, floor, and ceiling insulation in a suburban Washington, DC house, with a reduction of 60

"Williams proposes a large effort to capture 50 percent of the energy now used in heating residential and commercial buildings."

"Drilling for Oil and Gas in Our Buildings"

percent in the home's heating requirements at a cost of $650. The overall cost per barrel saved was $20, well beneath the cost of new domestic energy sources.

The Public Policy Challenge

The difficulty with such a large campaign as Williams proposes is that no one is leaping forward to do it. Utilities still charge average costs for energy sources, so that homeowners who go out and get their houses insulated can only expect to save at average rates, while the new marginal source is subsidized. As Williams notes, "Because present energy prices are so much lower than replacement costs, however, the typical consumer would usually invest only in [storm windows and wall insulation], and thereby capture less than the full savings which are justified on the basis of the replacement cost criterion. The challenge for public policy is to motivate the homeowner to make his investment decisions in accord with the national goal of minimizing overall costs."

"Other obstacles include the lack of technical capabilities to achieve highly efficient heating of residences, the low incentives for improving those rental buildings for which the fuel bill payer is someone other than the building owner, and the fact that the Department of Energy has set as its goal the capture of energy savings costing only $10 per barrel of oil equivalent energy in its Residential Conservation Service program."

In what is the meat of the report, Williams proposes a plan to short-circuit these "institutional barriers": retrofit pilot projects, train 'house doctors', and devise innovative financing measures to encourage a high level of implementation.

Pilot projects for home insulation seem an odd addition to Williams' list, but his argument is worth noting. Handbooks and computer models are currently available to measure heat losses from typical residential structures, but the models often tell the wrong story. Williams cites the example of models showing 20 percent of residential heat losses from the escape of warm air around insulation into the attics of houses; yet, in a test of 40 actual houses in the Northeast, heat losses to the attic were 3 to 7 *times* greater than sophisticated models predicted. Why? "Open shafts around a flue that extends from the basement to the attic, plumbing vent pipes, stairwells, spaces between walls, unsealed walls behind dropped ceilings, etc." What this generally means to homeowners is that they will get only one-half to two-thirds of the savings usually attributed to ceiling insulation; better that they find the holes in the house than add to the insulation that wasn't stopping much to begin with. Williams urges a 2-4 year program costing 50-100 million dollars to do pilot retrofits of all the nation's housing types to determine where the largest savings are and what they cost.

Is There a Doctor in the House?

Williams urges a return to the almost forgotten practice of doctors making house calls, only the ills to be cured are not aches and pains, but cracks and leaks. Builders, he adds, do not have the skills to determine what measures save

"*Williams urges a return to the almost forgotten practice of doctors making house calls, only the ills to be cured are not aches and pains, but cracks and leaks.*"

energy and money. These barefoot doctors would be certified to audit buildings for heat loss and perform simple on-the-spot repairs with tape, glue, plastic, and small quantities of insulation. "This remarkable possibility arises, Williams notes, because many bypass heat losses involve unwanted air flows through relatively small areas, the reduction of which is intelligence intensive but not nearly so labor and materials intensive as the more conventional conservation measures." Thus a four hour audit costing about $120 could probably save 15-20 percent of one's space heat bill.

"An all out effort," he notes, "to retrofit buildings heated by oil and gas to improve their energy performance could require on the order of $15 billion per year of capital investment. This level of investment is economically justified because the cost of saving energy even to the high degree proposed here would be less than the cost of the energy supply investments which would thereby be made unnecessary. This is a staggering investment rate, but it is only about 15 percent of the value of building construction contracts and about 25 percent of the level of energy supply investments in 1977."

Williams urges that utilities be used as the financing agent. In 1977, utilities "accounted for about one-fourth of

"Drilling for Oil and Gas in Our Buildings"

all new plant and equipment expenditures in the US economy.'' They also have the administrative structure to handle conservation loans, are doing it to some extent already, face marginal energy cost decisions on a daily basis, and don't have anything useful to do with their money anyway. On the debit side, we must point out, utilities are among the least-liked institutions in America, have a record of gold-plating all their investments so as to make the highest net return, give generally bad advice for building retrofits (especially in the commercial sector), and may not see a clear reason for moving conservation as quickly as it ought to go, thus setting back the entire effort.

Williams, nevertheless, argues strongly for utility financing, using the excellent Oregon plan as an example. In Oregon, electric utilities are allowed to finance conservation investments in electrically heated dwellings, of which there are many. Customers may request, at any time, a home audit. Subsequently, the utility will offer to finance and contract for the work the auditor recommends and the homeowner wants done. The cost of the conservation investment goes into the utilities' rate base and is allowed to earn a rate of return, which all customers pay. The customer pays for the cost of the conservation investment, without interest (because the rate base is designed to cover interest costs), when the house is sold. The criterion for making investments is that they cost less than the cost of a new electric source. Individuals choosing to insulate, and landlords too, get lowered energy costs and/or an interest-free home improvement loan. The

utilities' incentive is that investments are rate based, and so generate earnings. The energy bill for customers that do not insulate drops, too, because the cost of not insulating someone's home and finding a new electric source is less than the cost of rate-basing the conservation improvement. Thus, the cost of electric power to uninsulated customers would rise no faster than it would if supply investments were made.

Williams suggests a variant of this scheme for oil heated homes, using a specially chartered corporation to organize audits, arrange for installations, and provide financing. Insulated customers would pay nothing for installation,

"Obstacles include the lack of technical capabilities to achieve highly efficient heating of residences, the low incentives for improving those rental buildings for which the fuel bill payer is someone other than the building owner, and the fact that the Department of Energy has set as its goal the capture of energy savings costing only $10 per barrel of oil equivalent energy in its Residential Conservation Service Program."

the government would pay financing charges, and homeowners would pay the installation cost on sale of the home. Alternatively, financing could be covered by an initial charge of 1.5 ¢/gallon tax on heating oil, rising to 20 ¢/gallon at the end of ten years.

"The housing retrofit program we have outlined is very ambitious," Williams concludes. "It could not be realized if one or more of the three program components were slighted — the retrofit pilot projects, the house doctor training programs, or the financing measures. But the effort could in just a few years significantly reduce our dependence on insecure sources of foreign oil; it would help stabilize energy prices; it would lead to the creation of over 200,000 jobs for conservation installers and many more jobs for the manufacturing and marketing of the needed materials and hardware; it would be much less costly and much more effective in helping to rapidly close the oil supply/demand gap than a synthetic fuels strategy; and the effort would be far preferable to the imposition of draconian energy rationing measures."

Source: Marc Ross and Robert Williams, "Drilling for Oil and Gas in Our Buildings," 17 July 1979, 34 pp. Available from The Center for Energy and Environmental Studies, Princeton University, Princeton, New Jersey.

Jim Harding

"The challenge for public policy is to motivate the homeowner to make his investment decisions in accord with the national goal of minimizing overall costs."

Selling Savings

DESPERATELY SHORT OF CAPITAL for expansion and faced with a skeptical public and regulatory atmosphere, US electric utilities are beginning to examine their role in stimulating investments in alternative energy technologies and energy conservation equipment. Energy conservation or industrial power generation are hardly the kinds of investments salesmen of gas and electricity would seem to favor, but pinched finances and pervasive uncertainties have spurred innovation.

In many respects, the policies under review smack of deregulation. Under the federal Public Utilities and Regulatory Practices Act (PURPA), utilities and utility commissions are required to pay their "avoided costs" to an independent generator of electric power. Though definitional disagreements abound on what this avoided cost amounts to, the law is intended to create a competitive market for new investments in power generation. On the energy demand side, a number of utilities—representing about a fifth of US generating capacity—are developing "voucher" plans or "zero interest" loan programs to stimulate conservation investments that are competitive with new energy sources. The efforts suggest that the economies of scale that gave rise to service monopolies for electricity and gas may no longer apply.

In energy conservation, several notable programs exist. Motives differ, however, and there is tension within the utility industry over what is seen by some as capitulation to regulators. Says Ted Davenport, Assistant to the Chairman of Pacific Power and Light, a small Pacific Northwest utility: "We've almost been drummed out of the Edison Electric Institute (the major private utility trade association) for our conservation program." But to others, Davenport's program is a model. The utility has offered—

since 1978—"zero interest loans" for conservation investments in electrically heated homes. The utility "audits" a potential customer's home and authorizes loans that are competitive with the cost of power on the utility grid. The cost of the loan is treated as part of the utility's "rate base," until such time as the construction investment is repaid, at no interest, usually on sale of the property.

The plan has benefits for all participants, and even for non-participants. Homeowners get lower electric bills and a zero interest home improvement loan. Pacific Power & Light gets a rate of return on its investment (at the margin, higher than their cost of money), cash flow is improved because funds turn over in seven years or so, rather than the 30-40 years required for a power plant to repay its cost. This, in turn, makes it easier for the utility to earn its authorized return on traditional investments. The utility is saved the risk and trouble of raising large sums to finance uncertain and probably unnecessary power plants. Even landlords and tenants, whose incentives are normally at cross-purposes, share the benefits: the former gets a free loan and, possibly, higher occupancy rates, the latter gets reduced utility bills. Non-participants gain because the cost of the saved energy is less than the average cost of electricity in the system (3-4 ¢/kWh). Costs as high as 6.5-10.0 ¢/kWh (marginal cost) might be justified, but PP & L has not gone this far.

Following this lead, the California Public Utilities Commission directed the state's private utilities to develop "zero interest loan" programs for residential conservation and solar hot water heaters. Other utilities—among them, General Public Utilities, the Tennessee Valley Authority, Michigan Consolidated Gas, and New England Electric System—have impressive conservation financing programs. Together, utilities with innovative conservation financing programs under development represent about 20 percent of the electricity sold in the United States.

The effort is not without its problems, however.

First of all, power companies are not organized to pursue energy conservation. Their staffs are largely composed of construction engineers ready and willing to build plants.

Second, their demand projections are not broken down by end-uses, as the state Energy Commission's were, so they were unable to estimate the financial or energy impacts of conservation, let alone determine priorities among programs.

Third, a tremendous number of subsidies are given for new power plant construction, whereas few or none are given for energy conservation investments. Duane Chapman of Cornell has calculated that for an average utility in the fifty percent tax bracket, building a new nuclear plant will reduce the company's total tax obligation over a thirty year period. Investment credits, accelerated depreciation, and other book-keeping tricks make the investment look more attractive to the company than it is to taxpayers. (In the nuclear case, tax subsidies total nearly 75 percent of the construction cost.)

Fourth, utilities generally assume that consumers should pursue energy conservation by themselves, that they should respond to price signals. This perspective neglects the many subsidies present in utility regulation (lack of new hookup charges, average rather than marginal costing, equal urban and rural prices, disincentives for industrial power generation); the particular problems posed by landlords and tenants who have differing incentives (often a serious problem in the commercial sector as well as the residential one); and the particular advantages of utilities over, for example, banks in making a large number of carefully monitored, information intensive, conservation loans. What's more, utilities who want to project energy use in the future and adapt conservation techniques to the characteristics of energy supply will need to know how much conservation is going on and of what kinds, in each sector. Without that detailed knowledge, their ability to plan for the future—even assuming they do nothing to finance conservation—will be grossly diminished.

One Example

Legislation enacted by Congress in 1978 prevented utilities from entering the "conservation" business without a waiver from the Secretary of Energy. Most such waivers have passed the Secretary's desk without difficulty, but the impression of a skeptical Congress gives intransigent utilities or utility commissions a telling excuse. Most existing utility conservation programs address the residential sector; zero interest loans have not been offered to commercial or industrial customers. The reason for this is probably political: the residential sector has more votes. Nevertheless, the commercial and industrial sectors may yield more results per dollar than residences when it comes to saving electricity. Residential sector improvements—mainly insulation and weatherstripping—save scads of oil and gas, but they do little to relieve the financial crunch imposed by capital requirements for electricity.

In examining alternatives to the proposed Sundesert nuclear plant, the California Energy Commission found over 5000 megawatts of very cheap efficiency improve-ments available in southern California by 1985. The cost of the Sundesert project was 6.5 ¢/kWh plus 1-2 ¢/kWh for transmission and distribution; its size was 1900 megawatts; its capital cost, over $3 billion. The conservation improvements, available by expanding *existing* utility audit programs, cost an average of 0.5 ¢/kWh for nearly three times the total megawattage, at a program capital cost about a third that of the nuclear plant. Nearly 50 percent of the peak power savings was in the commercial sector. Twenty five percent was in residences; the remainder was in industrial cogeneration and voltage reductions to all customer classes. In energy terms, commercial programs saved five times the electricity of residential efforts (112 TWh at $308 million *vs.* 24 TWh at $703 million). Thus, commercial sector energy was saved at about a tenth the cost (0.3 ¢/kWh) of residential sector electricity (3 ¢/kWh).

It can certainly be argued that this review under-estimated the residential sector potential. No thought was given to a voucher program, like TVA's, that might reward buyers of efficient appliances with a check for the value of the difference between marginal and average cost energy displaced by the appliance. Window overhangs and coatings and motor retrofits got similar short shrift. But a counter argument also exists: residential sector savings are as big as they are partly because southern California has a high saturation of swimming pool filters, air conditioners, and electric water heaters—all of which can be cycled to more energy use "off peak." In any event, the costs and—perhaps more to the point—administrative difficulty associated with residential sector electricity improvements is high, probably beyond the current talents of utility staffs, and poorly evaluated by all.

The review also paid insufficient attention to industrial electricity use, where a similar voucher program could be instituted for improved electric motors. Efficiencies could probably be doubled using today's best technologies with 3-5 year paybacks on the investment. But the sector with impressive immediate payoff potential appears to be large office buildings.

A foundation of decentralized energy planning is the careful cataloging and understanding of individual energy "end-uses"—quantities and types of appliances in use, insulation levels in homes, processes in industry. With this information available, energy planners can project future energy needs, determine the impacts of new energy standards, evaluate the cost-effectiveness of programs, and select priority efforts for cost and energy-savings potential. This information is of utmost importance with respect to electric power, because of the need for long term planning, high standards of reliability, and great interaction between demand for energy and types of supply.

—*Jim Harding*

Communities Reduce Peak Electricity

According to PG&E, this voluntary community program is less expensive and requires less implementation time than other load management strategies.

WHEN CONGRESS PASSED the Public Utilities Regulatory Policies Act (PURPA) in 1978 to encourage electricity load management programs and rate reforms, California's Pacific Gas and Electric (PG&E) responded with $100,000 incentives for three communities. Under PURPA, utilities are to promote energy efficiency, reducing the need for peak power; and to increase conservation efforts among electricity consumers. PG&E's program in Chico, Merced, and Davis, California did just that.

For this first joint peak load reduction program, PG&E provided each community an incentive of $10,000 for each one percent reduction in peak electricity use, with a maximum incentive of $100,000. The three northern California test communities each obtained the maximum $100,000 for their voluntary efforts — to be spent on community energy conservation goods and services. From June through September 1980, Davis, Chico, and Merced reduced peak electricity use 22 percent, 17 percent, and 13 percent, respectively — a total reduction of 15.3 million peak kWh.

The program also brought a substantial reduction in off-peak electricity conservation (about fifteen to eighteen percent for the three cities).

According to PG&E, this program is less expensive and requires less time for implementation than other load management strategies such as time-of-use rates and metering, or radio-controlled appliance cycling. Moreover, in accord with PURPA, costly peak electricity use can be substantially reduced, as well as the need for new power plants to meet peak demand. Utilities can thus reduce the escalation in electricity costs to consumers.

In Davis, an eight-member technical advisory committee was appointed to work with a "community outreach committee" of 25 members. These two groups, assisted by a PG&E-supported coordinator were responsible for development and implementation of a joint utility-community agreement to establish the terms of the program and to develop residential as well as commercial peak load reduction strategies. Through this arrangement, the utility and the communities used complementary resources and experience, more effectively than could either separately.

Program Goals

The goals of the initial program were quite simple. First, the technical committee and the community outreach committee targeted peak electricity use activities and developed strategies to stimulate electricity consumers to avoid or reduce those uses. Second, the community outreach committee presented these peak reduction strategies to the community at large and provided feedback to the public on its load management efforts. Overall, these efforts aimed to educate the community about the benefits of and rationale for this program. In retrospect, the program was valuable not just for its immediate results in peak electricity reduction and conservation; it also provided community education about energy conservation and, ideally, will alter the way people think about energy in the long term.

Three Troubles

This pilot program was not, however, without its problems, particularly in measuring the success of the program. First, the initial index of energy use reduction worked *against* efforts to reduce peak consumption. Although the ratio of peak use over total use reflected the decrease in peak electricity use compared to 1979, any off-peak conservation served to reduce the ratio and thus the measure of progress. For example, if electricity consumers reduced their use equally during peak (noon to 6 pm weekdays) and off-peak hours, the performance index would reflect little or no progress.

Within a few weeks of the program's start, Davis adopted a new index. Weather variations are still a significant problem, however. In fact, the program's progress last year must be interpreted carefully in light of the state's unseasonally cool summer, which reduced peak energy use by air conditioners. In addition, the peak measurement index did not compensate for the dramatic rise in electricity prices that undoubtedly triggered some voluntary conservation. While measurement of the program's peak and overall electricity reduction progress presents difficult technical problems, PG&E and the com-

munities are optimistic: they expect that a more accurate measurement index will be derived.

Assessing which particular energy reduction strategies should be used was a second problem. Moreover, the technical and outreach committees found it difficult to determine which particular electricity uses and consumers should be targeted to maximize community success. Air conditioning was one obvious energy use that deserved attention — it constitutes the largest single use of peak power during summer months in California.

Third, in this, as in any infant program, coordination of the myriad activities was difficult. The program's success hinges on the cooperation and coordination of its committees.

Utilities Ambivalent

Many have argued that voluntary energy reduction is, at best, a losing proposition. This program, though, demonstrates the benefits from community-level efforts and indirect incentives to induce voluntary conservation and peak electricity reduction. In fact, PG&E has calculated that for each dollar spent on this program, the utility will gain $4.97 in benefits, and the ratepayers gain $2.11. It appears that the cooperative program is much more than cost-effective.

Implemented on a larger scale, this program could lead to substantial, cost-effective reductions in peak electricity use. Electricity consumers are more likely to embrace such a voluntary program than mandatory time-of-use metering and remote appliance cycling programs. Utilities, though, may have an ambivalent attitude. On the one hand, cooperative programs may dissuade some communities from creating their own municipal utility districts (see *Notes* vol. 3, no. 3), maintaining the utility monopoly on power production. On the other, utility revenues, under this program, are lower than they might be under the other load management options. However, a cooperative effort, with its long-term community education and overall energy conservation effects, seems far superior to the other more expensive, mandatory programs.

—*Eric Woychik*

Reference:

Pacific Gas & Electric

1980 *The Cooperative Electricity Management Program: A Joint PG&E-Community Effort.* PG&E, Load Management Section, Rate Department, San Francisco, California, March, 29 pp.

Eric Woychik was vice-chairman and previously chairman of the Davis Community Electricity Load Management Outreach Committee.

How to Finance the Solar Community

S OMEWHERE IN THE FILES of the Los Angeles Department of Water and Power there is a list of consumers waiting for solar system loans. Eighty miles south, in San Diego, the solar loans go begging. Why the difference? Not the interest rates, or the tax credits, but a novel institutional arrangement that in other California communities goes under the name of Municipal Solar Utility.

New technologies move through society only as rapidly as its institutions can carry them, and many soft energy advocates doubt whether our present-day utilities can proceed toward a low energy solar future at any great speed. Evolving in an era of cheap energy, low interest rates, and limited environmental awareness, the utilities organized around large, centralized facilities handling a limited number of special fuels. Dispersed and renewable energy supply systems call for very different kinds of service and management expertise. The response is a new institutional structure: the Municipal Solar Utility.

First proposed by the US Department of Energy, the Municipal Solar Utility (MSU) concept covers a collection of schemes for accelerating conservation efforts and bringing solar energy into commercial markets. Principally, the MSU helps to overcome two of the constraints on solar and conservation: high initial costs of the new technologies, and the public's lack of familiarity with them, even skepticism about their potential benefits. To date, MSUs have focused on residential retrofits for solar domestic hot water and swimming pool heating systems, but they are equally equipped to promote and manage other technologies such as district heating systems and conservation measures.

Financial Leverage

As a nonprofit entity, usually linked to a city government, the Municipal Solar Utility enjoys certain financing advantages over private utilities. The connection to government gives the MSU access to municipal bond financing, allowing them to borrow money at below-market interest rates. Because the return from municipal bonds is tax free, they readily attract high income bracket investors despite the low yield. And as long term investments, municipal bonds are well matched to the lengthy payback period of solar systems.

Conventional loans are often based on repayment in five years or less, leading to monthly installments on energy saving solar equipment that may exceed the utility bills they replace. This naturally discourages investment by consumers, who tend to emphasize short-term economies over solar equipment's lifetime savings.

MSUs, in contrast, are able to amortize solar systems over a period of up to 20 years. The utility is, in effect, extending a loan to the consumer which is repaid over the system's useful life. Monthly payments are so much lower than with conventional loans that MSU bills to consumers are less than those from the established utility.

The city of Santa Clara, California, near San Francisco, was among the first to explore the MSU concept for solar financing. Formed in 1976, the Santa Clara utility has expanded from swimming pool heating systems to include other solar applications such as domestic hot water and heating for a 2500 m² community recreation center. In this version of the MSU, the city owns and operates solar collectors, which are leased to on-site users. The utility shoulders the responsibility for choosing and servicing the equipment; all that is required of the user is an initial cash outlay, generally less than $300, to cover the labor and material costs of installation. Rental fees are calculated on a ten year amortization of the system's retail price, at an interest rate of 7%, and range from $13 to $17 monthly for domestic hot water. Charges for swimming pool systems accrue only during the season when they are in use, from April to September, and run about $28-30 per month, or 20-30% below the equivalent current cost of natural gas. Future energy price increases should have little effect on these rates, and consumers may expect only slight adjustments as operating and maintenance costs rise.

Consumer Confidence

The solar industry has shown high reliability for an infant enterprise, but with many small, new firms, problems are inevitable. A few failures can overshadow the industry's general success, making consumers reluctant to invest without some assurance of support should they encounter problems later on. Not even California's 55% tax credit for solar systems could induce very many consumers to take on the risk and responsibility of individual investment.

Because the credit did not totally overcome the problem of high initial system cost, one California bank — San Diego Federal Savings and Loan — developed an ingenious plan for full financing. Their loan program would add the cost of the solar system on to existing home mortgages, extending their duration to keep monthly payments constant. Homeowners could install domestic hot water systems under this arrangement for nothing down, with no increase in monthly payments, and without forfeiting any of the tax credit. High initial system cost would become a substantial first-year savings in tax and utility bills. The surprise is: still not many takers.

At the other extreme, a plan developed by the Los

Angeles Department of Water and Power offers consumers enough assurance to keep them waiting in line. The city provides interest free financing up to $2,000 for domestic hot water systems in homes with electric water heaters. Loans are repaid by adding $40, on the average, to the bimonthly water and power bill. As in the San Diego program, customers select their systems from a list of certified suppliers and receive the full 55% tax credit on their purchase. The difference between the programs is a strong backup warranty from the utility. In addition to one-year installation and two-year manufacturer's warranties from the solar firm, the Department of Water and Power agreed to inspect problem systems and negotiate with the seller/installer for three years subsequent to purchase. This added measure of consumer protection apparently reduces the risk of investing in a solar system to a level the public is willing to accept.

The Municipal Solar Utility concept should find even greater favor with consumers since the city owns the solar installation, is responsible for its maintenance, and provides long-term, low-cost financing. To explore the potentials of this concept, the California Energy Commission is currently sponsoring alternative institutional forms in six California cities. Some highlights of this program:

• The City of Los Angeles, expanding its Department of Water and Power plan, is developing financing packages through a non-profit corporation that will rely more on private equity.

• An existing community development corporation in Santa Monica is being transformed into an MSU tailored to the high local level (80%) of multi-family housing.

• Bakersfield is also using an existing redevelopment authority to explore self help options for solar installations.

• The City of Oceanside is extending the Santa Clara model to include both residential and governmental energy needs.

How these programs will complement the related activities of privately owned utilities remains to be seen.

—Craig Conley

Municipal utilities are, in theory, better equipped than private utilities to finance solar and conservation efforts in the community. They get faster local feedback, often elect their own boards, and have a lower cost of money. So why aren't there more of them?

References:

Braly, Mark
1979 *Public Power—Phase III.* Presented at the International Solar Energy Society Silver Jubilee Congress, May 29, 1979, Atlanta, Georgia.

Saunders, Robin
1978 "Santa Clara Pioneers with Solar Utility." *Public Power*, July-August p. 24-26.

White, Sharon Stanton
1979 *Municipal Bond Financing of Solar Energy Facilities.* December, SERI/TR 434-191.

More information on Municipal Solar Utilities and activities in California can be obtained from the California Energy Commission, 1111 Howe Ave., Sacramento, California 95825, Susan Garfield (916) 920-7393.

A Place In the Sun

Policy problems can be just as puzzling as technical questions in the development of solar energy systems. What do you do when your neighbor's tree, or new home addition, or even his solar collector blocks your solar collector?

SOLAR ACCESS ZONING is an extension of familiar land use regulation. Many states now authorize solar easements—private agreements in which a homeowner pays his neighbor for the freedom from shading. This private easement system becomes too complex in dense urban areas with multiple, overlapping easements. The cost to the solar owner is prohibitive, too.

Solar zoning shifts the burden of protecting solar access from the individual to the local government, lowering the costs of solar easements. We report here, in condensed form, on alternatives to the private easement approach with experience gained in communities across the US.

Albuquerque, New Mexico

Albuquerque, an early pioneer in solar access regulation, has perhaps gone the furthest in developing and implementing a comprehensive solar access program through zoning. In 1977, the city established a "buffer zone" between high-rise and lower-rise districts to minimize shading. In higher density zones, the Albuquerque ordinance sets a maximum height to build at any point on a lot. Any structure or vegetation over this height must lie within a pyramid defined by 45° planes rising from the four sides of the lot. However, to avoid neighborhoods with all the buildings in the absolute centers of their lots, structures may be shifted laterally, except to the north, within a set of planes set at 60° from the lot boundaries.

Variable setback requirements were adopted in 1980 to minimize shading of collectors, too. Front yard setback minimum is 20 feet (6 meters) and rear setback at least 15 feet (4.5 m), with a five-foot (1.5 m) minimum setback for side yards. To promote solar green-houses, convective air walls and Trombe walls, a homeowner can get a variance to that setback limit.

Sector plans can be re-evaluated every four years. *Sunpaper,* journal of the New Mexico Solar Energy Association, reports that one challenge to the sector plan came when a developer tore down some buildings to build townhouses. The structures would have been high enough to block neighbors' solar access, so the townhouse plans required some redesign, but not enough to cause any real hardship for the developer. The solar precedent in Albuquerque is now set.

Davis, California

Davis' street layout requirements ensure that every new rooftop will have a southern exposure appropriate for solar panels or photovoltaic arrays, and that every south-facing window will have direct solar access in winter. New streets are narrow and shaded by deciduous trees to cool neighborhoods a full ten degrees F (5.5° C) in summer. Zoning amendments permit flexible siting of fences and hedges for solar heating and greater use of shade control devices. Another zoning ordinance permits home businesses, eliminating the need for some residents to commute to work, which cuts transportation needs.

Lincoln, Nebraska

Wind rights are vital to the development of wind energy. Wind rights guarantee access to the wind for electric power generation in Lincoln. A new ordinance requires that applicants for wind machine operation guarantee covenants or easements from the abutting owners providing access to wind sufficient for the machine's adequate operation. There are currently no protected "rights" to the wind in any state, reports *A. T. Times,* July/August 1981. "The questions of whether and how to establish such rights, and determine their value once they are established, will need to be resolved before successful, widespread wind energy development can occur." A wind farm developer can now only ensure access to wind by acquiring sufficient land to preserve an unobstructed wind flow. This approach has limitations; first and foremost is the cost.

Current Oregon law suggests an alternative. Oregon law now permits the state and local governments to acquire by purchase, agreement, or donation "conservation or scenic easements." Such easements can be transferred to nonprofit corporations for protection. Analogous legislation, *A. T. Times* suggests, could authorize the state to acquire wind access easements which could be leased or assigned to wind system developers.

Los Alamos, New Mexico

Los Alamos has experience with a "solar-envelope-by-permit" approach to zoning. This solar amendment protects collectors against shading from vegetation and structures by a hypothetical 12-foot (3.7 m) fence at the lot line (between 10 a.m. and 3 p.m.) provided that the collector is smaller than one-half the heated floor area of its structure. Fifty solar access certificates are now on record in Los Alamos, and there have been no challenges filed. Martin Jaffe, writing for the *Solar Law Reporter*, adds: "Subsequent owners of restricted lots may not be overjoyed when later ordered to trim their trees or when a building permit application for an additional bedroom is desired, but present owners seem satisfied."

Sacramento County, California

Popular for its simplicity, the protection of solar access at the subdivision planning stage has been tried in several communities. In Sacramento, Albuquerque, San Diego in California, and Port Arthur in Texas, lot orientation is required to minimize summer heat gain and peak air conditioning energy use. Sacramento County has found an administrative problem: although lots were being oriented properly, buildings were not. An ordinance requiring that building lines follow lot lines would be an easy solution.

San Diego County, California

Plastic overlays, as an alternative to complicated drawings of building and vegetation shadow patterns, are being developed in San Diego County. The goal is a simple method of analyzing building layout and shadow problems caused by adjacent development.

Woodburn, Oregon

Solar access protection in Woodburn, a small city of 11,000, follows an

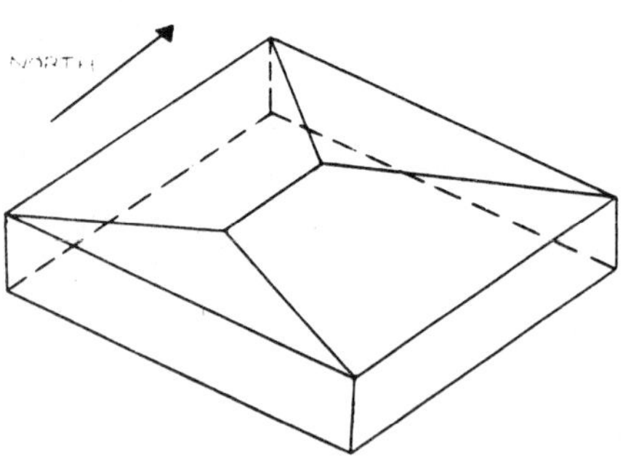

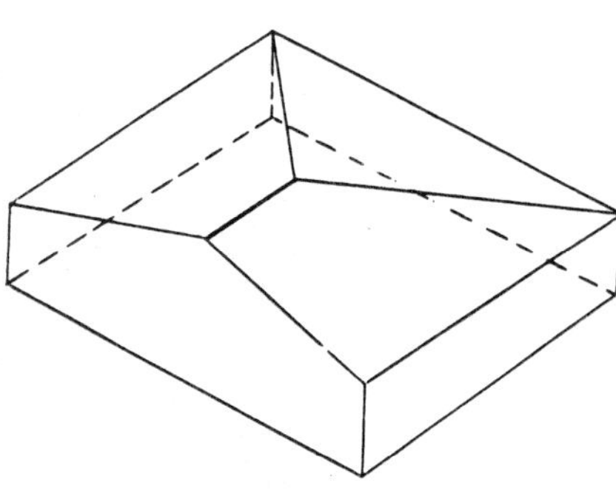

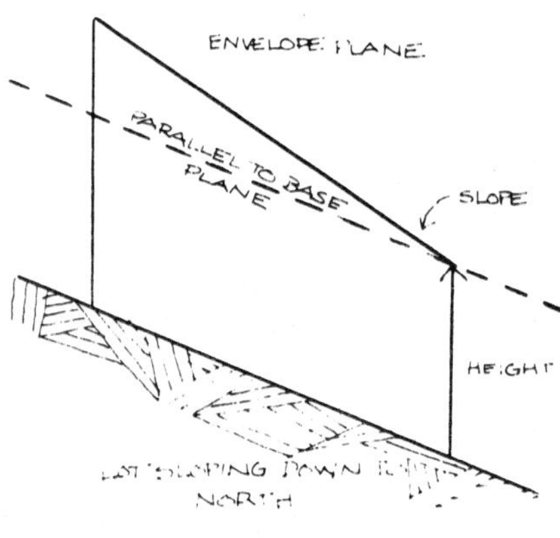

Solar envelopes *have restricted volumes within which a building can be sited yet not shade other nearby buildings within a specified time. Ashland, Oregon is one community that has simplified the solar envelope into a solar plane that slopes down the north side of a property to protect solar access to the adjacent property's south wall.*

from Gail Boyer Hayes, Solar Access Law, *Balinger, 1979.*

innovative ordinance based on the *status quo*. To qualify for access protection, the homeowner must first fully weatherize his home and then record with the city planning department how much sunlight is available to collectors at a certain location. New buildings and growing vegetation cannot later block those collectors. Community support for this detailed program ran high with 93 percent favoring solar access protection in new developments and 75 percent favoring access in existing neighborhoods, according to *Sunpaper*.

Wyoming State

The Wyoming Solar Rights Act is patterned after western water law: "First in time and first in right" to that water, over possible water users downstream. Enacted in 1981, the Solar Act protects a property owner's rights to the sun once he has purchased a building permit and a building inspector certifies that his solar system is using the sun's rays. Sun rights can be transferred to the next owner, too. The restrictions list that the collector must be set back from the property lines so that it would not be shaded by a ten-foot (3 m) wall on the winter solstice. The easements are protected only between 9 a.m. and 3 p.m.

—*Elyse Axell*

References:

Jaffe, Martin
1980 "A Commentary on Solar Access: Less Theory, More Practice" in *The Solar Law Reporter*, vol. 2, no. 4, November/December, pp. 769-779. Available from Superintendent of Documents, US GPO, Washington, DC 20402, $3.

New Mexico Solar Energy Association
1981 "Special Issue: Community solar access" in *Sunpaper*, vol. 6, no. 7, July/August. Available from NMSEA, PO Box 2004, Santa Fe, New Mexico 87501 for $2.

Noun, Bob and Peter Pollock
1981 "Legal Access: Getting your piece of the sun & the wind" in *A. T. Times*, July/August, p. 7. Available from *A. T. Times*, c/o National Center for Appropriate Technology, Butte, MT 59702.

SUN
1981 "Wyoming Solar Easements" in *SUN*, September, p. 13.

Feds Caution Solar Shoppers

'Let the buyer beware' is the best advice for energy-saving products, according to the US General Accounting Office (GAO) in a recently-released report to Congress.

GAO REPORTS that consumers have difficulty assessing the accuracy of the hundreds of energy performance and savings claims made by solar product manufacturers today. "According to the experts, many of these claims are inaccurate, do not tell the whole story, or are not comparable to claims of competing products." This government office believes "continued government efforts" are needed to assist consumers.

Responding to economic and patriotic pressure, consumers are buying energy-saving ceiling fans, insulation, gas-saving devices, and other efficiency products, GAO believes. Sales of energy-saving products have increased dramatically in recent years. Increases from $63,000 in 1978 to $7,170,000 in 1979 were reported by one manufacturer of a new space heater, advertised to be more energy efficient than conventional heaters, GAO notes. Production of solar collectors has increased ten times since

> Without accurate information, the consumer may decide to purchase one of these products, rather than add 6" of ceiling insulation. According to the DOE engineer, the additional insulation would provide about 50% more energy savings at far less cost—total cost about $300.

1974, while wood stove and fireplace insert sales grew sixfold from 1973 to 1979. By providing tax credits, federal and state governments also encourage consumers to purchase energy-saving devices. According to the GAO report, in 1978 and 1979, 10.8 million tax returns claimed tax credits for purchasing $7.6 billion in energy-saving products.

The Federal Trade Commission (FTC), responsible for protecting solar product consumers, works with industry to develop testing and advertising standards, but GAO believes the Commission could do more. According to GAO, the accuracy of most product's energy-saving claims has not been checked by the FTC: "We noted that the FTC was aware of several ads for solar and wood burning products with questionable claims.

FTC decided not to fully investigate these claims but to work with industry groups to develop standards."

The FTC responds that consumer problems with misleading claims "are not as bad" as the GAO report suggests. Fraudulent products, the FTC believes, tend to be sold door-to-door or mail-order: "Although there are a large number of companies marketing products in this way, we believe that it is not always cost-effective for a federal agency to spend considerable resources tracking down such small and elusive perpetrators."

DOE engineer, the additional insulation would provide about 50 percent more energy savings at far less cost (total cost about $300)."

GAO also concludes that, in general, solar product firms "are not responsive to consumer requests to support claims." GAO's test was to write, as consumers, to 97 firms asking for the information they use to support their claims. Nineteen percent did not reply, 51 percent of the replies were not considered responsive to the request, and only 30

SOLAR COLLECTOR CERTIFICATION	CALIFORNIA ENERGY COMMISSION
MODEL NO.	SERIAL NO.
GROSS COLLECTOR AREA (FT²)	COVER PLATE
DRY WT (LBS)	MAX OPERATING TEMP/PRESSURE (°F/PSI)
FLUID CAP (GAL)	MAX NO FLOW TEMP/PRESSURE (°F/PSI)
HEAT TRANSFER FLUID OR GAS	MAX LIVE LOAD (LB/FT²)
MAX FLOW RATE (GPM) (SCFM)	
THERMAL PERFORMANCE RATING (BTU/HR PER FT²) LOW TEMP.	MEDIUM TEMP.
THERMAL PERFORMANCE EFFICIENCY: SLOPE	Y-INTERCEPT
USE RESTRICTIONS:	

In its survey of the industry, GAO found "hundreds" of advertisements having questionable energy-saving claims. GAO reports: "the sellers generally do not provide consumers with information to support product claims; even when obtainable, the supporting information may be inaccurate or highly technical; and, the consumers often do not have the opportunity to learn through experience and then switch to more useful products." GAO relied on test reports from the Environmental Protection Agency, the Department of Defense, state agencies, and private consumer organizations as the basis for questioning an advertising claim's accuracy.

Solar Buyers Get Burned?

"For example," GAO reports, "one vendor told us that installing his double-pane windows (costing about $1,560) would save about 30 percent on the heating cost of a specific house, while three other dealers told us that installing their insulated siding (costing $3,000 to $5,000) on the same house would result in between 15 and 40 percent energy savings. Calculations by a DOE engineer, however, showed that either of these measures would probably achieve only about a five percent savings for this house. Without accurate and meaningful information, the consumer may decide to purchase one of these products, rather than to add six inches (15 cm) of ceiling insulation. According to the

percent of the replies did include information to support their claims. "Overall," GAO says, "more than two-thirds of the firms queried did not provide us, as consumers, with any basis for their claims. The most common type of nonresponsive reply contained only additional promotional literature, rather than information that supported the claim."

In making recommendations, GAO notes that according to FTC and state and local officials, many energy-related problems are localized and would be better resolved at the state or local level. Though most states have consumer protection laws that are aimed at eliminating unfair and deceptive trade practices, their efforts are generally limited by a lack of technical expertise and funds.

GAO does recommend that the FTC publish consumer factsheets about some of the difficulties with claims and ads, and then places most emphasis on the infant "Energy-Saving Device Clearinghouse," a project of the US Department of Energy. In 1980, the Clearinghouse received a two-year grant from DOE for an energy-saving device fraud prevention project to provide technical and legal support to federal, state, and local consumer protection agencies. The Clearinghouse manually maintains a data bank of names of products or companies making energy-saving claims, developed by monitoring ads in national and local periodicals. With the current Administration, this project's future looks doubtful.

A California Answer

Providing consumers with reliable, comparable performance data on different solar collectors is the goal of the California State Testing and Inspection Program for Solar Equipment (TIPSE). TIPSE is a voluntary solar collector certification program for solar manufacturers. Certification is awarded on the basis of collector testing done at state accredited labs following ASHRAE 93-77 testing procedures. The program aims to strengthen the solar industry through consumer protection. Certification provides the consumer with documented thermal performance data and durability assurance.

In California alone, TIPSE has certified 46 solar manufacturers of an estimated 75. Across the country, 183 solar collectors, representing 77 different solar manufacturers, have been certified (as of April 1981). To qualify for certification, the collector efficiency may not decrease more than ten percent during the test period. In addition, significant deterioration of collector materials, water damage or pressure losses must not occur. The tests simulate expected operating conditions but do not constitute a warranty.

Manufacturers have been receptive toward the program and think it should be expanded, according to a recent phone survey of solar manufacturers. Most of the California manufacturers surveyed were certified or in the process; three-quarters of them felt that TIPSE has benefited the solar industry by upgrading its image and providing quality assurance to the consumer. Nearly all of the manufacturers were in favor of requiring TIPSE certification for the state solar tax credit. One interesting utility tag-on: as of January, 1980, California utilities offering financial assistance to solar water heater buyers require that all solar collectors eligible must be TIPSE certified.

—*Elyse Axell*

References:
California Energy Commission
1981 *TIPSE Survey and Test Results.* Pamela L. Goetze, editor. California Energy Commission, Solar Office Development Division, 1111 Howe Avenue, Sacramento, CA 95825.

General Accounting Office
1981 *Consumer Products Advertised to Save Energy—Let the Buyer Beware.* July 24. Single copy available free of charge from US General Accounting Office, Document Handling and Information Services Facility, PO Box 6015, Gaithersburg, MD 20760.

TOTEM: The Little-Engine-That-Could

TOTEM—the TOTal Energy Module from Fiat—uses conventional automotive technology in an unconventional way. A Fiat 127 four-cylinder engine, parting company with its usual complement of transmission, axle, and wheels, drives a 380-volt, 200-amp, asynchronous generator producing 15 kWe. The heat from the engine's exhaust is not allowed to go to waste, but is converted to 38 kWt of hot water at a maximum temperature of 80°C. The entire unit is compact (1 m³), quiet (65 dbA), insulated (4.2 MJ/hr heat loss), and is easily adapted to run on natural gas, alcohol, liquid petroleum gas (LPG), or biogas.

The TOTEM generates electricity alone at 25% efficiency, and with hot water its overall First Law efficiency climbs to over 90%. Performance, as usual, comes at a price: $6,500-7,500 installed with a total operation and maintenance cost projected to be $0.30 per hour. This includes engine removal and overhaul every 3,000 hours, but the power unit is so easy to extract and replace that down time is kept to a minimum.

Given these features, how economical is the TOTEM? The answer depends on relative fuel costs and use. If electricity, now at $0.055/kWh, inflates at about the same rate as the unit's natural gas fuel (now $0.24/therm) and if the service life of the TOTEM reaches 15,000 hours, then the system attains a positive net present value with only four to eight hours of daily operation. Fiat designed the TOTEM for exactly such intermittent activity, as it reaches full power in just a few seconds. It was not intended to supply 100% of any application's electrical and heating needs, only a part of both. The current version of TOTEM is designed to respond to thermal demand; it cycles on and off to assure ample hot water and provides electricity as a bonus.

Connection to the grid is necessary, since startup requires an initial flow of current through the generator which doubles as a starter. Under these conditions, the TOTEM's economic performance is predicated on utility purchase of surplus electricity. Fiat has plans for a stand-alone unit, but applications independent of the grid are unlikely.

TOTEM would be most practical where electricity and heat demand follow roughly the same cycle, and where both exceed the unit's output. The application Fiat seemed to have foremost in mind was the large apartment complex, with multiple TOTEM units installed where demand levels warranted. Other possible applications would be in dairy facilities or food processing plants which need both hot water and electricity. In large residential or industrial installations, the TOTEM could be used as a peak shaving or load leveling device, since hot water storage would be adequate for any temporary excess accumulated during peak-hour operation. The benefits of the TOTEM are clear: primary energy savings up to 40% over contemporary peak power production techniques and no distribution losses which can amount to 10% of power output in centralized networks.

Utility rate structures, however, tend to discourage such small-scale production of electricity. Unless peak power comes at a premium price, apartment house owners and small industries have little incentive to invest in something like the TOTEM, since the utilities would capture most of the savings. Differential electricity rates would distribute these benefits more equitably and help commercialize the TOTEM. This has been the experience of utilities in Germany and Britain that popularized storage heating (another load leveler) by offering low off-peak rates through separately metered systems (Asbury, Geise, and Mueller, 1980). But until such time as multi-tiered rate structures become widespread, the most likely candidates for the TOTEM might well be utilities, particularly at the municipal level. They would face none of the problems of private owners, such as maintenance, grid integration, seasonal load matching and buy back rates, and could benefit from the TOTEM's high fuel efficiency.

Utility application is certainly on its way to being the primary use of the TOTEM in the United States. The most extensive tests to date have been conducted by the Brooklyn Union Gas Company on the TOTEM II, a version adapted to US voltage and frequency characteristics. After an initial shakedown period, three units performed reliably and at rated efficiencies over a nine-month, on-line run. Several non-utility sites for field tests are under consideration. The Office of Appropriate Technology in California has proposed to install two TOTEM II units at the Animal Health Facility in Fairfield. Since cooling needs there exceed the demand for space heat and hot water, the plan is to connect the system to an absorption chiller which would air condition the small animal hospital. Along with this modification, utility buy-back of surplus electricity will be needed for the units to be economically effective. Another field test is to be conducted in South Dakota under Department of Energy auspices.

Even after the utilities agree to attractive buy-back rates, the TOTEM will still have to find its way out of the morass of the United States Customs Service. Because it falls into the category of electrical generator (steam boiler might be more appropriate), the Italian-made TOTEM II will face a 7-8% import duty, although there are no domestic products with which it is in direct competition. The little-engine-that-could still has a few institutional barriers to overcome.

—*Craig Conley*
Charles Drucker

References:

Asbury, J.G., R.F. Geise, and R.O. Mueller
  1980 "Electric Heat: The Right Price at the Right Time." *Technology Review* Volume 82, Number 3, December/January 1980, 32-40.

Fiat of America
  n.d. Product brochure. Available through Fiat Motors of America, Inc., 155 Chestnut Ridge Road, Montvale, New Jersey 07645.

As Long as the Wind Shall Blow

Wind power, as a renewable source of electricity, offers stiff competition for fossil and nuclear electric power. Megawatt-rated aerogenerators are already operating in the US and Denmark, with more under construction in Sweden and West Germany. Rapid advances in hardware are only partly responsible for wind power's new-found popularity: recent analyses of wind in electric grids force a reappraisal of its economic value.

Traditionally, the only economic value assigned to wind power in a grid was as a fuel-saver for thermal power stations, or as a water-saver for hydroelectric generators. It was widely believed that a variable power source such as wind, without energy storage, could not reliably substitute for conventional capacity. Power plants would always have to be kept in reserve as back-up for the periods when the wind does not blow. In the jargon of electric power engineering, wind power was said to have no "capacity credit."

Redefining Reliability

Recent mathematical models and computer simulations of grids containing wind power have contested this traditional view. The main economic value of wind power is indeed the fuel saved, but there is also a useful economic contribution that wind power can make by substituting for some of the grid's conventional power plants. As a fuel saver, wind *energy* replaces a mixture of base, intermediate, and peak load energy sources. (In a fossil-fuel based grid, the latter is the most expensive.) But it turns out that wind power can be a capital saver, too. Provided that a small amount of extra peak load capacity (which is rarely operated) is also installed to maintain grid reliability, wind power can substitute for some conventional *base load capacity*. The plants thus replaced are those with the highest capital cost.

This insight discards the judgment that conventional plants are "reliable" while wind power generators are not. Both are unreliable to some extent, and both require some back-up capacity. More precisely, both types suffer from "forced outages"—periods when they are required to operate but unexpectedly fail. But the two "fail" for different reasons, and this affects how their respective reliabilities should be assessed. Forced outages of conventional plants result from breakdowns, while forced outages of wind power plants coincide with lulls in the wind. These lulls occur more frequently than conventional breakdowns, but their durations are shorter. (Aerogenerators break down, too, but random mechanical failures of individual units in a large array derates the average power output of the system only slightly.)

Calculating the reliability of wind power is complex because three separate probabilities are involved. First the wind is subject to lulls and other variations. In addition, the grid's conventional units must be assigned a forced outage probability. Finally, electricity demand varies. Early mathematical work on these probabilities was performed in 1978 at Lawrence Berkeley Laboratory by Ed Kahn, who indicated that wind power from dispersed sites does have capacity credit. Later work, moreover, suggests that even at a single site, an array of aerogenerators experiencing the same wind regime would have a significant capacity credit. Indeed, the single-site credit makes the principal contribution.

Simulating the Grid

The Electric Power Research Institute published, in 1979, a set of computer simulations for three separate, hypothetical electricity grids containing wind power capacity. EPRI found that the capacity credit of wind power increases with installed wind capacity only up to a point; at large wind penetrations, a constant limiting value is reached. However, the study could not explain the quantitative differences between the three grids in the capacity credit assigned to wind. As in the computer simulations of other complex systems, it was hard to see the forest for the trees.

Instead of a simulation, the explanation required a sensitivity analysis to show how the capacity credit of wind power depends on particular grid, aerogenerator, and wind characteristics. The proper research models, though, were not yet constructed.

Applied mathematicians and statisticians working in Canberra have taken an important step. They employ probability distributions or histograms to describe the three factors affecting capacity credit: wind power, forced outages of thermal power plants, and electricity demand. Brian Martin from the Australian National University, and John Haslett, Mark Diesendorf, and John Carlin from the Australian Commonwealth Scientific and Industrial Research Organisation (CSIRO) used these simplifying assumptions: there are no correlations between wind power and electricity demand; no relationship exists between wind power at different times;

and the grid system has no "memory" (e.g., hydroelectric storage). The capacity credit of wind power may then be determined by calculating the probability of bulk power demand exceeding available supply, or the "loss of load probability" (LOLP). This is expressed as a unit of time, such as one day in ten years.

Martin and Diesendorf derived empirical probability distribution functions of demand and wind power from hourly records spanning at least one year in the separate state grids of South Australia and Western Australia (where the data base covers a ten-year period). For both grids they performed a sensitivity analysis of how wind power capacity credit varies with several grid and aerogenerator parameters. Several of the conclusions of this *numerical probabilistic model* have since been confirmed with hour-by-hour *computer simulations* originally developed to investigate the operating problems of grids incorporating wind power.

In a separate paper, Haslett and Diesendorf assumed that all conventional units in the grid were identical, and that demand for electricity and the availability of conventional units could both be described by a normal (Gaussian) probability distribution. These approximations, though crude, enabled them to derive *analytical expressions*—mathematical equations instead of only numerical results—for the capacity credit of wind power with either small or large penetrations into the grid. All three methods—the numerical probabilistic model, the analytic probabilistic model, and the computer simulation—give generally consistent results.

More recently, John Haslett and John Carlin created simple statistical models to describe the probability distribution of wind power from arrays of aerogenerators at geographically dispersed sites experiencing different wind regimes. As expected from Kahn's original work, this leads to an enhanced capacity credit.

Research Results

The principal findings of the Canberra group's research follow:
• For small penetrations of wind power capacity into the grid (less than about ten percent of total annual grid energy generation) the capacity credit of wind power is approximately equal to the average wind power (measured in megawatts).
• With increasing wind penetration, capacity credit drops below the average wind power. In the Western Australia model, at a penetration of 20 percent by energy generation, the capacity credit falls to about 40 percent of average wind power.
• At large penetrations (40 percent and up), capacity credit tends to a constant whose value depends strongly on the grid LOLP in the absence of wind plants, and also on

---

Forced outages of conventional plants result from breakdowns, while forced outages of wind power plants coincide with lulls in the wind. These lulls occur more frequently than conventional breakdowns, but their durations are shorter.

---

the probability of zero wind power. The latter factor depends, in turn, on the cut-in speed of the aerogenerators.
• For all penetrations, the capacity credit of wind power is much higher when the conventional grid is composed of a small number of large units, as opposed to a large number of small units.
• In Western Australia, wind power correlations among one set of four well-dispersed sites average about 0.6 (a correlation of 1.0 would imply identical wind regimes). The effect of this dispersal is to raise the capacity credit about 20 percent above its single-site value.
• Computer simulations show that moderate positive correlations between wind power and electricity demand, as observed in W. Australia, also lift the capacity credit by about 20 percent.
• The economic value of capacity credit in this region is by no means negligible. At 20 percent penetration it amounts to roughly one-quarter of wind power's fuel-saving value.
• More recent work by the Canberra group suggests that the system can be optimised so that wind plants substitute for base load plants, which have the highest capital cost. But to maintain the same grid reliability (measured by LOLP) with base-loaded wind plants, some additional peak load must simultaneously be installed. Peak plants offer low capital cost, but

high fuel cost; they represent a kind of reliability insurance with low premiums since they are rarely operated.

Wind's Capacity Credit Compared

The numerical probabilistic model by Martin and Diesendorf can also be used to calculate the capacity credit of a single additional thermal power plant, permitting comparison with wind. For the same average power output, the capacity credit of an aerogenerator array (single-site) in W. Australia turns out to be at least 40 percent of the capacity credit of an additional thermal unit with a forced outage rate of eight percent.

Their analysis illustrates the importance of relative size. If the thermal unit is too big for the grid, its average power being greater than about 20 percent of average grid demand, its capacity credit drops. As an example, consider the proposal by the W. Australian government to install a 1,000 MW nuclear plant in its grid by 1995. Assume a 60 percent capacity factor, a (generous) forced outage rate of eight percent, and a doubling of 1978 average grid demand to 1,060 MW. The average power output of the nuclear plant would be 600 MW, but its capacity credit, measured in terms of its substitution for a hypothetical plant with zero forced outage, would be only 260 MW, or 43 percent of average output.

If a wind power penetration of 20 percent were chosen instead of an oversized nuclear plant, average wind power would be 212 MW (with a rated capacity of 650 MW). The capacity credit at a *single site* in W. Australia would be about 100 MW, or 47 percent of average power. Even a smaller nuclear plant rated at 600 MW would have a capacity credit of only 255 MW, or 71 percent of average power output.

Thus, in a small electricity grid without any storage, the reliability of wind power plus conservation is generally comparable with that of nuclear power, or, for that matter, any oversized thermal power plant. In an ordinary-sized grid, the capacity credit of wind power will still add significantly to its value as a fuel saver.

—*Mark Diesendorf*

Mark Diesendorf is an energy analyst with the Australian Commonwealth Scientific and Industrial Research Organization (CSIRO). For further information on his work: write to P.O. Box 1965, Canberra City, 2601 Australia.

Commuters Unsnarl Traffic

photo by Tom Turner

gasoline prices soar); and they may be unaware of the alternative forms of transportation provided by many communities and corporations, and the advantages they offer.

Encouraging efficient travel

Cities, businesses, and commuters all benefit from ridesharing. Urban congestion, air pollution, and noise abate as traffic diminishes, making the downtown area more habitable and attractive, and the freeways less jammed. Employers find that building parking for commuting workers is expensive — as much as $22,000 per space in parts of Los Angeles, and that they can save capital costs by reducing the number of incoming vehicles. The workers, in turn, save money on their gas bills and car costs — up to $1,500 per year each, according to a California Department of Transportation estimate.

Many programs now exist, or are planned, to discourage single-passenger vehicles and promote multiple ridership through financial and other incentives. The City of Los Angeles, for example, has under consideration a program whereby employers will be permitted to construct fewer parking spaces for their commuting workers if they organize and encourage a ridesharing program. Ridesharing there is already saving 2.3 million gallons of gasoline per year, and this is expected to increase.

Local and state governments can also provide incentives directly to the commuters, as illustrated by several programs in Washington State. The City of Seattle sets aside free downtown parking for ride pools, and park-and-ride lots on the peripheries. The number of vanpools commuting to the city center quadrupled in the first five months of 1980,

---

There is no better symbol of American energy waste than the lone commuter, crawling to work along clogged suburban freeways, encased in a shell of glass and steel that weighs twenty or thirty times as much as its solitary passenger.

O F THE 80 MILLION US workers, 65 percent drive to work alone, each consuming up to 11,000 Btu per passenger-mile (1,732 kCal/passenger-km). At this rate, a fifty-mile round trip commute uses more energy *per day* than the average Bangladeshi consumes in *two months* for all purposes combined.

The quickest and cheapest way to reduce this energy waste is by adding riders to the cars on the road. Twenty percent of the workforce now shares rides; if only half of the 52 million lone commuters were willing to double up, they would save about 350,000 barrels of oil each day, roughly five percent of our current imports. Public transportation, once the dominant means of commuting, is even more energy efficient than carpooling, but only six percent of the work force uses it. Mass transit ridership dropped steadily for the three decades after 1940, and despite recent gains, is still below the 1950 level.

Commuters resist travelling more than one to a vehicle for many reasons. The geography of American life — far-flung residential suburbs encircling a dense urban core with defined commercial/industrial districts — often makes public transportation commuting impractical, and ridesharing cumbersome. But not all lone commuters are victims of geographic circumstance; many, perhaps the majority, prefer to drive alone despite the cost. They cherish their private moments going to and from work (even when they are stalled in traffic and blanketed in exhaust fumes); they refuse to believe that their consumption is excessive (even when

though part of this success may be due to financial incentives from the State. Van purchasers who can demonstrate that the vehicle will be used for a pool are exempted from excise and sales tax, saving them up to $1,000 over four years. Encouragement comes, too, from the federal level: recent changes in gasoline regulations establish high priorities for vanpools in times of shortage.

Many employers have adopted a program called Transportation System Management — TSM — originally devised by the US Urban Mass Transportation Administration. Jon Twichell, who prepared TSM plans for San Francisco, describes its features: "TSM works by accumulating a variety of elements, rather than trying to provide a single answer to the commuting problem. It consists of a series of low-cost programs that include ride sharing, parking management, express transit service, and employee incentives. An administrator, usually called a transportation broker, is appointed to manage the TSM program."

The successes have been dramatic. The Continental Oil Company alone runs over 190 vanpools, with 900 of the 2,000 employees at their Houston plant signed up as members. The incentive they offer is that the driver rides free, and has personal use of the van for a nominal mileage charge. Ninety-three percent of the pool members surveyed found the vanpool at least as efficient as their previous commuting mode, and many of them noted that it eliminated the need for a second car.

The Tennessee Valley Authority, in another TSM program, contributes thirty-five percent of the cost of a vanpool, in parallel with a program of bus subsidies and carpooling. Since 1973, the number of single-occupant autos has dropped from 65 percent to only eighteen. The cost to TVA is about $50,000 per month, but they saved at least $20 million in captial outlay for parking and roads.

State-organized vanpools aren't cheap, either although they, too, save in the long run. Jack Derby, of CalTrans, estimates that each pool costs about $2,000 to organize. But, he notes, a single pool reduces the number of vehicle-miles travelled per year by 110,000, lowers the number of parking spaces needed by seven, and decreases the pollutant load by four tons. The total direct user benefits from each vanpool come to $17,500 annually, distributed among an average of 11.5 people. Since 1974, the number of vanpools in the state has jumped from 10 to 2,000, and is growing at about 35 percent per year. Of the state's 100,000 ridesharers, three-fourths carpool and the rest use vans. A vanpooling population three to four times that size would make economic sense right now, and as for future growth, Derby remarks that "the perimeter of feasible distance changes with the price of gas."

Bus comeback

Part of the Transportation System Management repertoire is a program of special express bus services, and employee discounts for commuter bus tickets. Long neglected, and in many areas allowed to deteriorate, bus systems are beginning to show new life. Public transit generally increased 13 percent in the three years prior to June 1980, and even in Los Angeles, the freeway capital of the world, bus ridership is up 48 percent over five years ago, to a record 1.2 million passengers per day. But the share of the total transportation demand served by buses is still well under two percent, and many urban transit systems find it extremely difficult to expand.

The problem, surprisingly, is not a scarcity of capital to finance growth. Funds from the Urban Mass Transportation Administration, supplemented with state and local taxes on sales and gasoline, are more than sufficient; *operating revenues*, though, are so low that many systems cannot afford to put on the streets all the vehicles they have the means to purchase. To improve their financial condition, transit operators are exploring ways to increase productivity, recognizing that between 70 and 80 percent of their expenses are labor costs.

Articulated and double-decker buses increase the number of passengers carried by a single driver, and many systems have invested in these high-load vehicles. Another way to lower labor costs is to match the number of drivers to the passenger load, allowing the size of the work force to follow the cyclical rush-hour demands. This requires a sizable fraction of part-time drivers — a notion which unions have strongly resisted until quite recently.

Improved bus systems will still need incentives to increase ridership and improve public awareness of transportation alternatives. Trenton, New Jersey, tried free bus service during off-peak hours, under a special federal grant. Ridership increased by 50 percent and, after the year-long experiment, remained 20 percent above the old figure. Though impractical on an indefinite basis, free service is an excellent way to introduce more people to public transportation.

Other "no-fare" experiments include free service in a number of downtown areas nationwide. In Albany, New York, downtown ridership tripled since the introduction of this program. And in Spokane, Washington State, bus use is encouraged by giving passengers tokens which downtown stores exchange for a discount.

Old world, new ploys

European city managers (and city dwellers) appear much more willing than their US counterparts to include disincentives in the effort to reduce urban auto traffic. Dutch cities employ several devices, including diverters and traffic bumps, to slow down cars, and computerized signal priority systems to speed up buses. This reduces the competitive advantage of the automobile *vis-a-vis* public transit. In the densely populated city of Delft, automobile use is controlled by an urban design principle called the *woonerf*, or "living yard," which the US Department of Transportation has hailed as "one of the most widely admired methods of living with and managing the automobile in modern urban neighborhoods... Streets, which in many cities occupy over 20 percent of the total land area, may be altered to allow pedestrians, bicycles, children, and leisure seekers to share the space with cars safely and without conflict." To create a *woonerf*, roadways are narrowed (though emergency vehicles may still pass), obstacles and speed bumps are installed, many parking places are eliminated, and streets are raised to sidewalk level, creating an open, courtyard-like effect. The addition of bicycle racks, plantings, and street furniture such as play objects adapt the *woonerf* to a variety of activities besides driving, and restore a human dimension to the urban environment.

Restricted access is another important European tactic. Several cities, notably Bremen in the Federal Republic of Germany, and Gothenburg in Sweden, employ a traffic cell system with great success. The urban center is divided into a number of cells, like the slices of a pie. The cells are connected by a ring road but, for purposes of automobile traffic, not to each other. Drivers can enter any cell they wish, but to reach another cell by car must first leave the city center and circle the periphery. Public transportation, or even walking, are more convenient. As transit ridership increased by eight percent between 1970 and 1975, and vehicle traffic dropped by 40 percent, the downtown area become much more habitable: noise levels on the main shopping street fell from 74 to 67 decibels; carbon monoxide declined from 65 to 55 ppm; and accidents were reduced by half.

The European experience with traffic management demonstrates what soft path advocates have maintained all along: that efforts to save energy need not diminish the quality of life, and can often improve it.

— *Charles Drucker*

References:
Twichell, Jon
1980 "Have van, will travel." *Planning*, February 1980, 16-18.   •

US Department of Energy.
1980 "Transportation and Energy." *The Energy Consumer*, September 1980. Available free from the US DOE Office of Consumer Affairs, 8G082, Washington DC 20585.

US Department of Transportation
1980 *Center City Environment and Transportation: Transportation Innovations in Five European Cities.* Urban Mass Transportation Administration, 400 7th Street, S.W., Washington, DC 20590.

Use Less, Pay More

New York City Examines Energy Use

photo by Donal F. Holway

High population density makes cities large consumers of energy. But this same density makes them energy efficient; in metropolitan New York City, per capita consumption, at 185 GJ per year, is only half the national average. And a recent report from the New York City Energy Office (NYCEO) calls for an overall improvement in energy efficiency of at least 20 percent.

When New York City officials recognized that high energy consumption, especially oil, was draining the local economy, and that there were substantial economic benefits in efficiency improvements, they began to take a closer look at the Big Apple's energy system. NYCEO's Robert M. Herzog and Project Director James DeMetro recently released the results of this examination. Their report, issued in conjunction with the Cooper Union Research Foundation scrutinizes local energy end uses and recommends efficiency measures to help the city reduce its heavy oil dependence.

Energy Consumption in New York City: Patterns and Opportunities details energy use in five consuming sectors: residential, commercial, industrial, transport, and public. Energy use in each sector was broken down by fuel type (oil, natural gas, electricity, steam, and coal), and within each fuel type by end use. The authors cite three reasons for this approach. First, energy efficiency improvements and the economic benefits that would accrue are all readily identifiable within this framework. Second, it permits comparison of a variety of energy options — conventional and unconventional — to identify the most cost-effective alternative. Finally, an analysis of this type also provides a convenient basis for developing fuel policy measures which might be instituted in the event of unanticipated supply interruptions. Each of these features is an essential part of any coherent, long-range energy policy for New York City.

The data for the NYCEO study were collected in an exhaustive survey of statistics on energy consuption by sector for the year 1979. Although it is the most detailed study of the city's energy use patterns ever undertaken, 1979 was not a typical year: weather was warmer than average; the gasoline shortage in the city was particularly acute; fuel oil prices also doubled during the observation period. Additional problems were encountered in characterizing end-uses within the industrial sector, forcing the relatively coarse-grained approach of a breakdown by fuel type and Standard Industrial Classification codes. Difficulties such as these underscore the necessity of continually updating the data base to develop an accurate picture of the energy consumption patterns.

A Picture of Urban Thrift

Energy consumption across all sectors during 1979 in New York City totalled 1,360 EJ (EJ = 10^{15} Joules). On a per capita basis, this amounted to 185 GJ per year, or approximately half the national average. This typifies the relative

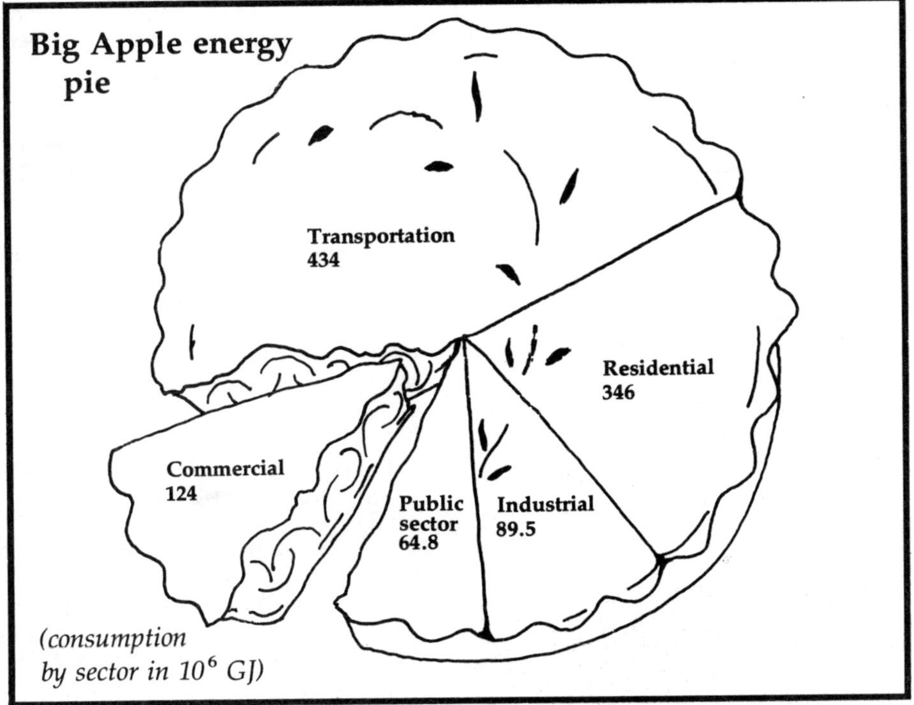

efficiency found in many of the world's urban areas.

Electricity consumption in New York City during 1979 came to 29 x 10⁹ kWh, principally from oil-fired and nuclear thermal generating stations. Only a small part of the city's electrical capacity is derived from hydropower, coal, or natural gas. As a result, New Yorkers pay *twice* the national average for their electricity canceling out their lower-than-average per capita consumption.

Residential energy consumption amounted to 345 x 10⁶ GJ for 1979, 126 GJ per household. Low-temperature end uses dominated this sector, with 69 percent of the total energy used for space heating and an additional 18 percent for domestic hot water. The remaining 13 percent was used for cooling, cooking, lighting, and household appliances. Although electricity supplied ten percent of the residential energy demand, only seven percent were actually electric-specific.

The NYCEO study identified several opportunities to reduce residential energy consumption. The major source of improvement is expected from reduced space heating demands as building envelopes are made more efficient and heating system performance increases. Raising the thermal integrity of residential buildings also lowers cooling requirements, and reduces high electricity peak load demand during summer. An initial target reduction of 25 percent has been set for this sector, recognized by the

study's authors as a conservative estimate. Given the relative inefficiency of the aged housing stock and the poor energy-conserving construction it incorporates, realistic energy savings are certainly well in excess of 25 percent.

Commercial activities consumed 122 x 10⁶ GJ during 1979, representing twelve percent of New York City's total energy budget. As with residential uses, the bulk of the end use demands — a full 86 percent — were for low temperatures space and water heating. Electric-specific needs took 4.7 x 10⁹ kWh, 13 percent of the sector's entire energy consumption. However, electricity use for all purposes amounted to 5.9 x 10⁹ kWh, 17 percent of the sectoral total.

Commercial activities and building types vary widely, and only close, individual attention can indicate which of the available energy-conserving options are appropriate. The study does, however, prescribe a series of general measures which can be implemented in most instances to reduce energy consumption. These include simple temperature and electrical load management techniques, insulation and weather-stripping, and relamping with energy-efficient flluorescent lamps. The use of Variable Speed Drives in commercial electric motors is also recommended (see "Efficient Electric Motors: Winding Up and Slowing Down " *Soft Energy Notes* 3:2:27. In all, it is expected that these simple measures can reduce energy consumption in the commercial sector

by 25 percent with only a modest capital investment.

Although the authors could not break down industrial energy consumption by end use, they did compile data on use by fuel type and industrial classification. Total industrial consumption was 89.5 x 10⁶ GJ, approximately eight percent of the city-wide total and less than one percent of the national total. Electricity provided 30 percent of the end-use energy (8.8 x 10⁹ kWh), and oil another 40 percent (5.5 million barrels). This sector's electricity use comprised 30 percent of the city's total.

Several factors serve to limit industrial energy consumption in New York City. In addition to the high price of oil and electricity, prime industrial space is scarce. High transportation costs and the distance to sources of raw materials also limit consumption.

> "Cities like New York have within their reach the potential to increase their efficiency and to use energy as the cornerstone for economic renewal."

Without detailed end-use data, appropriate energy conserving measures for industry are hard to identify. The study does propose obvious steps for the modification of existing capital stock for short-run improvements in energy efficiency. Heat recovery techniques can reduce energy consumption for both space and process heat. Load management, Variable Speed Drives, and operating electric motors at or near their full load are all suggested as measures which can reduce electricity use. The study also cites detailed industrial energy audits as the best approach to identify energy conservation measures.

New York City's transportation sector has several characteristics which distinguish it from other cities. New Yorkers depend heavily on an electrically-powered subway and commuter rail system for a large share of work-related trips. The area's two major airports skew energy consumption upward, as does the consumption of marine fuels by local port activities. Moreover, personal automobile use is limited by the high cost of parking, the relative inefficiency of city driving,

1979 New York City energy demand and fuel mix

(trillion Btu = 1.055 TJ)

	OIL	N. GAS	COAL	STEAM	ELEC.	NUCL.	HYDRO	TOTAL
Residential								
Space heating	156.80	63.12	.59	5.03	1.02			226.56
Water heating	32.12	26.63		1.55	.60			60.90
Cooling				.82	4.29			5.11
Cooking		7.89			.48			8.37
Drying		.99			.72			1.71
Lighting					5.74			5.74
Refrigs.					9.03			
Misc.					10.16			10.16
	188.92	98.63	.59	7.40	32.04			327.58
Commercial								
Space heating	64.19	11.86		9.22	.65			85.72
Water heating		4.87		1.79	.91			7.57
Cooling		.31		2.89	2.62			5.82
Lighting					10.72			10.72
Fans & motors					2.72			2.72
Misc.		.81			2.61			3.42
	64.19	17.65		13.90	20.23			115.97
Industrial	33.80	21.09			30.02			84.91
Transport								
Marine	72.55							72.55
Aviation	185.72							185.72
Cars & trucks	139.50							139.50
Rail	.30				7.32			7.62
Buses	5.94							5.94
	404.01				7.32			411.33
Public								
Space heating	20.83	10.87	3.70	5.31	.40			41.11
Water heating		1.96		.93	.29			3.18
Cooling		.51		1.96	1.43			3.90
Transport	3.75							3.75
Marine	.35							.35
Helicopters	.05							.05
Lighting					5.44			5.44
Cooking		.05						.05
Office equip.					.83			.83
Fans & motors					1.12			1.12
	24.98	13.39	3.70	8.20	9.51			59.78
Electric utility	163.95	44.78	12.04		(99.12)	108.1	50.86	280.61
Steam	40.66	6.86		(29.50)				18.02
TOTAL USAGE	920.51	202.40	16.37	(29.50)	(99.12)	108.1	50.86	1298.20

measures. In all, a reduction of consumption by approximately 20 percent is forecast.

Energy use in the public sector — those local, state, federal, and international governmental agencies located in New York City — amounted to 63 x 10^6 GJ during 1979, six percent of the total. As with the commercial sector, the demand for space heating and cooling, as well as water heating, comprised the largest portion of this sector's consumption (81 percent), while end uses specifically requiring electricity absorbed twelve percent.

Big Apple's Big Bill

Perhaps the most striking statistics to emerge from this report deal with annual energy expenditures. In 1979, New Yorkers spent a total of $7.7 billion (1979$) on energy. More than half of this amount was exported from the local economy, primarily to pay for oil. However, the relatively low-cost measures outlined in the study for reducing energy consumption could yield an annual savings of $1.6 billion.

The next phase of the study will be a more detailed analysis of a specific sector or sub-sector, aimed at understanding in greater depth the characteristics of urban energy use. It will also identify a wide range of feasible energy conserving measures. This will be followed by an implementation plan for achieving the savings outlined in the sectoral study in the most cost-effective manner. Finally, the multi-stage process will be repeated for each sector, making New York a model of urban energy efficiency.

Cities are, by their very nature, energy efficient. Yet cities like New York have within their reach the potential to increase their efficiency and to use energy as the cornerstone for economic renewal. New York has taken an important step in the transition to a renewable future, and the pursuit of an urban soft path.

— Michael Wade

Reference:
DeMetro, James, Elissa Udell, and Jason Zeller

1980 *Energy Consumption in New York City: Patterns and Opportunities.* New York City Energy Office, 49 Chamber St., Rm , 720, New York, NY 10007. Copies are available at no charge.

Michael Wade, formerly a petroleum geologist, is presently pursuing a graduate degree in energy policy at New York University, and working with the New York City Energy Office.

and the availability of mass transit.

Transportation in the city consumed 433 x 10^6 GJ in 1979, 41 percent of the five-sector total. Except for a two percent contribution from electricity for the subway and commuter rail systems, all of the energy used for transport was directly supplied by oil.

Unique demands require equally unique measures to reduce energy consumption for transport. In addition to the use of fuel-efficient cars and trucks, the study recommends the use of a flywheel energy storage system in the subway system. Traffic management techniques, proven successes during the transit strike of 1980, are also cited as feasible conservation

Urban Renewables: Garbage to Gas

Dotting the continental United States are more than 11,000 landfills—final disposal sites for over 90 percent of the nation's municipal solid waste (MSW). All the discarded bits and pieces of urban life find their way to the landfills, where they are bulldozed, compressed, and covered over. But out of sight is not out of mind, or at least not out of the atmosphere.

PEOPLE MAY BE DONE with what they throw away, but nature isn't; as the waste decomposes anaerobically, the landfill becomes a natural gas field, producing considerable quantities of a methane-carbon dioxide gas mixture. One pound of MSW can yield more than three standard cubic feet of landfill gas. Most of this vents harmlessly, if malodorously, into the air, but when the gas migrates into enclosed areas, it becomes potentially explosive. Landfill gas is also toxic to deep-rooted vegetation; its high concentration of carbon dioxide deprives a plant's root system of oxygen, and only shallow-rooted plants survive.

A Flare for Waste

At some landfills, gas is collected and flared off, to prevent migration and reduce the risk of uncontrolled combustion. In 1974, operators of several California landfills realized that a potentially valuable source of energy was going to waste. Since half the gas is methane, and only 49 percent inert carbon dioxide (one percent consists of trace materials), its average heating value is 550 Btu per standard cubic foot (20.5 MJ/m^3), or a bit more than half that of pipeline quality gas.

The two prototype landfill gas recovery projects—one in Mountain View, California, and the other in Los Angeles—based their recovery methods on technology originally developed to extract natural gas from shallow, on-shore deposits. They found that with only minimal treatment (water and particulate removal) landfill gas can be used as a supplemental boiler fuel, to generate heat or electricity. Alternatively, the gas may be purified completely and the saturated methane gas delivered to an existing utility pipeline, where it mixes with the stream of natural gas.

With the two California pilot plants serving as examples, there are now fifteen landfill gas projects underway across the country. Not every landfill site, however, has any realistic energy potential. Recovery is only economic if the landfill has at least one million tons (910,000 tonnes) of MSW in place—an average depth of forty feet (12.2 m) over at least forty acres (16.2 hectares)—or a filling rate of at least 150 tons (136 tonnes) per day. And even though decomposing MSW generates gas for at least a century, the quantity and quality of the gas can support economic recovery only during the first twenty years. But if the gas recovery potential of landfills could be completely utilized, they could supply one percent of the country's demand for natural gas.

New York's Garbage a Gas Bonanza

NEW YORK CITY runs one of the world's largest landfilling operations, handling over 19,000 tons (17,300 tonnes) per day of municipal solid waste. Under the direction of the Department of Sanitation, the city has been exploring ways to recover and use its landfills' gas. Efforts include economic and technical feasibility studies and tests, experimental projects, and leasing arrangements which open landfill acreage for project development by private companies.

Three principal options for managing gas recovery are under consideration: the city can develop recovery sites on its own, assuming all the risks and receiving all the revenues; private companies can assume the entire responsibility, funneling to the city a percentage of revenues in return for the use of the landfill; or, the city and a private company can enter a joint venture, sharing both risks and revenues.

Currently, New York is taking the second option, and leasing gas rights to a private developer. At the Fresh Kills landfill on Staten Island (the city's and perhaps the world's largest), a 400-acre (162 hectare) site has been leased to Getty Synthetic Fuels, Inc. A California-based company in the forefront of methane recovery and use, Getty is designing a purification facility to upgrade the landfill gas to pipeline quality. Brooklyn Union Gas Company will then purchase the gas—enough to heat 15,000 local homes, equivalent to 1,700 barrels of oil per day.

By July 1982, the plant should be fully on-line, and New York City will receive a royalty fee of 12.5 percent of gross revenues from the gas sale, about one million dollars per year. Or, the city retains the right to appropriate 12.5 percent of the purified gas for its own use.

Waste by the Kilowatt

NEARBY, ON A 40-ACRE (16.2 hectare) site within the same landfill, the Department of Sanitation is working in conjunction with the New York State Energy Research and Development Authority and the Brooklyn Union Gas Company to test raw landfill gas as an internal combustion engine fuel. Operating continuously six days a week for over six months during 1981, the test engine generated electricity which was used by the Fresh Kills landfill facility itself. Dismantled, the engine showed no signs of internal corrosion from the gas. The next phase of the project will study the feasibility of methanol conversion.

The city's second landfill gas extraction site will be the Pelham Bay Landfill in the Bronx, closed since 1979. Wehran Energy Corporation was selected from a pool of proposed developers, and a final agreement was approved by the city's Board of Estimate this past summer. Tests to ascertain the amount and quality of landfill gas are underway, and Wehran hopes to deliver medium-Btu gas to a 15,000-unit housing development one mile from the landfill.

The potential for energy and revenues from these projects has spurred New York to consider its other landfills as possible gas recovery sites. The Johns Hopkins Applied Physics Laboratory has been contracted to help determine the city's total landfill gas potential, and its development options. After an initial scoping and modeling study, a mobile laboratory will be constructed for field tests of landfill gas. Each landfill has unique gas characteristics, and this technical information, along with the other study results, will help the city to fully utilize its resources, either by soliciting outside developers, or through more direct involvement.

The value of landfill gas has convinced New York that its garbage is too valuable to go to waste.

—JoAnne Myers

JoAnne Myers is a Project Manager in Resource Recovery and Waste Disposal Planning with the New York City Department of Sanitation.

City Views

Washington, DC

The Anacostia Energy Alliance, a community group developing renewable energy in low-income neighborhoods, obtained $3,000 from the US Department of Energy to convert the southfacing brick wall of its office to a Trombe wall—one way of showing that passive solar retrofits work in urban settings.

More than two hundred local volunteers joined Alliance members in November 1979 to attach 512 square feet of glazing to the two-story wall. The glazing was mounted on two-by-four wooden risers to create an air space. The 12-inch thick wall was painted with flat black latex paint, and 32 vents were added to allow warm air to flow naturally into the building. (Each vent allows the air to flow in only one direction to prevent night-time heat loss.) The Trombe wall contributes about 60 percent of the heat for the Alliance's 2,100-square-foot office, saving about $300 a year.

The Alliance has a five-by-fifteen-foot sunspace nearly completed now. Using Kalwall fiberglass for glazing and bricks and water-filled drums for heat storage, the sunspace should make the building totally solar heated.

Idaho Falls, Idaho

Energy was plentiful when a flood caused by the collapse of the Teton Dam washed out three low-head hydroelectric plants on the Snake River near Idaho Falls, so the city never bothered to repair the plants. But times have changed. The city has decided to take advantage of cheap, reliable hydropower by installing new 7.2-megawatt bulb turbines at the three old sites and repairing some of the old turbines. When completed, the 24.6-megawatt system will produce 162 million kilowatt hours a year, enough electricity for the homes of Idaho Falls' 38,000 residents.

The Idaho Department of Energy contributed $7.3 million to the $53-million project, and the remainder of the capital was raised through the sale of revenue bonds. The first turbine will begin generating electricity by late 1981, and the other two sites will be operating by May 1982. Idaho Falls is considering building a fourth plant down river. If it does, the city could become a net exporter of electricity.

These descriptions of urban energy activities are reprinted with permission from the September/October 1981 issue of *Sun Times*, "Re-energizing the Cities," published by the Solar Lobby, 1001 Connecticut Avenue NW, Washington, DC 20036.

available through Robert Nathans, Institute for Energy Research, State University of New York, Stony Brook, New York 11794. The rural/small town scenario focuses on Franklin County, Massachusetts, and is being managed by the Future Studies Program at the University of Massachusetts, Amherst (see *Notes* 3:2:16-17 for a review of this project). Contact person is Mark Cherniak, Box 548, Greenfield, Massachusetts 01002.

Complementing these efforts, the American Institute of Architects Research Corporation has assembled a handbook to aid communities in reducing their dependence on depletable resources. The handbook includes: an explanation of building energy use; a guide for planners who wish to incorporate energy-saving designs; a guide for community members who wish to undertake the planning process; and a bibliography of publications and case studies on energy use and conservation strategies. A draft of this handbook is available from Charles Zucker, Design Arts Program, National Endowment for the Arts, 2401 E St. NW, Washington, DC 20506.

Various projects are now underway to construct solar futures tailored to the local conditions and needs of urban,

Planning from the Ground Up

Decentralized energy planning is alive and well in the United States, as communities in all regions look to meet their energy needs through conservation and the use of renewable resources. The following list of programs and studies provides a brief introduction to their activities. Much of this information is contained in the Solar Energy Research Institute publication, Decentralized Energy Studies, Compendium of U.S. Studies and Projects, *available from James Ohi, Community and Consumer Branch, Solar Energy Research Institute, 1617 Cole Boulevard, Golden, Colorado 80401.*

Decentralized Solar Energy: Community-Level Technology Assessment

This US Department of Energy program assesses the social, political, institutional, and life-style effects of deploying solar technologies at the local level. Its objective is to develop a process by which communities can conduct their own assessment efforts. The program manager is Jim Quinn, Office of Solar Energy, 600 E Street, NW, Washington, DC 20585.

As part of this program, three prototypical solar scenarios have been prepared for urban, suburban, and rural/small town communities. The urban scenario for Baltimore, Maryland, is being prepared by the Institute for Local Self-Reliance. It aims to show how solar energy combined with conservation can replace 50% of the city's fossil fuel requirements by the year 2000. A Draft Report, published in October, 1979, is available from David Morris, Institute for Local Self-Reliance, 1717 18th St. NW, Washington, DC 20009. The suburban scenario considers the Nassau/Suffolk region of Long Island in New York State, and examines the potential use of solar energy in the context of the regional land-use plan. Information is

suburban, small town and rural areas throughout the United States. Working teams that include both community leaders and technical consultants will develop plausible solar futures and assess the consequences. This work will lead to policy recommendations to federal, state, and local governments. One such program, in the Southern Tier Region of New York State, includes the three counties of Chemung, Schuyler and Steuben. The working team considered the resource potential of the area, and the technologies available for local energy development. Citizens were chosen at public meetings to represent industry, labor, community, and other interest groups in sketching a resource development plan. Project documents, including a resource inventory, a technology assessment workbook, and a renewable technology handbook are available from Steve Weisman, Energy Program Manager, Southern Tier Central Regional Planning and Development Board, 53½ Bridge St., Corning, New York 14830. Similar work is being done in Richmond, Kentucky and in Kent, Ohio. Reports are available through Sam Carnes, Oak Ridge National Laboratory, Energy Division, P.O. Box X, Oak Ridge, Tennessee 37880. The Franklin County study is also moving into a community involvement stage, and details may be obtained from David Pomerantz, University of Massachusetts, Amherst, Massachusetts 01003.

Research is also being conducted on the social impacts of energy decentralization. A workshop was held in Rensellaerville during June, 1979, and proceedings will soon be available through Gordon Enk, Institute on Man and Science, Rensellaerville, New York 12147.

Energy Conservation Project, Portland, Oregon

The objective of this project is to save over 30% of the energy that the people of Portland would need in 1985, or about 11 million barrels of oil per year. Toward this end, an Energy Policy Steering Committee was assembled with representatives from business, industry, utilities, and government. They developed energy policy for Portland which considers the extent of retrofits in various sectors, land use, renewable resource systems, transportation, and specifics of the role of city government. Reports on the project have been published and may be obtained from Marion Hemphill, City Energy Advisor, 620 S.W. Fifth, Room 610, Portland, Oregon 97204.

Energy Self-Sufficiency Study, Northhampton, Massachusetts

This study considers whether a small city can obtain all of its energy from renewable resources within its boundaries. The sources examined were solar ponds, on-site solar facilities, solid wastes, sewage, wood, hydropower, and wind. Cost-benefit analysis showed that even under pessimistic assumptions, the life-cycle revenues of these sources exceed maximum capital costs by $40 million. Final report for this study is available from Allan S. Krass, Associate Professor of Physics and Science Policy Assessment, Hampshire College, Amherst, Massachusetts 01002.

St. Louis Case Study: The Potential for Energy Conservation and Renewable Resources

The emphasis in St. Louis was on conservation options for the entire Standard Metropolitan Statistical Area (SMSA). Four renewable energy sources were also considered: active and passive solar energy, biomass, and wind. A draft report is now under review by the Department of Energy, and may be obtained from Mark Levine, Lawrence Berkeley Laboratory, Building 90, Room 3114, Berkeley, California 94720.

Planning for Energy Self-Reliance: A Case Study of the District of Columbia

A study of the energy picture in Washington, DC, and the potential for energy self-reliance has been prepared by the Institute for Local Self-Reliance. Its purpose is to bring a municipal perspective to energy planning. The contact person is David Morris, Project Director, Institute for Local Self-Reliance, 1717 18th Street NW, Washington, DC 20009.

Integrated Community Energy Systems

The objective of this program is to include energy planning in the land development process. Case studies will be done of new towns that incorporate state-of-the-art methods of energy conservation into community design. The sites selected are Burke Center, Virginia; Greenbriar, Virginia; Shenandoah, Georgia; Radisson, New York; and Woodlands, Texas. Information is available from Jacob Kaminsky, Division of Buildings and Community Systems, Assistant Secretary for Conservation and Solar Applications, Department of Energy, 20 Massachusetts Avenue NW, Washington, DC 20585.

Philadelphia Solar Planning Project

This project seeks to plan and organize a comprehensive program for assessing and achieving the maximum local use of solar energy. It is supported by a grant from the National Endowment for the Arts to the University City Science Center, with subcontractors in universities, local governments, and private consulting firms. The Philadelphia project is a pilot study for SUNACT (Solar Utilization Applied to Cities and Towns), a Department of Energy program managed by Frank DeSerio, Office of Solar Applications for Buildings, Room 5G-070, Forrestal Building, Washington, DC 20585.

Soldiers Grove, Wisconsin Project

The central business district of Soldiers Grove is frequently flooded by the Kickapoo River, and a plan to relocate the district provides an opportunity to develop and demonstrate the use of alternative energy technologies. Energy options include district heating, cogeneration, solar ponds, wind generators, biomass, passive solar design, and low-head hydro. The village is working with Argonne National Laboratory, the University of Wisconsin, the Department of Energy, and the Wisconsin Division of State Planning and Energy to coordinate efforts in technology development, transportation, architecture, community design, and flood prevention. The project coordinator is Tom Hirsch, Community Development Office, Soldiers Grove, Wisconsin 54655.

Energy Planning Framework, Corvallis, Oregon

The City of Corvallis has prepared an Energy Planning Framework to reduce energy use over the next few years by more than 30 percent. The framework examines end-use efficiencies, land use planning, transportation services, economic development, building standards, and means of implementing alternative and renewable energy technologies. A copy may be obtained from the Planning Department, City of Corvallis, 180 N.W. 5th St., Corvallis, Oregon 97330.

Alternative Energy Feasibility Study, Pulaski, New York

This study estimates the possibility of using energy from solar, wind, solid waste, hydropower, and local natural gas to meet the needs of ten public and institutional facilities in the village of Pulaski. Among the issues addressed are technical, legal and institutional considerations, and social and environmental impacts. Information is available from Leonard F. O'Reilly, O'Reilly Associates, Suite 9B, 828 Bloomfield Avenue, Montclair, New Jersey 07042.

Grassroots Energy

"Decentralization, in attacking energy problems, has brought together a diversity of solutions with a unity of purpose," writes Solar Energy Research Institute Director Denis Hayes. These solutions are appearing faster than researchers and interested members of other communities can keep up with them. An excellent overview of community-level energy activities, though, appears in a recent issue of *The Energy Consumer*, a bimonthly publication of the Department of Energy's Office of Consumer Affairs. Along with brief descriptions of locally-based initiatives around the country, this issue contains a list of resources for community energy projects. It is available free from the Office of Consumer Affairs, Department of Energy, Room 8G082, Washington, DC 20585.

Jobs and Energy on Long Island

Employment Potential of Energy Scenarios Compared

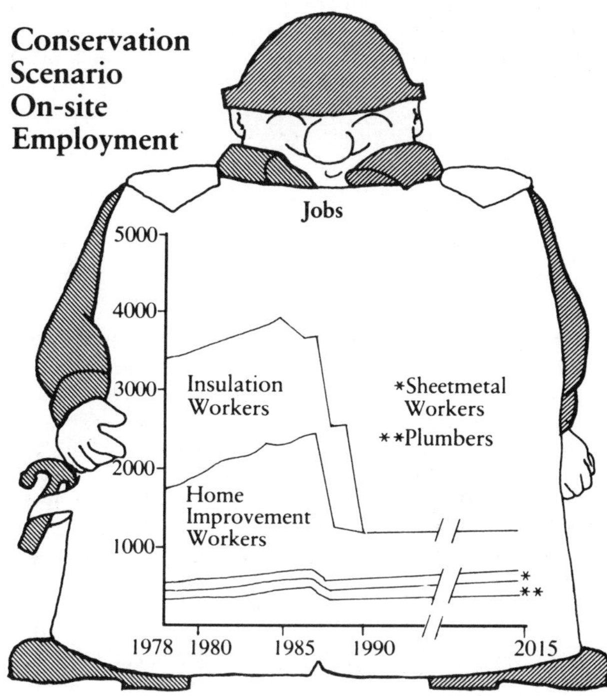

After the first 7-12 years of the Conservation/Solar Scenario, on-site employment drops sharply. Retrofits to existing housing stock have been completed, and the on-going labor demand level of new construction and maintenance is much lower.

CONSERVATION is not only the least expensive "new energy source" presently available, it is also one of the most effective ways to create new jobs. Energy savings and lower rates of unemployment combine to make well-devised conservation programs attractive alternatives to energy from fossil fuels and nuclear power. These are the main conclusions of *Jobs and Energy,* a study of energy options for Nassau and Suffolk Counties on New York's Long Island, conducted by the Council on Economic Priorities (CEP), a non-profit public interest research group. CEP recommends that the Jamesport nuclear plants in Suffolk County (now canceled) not be built; they would cost more and provide fewer jobs per dollar invested than a combination of conservation and solar measures equally effective in meeting the area's energy needs.

CEP's research was designed to examine claims that nuclear power plants are economically beneficial and merit union support. "Employment has been among the most compelling arguments for nuclear power; the construction and operation of these plants creates many jobs, and the nuclear industry has not been slow to document them. Proposals to build nuclear facilities are often backed by organized labor, especially the construction and manufacturing trade unions whose workers stand to benefit most directly. Yet to place these employment benefits in perspective, it is necessary to compare them with the number and type of jobs which would be created if conservation and solar measures were installed in the buildings which would otherwise be supplied by the proposed plant."

CEP selected two eastern Long Island counties as the site for their comparison of the employment levels generated by various energy scenarios. The area's severe climate, high energy prices and serious unemployment problem typify conditions throughout New England, making it an interesting and sensitive site for a soft path study. In addition, high quality data on key research items were readily available. A previous study for the Suffolk County Legislature (by the architectural engineering firm of Dubin-Bloome Associates) assessing the regional potential for increasing residential energy efficiency was the starting point for CEP's analysis of employment effects. Parallel information on labor requirements for nuclear plant construction and operation came from permit applications by the Long Island Lighting Company (LILCO) to build two 1150 megawatt reactors at Jamesport, in Suffolk. To allow a comparison between these two energy options, the scenario was set within a 38-year period—1978 through 2015—corresponding to the construction period and service lifespan of the Jamesport nuclear plants. Unfortunately, the commercial and industrial sectors proved impossible to include, and one of the study's limitations is that it is a purely residential conservation scenario.

The Conservation Package

CEP began with a list originally developed by Dubin-Bloome of more than 100 possible measures for residential conservation. Cost/benefit analysis trimmed this to a set of 34 highly cost-effective measures, which CEP incorporated into their Conservation/Solar Scenario. The most important group of conservation measures included seven *building envelope improvements* to increase the structure's thermal integrity. Seven *space heating measures* and four *cooling measures* were selected for their substantial contributions to household energy savings. Four *water heating measures* (including 50 square feet of solar collector panels) and twelve

Conservation Scenario
Gross Energy Savings by Fuel Type

TOTAL = $16.27 BILLION

Fuel Type	Unit Cost	Total Energy Savings	$ Value
Electricity	$.0576/kWh	110,968 $\times$ 10^6 kWh	$6.39 billion
Natural Gas	$3.40/M ft.^{3}	532 $\times$ 10^6 M ft.^{3}	$1.81 billion
Oil	$.467/gallon	17,274 $\times$ 10^6 gallons	$8.07 billion

efficient appliances complete the conservation package. All of these measures are purely technical fixes. They employ only currently available technology, they have no significant impact on individual style of life or creature comforts, and they are generally sensible economic investments by today's energy prices.

CEP tried to make their projection of energy savings under the Conservation/Solar Scenario as realistic as possible by incorporating many other real-life assumptions. For example, they introduce conservation measures gradually, with full implementation in 10 to 15 years. Their calculation of yearly energy savings reflects the gradual nature of this process. Similarly, the total costs of the Conservation/Solar Scenario include re-investment in short-lived goods, such as energy-efficient refrigerators, which would be purchased a second time after the first ones wore out.

Conservation/Solar Scenario Savings

The cost of the conservation package, though varying somewhat from household to household, would average $2,200 (1976 $) for the original installation, with an additional $1,000 needed for maintenance and replacement. The total investment in conservation and solar measures by the year 2015 would be about $4.04 billion. Based on present costs and consumption rates for Long Island, these measures would, over 38 years, displace energy valued at $16.27 billion, for a net savings of $12.23 billion.

Estimating Effects on Employment

CEP calculated that the Conservation/Solar Scenario would generate an average of 10,400 to 12,700 *more jobs* than would traditional patterns of energy consumption. Almost all of the employment would be local to eastern Long Island, reducing the area's unemployment rate from its 1979 level of 6.3% to 5.5%. Considering only those measures which displace electricity, conservation is *over* 40% *more effective* at creating employment opportunities than the same amount spent on the Jamesport nuclear power plant.

Employment effects fall into four basic categories:

- *Direct Effects.* The Conservation/Solar Scenario provides direct, on-site employment for insulation installers, solar equipment specialists, heating and ventilation workers and plumbers, among others.
- *Indirect Effects.* The industries that supply on-site workers with materials and services experience indirect employment effects when they step up production to satisfy an increasing demand. Their greater activity, in turn, diffuses employment effects to other economic sectors, such as mining and transportation.
- *Induced Effects.* The workers and businesses directly and indirectly affected by increased economic activity receive wages or profits, some portion of which returns to the economy as further spending. This induces a second round of employment effects.
- *Responding Effects.* The Conservation/Solar Scenario would lower fuel costs, thereby increasing household discretionary income. Diverting this money to consumer goods rather than energy purchases would generate a large number of additional jobs. [cf. Schachter, 1979 for a review of these concepts.]

Two separate computer models assisted CEP in calculating employment effects of various energy scenarios. One model determined how investments in energy conservation and solar energy would affect the national economy, and the other focused exclusively on the economy of the Nassau/Suffolk region. Though clearly the best available, these models are not without problems. They are built around 1972 and 1967 economic data, respectively, and require conversions to present prices and costs. Inter-industry connections are also different today from those 8 to 13 years ago, so the models are to some extent functionally out of date. Using them to project employment nearly 40 years into the future implies a static economic structure, unresponsive to technological change. [See Boland, 1979 for further critique of these models.] Perhaps the most serious difficulty for the present study is that the regional and national models are not strictly comparable. The regional model includes induced effects from wage and contractor profit spending, while the national model does not. Consequently, as the authors note, "one cannot subtract the regional employment from the national to derive employment outside the Nassau/Suffolk region."

Conservation/Solar Scenario vs. Continued Consumption

Using the national employment model, CEP determined that Long Island's present residential energy mix (49.6% fuel oil, 39.3% electricity and 11.1% natural gas) generates 17.7 labor years nationally per million dollars of expenditure. Over the 38 years of the scenario, assuming continued consumption at present rates and proportions, $16.27 billion would be spent on energy, supporting an average of 7,600 jobs. The Conservation/Solar Scenario is almost three times more effective (48.8 to 17.7), dollar for dollar, than traditional energy purchases in creating jobs, but there is still a net loss of jobs because the total investment in conservation measures is only one-quarter the value of the energy it displaces. On-site and 'multiplier' employment for conservation together produce only about 5,200 jobs, 2,400 fewer than with continued consumption patterns. This difference, though, is only a fraction of the employment generated by twelve and a quarter billion dollars in energy savings percolating through the national economy.

A large portion of this amount may be pre-empted by the

TOOLS for the SOFT PATH

Long Island Lighting Company. As electricity is displaced by conservation and solar measures, utilities are likely to raise rates to meet their fixed costs. CEP estimates that 25% to 75% of the dollar savings from lower electricity consumption will be captured as rate increases. This estimate is much too high, according to other energy analysts, particularly if the utility finances solar or conservation measures (see Lovins, 1979). The rate increases assumed by CEP would fund between 1,000 and 3,000 jobs within the utilities. The remainder of the savings from conservation would move into consumer expenditures, which produce 50.3 labor years per million dollars spent. The net employment gain from conservation will thus average between 10,400 and 12,700 jobs over 38 years, and most of these employment effects will be local.

The Jobs from Jamesport

Jobs and Energy goes on to consider the employment associated with construction and operation of the two 1150 megawatt nuclear power plants planned for Jamesport. The nuclear option on Long Island, CEP notes, would lead to 3,180 jobs nationally, at an estimated cost of $4 billion. Each million dollars invested would generate 30.2 labor years. To compare the employment effects of nuclear power and conservation, CEP devised a "Conservation Electric Scenario," consisting of a subset of their 34 original conservation measures. Only those items that saved electricity in cost-effective fashion were included. In this way, CEP avoided an asymmetrical comparison between the employment effects of producing one kind of energy and the jobs created by conserving another.

Of the $4.01 billion Conservation/Solar Scenario, only $620 million represented cost-effective displacement of electric power. This investment, though dwarfed by the $4 billion Jamesport plants, would still save one-quarter of the two reactors' total electrical output, and provide about that fraction of Jamesport's associated employment. Though smaller in scale than the nuclear project, the Conservation Electric Scenario creates jobs more efficiently, providing 40% more labor years per million dollars of investment.

CEP found that "improvements in the efficiency of residential electricity use can satisfy increases in regional end-use needs for electricity more cheaply than can nuclear power. This suggests that, at least up to a certain point, investing in electricity conservation measures would allow the economy to increase total productivity at a higher rate than would investing in nuclear power plants."

Notes on Methodology

Jobs and Energy is an important addition to the growing body of soft path studies. It combines within a single analytical frame a realistic, detailed energy scenario and a state-of-the-art assessment of employment effects. Beginning as a search for reasonable alternatives to the Jamesport nuclear project, and building upon the results of related conservation studies, *Jobs and Energy* emerges as "what is perhaps the most comprehensive analysis of this kind" (Schachter, 1979).

However the results of their study are interpreted by labor economists, or acted upon by policy makers, CEP intended *Jobs and Energy* as a prototype, and included detailed methodological accounts in the report. At least one other

National and Regional Employment Comparison
Conservation Scenario and Continued Energy Consumption

Energy Future	Dollars of expenditure (billions 1976$)	Labor years per million dollars expenditure		Average number of jobs	
		Regional	National	Regional	National
Continued Consumption	16.27	8.0	17.7	3,400	7,600
Conservation Scenario					
On-site Implementation plus multiplier	4.04	45.4	48.8	4,800	5,200
Increased Discretionary Spending	7.43-10.63	34.9	50.3	6,800-9,800	9,800-14,100
Utility Fixed Costs	4.79-1.60	15.2	24.0	1,900-600	3,000-1,000
Total	16.27			13,500-15,200	18,000-20,300

study has been modeled directly after *Jobs and Energy,* running essentially the same computer programs with assumptions appropriate to all of New England (Rosen & Stutz, 1978). In support of further efforts such as these, CEP included a brief chapter on the methodological refinements they considered most worthwhile. They pointed to additional renewable energy sources and conservation measures as possible scenario components, and the tremendous potential for conservation in commercial and industrial sectors, which they were forced to exclude from their own analysis.

To this list of future research needs might be added a systematic examination of the "responding effect." It is standard for employment studies to assume that energy savings are funneled into discretionary consumer purchases, where they induce employment at a much higher rate. This is a critical assumption, since, at least in *Jobs and Energy,* much of the Conservation/Solar Scenario's employment advantage over Continued Consumption stems from the labor required to meet increased consumer demand.

CEP's authors note their reservations on this issue: "Ideally, the effects of increased discretionary income should be measured as marginal, rather than average, consumer purchases, but this was not possible. . ." The problem with this assumption, however, may go beyond the difference between the two types of purchase. Marginal phenomena are quite complex, and of the 7 to 11 billion dollars' worth of energy savings, major chunks may not go into consumer purchases at all, but move through the economy in ways that generate employment at a much lower rate.

Inflation is one of the primary economic forces reducing the value and the purchasing power of savings. *Jobs and Energy* tries to eliminate this factor by assuming constant 1976 dollars throughout the scenario. This method controls for inflationary increases in the *volume* of revenue flowing through a household, but not for the accompanying changes in relative *proportions* assigned to key items on the domestic budget. Inflation affects some costs more than it does others; leading the pack, for the past few years, are the costs for food, clothing, housing, interest charges and taxes (as cost-of-living increases propel wage earners into higher percentage brackets). Some of these items, notably housing and food, generate more jobs per dollar than the average; the others have a much lower job creation potential (Hannon, 1975).

Finance charges will reduce even further the sum available for discretionary consumer purchases. The Conservation/Solar Scenario requires an immediate investment of about $2,200 per residence, and obviously many times that amount for an apartment house with multiple units. Much, if not all, of this investment will have to be financed, and innovative programs such as the Oregon and TVA plans have demonstrated that this is possible at relatively low costs to the consumer. By comparison, the terms of the 1977 New York State Energy Conservation Act—9.83% on a seven-year loan with perhaps 20% of the interest charges returned as a tax credit—constitute a 'worst case.' Assuming these terms (as CEP does), total scenario costs will increase by 31%, from $4.01 to $5.26 billion. This should not affect employment, according to CEP, if finance charges have the same employment impacts as average consumer purchases, but Hannon (1975:99) presents data showing that they may be considerably lower.

Higher levels of consumer debt also serve to shift revenue away from average consumer purchases. For roughly the first third of the scenario, (10-12 years depending on financing) cumulative costs exceed total savings, using a worst-case assumption of high utility rate increases. During this period, households may enjoy less in the way of discretionary income than previously. Their debt burden will almost certainly increase: partly to finance the conservation measures, and partly to maintain their standard of living in the face of temporarily higher energy-related fixed costs. Consumers may emerge from this initial phase of the scenario with a higher ratio of debt to income than is comfortable for most middle-class families. Many may use their first personal savings from reduced energy bills not to acquire consumer goods but to service and retire some portion of their indebtedness. Older couples and others with lower debt burdens are likely to channel energy savings into consumer spending much as CEP predicts. But younger families at earlier phases of the domestic debt cycle may be more affected by the higher credit load. Local demographic conditions, then, play a role in determining marginal consumer spending patterns, which in turn control the number of jobs created by energy savings.

Because the respending effect accounts for so much of the employment advantage of conservation and solar measures over traditional energy use, further research along the lines suggested here appears to be of merit. However, even at this point, certain policy implications are clear. For energy savings from conservation to produce the maximum employment benefits, it is essential to institute financing arrangements that prolong the payback period and keep cumulative costs below total savings. Even though finance charges may rise as a result of this policy, consumer debt is unlikely to increase. The employment generated by respending of energy savings could ultimately be greater, and more evenly distributed over the life of the scenario.

References: —*Charles Drucker*

Boland, Tom
1978 *Jobs and Energy in New England and the United States: Overviews of Some of the Recent and Forthcoming Studies, Hearings, Proposals, Literature and Activities.* Prepared for the New England Energy Congress. Available from the author, 35-R Jaques St., Somerville, Massachusetts 02145.

Buchsbaum, Steven, James W. Benson, *et al.*
1979 *Jobs and Energy: The Employment and Economic Impacts of Nuclear Power, Conservation, and other Energy Options.* Council on Economic Priorities, 84 Fifth Avenue, New York 10011. $12.

Hannon, Bruce
1975 "Energy Conservation and the Consumer," *Science,* Volume 189, Number 4197, pp. 95-102.

Lovins, Amory B.
1979 "Electric Utility Investments: *Excelsior* or Confetti?" Paper delivered to the E.F. Hutton Fixed Income Research Conference on Public and Investor Owned Electric Utilities. Available for $1.50 from IPSEP.

Rosen, Richard A., and John Stutz
1978 *The Employment Creation Potential of Energy Conservation and Solar Technologies: The Implications of the Long Island Jobs Study for New England* 1978-1993. Prepared for the Low-Income Caucus of the New England Energy Congress. Available through Energy Systems Research Group, 120 Milk St., Boston, Massachusetts 02109.

Schachter, Meg
1979 *The Job Creation Potential of Solar and Conservation: A Critical Evaluation.* US Department of Energy, Policy and Evaluation, Advanced Energy Systems Division. Available through DOE, San Francisco, 111 Pine Street, San Francisco, California 94111.

Conservation In the Corn Belt

SMALL FARM ENERGY PROJECT ENCOURAGES LOCAL SELF-RELIANCE

THE RISING COST OF ENERGY affects farms regardless of size. No enterprise is too small-scale to benefit from an innovative approach to energy technologies and management practices. Two dozen farmers, working relatively small family holdings in Cedar County, Nebraska, have reduced their annual energy consumption by 15 percent. Each saved an average $1,100 in energy bills during 1979, with an average initial investment of only $1,200 per farm. The low-cost, locally-adapted technologies they used, however, tell only a small part of the story. Nearly 70 percent of the total energy savings came from the more effective use of existing farm machinery, indicating the importance of energy-conserving attitudes and management practices.

They did it with the Small Farm Energy Project. Established in November 1976 by the Center for Rural Affairs, the Project employs energy-saving technologies to increase the incomes from small farms in Cedar County. A three-year research and demonstration program involving 48 low-income grain and livestock farm families, the Project emphasizes on-farm innovation in the face of growing resource scarcity. Conventional agricultural research, in contrast, is typically oriented toward specialized and expanding enterprises. It tends to under-serve the small independent operator.

Energy Use on a Small Farm

Small Farm Energy Project participants ran operations typical of the small, well-kept farms in the rolling hills of Cedar County. In 1974, the average Cedar County farm had 354 acres, considerably below the 683 acre average for the State of Nebraska. Unlike the large farms found elsewhere in the region, the small Cedar County farmers do not rely entirely upon grain as their cash crop. In the tradition established there by German and Czech settlers in the last century, they maintain livestock as the principal final product for market. Hogs and dairy cattle are their mainstay, and Cedar County often leads the state in the number of pigs farrowed. Bulk tank dairying was established in the 1950s to diversity the local economy and provide small farmers with another sound enterprise.

To support their livestock and provide their families with income, local farmers grow a range of crops: corn, oats, alfalfa, and soybeans on an average of 240 acres, or two-thirds of their land. Crop rotation is the primary means of maintaining soil fertility, though some commercial fertilizer is applied. The farms provide most of the family income, and take the labor of the whole family (average size of five), even though they are fully mechanized, with an average of 3.4 tractors per farm.

Project participants were identified by income guidelines set by the Community Services Administration, with final selection by a community advisory committee. The Project design called for data collection on

Average farm energy use, 1976-1979
(In millions of Btu's)

	Cooperating Farms				Control Group			
	1976	1977	1978	1979	1976	1977	1978	1979
Space Heating	155	156	128	131	157	131	151	181
Electricity	313	325	318	346	268	280	322	348
Motor Fuels	408	495	488	466	438	522	545	529
Fertilizer	100	103	107	118	115	123	167	160
Total	976	1,079	1,041	1,061	978	1,056	1,185	1,218

Record-keeping farms purchased 15% more energy than cooperating farmers after cooperators worked three years on reducing energy expenses.

two groups of 24 farms each, closely matched in their acreage, machinery and livestock operations. Both groups agreed to keep careful records of their energy expenses, farm operations, and production levels. One group consented, in addition, to participate in educational workshops and consider adopting energy innovations; the other, a control group, provided a basis for comparison.

Energy purchases by project participants were typical of the area's small farmers. Electricity accounted for 27 percent of the total energy budget and was used for heating, lighting, and running the farm's numerous electric motors. In Cedar County, a regional supplier provides the electricity to a rural electrical cooperative for distribution to end-users. Though service is generally reliable, most dairy farms have a small standby generator for milking machines if power is interrupted. Motor fuels for transportation, field operations, and other chores make up 46 percent of total energy consumption, with half of this going to non-farm transportation. This high demand level forces farmers into competition with other major users of petroleum. Supply problems occasionally arise, such as the rationing of diesel fuel by dealers following the Iranian fuel embargo in the spring of 1979.

Heating fuels total 14 percent of energy consumption, with crop drying and space heating the two major end-uses. Propane has been a readily available substitute during fuel oil supply interruptions. Though some farmers do not apply commercial fertilizer to their fields, the average energy budget for fertilizer runs 13 percent. Long-term availability and cost of nitrogen fertilizer will likely reflect that of the natural gas used in its manufacture.

Energy use among the control group of farmers increased during the project period by nearly 25 percent over 1976 levels. A three-year drought, which ended the following year, may have depressed 1976 use levels. The subsequent three growing seasons enjoyed good to excellent weather, producing better crops and requiring more energy for field operations. Livestock operations expanded as well.

Just as yearly variations in weather influence the general level of farm activity and energy use, seasonal fluctuations in farm energy consumption are also quite marked. Purchases are heaviest in the spring quarter, led by high demands for diesel fuel and fertilizers. During fall and winter quarters, with the onset of the heating season, electricity and heating fuels predominate. Household uses of these and other fuels compose 40 percent of the total small farm energy consumption.

Average farm energy purchases and expenditures *(Control Group)*

	1976	1977	1978	1979
Electricity				
kWh	25,943	26,990	31,333	33,809
$	788	830	1,027	1,175
Fuel Oil				
gal.	496	414	413	497
$	191	178	189	322
Propane				
gal.	871	734	898	1,013
$	336	249	357	439
Diesel				
gal.	829	951	1,146	1,309
$	318	421	515	893
Tractor Gas				
gal.	1,460	1,554	1,776	1,590
$	792	903	1,051	1,316
Car Gas				
gal.	1,092	1,528	1,279	1,294
$	592	903	789	1,069
Fertilizer				
10^6 Btu	115	123	167	160
$	1,220	1,252	1,272	1,674
Total				
10^6 Btu	978	1,056	1,185	1,218
$	4,237	4,736	5,200	6,888

Sound investments for expanding farmers included farm chemicals, large machinery, and sophisticated livestock facilities, back in the days of cheap energy. Though energy-intensive technical innovations such as these increased the productivity of land and labor, the labor they "saved" was too often that of the frugal small farmer down the road—the one who resisted, or could not afford, energy-intensive farm practices. Combines and other machines improved the farmer's competitive position, and allowed him to outbid his more conservative neighbors when farmland came up for rent. And with government policies supporting farm specialization and energy-intensive expansion, the small family farmer has frequently faced the ultimatum posed by former Secretary of Agriculture Earl Butz: "Get bigger or get out."

The Small Farm Energy Project took an alternative approach. The cooperating farmers employed a variety of techniques for conserving, recycling, and producing energy while enhancing self-reliance. The Project stressed simple but proven technologies making fuller use of on-farm resources, and involving the farmer's own construction labor with locally available materials. The resourcefulness and the skills of the participants helped shape the technologies to meet the farmers' individual needs and the management constraints of their operations. As it reduced dependence on farm inputs, this approach appealed to the conservative north-central Plains farmers. They choose a risk-averse management style.

Cooperating farmers attended a series of winter workshops and heard some 40 speakers describe a range of energy alternatives. The decisions to develop particular energy innovations came, however, from the farmers themselves. Project staff supplied design guidance and assistance, but principal responsibility for construction and performance monitoring rested with the farmers. All farms adopted at least two of the energy conservation technologies, and nearly all took on one or more renewable energy innovations.

The Project had no energy axe to grind. It did not set out to prove that a particular energy alternative was technically and economically effective. Rather, the participatory process allowed farmers to find technologies suited to their individual needs. Skeptical neighbors and friends meant community values were kept in the forefront.

Energy Conserving Ethic

Exposed to energy innovations, and suddenly aware of the multitude of opportunities for savings, the cooperating farmers became more aware of their energy use and found ways to manage more efficiently. They reduced domestic and other non-farm energy consumption through the same behavioral changes that urban dwellers often make: less automotive travel, and shutting off unused rooms for zone space heating. Farm fuel use was reduced by more frugal use of tractors and other energy-hungry equipment. A soil testing program helped reduce energy consumption via fertilizers to a bare minimum.

The 24 cooperating farmers invested a total of $29,699 in 148 energy innovations during the three-year project. By 1979, they were saving $27,312 on energy per year compared to the 24 farmers of the control group, with no difference in the production levels of the two. After three years of involvement in the Energy Project, conservation participants were purchasing 15 percent less energy than their uninvolved counterparts.

Specifically, solar technologies accounted for approximately six percent of the realized energy savings. While weatherization and conservation measures were responsible for an additional 25 percent, the bulk of the savings may be attributed to more careful energy expenditures with conventional farm technologies.

The Small Farm Energy Project must be regarded as a double success. It increased the incomes of local farmers, its initial goal, by significantly cutting their agricultural energy costs. Yet it also helped to imbue the project's participants with an ethic of energy conservation that stimulated a search for additional savings and a new sense of self-reliance.

—Rob Aiken

Rob Aiken is a research assistant on the Small Farm Energy Project, and author of the Project's Final Report, *which is available for $5 from the Center for Rural Affairs, P.O. Box 736, Hartington, Nebraska 68739.*

IPSEP'S NATIONAL AND REGIONAL CONTACT LIST

To reduce mailing and transaction costs, we have asked certain individuals to act as national or regional contacts, distributing *Soft Energy Notes* to other soft energy workers in their areas. If this system is not working, or if you have a better idea, please let us know.

We hope to receive as much information as we send out. The sustained quality of this service depends upon accuracy, speed, and breadth. We are therefore placing a double burden on regional and national contacts: to bring to our attention key developments that we might not otherwise hear of and to circulate *Soft Energy Notes* to appropriate recipients. Here we list contacts and their addresses for your information.

Australia: Mark Diesendorf, P.O. Box 1965, Canberra City. 2601.
Jeff Nicholls, Glenwarrin Mill, ELANDS 2429. (065) 504 518.

Austria: Peter Weish, Institut fur Umweltwissenshaften, Messepalast, Stiege 14, 1070 Wien. 0 22 2 93 64 78.

Bangladesh: Dr. M.N. Islam, Chemical Engineering Dept., Bangladesh University of Engineering and Technology, Dacca-2.

Belgium: Luc De Brabandere, Rue du Puison 13, 1392 Hoves. (02) 395 3996.

Brazil: José Goldemberg, Instituto de Fisica, CIDADE Universitaria-Caixas, Postais 20516-8219 Sao Paulo.

Canada: David Brooks, 54-53 Queen Street, Ottawa, Ontario KIP 5C5, 613 233 0260.

Denmark: Jørgen Nørgaard, Physics Lab III, Technical University of Denmark, Bldg. 309C, DK-2800, Lyngby. (02) 88 16 11.

Ecuador: Eugene R. Braun. Asesoria Para Desarollo En Latinoamerica, P.O. Box 498, Quito, Ecuador. 234 402.

Federal Republic of Germany: Martin Kuenzlen, Freunde der Erde, Wizlebenstrasse 32, 1 Berlin 19. 45 56 16.
Florentin Krause (current address) 2646 Dana, Berkeley, California USA.

Fiji: Peter Johnson, Ministry of Energy, Government Bldg., PO Box 2256, Suva, Fiji.

France: Pierre Samuel, 14 bis rue de l'Arbalete, 75005 Paris.

Greece: Georges Rallis, National Energy Council, 42 Academias Street, Athens, Greece.

Hong Kong: Donna M. Liu, Asia 2000, 146 Prince Edward Road, W., Kowloon, Hong Kong.

India: N.K. Gopalakrishnan, TATA Energy Research Institute Documentation Centre, Bombay House, 24 Homi Mody St., Bombay-400 023.
Amulya Reddy, Indian Institute of Science, Bangalore 560012.

Ireland: Brian Hurley, 3 Larkfield Gardens, Dublin 6, 01-960653.

Israel: Avram Kalisky, Hahagana 26/18, 97825 Jerusalem.

Italy: Pier Luigi Lombard, Amici della Terra, Piazza Sforza Cesarini 28, 00186 Roma. 06 655308.

Japan: Yukio Tanaka, Chikyu-no-Tomo (Friends of the Earth), 1-51-8 Yoyugi, Shibuya-ku, Tokyo 151, Japan.

Kenya: P.M. Githinji, Dept. of Mechanical Engineering, University of Nairobi, Box 30197, Nairobi.

Malaysia: Gurmit Singh, Environmental Protection Society, P.O. Box 382, JLN Sultan, Petaling Jaya, Selangor.

Nepal: Deanna Donovan, P.O. Box 1615, Katmandu.

Netherlands: Eric-Jan Tuininga, Morgenster 5, Leusden. 055 77 33 44.

New Guinea: Ken Newcombe, P.O. Box 2352, Konedobu, Papua.

New Zealand: Denis Hocking, Rangitoto Farm, RD 2 Bulls.

Norway: Paul Hofseth, Energy Office, Royal Ministry of the Environment, Myntgaten 3, Oslo Dept., Oslo. 2-117521.

Puerto Rico: Jeff Malley, Energia Verde, Box 40612, Minillas Station, Santurce 00940.

Romania: Adrian Gheorghe, Spl. Independentei 313, Sector VI, Bucharest.

Somalia: Prof. Bixi, c/o SRUERD, PO Box 2962, Mogadishu, Somalia.

Spain: Santiago Abad, c/Martin de los Heros, 17, Madrid 8.

Sweden: Erika Daleus, Jordens Vanner, Box 7331, S-103 90, Stockholm. 08 11 42 32.

Switzerland: Theo Ginsburg, SES, Auf der Mauer 6, CH-8001, Zurich.

Thailand: Dr. Nart Tuntawiroon, Mahidol University, c/o R.S. Hotel, Larn Luang Road, Bangkok, 281 3047.

UK: Mike Flood, Friends of the Earth, Ltd., 9 Poland St., London W1V 3DG, (01) 434 1684.

Venezuela: Gareth Price, Marven SA, Spartado 809, Caracas 101.

West Africa: David L.B. Kamara, Dept. of Electrical Engineering, Fouran Bay College, Univ. of Sierra Leona Private Mail Bag, Freetown, Sierra Leone.

UNECE: Claude Ducret, UNECE, Palais des Nations, CH-1211 Geneve 10, Switzerland 34 60 11.

Other recipients of *Soft Energy Notes* are in the following nations: Philippines, Indonesia, Ethiopia, Zambia, Mexico, People's Republic of China, Niger, Sudan, Jordan, Somalia, Ghana, Korea, Tanzania, Turkey, Mozambique, Netherlands, Nicaragua, New Caledonia, Jamaica, Sri Lanka, Argentina, Pakistan, Thailand, Costa Rica, Senegal, Botswana, Congo, Lesotho, Mali, Taiwan, Chile, Colombia, Guatemala, El Salvador, Iraq, Saudi Arabia, Trinidad, Iran, Yugoslavia, Finland, Barbados, Portugal, Romania, Bolivia, Cyprus, Syria, Guyana, French West Indies, Nigeria, Korea, Wales, Scotland, Union of Soviet Socialist Republics, Bahamas, Dominican Republic.

Earth Needs More Friends

And so does Friends of the Earth. Not that the friends we've got haven't performed near-miracles. They have. They've enabled FOE to play a leading role in unmasking the "peaceful atom," in espousing a soft energy path, in strengthening the Clean Air Act, in getting (after lo, these many years) stripmine control legislation through Congress, in rallying support for endangered marine mammals, in debunking the SST, in developing and pushing proposals to save Alaskan wilderness, in opposing public works boondoggles, in exposing the undesirability (not to mention the impossibility) of incessant material growth, and in striving for the preservation, restoration, and rational use of Earth and its resources.

FOE has never been large, as membership organizations go. This has made it easier, no doubt, for us to be quick-reacting and maneuverable. But there are times when there's no substitute for sheer muscle—and organizational muscle, in the view of many politicians and publicists, is strictly proportional to membership size. More members, more muscle. (More money, too, a fact we can't afford to ignore.) So one of the most crucial ways you can help FOE defend our planetary habitat is by helping us to enroll new members. Please bring the membership coupon, below, to the attention of someone who shares our concerns. Thank you.

Friends of the Earth ● 1045 Sansome, #404 ● San Francisco, Calif. 94111 ● (415) 433-7373

Please enroll me as checked, including **Not Man Apart** and discounts on selected FOE books.

☐ Regular, $25 a year; ☐ Contributing, $60 a year; ☐ Life, $1,000; ☐ Retired, $12 a year;
☐ Supporting, $35 a year; ☐ Sustaining, $250 a year; ☐ Student, $12 a year; ☐ Patron, $5,000 or more
☐ Sponsor = $100 ☐ Spouse = add $5 to any membership

(Name) ____

(Address) ____

(City, State, Zip) ____

(Contributions to FOE are not tax-deductible.)